Lecture Notes in Mathematics

Edited by A. Dold and B. Eckmann



1348

F. Borceux (Ed.)

# Categorical Algebra and its Applications

Proceedings of a Conference, held in Louvain-La-Neuve, Belgium, July 26 – August 1, 1987

Springer-Verlag
Berlin Heidelberg New York London Paris Tokyo

**Editor**

Francis Borceux
Département de Mathématiques, Université Catholique de Louvain
2 chemin du Cyclotron, 1348 Louvain-La-Neuve, Belgium

Mathematics Subject Classification (1980): 18-06

ISBN 3-540-50362-5 Springer-Verlag Berlin Heidelberg New York
ISBN 0-387-50362-5 Springer-Verlag New York Berlin Heidelberg



Printing and binding: Druckhaus Beltz, Hemsbach/Bergstr.
2146/3140-543210

# INTRODUCTION

The first Louvain-la-Neuve Conference on Categorical Algebra and its Applications has been held in Louvain-la-Neuve (Belgium) from July 26 to August 1, 1988. The organizing committee was composed of:

J. Adamek (Prague, Czecoslovakia)
A. Kock (Aarhus, Denmark)
F.W. Lawvere (Buffalo, USA)
R. Walters (Sydney, Australia).

The local organizers were:

F. Borceux (Louvain-la-Neuve, Belgium)
F. Warrinier (Leuven, Belgium)

128 persons did take part in the Conference and 96 of them did deliver a talk.

The organizers would like to express their gratitude to the various organisms who did support financially the meeting and therefore made it accessible to a very large international audience:

Le Fonds National de la Recherche Scientifique
L'Université Catholique de Louvain
De Katholieke Universiteit te Leuven
Le Centre Belge de Recherches en Mathématiques
Le Centre National de Recherches en Logique
Apple Computer
Les commerçants de Louvain-la-Neuve.

The proceedings volume has been realized with the competent collaboration of 46 referees who did study carefully as many papers; the editor expresses his gratitude to all of them. The material submitted for these proceedings represented not far from 800 pages. Since just half of this material could be published, some very painful choices had to be made and it is certainly a pity that some good papers had to be rejected; the editor hopes they will find a good home somewhere else.

TABLE OF CONTENTS



---

‡ This paper has been edited by A. Kock

LIST OF PARTICIPANTS

**J. Adamek**; Prague, Czecoslovakia.
**I. Aitchison**; Melbourne, Australia.
**C. Anghel**; Dingolfing, Germany.
**E. Aznar**; Granada, Spain.
**B. Banaschewski**; Hamilton, Canada.
**L. Barbieri**; Genova, Italy.
**J. Barja**; Malaga, Spain.
**M. Barr**; Montréal, Canada.
**J. Bénabou**; Paris, France.
**H. Bentley**; Toledo, USA.
**F. Borceux**; Louvain-la-Neuve, Belgium.
**R. Börger**; Hagen, Germany.
**D. Bourn**; Amiens, France.
**D. Brandt**; Garbsen,Germany.
**S. Breitsprecher**; Tübingen, Germany.
**H. Brinkmann**; Konstanz, Germany.
**M. Bullejos**; Granada, Spain.
**A. Carboni**; Milano, Italy.
**P. Carrasco**; Granada, Spain.
**G. Castellini**; Mayaguez, Puerto Rico.
**P. Cherenack**; Cape Town, South Africa.
**E. Colebunders**; Brussel, Belgium.
**R. Cruciani**; Roma, Italy.
**V. De Paiva**; Cambridge, United Kingdom.
**F. De Vries**; Utrecht, Netherlands.
**Y. Diers**; Valenciennes, France.
**E. Dubuc**; Buenos Ayres, Argentine.
**P. Dupont**; Louvain-la-Neuve, Belgium.
**S. Eilenberg**; New York, USA.
**L. Español**; Logrono, Spain.
**A. Frei**; Vancouver, Canada.
**A. Frölicher**; Genève, Switzerland.
**F. Gago**; Santiago de Compostela, Spain.
**S. Ghilardi**; Bergamo, Italy.
**M. Grandis**; Genova, Italy.
**J. Gray**; Urbana, USA.
**R. Guitart**; Paris, France.
**H. Herrlich**; Bremen, Germany.
**P. Higgins**; Durham, United Kingdom;
**R. Hoffmann**; Bremen, Germany.
**H. Hušek**; Prague, Czecoslovakia.
**J. Isbell**; Buffalo, USA.
**G. Jarzembski**; Torun, Poland.
**J. Johnson**; New Brunswick, USA.
**M. Johnson**; Sydney, Australia.
**P. Johnstone**; Cambridge, Un. Kingdom.
**A. Joyal**; Montréal, Canada.
**K. Kamps**; Hagen, Germany.
**S. Kasangian**; Milano, Italy.
**G. Kelly**; Sydney, Australia.
**H. Kleisli**; Fribourg, Switzerland,
**A. Kock**; Aarhus, Denmark.
**M. Korostenski**; Johannesburg, S. Africa.
**J. Koslowski**; Nashville, USA.
**I. Kříž**; Prague, Czecoslovakia.
**A. Labella**; Roma, Italy.
**J. Lambek**; Montréal, Canada.
**R. Lavendhomme**; Louvain-la-N., Belgium.
**F. Lawvere**; Buffalo, USA.
**P. Lecouturier**; Kinshasa, Zaïre.
**D. Lever**; Nova Scotia, Canada.
**F. Linton**; Middletown, USA.
**A. Lirola Terrez**; Almeira, Spain.
**B. Loiseau**; Louvain-la-Neuve, Belgium.
**R. Lowen**; Antwerpen, Belgium.
**L. Lubikulu**; Louvain-la-Neuve, Belgium.
**T. Lucas**; Louvain-la-Neuve, Belgium.
**J. Mac Donald**; Vancouver, Canada.
**K. Mackenzie**; Durham, United Kingdom.
**S. Mac Lane**; Chicago, USA.
**R. Macleod**; Sackville, Canada.
**S. Mantovani**, Parma, Italy.
**S. Mawanda**; Louvain-la-Neuve, Belgium.
**G. Meloni**; Milano, Italy.
**M. Minguez**; Logrono, Spain.
**I. Moerdijk**; Amsterdam, Netherlands.
**T. Mubanga**; Louvain-la-Neuve, Belgium.
**J. Navaro**; Badajos, Spain.

**L. Nel**; Ottawa, Canada.
**S. Niefield**; Schenectady, USA.
**A. Obtułowicz**; Warsawa, Poland.
**D. Pape**; Berlin, Germany.
**R. Paré**; Halifax, Canada.
**D. Pavlovic**; Utrecht, Netherlands.
**M. Pedicchio**; Trieste, Italy.
**J. Pelletier**; Toronto, Canada.
**M. Pfender**; Berlin, Germany.
**A. Pitts**; Brighton, United Kingdom.
**T. Porter**; Bangor, United Kingdom.
**J. Power**; Sydney, Australia.
**J. Pradines**; Toulouse, France.
**A. Pultr**; Prague, Czecoslovakia.
**D. Pümplun**; Hagen, Germany.
**J. Reiterman**; Prague, Czecoslovakia.
**G. Reyes**; Montréal, Canada.
**J. Riguet**; Paris, France.
**E. Robinson**; Cambridge, Un. Kingdom.
**E. Rogora**; Milano, Italy.
**J. Roisin**; Louvain-la-Neuve, Belgium.
**R. Rosebrugh**; Sackville, Canada.
**K. Rosenthal**; Schenectady, USA.
**J. Rosický**; Brno, Czecoslovalia.
**G. Rosolini**; Parma, Italy.
**F. Rossi**; Trieste, Italy.
**S. Salbany**; Harare, Zimbabwe.
**J. Sancho de Salas**; Badajos, Spain.
**M. Sancho de Salas**; Badajos, Spain.
**L. Schinaia**; Roma, Italy.
**D. Schumacher**; Wolfville, Canada.
**F. Schwarz**; Toledo, USA.
**S. Semadeni**; Warsawa, Poland.
**H. Simmons**; Aberdeen, United Kingdom.
**R. Street**; Sydney, Australia.
**P. Taylor**; London, United Kingdom.
**M. Thiebaud**; Thônex, Switzerland.
**W. Tholen**; Toronto, Canada.
**M. Tierney**; New Brunswick, USA.
**V. Topencharov**; Sofia, Bulgaria.
**V. Trnková**; Prague, Czecoslovakia.
**J. Turpin**; Louvain-la-Neuve, Belgium.
**K. Ulbich**; Hamburg, Germany.
**G. Van den bossche**; Louvain-la-N., Belgium.
**B. Veit**; Roma, Italy.
**J. Vermeulen**; Pretoria , South Africa.
**R. Walters**; Sydney, Australia.
**F. Warrinier**; Leuven, Belgium.
**S. Weck**; Toledo, USA.
**W. Weiss**; Darmstadt, Germany.

# ARE ALL LIMIT-CLOSED SUBCATEGORIES OF LOCALLY PRESENTABLE CATEGORIES REFLECTIVE?

J. Adámek, J. Rosický and V. Trnková

Tech. Univ.
Suchbátarova 2
16627 Praha 6,
CSSR

Purkyne Univ.
Janáckovo nám. 2a
66295 Brno, CSSR

Math. Inst. of the
Charles University
Sokolovská 83
18600 Praha 8, CSSR

Introduction. It is not surprising that the answer to the question in the title is negative, under "reasonable" set-theoretical hypothesis (e.g., assuming the non-existence of measurable cardinals). We even construct two reflective subcategories*) of the locally presentable category *Graph* of graphs whose intersection is not reflective in *Graph* (although it is, of course, closed under limits). What *might* be surprising is that under other set-theoretical hypothesis, the answer is affirmative. We introduce a condition called Weak Vopěnka Principle, consistency of which follows by the existence of huge cardinals and, as some set-theorists believe, may be added to the usual axioms of set theory. We prove that assuming Weak Vopěnka Principle, then each locally presentable category $K$ has the following properties:

(i) Every subcategory of $K$ closed under limits is reflective in $K$;

(ii) All reflective subcategories of $K$ form a large complete lattice (ordered by inclusion);

(iii) The intersection of two reflective subcategories of $K$ is reflective.

Conversely, assuming the negation of Weak Vopěnka Principle, none of the above statements holds in $K$ = *Graph*.

*Weak Vopěnka Principle* is the following statement:

$Ord^{op}$ does not have a full embedding into *Graph*.

(Here $Ord^{op}$ is the dual of the well-ordered category of all ordinals.) We have chosen that name because the principle is weaker than the well-known *Vopěnka Principle* which, as we show below, can be formulated as follows:

---

*) Subcategories are understood to be full throughout our paper.

*Ord* does not have full embedding into *Graph*.

The position of Vopěnka Principle in set theory is discussed e.g. in [J]; there are good reasons to believe that Vopěnka Principle does not contradict the usual axioms of set theory. If so, then we can add Weak Vopěnka Principle to the usual axioms, and in the resulting set theory the answer to our title question is affirmative.

On the other hand, by results to be found in [PT], every concrete category has a full embedding into *Graph* provided that we assume the following

(M) There does not exist a proper class of measurable cardinals.

Thus, (M) implies the negation of both of the above principles. Since the non-existence of measurable cardinals is certainly not contradictory to the usual axioms of set theory, we conclude that we can add the negation of Weak Vopěnka Principle to those axioms. In the resulting set theory, the answer to the above question is negative. A closely related result has been proved in [AR] where, under (M), a proper class of reflective subcategories of *Graph* was presented whose intersection is not reflective in *Graph*. The present result is a refinement of the previous one: the set-theoretical hypothesis is weaker (viz., as weak as possible) and we have two subcategories in place of a proper class. The price of that refinement is a much deeper delving into the results of Prague School. This has made our main construction technically quite involved. For a reader interested only in the equivalence of our title question and Weak Vopěnka Principle, it is not necessary to study our construction (i.e., he can skip parts II and III of our paper) since J. R. Isbell has made the observation in [$I_1$] that the affirmative answer easily implies Weak Vopěnka Principle and our Theorem 2 states the converse. This converse also follows by a result of E. R. Fisher announced in [$F_1$]. He unfortunately has not published the proof yet but he kindly sent it to us [$F_2$].

We finally mention a recent result of M. Makkai and A. Pitts [MP] which holds without set-theoretical restrictions: let $K$ be a locally finitely presentable category, then each subcategory of $K$ closed under limits and filtered colimits is reflective in $K$.

In contrast to locally presentable categories, in the category *Top* of topological spaces we have found, without additional set-theoretical hypothesis, two reflective subcategories with non-reflective intersection, see [TAR].

## I. Assuming Weak Vopěnka Principle

Convention. We work, throughout our paper, within the usual Gödel-Bernays theory of sets with AC.

We will heavily use the category *Graph* of graphs (directed), i.e., pairs $(V,E)$ where $V$ is a set (of vertices) and $E \subseteq V \times V$ is a set (of edges). Morphisms $f:(V,E) \to (V',E')$, called homomorphisms, are maps $f:V \to V'$ with $(f\times f)(E) \subseteq E'$. An important property of that category is that every graph is a coproduct of its connected components (=maximal indecomposable subobjects). The following result explains why *Graph* can be used as a representative locally presentable category:

Theorem 1. Every locally presentable category can be fully embedded into *Graph*.

Proof. For each small category $A$, the category $Set^A$ has a full embedding into *Graph*, see [PT] (II.5.3 and Ex. I.7.1). Consequently, every locally presentable category is equivalent to a subcategory of *Graph*, see [GU]. It is obvious that a category equivalent to a subcategory of *Graph* is also isomorphic to (another) subcategory of *Graph*.

A *large discrete category* is defined by having a proper class of objects and no other morphisms than identities.

Lemma 1. (Formulations of Vopěnka Principle.) The following statements are equivalent:

(i) Vopěnka Principle: for each first-order language, every class of models such that none of them has an elementary embedding into another one is a set;

(ii) No locally presentable category has a large discrete subcategory;

(iii) *Graph* does not have a large discrete subcategory;

(iv) *Ord* cannot be fully embedded into *Graph*.

Proof. i → iv: Suppose that, on the contrary, *Ord* has a full embedding into *Graph* (denoted by $(V_i,E_i)$ on objects and by $\alpha_{ij}:(V_i,E_i) \to (V_j,E_j)$ on morphisms, $i \le j$). There clearly exists a proper class $K \subseteq \mathrm{Ord}$ such that for each $j \in K$ the maps $\alpha_{ij}$ for $i \in K$, $i < j$, are not collectively onto; let us choose a vertex $x_j \in V_j - \bigcup_{i\in K, i<j} \alpha_{ij}(V_i)$. Now consider the language of one binary relation and one nullary operation (and no axioms). Then each $(V_i,E_i,x_i)$, $i \in K$, is a model, and there are no elementary extensions in the resulting

class. This contradicts (i).

iv → iii: The category $K$ of connected graphs has the property that *Graph* can be fully embedded into $K$, see [PT](I.1.11). Thus, if (iii) would be false, there would exist a large discrete category of connected graphs $G_x$, $x \in X$. For any ordinal $i$, $X_i = \{x \in X \setminus$ the rank of $x$ is smaller than $i\}$ is a set (see [J]). (Using the rank, which is based on the axiom of regularity, we avoid the axiom of choice for classes.) There is an isotone map $F:\mathrm{Ord} \to \mathrm{Ord}$ such that $X_{F(i)} \neq X_{F(j)}$ for $i \neq j$. The functor $A:\mathit{Ord} \to \mathit{Graph}$ defined on objects by $A_i = \coprod_{x \in X_{F(i)}} G_x$ and on morphisms by the obvious coproduct injections is full. (In fact, given a homomorphism $f:A_i \to A_j$ then, since $f$ preserves connected components, for each $x \in X_{F(i)}$ there is $y \in X_{F(j)}$ such that $f$ maps $G_x$ inside $G_y$. It follows that $x = y$ and $f(v) = v$ for each vertex $v$ of $G_x$. Thus, $i \leq j$ and $F(v) = v$ for each vertex $v$ of $A_i$.) This contradicts (iv).

iii → ii by Theorem 1.

ii → i : Suppose that, on the contrary, there is a first-order language $L$ and a proper class $M_i$, $i \in X$ of $L$-models such that none of them has an elementary embedding into another. Add a new binary relation symbol to $L$ and denote by $\bar{L}$ the resulting first order language. By [PT] II.3.12., the underlying set $|M_i|$ of $M_i$ always carries a rigid binary relation $r_i$. Hence $(M_i,r_i)$, $i \in X$ form a large discrete subcategory of the category $\mathrm{Mod}(\bar{L})$ of $\bar{L}$-models and elementary embeddings. By the downward Löwenheim-Skolem theorem (cf. [CK], 3.1.6.), $\mathrm{Mod}(\bar{L})$ has a small dense subcategory. Hence $\mathrm{Mod}(\bar{L})$ can be fully embedded into a category of algebras (cf. $[I_2]$, 4.2.), i.e. into a locally presentable category. It contradicts (ii).

<u>Lemma 2</u>. Vopěnka Principle implies Weak Vopěnka Principle (in other words, if $\mathit{Ord}^{op}$ has a full embedding into *Graph* then also *Ord* has such an embedding).

<u>Proof</u>. Let $A:\mathit{Ord}^{op} \to \mathit{Graph}$ be a full embedding denoted by $A_i = (V_i,E_i)$ on objects and $\alpha_{ij}:A_i \to A_j$ $(j \leq i)$ on morphisms. There clearly exists a proper class $K \subseteq \mathrm{Ord}$ such that $\alpha_{ij}$ is not one-to-one for any $i,j \in K$, $i \neq j$. Consider the theory of two binary relations (and no axiom). For each $i \in K$ we have a model $(V_i,E_i,E_i')$ where $E_i' = \{(x,y) \mid x,y \in V_i,\ x \neq y\}$ and there exist no homomorphisms from one of those models into another one - thus, Vopěnka Principle does not hold.

<u>Open problem</u>: Are the two principles equivalent?

Theorem 2. Assuming that Weak Vopěnka Principle is true, each subcategory of a locally presentable category $K$ which is closed under limits is reflective in $K$.

Proof. We first observe that for each set $H$ of objects of $K$, the least subcategory $H^*$ of $K$ closed under limits and containing $H$ is reflective in $K$. This follows from the Special Adjoint Functor Theorem [M]: $H$ is a cogenerating set, and since $K$ is wellpowered and complete (being locally presentable), so is $H^*$.

Let $L$ be a subcategory of $K$ closed under limits. For $j \in$ Ord, let $L_j$ be the set of all objects of $L$ having the rank smaller than $j$ (compare the proof of Lemma 1; we are again avoiding the axiom of choice for classes). The categories $L_j^*$ are reflective. Thus, for each object K of $K$ we have reflections $r_j : K \to K_j$ of K in $L_j^*$ for all $j \in$ $\in$ Ord. Since $i \le j$ implies $L_i^* \subseteq L_j^* \subseteq L$ we have the unique $K$-morphism $e_{ji}: K_j \to K_i$ with $r_i = e_{ji}.r_j$. We are going to prove that the chain $e_{ji}$ is stationary, i.e., that there exists an ordinal $i_0$ such that all $e_{ji}$ with $i_0 \le i \le j$ are isomorphisms. It then follows that $r_{i_0}:K \to K_{i_0}$ is the reflection of K in each $L_i^*$, $i \ge i_0$, and hence, the reflection in $L$; this will conclude the proof.

Suppose that, on the contrary, there is an object K such that $e_{ji}$ is not stationary. From the wellpoweredness of $K$ it follows that there is a transfinite sequence $t_0 < t_1 < \dots < t_i < \dots$ of ordinals such that for $i < j$ the morphism $e_{t_j t_i}$ is not a monomorphism. Now consider the comma-category K↓K (of all $K$-arrows with domain K, see [M]): we have a functor

$$E: \mathcal{O}rd^{op} \to K{\downarrow}K$$

defined on objects by $E_i = (K \xrightarrow{r_{t_i}} K_{t_i})$ and on morphisms by $(i \to j) \mapsto e_{t_j t_i}$. This functor is a full embedding. In fact, let $f:E_j \to E_i$ be a morphism, i.e., let $f:K_{t_j} \to K_{t_i}$ fulfil $f.r_{t_j} = r_{t_i}$. If $i \le j$, then $f.r_{t_j} = e_{t_j t_i}.r_{t_j}$ and this implies $f = e_{t_j t_i}$ (by the uniqueness requirement on reflections). And the case $i > j$ cannot occur since we would have $(f.e_{t_i t_j}).r_{t_i} = f.r_{t_j} = r_{t_i}$, and hence, $f.e_{t_i t_j} = id$, although $e_{t_i t_j}$ is not a monomorphism. It is well-known that since $K$ is locally presentable, so is K↓K (see, e.g., [MPa]6.1.1.), and hence, there is a full embedding of K↓K in $\mathcal{G}raph$ by Theorem 1. This embedding composed with E above, shows that Weak Vopěnka Principle is false - a contradiction. (There is, a more straightforward way to finish the proof because E induces a full embedding of $K\downarrow\mathcal{K}$ in $G\downarrow\mathcal{G}raph$ for $G = EK$ and $G\downarrow\mathcal{G}raph$ has a full embedding into a locally presentable category of graphs with constants $c_x$ indexed by vertices of G (sending $f:G \to X$ to $(X,f(v))_{v\in V(G)}$)).

Corollary. Assuming Weak Vopěnka Principle, each intersection of reflective subcategories of a locally presentable category $\mathcal{K}$ is reflective in $\mathcal{K}$. Thus, the lattice of all reflective subcategories of $\mathcal{K}$ is large-complete.

## II. Assuming Negation of Weak Vopěnka Principle

Theorem 3. Assuming that Weak Vopěnka Principle is false, the category *Graph* has two reflective subcategories whose intersection is not reflective.

The aim of the following two sections is to prove the theorem. We shall construct a collection $D_i$ ($i \in$ Ord) of graphs with the following properties:

(a) The subcategory $\mathcal{L}$ of *Graph* consisting of those graphs G such that for each ordinal i there is precisely one homomorphism from $D_i$ to G is not reflective in *Graph*;

(b) Let $\mathcal{L}_e$ and $\mathcal{L}_o$ denote the subcategories of *Graph* consisting of those graphs G satisfying the following conditions:

(b1) Given an even ordinal i (odd ordinal i, respectively) such that for each $t < i$ there is a homomorphism from $D_t$ to G, then there is also a homomorphism from $D_i$ to G,

(b2) For each ordinal i there is at most one homomorphism from $D_i$ to G,

(b3) There is a homomorphism from $D_0$ to G.

Then $\mathcal{L}_e$ and $\mathcal{L}_o$ are both reflective subcategories of *Graph*. Since, obviously, $\mathcal{L} = \mathcal{L}_e \cap \mathcal{L}_o$, this will prove Theorem 3.

We will now present a condition guaranteeing that (b) is true. In the last section we will construct, assuming the negation of Weak Vopěnka Principle, a collection $D_i$ satisfying both that condition and (a) above. Recall that a collection of objects is *rigid* if the corresponding subcategory is discrete.

Proposition 1. Let $D_i$ ($i \in \mathrm{Ord}$) be a rigid collection of connected graphs such that for each graph G the following holds:

if there is a homomorphism from $D_{i+1}$ to G and a surjective homomorphism from $\coprod_{k<i} D_k$ to G, then there is also a homomorphism from $D_i$ to G.

Then the above subcategories $L_e$ and $L_o$ are reflective in *Graph*.

Proof. We present a proof that $L_o$ is reflective, the case of $L_e$ is analogous. Let $\hat{L}$ denote the category of all graphs satisfying (b2) above. Observe that $\hat{L}$ is obviously closed under products and subobjects in *Graph*, and therefore, $\hat{L}$ is epireflective in *Graph*. It is thus sufficient to prove that $L_o$ is reflective in $\hat{L}$. We denote, for short, by $[P,Q]$ the set of all homomorphisms from P to Q.

We first show that for each graph $G \in \hat{L}$ with $[D_i,G] \neq \emptyset$ for all even ordinals i there exists an ordinal $i_o$ with $[D_i,G] \neq \emptyset$ for all $i \geq i_o$. Without loss of generality, we can assume that each vertex of G lies in the image of some homomorphism from $D_i$, $i \in \mathrm{Ord}$. (For else we work with the subgraph G' of G consisting of all vertices lying in those images: proving the statement for G' would clearly prove it for G too.) We can, then, choose an ordinal $i_o$ such that each vertex of G lies in the image of some homomorphism from $D_i$, $i < i_o$. Let us show that for each ordinal $j \geq i_o$ we have $[D_j,G] \neq \emptyset$. Suppose, on the contrary, that $[D_j,G] = \emptyset$. It follows that j is odd, and hence $j + 1$ is even and therefore, $[D_{j+1},G] \neq \emptyset$. Put

$$A = \{i \in \mathrm{Ord} \mid i < j, \text{ and } [D_i,G] = \emptyset\}.$$

There is a surjective homomorphism from $\coprod_{k<j} D_k$ to $G + \coprod_{i\in A} D_i$: for each $k < j$ we either have a unique homomorphism from $D_k$ to G (if $k \notin A$) or $D_k$ is a summand of $\coprod D_i$ (if $k \in A$) and the resulting homomorphism $f: \coprod_{k<j} D_k \to G + \coprod_{i\in A} D_i$ is surjective by the choice of $i_o$, since $j \geq i_o$. We now use the above property of the graphs $D_i$ to conclude that there is a homomorphism from $D_j$ to $G + \coprod_{i\in A} D_i$. But this is a contradiction since $j \notin A$ and $[D_j,G] = \emptyset$: each homomorphism $g: D_j \to G + \coprod_{i\in A} D_i$ would, by the connectedness of $D_j$, map $D_j$ either to G, or to some $D_i$ (but then $i = j$).

We are ready to describe the reflection of any graph $G \in \hat{L}$. We denote by $i_o$ either the least even ordinal with $[D_{i_o},G] = \emptyset$, or (if no such even ordinal exists) the least ordinal with $[D_i,G] \neq \emptyset$ for all $i \geq i_o$. We will prove that the reflection of G in $L_o$ is the following coproduct injection

$$r : G \to G + \coprod_{j \in B} D_j$$

where

$$B = \{j \in \mathrm{Ord} \mid j < i_0 \text{ and } [D_j, G] = \emptyset\}.$$

(A) $G + \coprod_{j \in B} D_j$ lies in $L_0$. First, this graph obviously lies in $\hat{L}$. Now let i be an odd ordinal with $[D_t, G + \coprod D_j] \neq \emptyset$ for all $t < i$. We are to show that $[D_i, G + \coprod D_j] \neq \emptyset$. This is obvious whenever either $[D_i, G] \neq \emptyset$, or $i \in B$. Thus, we can suppose that $i \geq i_0$.

(A1) If $i_0$ is even with $[D_{i_0}, G] = \emptyset$ then, since i is odd, we have $i > i_0$ and then $i_0$ is one of the t's above – thus, $[D_{i_0}, G + \coprod_{j \in B} D_j] \neq \emptyset$. This is a contradiction since $[D_{i_0}, G] = \emptyset$ and $i_0 \notin B$.

(A2) If $i_0$ is such that $[D_k, G] \neq \emptyset$ for all $k \geq i_0$, we are ready since $i \geq i_0$.

(B) Any homomorphism $h: G \to H$ with $H \in L_0$ factors uniquely through r. Since $H \in \hat{L}$, it is sufficient to show that $[D_j, H] \neq \emptyset$ for all $j \in B$: then we have a (necessarily unique) extension of h to $G + \coprod_{j \in B} D_j$. We will prove by induction on $i < i_0$ that $[D_i, H] \neq \emptyset$. The case $i = 0$ follows from the definition of $L_0$. Suppose $i < i_0$ is such that $[D_t, H] \neq \emptyset$ for all $t < i$. Since $H \in L_0$, if i is odd we conclude $[D_i, H] \neq \emptyset$. If i is even then $i < i_0$ implies $[D_i, G] \neq \emptyset$, and hence, $[D_i, H] \neq \emptyset$.

## III. Main Construction

The construction of the graphs $D_i$ which would prove Theorem 3 has several levels, and we start with some auxiliary constructions. We denote by

$$\mathit{Graph}_0$$

the subcategory of *Graph* formed by all non-trivial, acyclic, strongly connected graphs, i.e., graphs of at least two vertices and such that for arbitrary two distinct vertices x and y there exists either a directed path from x to y, or a directed path from y to x, but not both. The sets of vertices and edges of a graph G are denoted by V(G) and E(G), respectively. When speaking about subgraphs and quotient graphs, we understand the strong sense, i.e., regular monos and regular epis, respectively. A pair (G,a) consisting of a graph G and a vertex a of G is called a pointed graph; homomorphisms of pointed graphs are those graph homomorphisms which preserve the chosen vertices.

Construction $\mapsto$. Let G and H be graphs, and let a be a vertex of H. We denote by

$$H \overset{a}{\mapsto} G$$

the graph obtained from G by gluing onto each vertex a copy of the graph H using the gluing vertex a:

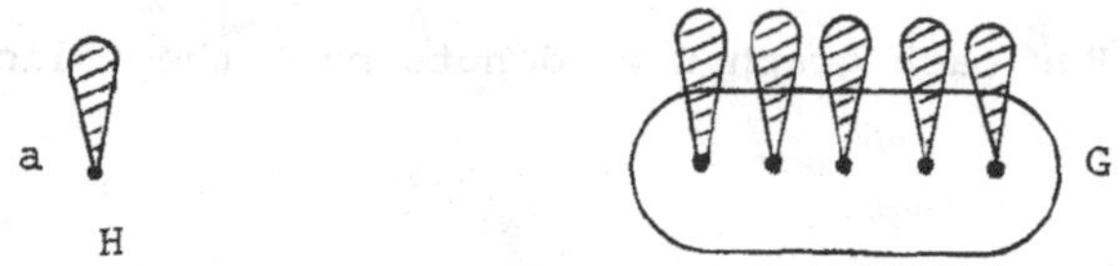


$H \overset{a}{\mapsto} G$

Formally, $H \overset{a}{\mapsto} G$ is the quotient of the sum $G + \coprod_{x \in V(G)} H \times \{x\}$, where $H \times \{x\}$ is a copy of the graph H, under the equivalence $\sim$ with $x \sim (a, x)$ for each $x \in V(G)$ and with all other equivalence classes singleton sets.

Construction #. For each pointed graph (H,b) denote by $(H,b)^{\#}$ the graph obtained from H by iteratively gluing a copy of H onto vertices (using b as the gluing point) in such a way that, in the end, each vertex x has a separate copy of $(H,b)^{\#}$ glued to x.

Thus, $(H,b)^{\#}$ is the direct limit of the following $\omega$-sequence $H_0 \subseteq H_1 \subseteq H_2 \subseteq \ldots$ of subgraphs:

$H_0$ is the graph with single vertex b and no edge,

$H_1 = H$,

$H_2$ is obtained by gluing a copy of H on each vertex in $V(H) - \{b\}$, i.e., $H_1$ is the quotient of $H + \coprod_{x \in V(H), x \neq b} H \times \{x\}$ (where $H \times \{x\}$ is a copy of H disjoint with H) under the least equivalence $\sim$ with $x \sim (b,x)$ for all $x \in V(H) - b$: and, in general, $H_{n+1}$ is obtained by gluing a copy of H on each vertex in $V(H_n) - V(H_{n-1})$, i.e., $H_{n+1}$ is the quotient of

$$H_n + \coprod_{x \in V(H_n) - V(H_{n-1})} H \times \{x\}$$

(where $H \times \{x\}$ is a copy of H disjoint with $H_n$) under the least equivalence $\sim$ with $x \sim (b,x)$ for all $x \in V(H_n) - V(H_{n-1})$.

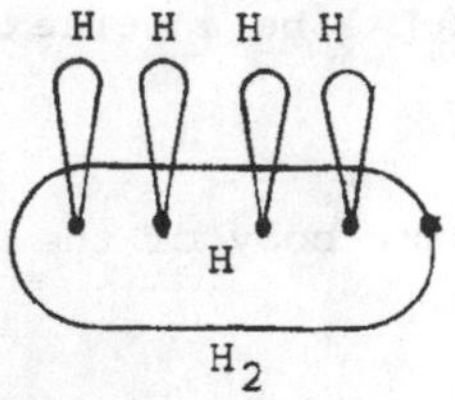


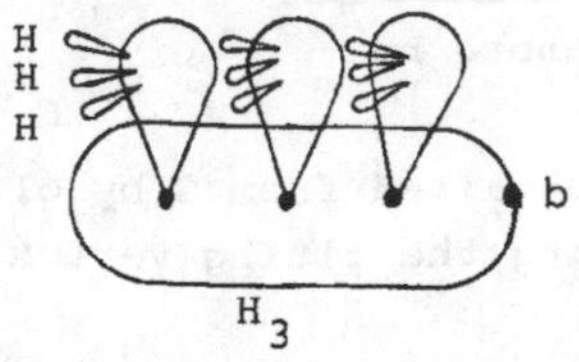


Construction ~. For each graph G we denote by $\widetilde{G}$ the following graph.

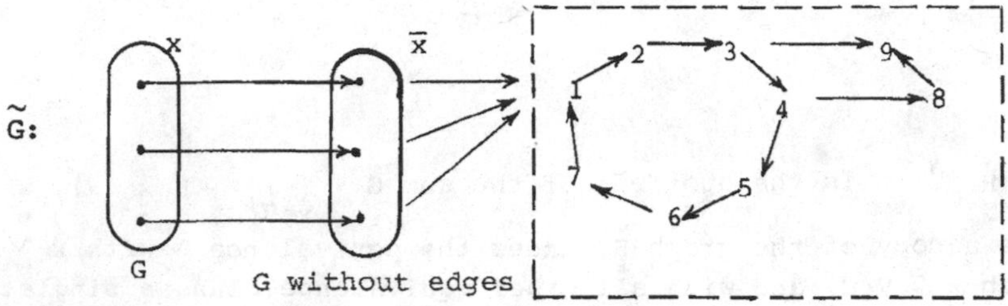


Formally, $V(\widetilde{G})$ is a disjoint union of $V(G)$, $\overline{V(G)} = \{\overline{x} \mid x \in V(G)\}$ and $\{1,2,\ldots,9\}$ (we suppose, for simplicity, that the sets are disjoint), and $E(\widetilde{G}) = E(G) \cup E(K) \cup \{x,\overline{x})\}_{x\in V(G)} \cup \{(\overline{x},1)\}_{x\in V(G)}$, where K denotes the above graph on the vertices $1,2,\ldots,9$.

Construction *. Given graphs G and H, we denote by

$$G * H$$

the graph obtained from H by first gluing a copy of the following graph $C_{3,5}$

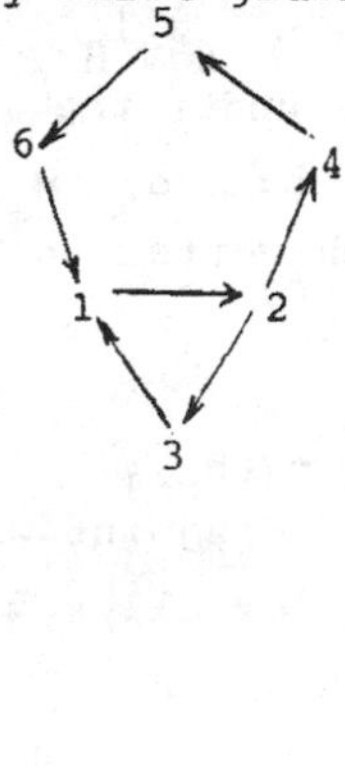

onto each vertex (with the gluing point 1) and, in the resulting graph, gluing a copy of $(\widetilde{G},9)^{\#}$ onto each vertex. Formally,

$$G * H = (\widetilde{G},9)^{\#} \overset{9}{\leftarrow} (C_{3,5} \overset{1}{\mapsto} H).$$

All the constructions above can be performed on homomorphisms too (giving rise to naturally defined bifunctors of the category *Graph*). For example, let $h:(H,a) \to (H',a')$ be a homomorphism of pointed graphs and $g:G \to G'$ a homomorphism of graphs. Then

$$h \mapsto g:(H \overset{a}{\leftarrow} G) \to (H' \overset{a'}{\leftarrow} G')$$

denotes the homomorphism defined as g on the "basic" subgraph G of $H \overset{a}{\leftarrow} G$, and defined as h on each copy $H \times \{x\}$, sending it to the copy $H' \times \{g(x)\}$ of $H' \overset{a'}{\leftarrow} G'$.

Analogously, $h^{\#}:(H,a)^{\#} \to (H',a')^{\#}$ is the homomorphism iteratively defined by using $h:h^{\#}$ sends $H_0$ to $H_0'$ as h, $h^{\#} = h \overset{a}{\leftarrow} h$ on the copy $H_1 = H \overset{a}{\leftarrow} H$, and having defined $h^{\#}$ on $H_n$, each of the new copies $H \times \{x\}$ is sent (as with h) to the new copy $H' \times \{h^{\#}(x)\}$.

Furthermore, $\widetilde{g}:\widetilde{G} \to \widetilde{G}'$ denotes the homomorphism given by $x \mapsto g(x)$, $\overline{x} \mapsto \overline{g(x)}$ and $i \mapsto i$ ($x \in V(G)$ and $i = 1,\dots,9$). Finally, put

$$g * h = \widetilde{g}^{\#} \mapsto (id_{C_{3,5}} \mapsto h) : G * H \to G' * H'.$$

Lemma 3. Let $f:G * H \to G' * H'$ be a homomorphism (G,G',H and H' in $\mathcal{G}raph_0$). Then

(i) both $[G,G'] \neq \emptyset$ and $[H,H'] \neq \emptyset$

and

(ii) assuming that there is just one homomorphism $g:G \to G'$, then $f = g * h$ for some homomorphism $h:H \to H'$.

Proof. (i)(a) Since G is acyclic, the only cycles of $\widetilde{G}$ are the subgraphs $C_7$(= 7 - cycle), and hence, all 3-cycles of $G * H$ lie in $C_{3,5} \mapsto H$. The same is true about $G' * H'$. Every homomorphism maps 3-cycles onto 3-cycles. Consequently, $f(C_{3,5} \overset{1}{\mapsto} H) \subseteq C_{3,5} \overset{1}{\mapsto} H'$. Moreover, since $C_{3,5}$ is a rigid graph, there clearly exists a homomorphism $h:H \to H'$ such that the restriction of f to $C_{3,5} \overset{1}{\mapsto} H$ is equal to $id_{C_{3,5}} \mapsto h$.

(b) Each 7-cycle of $G * H$ containing a vertex joined with $C_{3,5} \overset{1}{\mapsto} H$ lies in some copy $G \times \{x\}$ of the graph G (x a vertex of $(C_{3,5} \overset{1}{\mapsto} H)$). The point (3,x) of that 7-cycle is the initial vertex of an edge terminating in x. Since f maps 7-cycles onto 7-cycles, and it maps

$C_{3,5} \mapsto H$ identically, f(3,x) has the analogous properties w.r. to the vertex f(x) in $G' * H'$. Consequently, f(3,x) = (3,h(x)). It is easy to conclude that f(i,x) = (i,h(x)) for each i = 1,...,8 and each x in $C_{3,5} \overset{1}{\nleftrightarrow} H$.

(c) We want to show that for each vertex x of $C_{3,5} \overset{1}{\leftrightarrow} H$ there exists a homomorphism $g_x : G \to G'$ with $f(z) = g_x(z)$ for each vertex z of the subgraph $G \times \{x\}$ of the copy $\tilde{G} \times \{x\}$ of $\tilde{G}$ glued to x.

In fact, let $D_x$ be the subgraph of $(\tilde{G},9)^{\#} \times \{x\}$ (i.e., of the copy of $(\tilde{G},9)^{\#}$ glued to x) formed by all vertices from which a path of length 2 leads to (1,x). By inspecting $(\tilde{G},9)^{\#} \times \{x\}$:

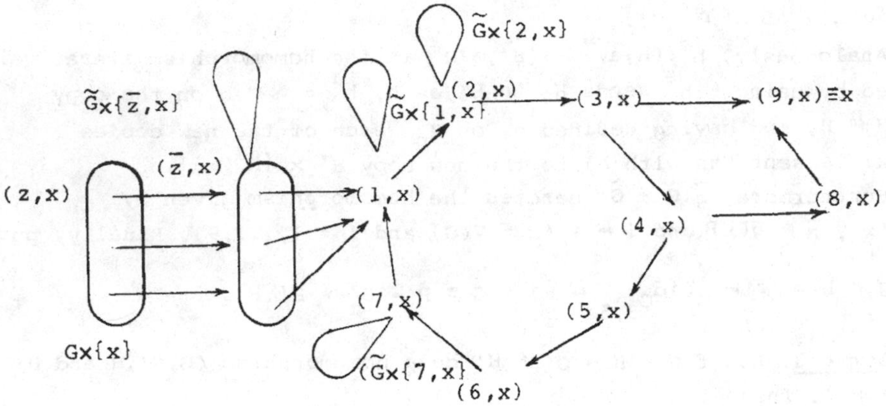


we see that $D_x$ consists of

(α) all vertices (z,x) of the copy $G \times \{x\}$ of G (i.e., $z \in V(G)$),

(β) all vertices $(3,\bar{z},x)$ and $(8,\bar{z},x)$ of the copy of $\tilde{G}$ glued to $(\bar{z},x)$ for $z \in V(G)$

(γ) the vertices (6,x), (3,7,x), (8,7,x), (2,1,x) and (4,1,x), where the copy of $\tilde{G}$ glued to (i,x) is denoted by $\tilde{G} \times \{i,x\}$ (i = 1,7).

We observe that each of the vertices listed in (β) and (γ) above is isolated in $D_x$. In contrast, none of the vertices listed in (α) is isolated in $D_x$ because the graph G has no isolated vertices (due to the choice of the subcategory *Graph*₀ above) - now, if $(z_1,z_2)$ is an edge of G, then $((z_1,x), (z_2,x))$ is an edge of $D_x$.

We can form the analogous subgraph $D'_x$ of all vertices with a path of length 2 into $(1,f(x))$ in the graph $G' * H'$. Obviously, $f(D_x) \subseteq \subseteq D'_{f(x)}$ and moreover, the non-isolated vertices of $D_x$ are mapped onto non-isolated vertices of $D'_{f(x)}$. In other words, for each vertex $z \in V(G)$ there is a vertex $\hat{z} \in V(G')$ with $f(z,x) = (\hat{z},f(x))$. It is obvious that the resulting map $g_x: z \mapsto \hat{z}$ is a graph homomorphism $g_x: G \to G'$.

(d) The assertions (a) and (c) prove (i).

(ii) Suppose that $g: G \to G'$ is the unique homomorphism in $[G,G']$. As proved above, there is a homomorphism $h: H \to H'$ such that f coincides with $id \mapsto h$ on the subgraph $C_{3,5} \overset{1}{\mapsto} H$, and f coincides with g on each of the subgraphs $G \times \{x\}$ for all x in $C_{3,5} \overset{1}{\mapsto} H$. We are going to prove that $f = g * h$ by induction on the "depth" of a vertex t of $G * H$, where depth 0 means that t lies in $C_{3,5} \overset{1}{\mapsto} H$, depth 1 that t lies in a copy of $\widetilde{G}$ glued to a vertex of depth 0, depth 2 that t lies in a copy of $\widetilde{G}$ glued to a vertex of depth 1, etc.

The claim that $f(t) = (g * h)(t)$ was established above for all vertices of depth 0, and for all vertices $t = (z,x)$ $(z \in V(G))$ or $t = (i,x)$ $(i = 1,\dots,8)$ of depth 1. The remaining vertices $t = (\bar{z},x)$ of depth 1 make no difficulties because t is the unique vertex lying on a path from $(z,x)$ to $(1,x)$, and hence, $f(t)$, having the analogous property in $G' * H'$, must be equal to $(\overline{g(z)},h(x))$. Let us proceed to depth 2 (the step from n to n+1 being analogous for all $n \geq 1$). We have $t = (\alpha,\beta,x)$ for $\alpha,\beta \in V(G) \cup \overline{V(G)} \cup \{1,\dots,8\}$, and we are to show that $f(t) = (\tilde{g}(\alpha),\tilde{g}(\beta),h(x))$. We have established $f(\beta,x) = (\tilde{g}(\beta),h(x))$ above. Observe that the vertex $(3,\beta,x)$ is the initial vertex of an edge terminating in $(\beta,x)$, and moreover, the following path of length 3.

$$(3,\beta,x),\ (4,\beta,x),\ (8,\beta,x),\ (\beta,x)$$

connects $(3,\beta,x)$ with $(\beta,x)$ too. The first edge of the above path lies on a 7-cycle. Now, f preserves all of the mentioned properties of $(3,\beta,x)$, and the unique vertex of $G' * H'$ fulfilling them (with respect to $f(\beta,x) = (\tilde{g}(\beta),h(x))$ is clearly $(3,\tilde{g}(\beta),h(x))$. Thus, $f(3,\beta,x) = (3,\tilde{g}(\beta),h(x))$. This proves our claim for $t = (3,\beta,x)$. Consequently, the claim clearly holds for all $t = (i,\beta,x)$, $i = 1,\dots,8$. The extension of the claim to all $t = (z,\beta,x)$, $z \in V(G)$ is analogous to (c) above, and the case $t = (\bar{z},\beta,x)$ is then clear.

<u>An Arrow Construction</u>. We are going to use the arrow construction introduced in [PT] (pp. 106-109) which consists of substituting each edge

(a,b) of a graph by the following subgraph M.

M:

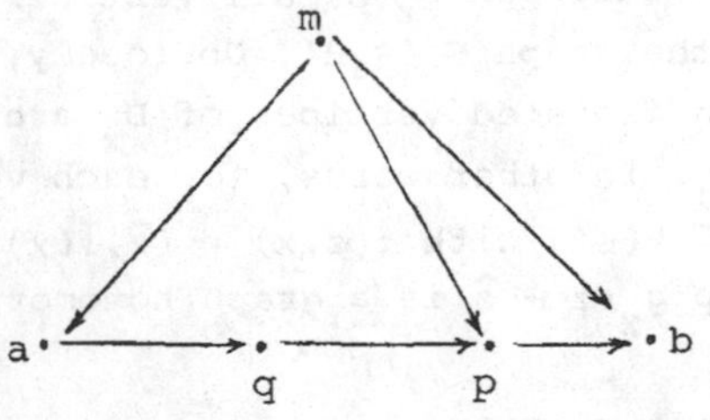


Formally, for each graph G in *Graph*$_0$ we denote by FG the quotient of the sum $\coprod_{r \in E(G)} M \times \{r\}$ (of copies $M \times \{r\}$ of M) under the least equivalence $\sim$ with the following properties:

$(a,r) \sim (a,r')$ for edges $r = (x,y)$ and $r' = (x,y')$ of G,

$(b,r) \sim (b,r')$ for edges $r = (x,y)$ and $r' = (x',y)$ of G,

and

$(a,r) \sim (b,r')$ for edges $r = (x,y)$ and $r' = (y,z)$ of G.

For each homomorphism $g:G \to G'$ in *Graph*$_0$, we denote by $Fg:FG \to FG'$ the naturally arising homomorphism which, for each edge $r = (x,y)$ of G, maps the copy $M \times \{r\}$ as the identity map onto the copy of M in FG' corresponding to the edge $(f(x),f(y))$.

Example:

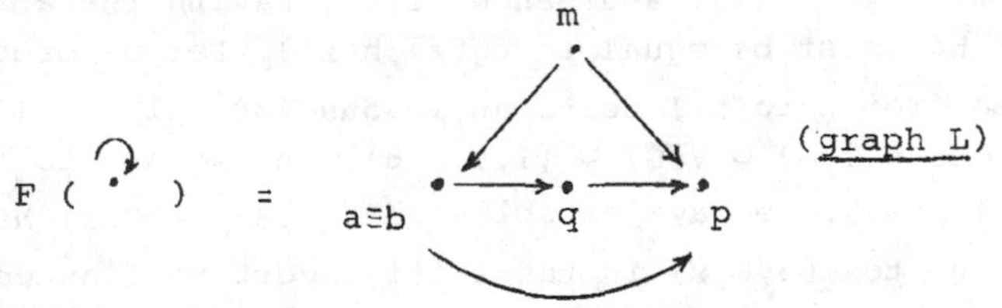


Proposition 12. The above rule F defines a full functor from the category of strongly connected graphs into *Graph*$_0$.

Proof. (a) We first prove that given a strongly connected graph G, then for each homomorphism $f:M \to FG$ there is an edge r in G with f the canonical injection onto the copy $M \times \{r\}$ of M in FG. Observe that M is a rigid graph, and hence, it is sufficient to prove that there is an edge $r \in E(G)$ with $f(m) = (m,r)$ - it is easy to conclude that, then, $f(x) = (x,r)$ for all $x \in V(M)$.

The vertex m in M has two outcoming edges with endvertices connected by an edge, viz., b and p. Since f(m) has the same property in FG, it is clear that either $f(m) = (m,r)$ for some edge r (and we conclude (a)), or, $f(m) = (a,r)$ for some loop $r = (x,x)$ of G. In the latter case, which we are to contradict, the outcoming edges of f(m) force us to conclude that f(b) = (q,r) and f(p) = (p,r). The edge (q,p) of M is mapped onto an edge with the endvertex f(p) = (p,r), i.e., f(q) is either (q,r), or (m,r). If f(q) = (q,r), then the edge (a,q) of M is mapped onto an edge with the endvertex f(q) = (q,r), and hence, f(a) = (a,r) [= f(m)] - this is a contradiction because (a,m) is an edge of M but (f(a),f(m)) is a loop which cannot be an edge of FG. If f(q) = (m,r) then the edge (a,q) of M is not mapped on an edge of FG because no edge of FG has the endvertex f(q) = (m,r) - a contradiction.

(b) Let $h:FG \to FG'$ be a homomorphism in $\mathit{Graph}_o$. By (a), h maps each copy $M \times \{r\}$ in FG identically onto a copy of $M \times \{r'\}$ in FG'. Let $g:V(G) \to V(G')$ be the unique map with $h(M \times \{x,y\}) = M \times \{g(x),g(y)\}$; the fact that g is unique follows from the choice of the category $\mathit{Graph}_o$. Then $g:G \to G'$ is clearly a homomorphism with $h = Fg$.

Remark. The above proposition is proved in [PT] for graphs without loops.

The rigid collection $D_i$. We are now prepared to produce the collection $D_i$ ($i \in$ Ord) of graphs which will prove Theorem 3 in the manner explained in Section II. We assume that Weak Vopěnka Principle does not hold, i.e., that $\mathit{Ord}^{op}$ has a full embedding to $\mathit{Graph}$. As proved in [PT] (IV.3.5), there is a full embedding of $\mathit{Graph}$ to $\mathit{Graph}_o$. Consequently, there is a full embedding

$$U:\mathit{Ord}^{op} \to \mathit{Graph}_o$$

(denoted by $U_i$ on objects and $u_{ij}:U_i \to U_j$, $i \geq j$, on morphisms). By Lemma 2, $\mathit{Graph}$ contains a large discrete full subcategory, and hence, the same is true about $\mathit{Graph}_o$. Let $V_i$ ($i \in$ Ord) be a rigid collection in $\mathit{Graph}_o$. We assume without loss of generality that the vertices $1,\dots,6$ of $C_{3,5}$ do not lie in any $V_i$.

Denote by $\overline{V_i}$ the graph obtained from $\coprod_{j \leq i} V_j + C_{3,5}$ by adding all edges with the initial vertex in some $V_j$ and the final vertex 1: see the next Figure.

Now put

$$D_i = FU_i * F\overline{V_i}.$$

Lemma 5. The collection $D_i$($i \in$ Ord) is rigid.

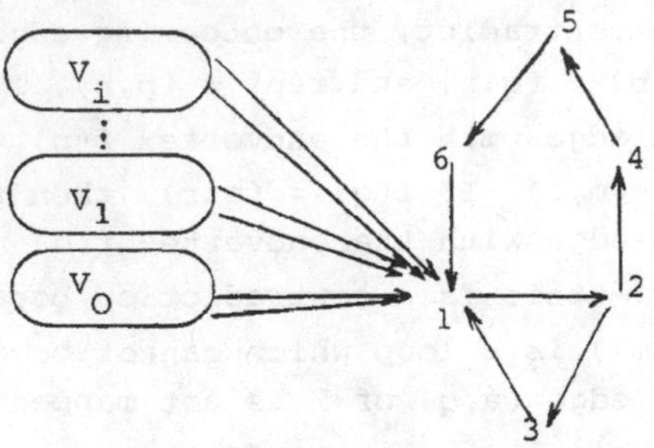


Proof. For $i < j$ we have $[D_i, D_j] = \emptyset$ since, by Proposition 2, $[U_i, U_j] = \emptyset$ implies $[FU_i, FU_j] = \emptyset$, and we can apply Lemma 3.

For $i > j$ there is no homomorphism from $\overline{V_i}$ to $\overline{V_j}$ : any homomorphism would have to be the identity map on $C_{3,5}$ and this implies that the subgraph $\coprod_{k<i} V_k$ of $\overline{V_i}$ is mapped into the subgraph $\coprod_{k<j} V_k$ which is impossible since $[V_j, \coprod_{k<j} V_k] = \emptyset$. Thus, $[\overline{V_i}, \overline{V_j}] = \emptyset$, and by Proposition 2, this implies $[F\overline{V}_i, F\overline{V}_j] = \emptyset$. We conclude, by Lemma 3, that $[D_i, D_j] = \emptyset$.

Finally, for $i = j$ we see, as above, that $[F\overline{V_i}, F\overline{V_i}] = \{id\}$. By Lemma 3, it follows that $[D_i, D_i] = \{id\}$.

Lemma 6. For the above graph L we have

$$[D_i, FU_j * L] = \begin{cases} \emptyset & \text{whenever } i < j \\ \{\delta_{ij}\} & \text{whenever } i \geq j \end{cases}$$

where $\delta_{ij} = Fu_{ij} * Fv_i$ for the homomorphism $v_i$ collapsing $\overline{V_i}$ to a single loop.

Proof. For $i < j$ we have $[U_i, U_j] = \emptyset$ and thus $[FU_i, FU_j] = \emptyset$. By Lemma 3, $[D_i, FU_j * L] = \emptyset$.

For $i \geq j$, $Fv_i : F\overline{V_i} \rightarrow L$ is the unique morphism in $[F\overline{V_i}, L]$, and $Fu_{ij} : FU_i \rightarrow FU_j$ is the unique morphism in $[FU_i, FU_j]$. By Lemma 3, this proves that $\delta_{ij}$ is the unique morphism in $[D_i, FU_j * L]$.

The proof of Theorem 3. We have defined above the rigid collection of connected graphs $D_i$ ($i \in \mathrm{Ord}$). It remains to prove that the collection satisfies the condition of Proposition 1 (whereby the categories $\mathcal{L}_o$ and $\mathcal{L}_e$ are both reflective) and that the category $\mathcal{L}$ of Section II ($\mathcal{L} = \mathcal{L}_e \cap \mathcal{L}_o$) is not reflective in *Graph*.

(a) For each graph G and each ordinal i, given a homomorphism $g:D_{i+1} \to G$ and a surjective homomorphism $e: \coprod_{k<i} D_k \to G$, we shall present a homomorphism $f:D_i \to G$.

The "basis" of $D_i = FU_i * F\overline{V_i}$ is the graph $C_{3,5} \overset{1}{\mapsto} F\overline{V_i}$ which is a subgraph of $C_{3,5} \mapsto F\overline{V_{i+1}}$ (since, by the definition of $\overline{V_i}$, we have $\overline{V_i}$ as a subgraph of $\overline{V}_{i+1}$) and the latter is the "basis" of $D_{i+1}$. We define f on the vertices of $C_{3,5} \overset{1}{\mapsto} F\overline{V_i}$ as the restriction of g. Next, we must define f on the copy of $(\widetilde{FU_i},g)^\#$ glued to x (for each vertex x of $C_{3,5} \mapsto F\overline{V_i}$). We choose a vertex $\overline{x}$ of $\coprod_{k<i} D_k$ with $e(\overline{x}) = x$; $\overline{x}$ lies in $D_{j(x)}$ for some ordinal $j(x) < i$. We lift the homomorphism $u_{i,j(x)}:U_i \to U_{j(x)}$ to the homomorphism $(\widetilde{Fu}_{i,j(x)})^\#:(\widetilde{FU}_i,9)^\# \to (\widetilde{FU}_{j(x)},9)^\#$ which we can compose with the embedding $(FU_{j(x)},9)^\# \to D_{j(x)} \to \coprod_{k<i} D_k$ (mapping 9 to $\overline{x}$). The resulting homomorphism will be denoted by $h_x:(\widetilde{FU}_i,9)^\# \to \coprod_{k<i} D_k$. Now define

$$f(y) = e(h_x(y)) \text{ for each vertex y of the copy of } (\widetilde{FU}_i,9)^\# \text{ glued to x.}$$

It is obvious that the restriction of f to the "basis" $C_{3,5} \overset{1}{\mapsto} F\overline{V_i}$ is a homomorphism (since g is a homomorphism), and each restriction of f to a copy of $(\widetilde{FU}_i,9)^\#$ glued to some vertex of the "basis" is also a homomorphism. Since, moreover, the gluing points are correctly mapped ($e(h_x(9)) = e(\overline{x}) = x$ for each x), it follows that f is a homomorphism.

(b) The category $L$ (of all graphs G such that each $[D_i,G]$ is a singleton set) is not reflective in *Graph*.

In fact, assuming that $D_0$ has a reflection $r_0:D_0 \to D_0^*$ in $L$, we will derive a contradiction. Since $D_0^*$ lies in $L$, we have unique homomorphisms $r_i:D_i \to D_0^*$ for all $i \in \mathrm{Ord}$, and to derive a contradiction, we will prove that each $r_i$ is one-to-one.

The following graph

$$B_j = FU_j * L + \coprod_{k<j} D_k$$

lies in $L$ for each ordinal j. In fact, for every ordinal i we have a homomorphism

$$f_{ij}:D_i \to B_j,$$

viz., the appropriate coproduct injection if $i < j$, and the homomorphism $\delta_{ij}$ of Lemma 6 composed with the first coproduct injection if $i \geq j$. It follows from the rigidity and connectedness of $D_i$'s, and from Lemma 6, that $[D_i,B_j] = \{f_{ij}\}$. Consequently, for the morphism $f_{0j}:D_0 \to B_j (\in L)$ there is a unique morphism $f_{0j}^*:D_0^* \to B_j$ with

$f_{0j} = f^*_{0j}.r_0$. Since for such ordinal i we have $[D_i, B_{i+1}] = \{f_{i,i+1}\}$, it follows that

$$f^*_{0,i+1} \cdot M_i = f_{i,i+1}: D_i \to B_{i+1}$$

Since $f_{i,i+1}$ is one-to-one, this proves that $r_i$ is one-to-one, concluding the proof.

References:

[AR] J. Adámek and J. Rosický. Intersections of reflective subcategories. Proc. Amer. Math. Soc. (to appear).

[CK] C. C. Chang and H. J. Keisler, Model Theory, North-Holland, Amsterdam 1973.

[$F_1$] E. R. Fisher, Vopěnka's Principle, Category Theory and Universal Algebra, Notices Amer. Math. Soc. 24 (1977), A - 44.

[$F_2$] E. R. Fisher, Vopěnka's principle, universal algebra, and category theory, preprint 1987.

[FK] P. J. Freyd and G. M. Kelly, Categories of continuous functors I, J. Pure Appl. Alg. 2(1972), 169-191.

[GU] P. Gabriel and F. Ulmer, Lokal präsentierbare Kategorien, Lect. Notes in Math. 221, Springer-Verlag, Berlin 1971.

[$I_1$] J. R. Isbell, Math. Reviews 41 (1971), # 5444.

[$I_2$] J. R. Isbell, Subobjects, adequacy, completeness and categories of algebras, Rozprawy Matem. XXXVI, Warszawa 1964.

[J] T. J. Jech, Set Theory, Academic Press, New York 1978

[M] S. Mac Lane, Categories for the Working Mathematician, Springer-Verlag, New York 1971.

[MPa] M. Makkai and R. Paré, Accessible categories: The foundations of categorical model theory, to appear in Contemp. Math.

[MṖ] M. Makkai and A. M. Pitts, Some results on locally finitely presentable categories, Trans. Amer. Math. Soc. 299 (1987), 473-496.

[O] O. Ore, Theory of graphs, Amer. Math. Soc., Providence 1962.

[PT] A. Pultr and V. Trnková, Combinatorial, algebraic and topological representations of groups, semigroups and categories, North-Holland, Amsterdam 1980.

[TAR] V. Trnková, J. Adámek and J. Rosický, Topological reflections revisited, to appear.

This paper is in final form and will not be published elsewhere.

# ON CATEGORIES WITH EFFECTIVE UNIONS

Michael Barr
Department of Mathematics and Statistics
McGill University
805 Sherbrooke St. West
Montréal, Québec
Canada H3A 2K6

**Abstract**

We study an exactness condition that allows us to treat many of the familiar constructions in abelian categories and toposes in parallel. Examples of such constructions include Grothendieck's theorem on the existence of injective cogenerators, the exactness of right exact functors and torsion theories/topologies.

## 1 Introduction

One of the most surprising results in the early days of category theory —the one that demonstrated that there was power in the notion —was Grothendieck's theorem that what we now call a Grothendieck abelian category has an injective cogenerator, [Grothendieck, 1957]. The hypotheses of that theorem include two fundamental ideas. The first is that of an abelian category [Buchsbaum, 1956] that people had been zeroing in on since Mac Lane's influential paper, [Mac Lane, 1950]. The second was the infinite exactness condition that characterizes a Grothendieck category and, for abelian categories, goes under the name AB5.

It is easy to divorce the latter condition from that of additivity. One simply supposes that finite limits commute with filtered colimits. And indeed, this has been a very important concept in equational categories and many other places. Not so successful have been attempts to translate the finite exactness conditions into a non-additive setting. For example, it has been difficult to find a finite exactness condition that, together with exactness of filtered colimits, guarantees the existence of enough injectives. Of course, toposes have enough injectives, but the conditions defining toposes are much more than exactness conditions. (By an exactness condition, I mean any condition that says a limit and a colimit commute.)

Another interesting property that abelian categories share with toposes is that a functor between abelian categories is left exact if it preserve finite products and monos and is right exact.

One would hope that results like these could be shown to follow from well chosen exactness conditions. Unfortunately, insufficiently many exactness conditions are known that would permit one to prove the results above. In fact, the principal exactness property of toposes —the universal effective sums—is not enjoyed by abelian categories, while they in turn have exactness properties not possessed by toposes.

The main purpose of this paper is to explore an exactness condition which is satisfied by both abelian categories and toposes and which is strong enough to prove (under relatively straightforward additional conditions) both the theorem on the existence of injective envelopes and the left exactness of certain functors. For example, this gives a new proof of the left exactness of the associated sheaf functor as a special case of reflectors for certain kinds of topologies on categories.

We apply these results to topologies, also known as torsion theories, by proving that, under appropriate additional conditions, the sheaf, that is torsion-free divisible, reflector exists and is exact. Although it was previously known that topologies were a non-abelian analog of torsion theories (see [Barr, 1973], the analogy has rarely, if ever, been fully exploited.

The author of this paper was supported by the Ministère de l'Education du Québec through FCAR grants to the Groupe Interuniversitaire en Etudes Catégoriques and by an individual operating grant from the National Science and Engineering Research Council.

## 2 Effective unions

**Definition.** A category is said to have **effective unions** if in any pullback diagram

$$\begin{array}{ccc} A & \rightarrowtail & B \\ \downarrow & & \downarrow \\ C & \rightarrowtail & D \end{array} \qquad (1)$$

if all the indicated arrows are regular monomorphisms, then the pushout $B +_A C$ exists and the arrow $B +_A C \to D$ is also a regular mono. This condition is satisfied in abelian categories and toposes. For the former, it follows from the fact that when 1 is a pullback, then

$$0 \to A \to B \oplus C \to D$$

is exact and then that $(B \oplus C)/A \rightarrowtail D$. For toposes, it is proved in [Johnstone, 1977], p. 41. It is readily inferred from the following:

**2.1 Proposition.** *A regular category with finite limits and colimits in which finite sums are disjoint and universal and every mono is regular has effective unions.*

**Proof.** Let $E$ be the union in the subobject lattice of $D$ of the subobjects $B$ and $C$. Then

$$(B + C) \times_E (B + C) \rightrightarrows B + C \to E$$

is a kernel pair/coequalizer diagram. Since $E \subseteq D$, the pullback over $E$ is the same as that over $D$. Because of the universality of sums,

$$(B+C)\times_D (B+C) \cong B\times_D B + B\times_D C + C\times_D B + C +_D C.$$

But with $B \rightarrowtail D$ and $C \rightarrowtail D$, the first and last terms add up to the diagonal and the two middle terms are simply the symmetric versions of each other. In other words, $B\times_D C = C\times_D B = B\cap C$. Thus,

$$B\cap C \rightrightarrows B+C \to E$$

is a coequalizer which implies that

$$\begin{array}{ccc} A & \rightarrowtail & B \\ \downarrow & & \downarrow \\ C & \rightarrowtail & E \end{array} \qquad (2)$$

is a pushout. □

**2.2** The condition is inherited by slices, coslices, products and disjoint unions of categories and the formation of functor categories. It is also satisfied by the category of sheaves for a topology. (For the definition of topology, see Section 5 below.) The reason is that in all these cases connected finite limits and connected finite colimits are created in the original category.

Other examples include the category of compact Hausdorff spaces as well as by various full subcategories, like Stone spaces. The reason is that subspaces in those categories are closed and a function on a union of closed subspaces is continuous if its restriction to the subspaces is. John Isbell has observed that this condition will be satisfied in any variety or even quasi-variety whose theory includes no non-trivial finitary operation. The reason is that in those cases, the underlying set functor creates finite limits and colimits. Nevertheless, effective unions remain a relatively rare property.

In various parts of this paper, there will other exactness conditions required. In order to avoid having to impose different and quite technical conditions, we will simplify the presentation by supposing our categories to be *biregular* by which we mean that every morphism factors as a regular epimorphism followed by a regular monomorphism and that these factorizations are preserved by pushouts and pullbacks.

# 3 Injectives

**Definition.** Let $\mathcal{G}$ be a full subcategory of $\mathcal{C}$. We say that an object $Q$ is **$\mathcal{G}$-injective** if whenever $H \rightarrowtail G$ is a regular monic in $\mathcal{C}$ between objects of $\mathcal{G}$, then $\mathrm{Hom}(G,Q) \to \mathrm{Hom}(H,Q)$ is surjective. An object is called **injective** if it is $\mathcal{C}$-injective.

**3.1 Lemma.** *Suppose the category $\mathcal{C}$ and full subcategory $\mathcal{G}$ satisfy the following conditions:*

(a) *$\mathcal{C}$ has finite limits and exact filtered colimits;*

(b) *$\mathcal{C}$ has pushouts of regular monos and they are regular monos;*

(c) *$\mathcal{G}$ is small;*

(d) *$\mathcal{C}$ is well-powered with respect to regular subobjects.*

*Then each object of $\mathcal{C}$ can be embedded by a regular mono into a $\mathcal{G}$-injective.*

**Proof.** Fix an object $C$ of $\mathcal{C}$. Begin by well-ordering the set of all diagrams

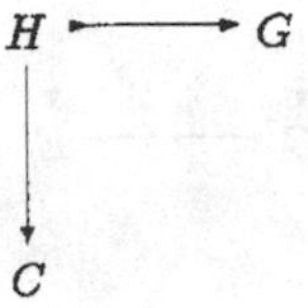


in which $H \rightarrowtail G$ is a regular monomorphism between objects of $\mathcal{G}$. We will take this to mean that a one-one correspondence has been chosen between a set of ordinals $\alpha < \lambda$ and diagrams

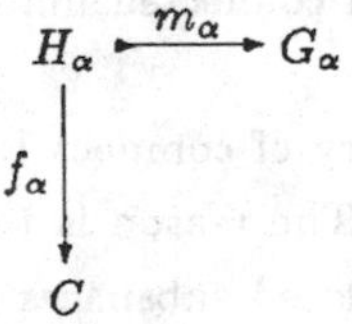


We will construct an ordinal sequence of objects $C_\alpha$ and regular monos $m_{\alpha\beta} : C_\beta \to C_\alpha$, for $\beta \leq \alpha$, subject to the usual commutativity conditions as follows. Begin by letting $C_0 = C$. If $\alpha$ is a limit ordinal and $C_\beta$ has been constructed for all $\beta < \alpha$, we let $C_\alpha = \operatorname{colim}_{\beta<\alpha} C_\beta$. For $\beta \leq \alpha$, we define $m_{\alpha\beta}$ to be the transition map to the colimit. We leave it to the reader to show, using the exactness of the filtered colimits, that it is a regular monic. We define $C_{\alpha+1}$ so that

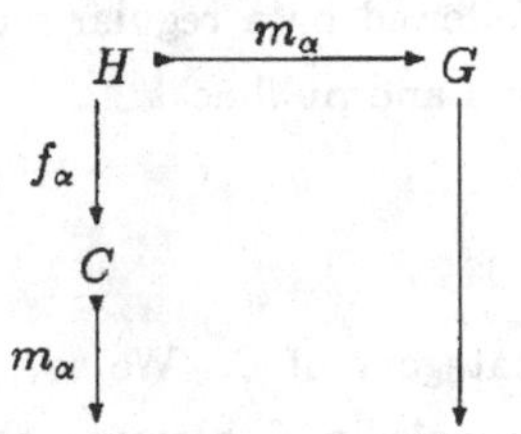

is a pushout and $m_{\alpha+1,\beta} = f_{\alpha+1,\alpha} \circ m_{\alpha\beta}$. We let $Q_1(C) = C_\lambda$. It is evident that each diagram

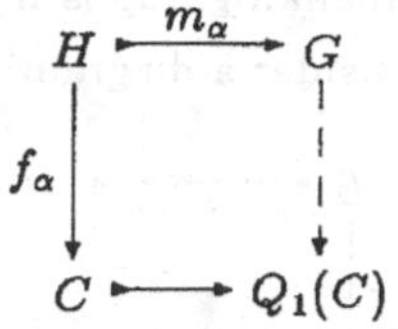


can be completed as shown. Next we choose a regular cardinal $\mu$ with the property that each object of $\mathcal{G}$ has a regular subobject lattice of cardinality less than $\mu$. Then we define an ordinal sequence $Q_\alpha(C)$ and regular monics $n_{\alpha\beta} : Q_\beta(C) \to Q_\alpha(C)$ for $\beta \leq \alpha < \mu$ by letting $Q_\alpha(C) = \operatorname{colim}_{\beta<\alpha} Q_\beta(C)$ when $\alpha$ is a limit ordinal and $Q_{\alpha+1}(C) = Q_1(Q_\alpha(C))$. Finally, $Q(C) = Q_\mu(C)$. I claim that any arrow $f : G \to Q(C)$ with $G$ an object of $\mathcal{G}$ factors through $Q_\alpha(C)$ for some $\alpha < \mu$. In fact, filtered colimits commute with finite limits so if we let $G_\alpha = f^{-1}(Q_\alpha(C))$, we have that $\operatorname{colim} G_\alpha = G$. However, this colimit is taken over a set of regular subobjects of $G$ of cardinality larger than that of the whole subobject lattice of $G$. If we eliminate repetitions, we get $G$ as the colimit of a set of subobjects $G_\alpha$ where now the index set runs over a set of ordinals smaller than $\mu$. If we denote this index set by $I$, the fact that $\mu$ is regular implies that $\sup\{\alpha \mid \alpha \in I\} = \nu < \mu$. But $\operatorname{colim}_{\alpha \in I} G_\alpha = G$ which means that $f$ factors through $Q_\nu(C)$. Now if we have a diagram

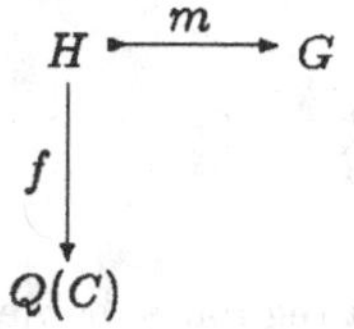


it can be factored and then filled in as indicated

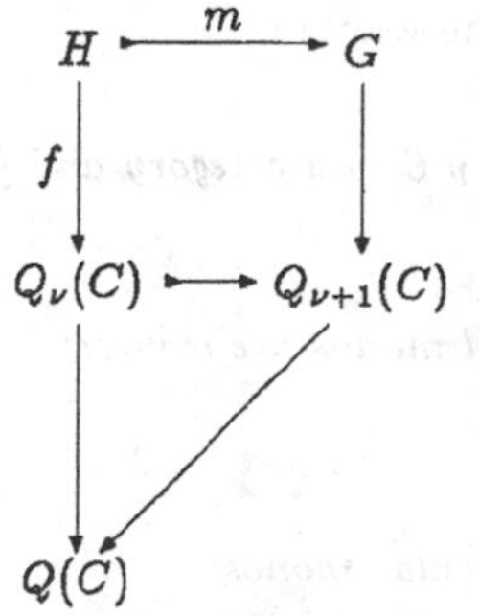


**3.2 Proposition.** *Suppose that in addition to the hypotheses of 3.1, C has effective unions and cokernel pairs and $\mathcal{G}$ is closed under epimorphic images and regular subobjects and includes a set of strong generators. Then a $\mathcal{G}$-injective is injective.*

This argument is adapted from the one found in [Grothendieck, 1957] and really goes back to Baer's proof that a divisible abelian group is injective.

**Proof.** Let $Q$ be a $\mathcal{G}$-injective and consider a diagram

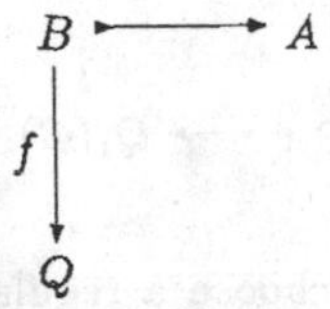

in which $B \rightarrowtail A$ is a proper regular mono. There is an object $G$ of $\mathcal{G}$ and an arrow $G \to A$ that does not factor through $B$. Under the hypothesis 3.1(b) and the existence of cokernel pairs every arrow factors as an epi followed by a regular mono and $\mathcal{G}$ is closed under quotients, we can suppose that $G \to A$ is a regular mono. Then $H = G \times_A B$ is a proper regular subobject of $G$. We have the diagram

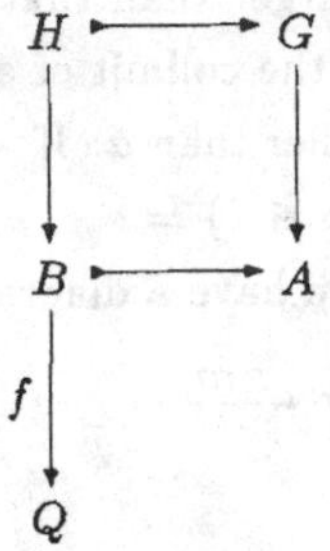

The pushout $G +_H B$ is by hypothesis a regular subobject of $A$ that properly contains $B$ and the mapping property of the pushout implies that the map extends to it. This shows that $f$ can be extended a little and a standard Zorn's lemma argument (using the exactness of filtered colimits) takes us the rest of the way. □

**3.3 Theorem.** *Suppose the category $\mathcal{C}$ is a category and $\mathcal{G}$ a full subcategory that satisfy the following conditions:*

(a) *$\mathcal{C}$ is regular and coregular and all monos are regular;*

(b) *$\mathcal{C}$ has exact filtered colimits;*

(c) *pushouts of regular monos are regular monos;*

(d) *$\mathcal{C}$ has effective unions;*

(e) *$\mathcal{C}$ is co-well-powered;*

(f) *$\mathcal{G}$ includes a set of strong generators.*

*Then $\mathcal{C}$ has enough injectives.*

**Proof.** Since $\mathcal{C}$ has a set of generators, it is also well-powered and so we can suppose that $\mathcal{G}$ is closed under subobjects and regular quotient objects. Then the hypotheses of 3.1 and 3.2 are satisfied and the conclusion follows. □

**3.4 Injective cogenerators.** We have shown that there are enough injectives. We wish to show that Grothendieck's argument about the existence of injective cogenerators can also be adapted to this setting. We begin by closing up the set of generators under finite sums and also under quotients. Then every object is a monomorphic filtered colimit of generators. For each generator $G$, let $Q(G)$ denote an injective container of $G$. Let $\mathcal{Q}$ denote the set of the $Q(G)$.

**3.5 Theorem.** *A locally finitely presentable category in which all epis are regular and which has exact filtered colimits, a small generating subcategory, and sufficient injectives has a small injective cogenerating family.*

**Proof.** We can suppose that the generating subcategory is small and closed under finite sums, finite products, subobjects and quotient objects. Let $\mathcal{C}$ and $\mathcal{G}$ denote the category and the small generating subcategory, respectively.

Given a proper epimorphism $B \twoheadrightarrow A$, let $K \rightrightarrows B$ be the kernel pair. Write $B = \operatorname{colim} G_i$ a monomorphic filtered colimit with the $G_i$ objects of $\mathcal{G}$. Since $K = B \times_A B$, it follows that $K$ is the colimit of the diagram

$$(\operatorname{colim} G_i) \times_A (\operatorname{colim} G_i) \cong \operatorname{colim}(G_i \times_A G_i)$$

Since the two arrows from $K$ to $B$ are distinct (otherwise $B \twoheadrightarrow A$ is not a proper epi), it follows that for some index $i$ the two composite arrows

$$G_i \times_A G_i \to B \times B \rightrightarrows B \twoheadrightarrow A$$

are distinct. Now consider the serially commutative diagram

$$\begin{array}{ccccc}
G_i \times_A G_i & \overset{e^0}{\underset{e^1}{\rightrightarrows}} & G_i & \rightarrowtail & Q(G_i) \\
\downarrow & & \downarrow & & \\
B \times_A B & \overset{d^0}{\underset{d^1}{\rightrightarrows}} & B & \twoheadrightarrow & A
\end{array}$$

The injectivity forces the existence of an arrow $g : B \to Q(G_i)$ making the triangle commutative. Since $e^0 \neq e^1$ and $q$ is mono, it follows that $q \circ e^0 \neq e^1$ hence $g \circ d^0 \neq d^1$ whence $g$ cannot factor through $A$. This shows, by a standard argument, that $B$ can be embedded into a product of the $Q(G_i)$. □

**3.6 injective envelopes.** By combining our results with those of Banaschewski [1970], we easily obtain,

**3.7 Corollary.** *Under the same hypotheses, every object can be embedded into an injective envelope.*

## 4 Left exact functors

It is a familiar fact of abelian categories that a right exact functor is left exact as soon as it preserves monos. In some ways this is a very surprising fact. Despite the example of the exact sheaf reflector, no non-additive version of this has ever been stated. It is interesting that we have such a theorem for categories with effective unions. A similar theorem has been discovered by Borceux and Veit [unpublished manuscript].

**4.1 Theorem.** *Suppose $F : \mathcal{C} \to \mathcal{D}$ is a functor such that $\mathcal{C}$ has finite limits, cokernel pairs and effective unions and $F$ preserves finite products, regular monomorphisms and cokernel pairs. Then $F$ preserves finite limits.*

In the case of an abelian category, a right exact functor automatically preserves finite products and cokernel pairs, so this implies that every right exact functor that preserves (regular) monos is left exact.

**Proof.** It is sufficient to show that $F$ preserves equalizers. The diagram

$$C \to C' \rightrightarrows C'' \qquad (3)$$

is an equalizer if and only if

$$C \to C' \rightrightarrows C' \times C'' \qquad (4)$$

is and in 4, the two arrows have a common left inverse. Since also

$$FC \to FC' \rightrightarrows FC''$$

is an equalizer if and only if

$$FC \to FC' \rightrightarrows F(C' \times C'') \cong FC' \times FC''$$

is, it is sufficient to show that an equalizer of type 4 is preserved. Hence we can suppose without loss of generality that 3 is an equalizer of two split monos with a common left inverse. We observe that if 4 is an equalizer of that sort, then

$$\begin{array}{ccc} C & \longrightarrow & C' \\ \downarrow & & \downarrow \\ C' & \longrightarrow & C'' \end{array}$$

is a pullback. The arrows are all regular monos, two of them by hypothesis and the other two being split monos. It follows from the effective unions that the induced arrow $C' +_C C' \rightarrowtail C''$ is a regular mono. If we apply $F$ to the diagram

$$\begin{array}{ccccc} C & \rightarrowtail & C' & \rightrightarrows & C' +_C C' \\ \| & & \| & & \downarrow \\ C & \rightarrowtail & C' & \rightrightarrows & C'' \end{array}$$

we get the diagram

$$\begin{array}{ccccc} FC & \rightarrowtail & FC' & \rightrightarrows & FC' +_{FC} FC' \\ \| & & \| & & \downarrow \\ FC & \rightarrowtail & FC' & \rightrightarrows & FC'' \end{array}$$

We have used the fact that $F$ preserves cokernel pairs. The arrow $FC \rightarrowtail FC'$ is a regular mono, hence is the equalizer of its cokernel pair. Moreover, since $F$ preserves regular monos, the induced arrow $FC' +_{FC} FC' \to FC''$ is also a (regular) mono. It is now a trivial diagram chase to see that the bottom line is also an equalizer. □

## 5 Topologies

Not surprisingly, Theorem 4.1 can be used to give a "soft" proof of the fact that the sheaf reflector is left exact. At that the same time, it shows that the torsion-free divisible reflector with respect to a torsion theory on an abelian category is exact. In fact, categories with effective unions and sufficient injectives seem to be the right kind of category in which to study topologies in general. We begin with a definition.

**5.1** A **topology**, also known as a **torsion theory**, on a category with finite limits is a natural endomorphism of the subobject functor which is monotone, inflationary and idempotent. This means that if $\mathrm{j} : \mathrm{Sub} \to \mathrm{Sub}$ is the endomorphism, that with each subobject $A_0$ of an object $A$, there is associated a subobject $\mathrm{j}\, A_0 = \mathrm{j}_A\, A_0$, which is

(a) (natural): If $f : B \to A$ and $A_0 \subseteq A$, then $f^{-1}(\mathrm{j}_A\, A_0) = \mathrm{j}_B(f^{-1}(A_0))$.

(b) (monotone): If $A_0 \subseteq A_1 \subseteq A$, then $\mathrm{j}_A\, A_0 \subseteq \mathrm{j}_A\, A_1$.

(c) (inflationary): $A_0 \subseteq \mathrm{j}\, A_0$.

(d) (idempotent): $\mathrm{j}\,\mathrm{j}\, A_0 = \mathrm{j}\, A_0$.

**5.2 Sheaves for a topology.** Let j be a topology on the category $\mathcal{C}$. A subobject $C_0 \subseteq C$ is called **j-closed** if $j_C C_0 = C_0$ and **j-dense** if $j_C C_0 = C$. An object $C$ will be called **j-separated** or **j-torsion free** if the diagonal $C \to C \times C$ is j-closed. It is called a **j-sheaf** or **j-torsion free and divisible** if it is j-separated and if it is injective with respect to j-dense monos. It is normal to omit the j when there is no doubt of the topology in question.

Note that we use the terminology of topologies on a category, although there is a parallel terminology developed in the context of abelian categories.

**5.3 Lemma.** *Let $\mathcal{C}$ be a category with a topology* j. *Let $A$ be an object of $\mathcal{C}$ and $B$ be a subobject of $A$. Then*

(a) *If*

$$\begin{array}{ccc} B' & \rightarrowtail & A' \\ \downarrow & & \downarrow f \\ B & \rightarrowtail & A \end{array}$$

*is a pullback, then so are both squares of*

$$\begin{array}{ccccc} B' & \rightarrowtail & j_{A'} B' & \rightarrowtail & A' \\ \downarrow & & \downarrow & & \downarrow \\ B & \rightarrowtail & j_A B & \rightarrowtail & A \end{array}$$

(b) *If $C \subseteq B$, then $j_B C = B \cap j_A C$.*

(c) *If*

$$\begin{array}{ccc} B' & \rightarrowtail & A' \\ \downarrow & & \downarrow \\ B & \rightarrowtail & A \end{array}$$

*is a commutative square with the top arrow dense and the bottom arrow closed, then there is a unique arrow $A' \to B$ making both triangles commute.*

(d) *$A \subseteq j_A B$ is dense and $j_A B \subseteq A$ is closed; moreover these properties characterize $j_A B$ uniquely.*

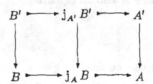

(e) *If $f : A' \to A$ is any map in $\mathcal{C}$ and $B$ is dense (respectively closed) in $A$, then $B' = f^{-1}(B)$ is dense (respectively closed) in $B$.*

(f) *If $C \subseteq B \subseteq A$ and both inclusions are dense (respectively closed), then $C$ is dense (respectively closed) in $A$.*

(g) *If $B$ and $C$ are both dense (respectively closed) in $A$ then $B \cap C$ is dense (respectively closed) in $A$.*

**Proof.** (a) That the right hand square is a pullback follows from the fact that j is a natural endomorphism of the subobject functor means that

$$\begin{array}{ccc} \mathrm{Sub}(A) & \xrightarrow{\mathrm{Sub}(f)} & \mathrm{Sub}(A') \\ \downarrow{\mathrm{j}} & & \downarrow{\mathrm{j}} \\ \mathrm{Sub}(A) & \xrightarrow[\mathrm{Sub}(f)]{} & \mathrm{Sub}(A') \end{array}$$

commutes. For $\mathrm{Sub}(f)(B) = B'$ by assumption and so $\mathrm{Sub}(f)(\mathrm{j}_A\, B) = \mathrm{Sub}_{A'}\, B'$.

(b) This follows from (a) and the observation that

$$\begin{array}{ccc} C & \rightarrowtail & B \\ \downarrow & & \downarrow \\ C & \rightarrowtail & A \end{array}$$

is a pullback.

(c) It follows from (a) that we have a commutative diagram

$$\begin{array}{ccccc} B' & \rightarrowtail & \mathrm{j}_{A'}\, B' & \stackrel{=}{\rightarrowtail} & A' \\ \downarrow & & \downarrow & & \downarrow \\ B & \underset{=}{\rightarrowtail} & \mathrm{j}_A\, B & \rightarrowtail & A \end{array}$$

which gives existence. The uniqueness follows from the fact that $\mathrm{j}_A\, B \rightarrowtail A$.

(d) $\mathrm{j}_{\mathrm{j}\,B}\, B = \mathrm{j}_A\, B \cap \mathrm{j}_A\, B = \mathrm{j}_A\, B$ from (b) which gives the density. Since j is idempotent, $\mathrm{j}_A(\mathrm{j}_A\, B) = \mathrm{j}_A\, B$, which gives the closedness. The uniqueness is assured by (c).

(e) This is immediate from (a) since a pullback of an equality is an equality.

(f) Suppose both inclusions are dense. From (b) we have that $B = \mathrm{j}_B\, C = B \cap \mathrm{j}_A\, C$ so that $B \subseteq \mathrm{j}_A\, C$, whence $A = \mathrm{j}_A\, B = \mathrm{j}_A \mathrm{j}_A\, C = \mathrm{j}_A\, C$. Next suppose both are closed. Then $\mathrm{j}_A\, C \subseteq \mathrm{j}_A\, B = B$. Thus $C = \mathrm{j}_B\, C = B \cap \mathrm{j}_A\, C = \mathrm{j}_A\, C$.

(g) This is immediate from (e) and (f). □

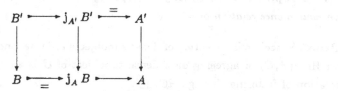

**5.4** Let j be a topology on $\mathcal{C}$. Then for $C$ an object of $\mathcal{C}$, we let $RC$ denote the j-closure of the diagonal of $C$ in $C \times C$.

**5.5 Proposition.** *Let* j *be a topology on* $\mathcal{C}$. *Then for any arrows* $f, g : A \to C$, *the equalizer of* $f$ *and* $g$ *is dense in* $A$ *if and only if* $< f, g >: A \to C \times C$ *factors through* $RC$.

**Proof.** The equalizer $E$ of $f$ and $g$ is characterized by the fact that

$$\begin{array}{ccc} E & \longrightarrow & A \\ \downarrow & & \downarrow \\ C & \longrightarrow & C \times C \end{array}$$

is a pullback. Applying j, we get

$$\begin{array}{ccccc} E & \longrightarrow & \mathrm{j}_A E & \longrightarrow & A \\ \downarrow & & \downarrow & & \downarrow \\ C & \longrightarrow & RC & \longrightarrow & C \times C \end{array}$$

If $E$ is dense, we see the required factorization immediately. If $< f, g >$ factors through $RC$, the fact that the right hand square is a pullback implies that $\mathrm{j}_A E = A$ and then $E$ is dense. □

**5.6 Corollary.** *Two maps to a separated object that agree on a dense subobject are equal.*

**5.7 Proposition.** *Suppose* j *is a topology on a category* $\mathcal{C}$. *Then for any object* $C$, $RC$ *is an equivalence relation on* $C$.

**Proof.** Since the intersection of dense subobjects is dense, one easily shows that the relation on $\mathrm{Hom}(A, C)$ of agreeing on a dense subobject of $C$ is an equivalence relation and is the relation of factoring through $RC$. □

**5.8** We will call a topology j on $\mathcal{C}$ **effective** if for each object $C$ of $\mathcal{C}$, $RC$ is an effective equivalence relation on $C$.

**5.9 Proposition.** *Let* j *be an effective topology on the regular category* $\mathcal{C}$ *and* $SC = C/RC$ *denote the quotient functor. Then for any object* $C$, $SC$ *is separated and is the separated reflection of* $C$.

**Proof.** In the pullback diagram

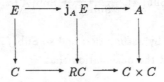

$$\begin{array}{ccccc} RC & \xrightarrow{=} & RC & \longrightarrow & C \times C \\ \downarrow & & \downarrow & & \downarrow \\ SC & \longrightarrow & RSC & \longrightarrow & SC \times SC \end{array}$$

the fact that $RC \to RSC$ is a regular epi implies that $SC \to RSC$ is as well and hence, being a mono, is an isomorphism.

This shows that $SC$ is separated. If $A$ is separated and $f : C \to A$ is an arrow, we have

$$\begin{array}{ccc} C & \longrightarrow & RC \\ \Big\downarrow & & \downarrow \\ & & C \times C \\ \Big\downarrow & & \downarrow \\ A & \longrightarrow & A \times A \end{array}$$

whose diagonal fill-in shows that the arrow $C \to A$ coequalizes the kernel pair and hence that there is an induced map $SC \to A$. The uniqueness is evident. □

**5.10 Proposition.** *Under the same hypotheses, $S$ preserves monos and finite products.*

**Proof.** It follows from Lemma 5.3(e) that $RA \times B$ and $A \times RB$ are dense in $RA \times RB$ and hence by part (g) that their intersection $A \times B$ is dense in $RA \times RB$. It similarly follows that $RA \times RB$ is closed in $A \times A \times B \times B$ and hence by (d) that the j-closure of $A \times B$ is $RA \times RB$, in other words that $R(A \times B) = RA \times RB$. In a regular category, a product of coequalizers is a coequalizer so that

$$RA \times RB \rightrightarrows A \times B \to SA \times SB$$

is a coequalizer and hence $S(A \times B) \cong SA \times SB$.

As for monos, if (and only if) $A \to B$ is monic,

$$\begin{array}{ccc} A & \longrightarrow & A \times A \\ \downarrow & & \downarrow \\ B & \longrightarrow & B \times B \end{array}$$

is a pullback and hence so is

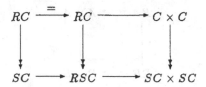

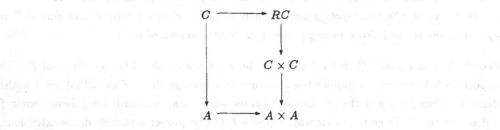

$$\begin{array}{ccc} RA & \longrightarrow & A \times A \\ \downarrow & & \downarrow \\ RB & \longrightarrow & B \times B \end{array}$$

by Lemma 5.3(a). It is then a simple exercise, using the regular embedding of [Barr, 1971] or [Barr, 1986] to see that the induced $SA \to SB$ is monic. □

**5.11 Proposition.** *Suppose that C is regular, that pushouts of monos exist and are mono and that* j *is an effective topology on C. Then an object F is a* j*-sheaf if and only if F is separated and is not* j*-dense in any properly containing separated object.*

**Proof.** If $F$ is a sheaf, then let $f : F \rightarrowtail C$ be a j-dense monic. The injectivity of $F$ with respect to j-dense monics implies the existence of a map $g : B \to F$ such that $g \circ f$ is the identity. Then $f \circ g$ and the identity of $C$ agree when composed with the j-dense monic $f$ and so are equal. To go the other way, suppose $F$ has no proper separable dense extensions. Then supposing we have

$$\begin{array}{ccc} A & \overset{\text{dense}}{\rightarrowtail} & B \\ \downarrow & & \\ F & & \end{array}$$

we form the pushout:

$$\begin{array}{ccc} A & \rightarrowtail & B \\ \downarrow & & \downarrow \\ F & \rightarrowtail & G \end{array}$$

With $F$ closed in $G$, Lemma 5.3(c) gives the required map. □

**5.12 Theorem.** *Suppose that C is regular with every mono regular and that* j *is an effective topology on C. Suppose each object can be embedded in an injective. Then each separated object is embedded in a sheaf.*

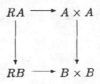

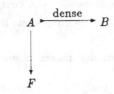

**Proof.** Let $C$ be separated and $f : C \rightarrowtail Q$ an injective container. Then $f$ induces a mono $Sf : SC \rightarrowtail SQ$. Since $C$ is separated, $C = SC$ and we have $Sf = g \circ f : C \to SQ$ where $g : Q \to SQ$ is the canonical map. A subobject of a separated object is separated (an easy consequence of Lemma 5.3(e)) so that if we factor $Sf = r \circ m$, where $m : C \rightarrowtail F$ is dense and $r : F \rightarrowtail SQ$ is closed, then $F$ is separated. Now we claim that $F$ is a sheaf. In fact, let $h : A \rightarrowtail B$ be a dense mono and $k : A \to F$ an arrow. Define $A_0$ so that the upper left square in the diagram

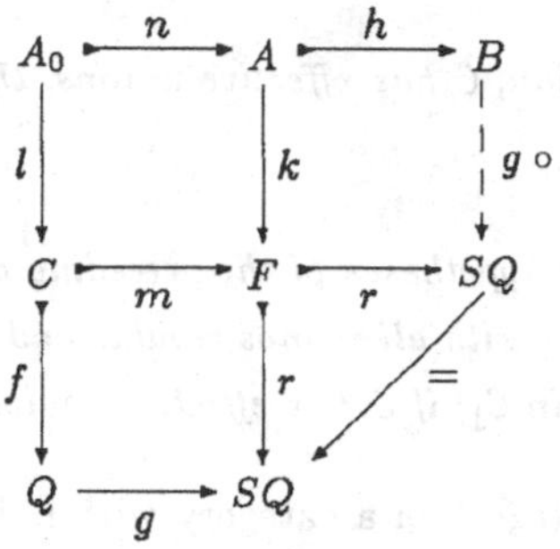

is a pullback. Now the injectivity of $Q$ implies the existence of a map $s : B \to Q$ such that $t \circ h \circ n = f \circ l$. Then

$$g \circ s \circ h \circ n = g \circ f \circ l = r \circ m \circ l = r \circ k \circ n.$$

But the target, $SQ$, of that map is separated and $n$ is dense so that $g \circ s \circ h = r \circ k$ and the right hand square commutes. Finally, $m$ is assumed dense and $s$ is closed by the definition of $F$ and so 5.3(c) gives the desired conclusion. □

**5.13 Theorem.** *Under the same hypotheses, a subobject of a sheaf is separated and its closure is a sheaf, in fact, the sheaf reflector of that object.*

**Proof.** Any subobject of a sheaf (or of any separated object), is readily seen to be separated. Let $A$ be such an object and let $A \to F$ be dense with $F$ a sheaf. Consider a diagram

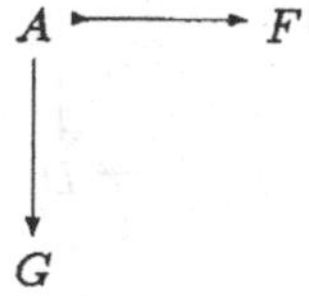

with $G$ a sheaf. Since $G$ is a sheaf and the arrow $A \rightarrowtail F$ is dense, it follows there is a map $F \to G$ making the triangle commute. It is unique because two maps to a separated object that agree on a dense subobject are equal. □

**5.14 Theorem.** *Suppose that $\mathcal{C}$ is a category and $\mathrm{j}$ a topology on $\mathcal{C}$ that satisfy the hypotheses of Theorem 5.12. Then the inclusion of the full category of $\mathrm{j}$-sheaves has a left adjoint that preserves monos and finite products.*

**Proof.** Only the preservation properties need be verified. The preservation of monos is easy since if $A \rightarrowtail B$, we have $SA \rightarrowtail SB$ and if $FB$ is the sheaf associated to $SB$, the sheaf associated to $A$ is the j-closure of $SA$ under the inclusion $SA \rightarrowtail SB \rightarrowtail FB$. As for products, the product (including empty product) of sheaves is a sheaf because the inclusion is a right adjoint. Then $S(A \times B) \cong SA \times SB \rightarrowtail FA \times FB$ is dense from Lemma 5.3(e) and (g). □

**5.15 Corollary.** *If, in addition, $C$ has effective unions, then the associated sheaf functor is left exact.*

**5.16 Theorem.** *Suppose the hypotheses of the preceding corollary are satisfied. Then the category $C_j$ of j-sheaves is regular with all monos regular and effective unions. If pushouts of monos are mono in $C$, they are in $C_j$; if $C$ has effective equivalence relations, so does $C_j$.*

**Proof.** Being a reflective subcategory of a category with finite limits, $C_j$ has them too.

If

$$\begin{array}{ccc} A' & \longrightarrow & B' \\ \downarrow & & \downarrow \\ A & \twoheadrightarrow & B \end{array}$$

is a pullback in the category of sheaves, it is in $C$ since the inclusion is a right adjoint. Then the arrow $A' \rightarrow B'$ is a regular epi in $C$, *a fortiori* in $C_j$.

A mono in $C_j$ is regular in $C$ and by applying the associated sheaf functor to the corresponding equalizer diagram, we conclude that it is regular in $C_j$ as well.

To see that unions are effective, we consider a pullback of monos in $C_j$.

$$\begin{array}{ccc} B \cap C & \rightarrowtail & B \\ \downarrow & & \downarrow \\ C & \rightarrowtail & D \end{array}$$

This is also an intersection in $C$. If $D' \rightarrowtail D$ is the pushout in $C$, then $FD' \rightarrowtail FD = D$ is the pushout in $C_j$. The universal mapping property of the adjoint insures that any subsheaf of $D$ that includes both $B$ and $C$ also includes $FD'$, so it is also their union in the subobject lattice.

Next, consider a pushout in $C$

$$\begin{array}{ccc} A & \rightarrowtail & B \\ \downarrow & & \downarrow \\ A' & \longrightarrow & B' \end{array}$$

in which $A$, $B$ and $A'$ are sheaves and the upper arrow is monic. Since left adjoints preserve pushouts, the pushout in $C_J$ is the sheaf associated to $B'$. Since the reflector preserves monos, the arrow $A' \rightarrowtail FB$ is still monic.

Suppose that equivalence relations in $C$ are effective. Let $A$ be a sheaf and $E \subseteq A \times A$ an equivalence relation which is also a sheaf. Then we have a kernel pair diagram in $C$

$$E \rightrightarrows A \to B$$

Applying the associated sheaf functor $F$, we get that

$$E \rightrightarrows A \to FB$$

is also a kernel pair since the associated sheaf functor preservers coequalizers. The conclusion now follows since we have shown that all monos in $C_J$ are regular. □

## References


B. Banaschewski, *Injectivity and essential extensions in equational classes of algebras.* **Proc. Conf.. on Universal Algebra, (1969). Queen's Series Pure Applied Math., 25** (1970).

M. Barr, *Exact categories.* **In Exact Categories and Categories of Sheaves, Springer Lecture Notes in Mathematics 236** (1971), 1-120.

M. Barr, *Non-abelian torsion theories.* **Canad. J. Math., 25** (1973), 1224-1237.

M. Barr, *Representations of categories.* **J. Pure and Applied Algebra, 41** (1986), 113-137.

F. Borceux & B. Veit, *On the left exactness of orthogonal reflections.* Unpublished manuscript.

D. Buchsbaum, *Exact categories.* Appendix to H. Cartan & S. Eilenberg, **Homological Algebra**, Princeton University Press, Princeton, N. J., 1956.

A. Grothendieck, *Sur quelques points d'algèbre homologique.* **Tohôku Math. Journal 2** (1957), 199-221.

P. M. Johnstone, Topos Theory. **Cambridge University Press**, 1977.

S. Mac Lane, *Duality for groups.* **Bull. Amer. Math. Soc. 56** (1950), 485-516.


This paper is in final form and will not be published elsewhere.

# DESCENT THEORY FOR BANACH MODULES

Francis Borceux*
Institut de Mathématique
Pure et Appliquée
Université Catholique de Louvain
B-1348 Louvain-la-Neuve, Belgium

Joan Wick Pelletier*
Department of Mathematics
York University
North York, Ontario
M3J 1P3, Canada

In the context of rings, the question of descent theory of modules can be described briefly. When R and S are two commutative rings with unit connected by a homomorphism $f:R \to S$, there is an obvious way of viewing an S-module as an R-module. Moreover, given an R-module N, there is a naturally associated S-module, $N \otimes_R S$, which is called the S-module induced by N. This association of S-modules with R-modules is an expression of an adjointness relation which is a theme common to many branches of mathematics and in particular to many algebraic constructions. Descent theory is the study of which S-modules arise as induced R-modules via the R-tensor product with S.

As well as being a topic of interest in ring theory, descent theory has been studied in other settings. Its generalization in the case of $C^*$-algebras has lead to some deep work of Rieffel on the theory of induced representations [10] and on the Morita theory of $C^*$-algebras [11]. Recently, Joyal and Tierney have looked at descent theory in the context of locales [5] and characterized the type of morphism f for which a satisfactory descent theory holds.

In this article we study the category of Banach modules over commutative Banach algebras with unit and develop a descent theory for modules related by an algebra homomorphism $f:R \to S$. We find necessary and sufficient conditions for f to be a descent morphism and sufficient conditions for it to be an effective descent morphism (in the terminology of [5]). Although the results obtained do not completely characterize effective descent morphisms, they are adequate to yield as an application a Morita theorem for Banach modules.

---

The paper is in final form and will not be published elsewhere.
*Both authors acknowledge the support of the Université Catholique de Louvain and of the National Sciences and Research Council of Canada.

## 1. Preliminaries

The theory of modules over a commutative monoid in a symmetric monoidal category has been well presented by Joyal and Tierney [5]. The module categories we are interested in, namely categories of Banach $\mathbb{R}$-modules, where $R$ is a commutative Banach algebra with unit, are special cases of this theory, and we will refer the reader to this source for details of much of our preliminary set-up.

In the applications that follow it will be necessary to consider categories of Banach modules based on two categories of Banach spaces, $\mathrm{Ban}_1$ and $\mathrm{Ban}_\infty$. The objects of both categories are all Banach spaces over $K$ ($K = \mathbb{R}$ or $\mathbb{C}$). The morphisms of $\mathrm{Ban}_1$ are all linear transformations bounded by $1$, while those of $\mathrm{Ban}_\infty$ are all bounded linear transformations. $\mathrm{Ban}_1$ is well known to be complete and cocomplete, while $\mathrm{Ban}_\infty$ is merely finitely complete and cocomplete. Other major differences between the two categories show up in the concepts of isomorphism and regular monomorphism. A $\mathrm{Ban}_1$-isomorphism is an isometric isomorphism, which implies a strong identification of Banach spaces in the norm; a $\mathrm{Ban}_\infty$-isomorphism is merely a topological isomorphism. The class of regular monomorphisms in $\mathrm{Ban}_1$ coincides with the class of isometric inclusions; on the other hand, in $\mathrm{Ban}_\infty$ a regular monomorphism is the same thing as a monomorphism with closed image. With these distinctions in mind, we make the following definition:

1.1 DEFINITION. Let $R$ be a commutative Banach algebra with unit. A Banach space $N$ is said to be a $\mathrm{Ban}_1$-module ($\mathrm{Ban}_\infty$-module) over $R$ if $N$ is equipped with a bilinear map $\mu: N \times R \to N$ such that $\mu(n,1) = n$, $\mu(n,r_1r_2) = \mu(\mu(n,r_1),r_2)$, and such that $\mu_n/\|n\|: R \to N$ and $\mu_r/\|r\|: N \to N$ are $\mathrm{Ban}_1(\mathrm{Ban}_\infty)$-morphisms, where $\mu_n$ and $\mu_r$ denote the restriction maps, for all non-zero $n \in N$, $r \in R$.

We let $\mathrm{Mod}_R{}^1$ denote the category of all $\mathrm{Ban}_1$-modules over $R$ together with $\mathrm{Ban}_1$-morphisms $g: N \to M$ which are $R$-linear and let $\mathrm{Mod}_R{}^\infty$ denote the category of $\mathrm{Ban}_\infty$-modules over $R$ with $\mathrm{Ban}_\infty$-morphisms which are $R$-linear. We will use the notation $\mathrm{Mod}_R$ to denote indifferently the categories $\mathrm{Mod}_R{}^1$ or $\mathrm{Mod}_R{}^\infty$ in all results which are common to both categories, and $\mathrm{Ban}$ to denote either $\mathrm{Ban}_1$ or $\mathrm{Ban}_\infty$.

Given $N, M \in \mathrm{Mod}_R$, a Banach module tensor product over $R$, denoted $M \otimes_R N$, can be defined. Explicitly, $M \otimes_R N$ is the quotient

$M\hat{\otimes}N/P$ , where $M\hat{\otimes}N$ is the (projective) Banach space tensor product and $P$ is the closed linear span in $M\hat{\otimes}N$ of all elements of the form $mr\otimes n - m\otimes rn$ . Clearly, $M\otimes_R N \in Mod_R$ by the definition $(m\otimes n)r = m\otimes nr$ . Similarly, given $N,M \in Mod_R$ , one can form $L_R(N,M) \in Mod_R$ , the R-linear Ban-morphisms from $N$ to $M$ with the usual operator norm. The R-module structure $L_R(N,M) \times R \to L_R(N,M)$ is given by $\alpha r(n) = \alpha(nr)$ for $\alpha \in L_R(N,M)$ , $r \in R$ , $n \in N$ .

We have the following standard and useful results, e.g. see [6].

1.2 PROPOSITION. Let $M,N,P \in Mod_R$ . Then

(1) $M\otimes_R N \cong N\otimes_R M$ ,
(2) $L_R(M\otimes_R N, P) \cong L_R(M, L_R(N,P))$ ,
(3) $M\otimes_R R \cong M$ ,
(4) $L_R(R,M) \cong M$ .

Now let $R$ and $S$ be commutative Banach algebras with unit (this will be our standard assumption throughout this article) and let $f:R \to S$ be a norm-decreasing algebra homomorphism. As in the general setting described in [5], we have an obvious functor $()_f : Mod_S \to Mod_R$ given by restricting scalars along $f$ , that is, for $M \in Mod_S$ , we define the module action $M \times R \to M$ by $mr \equiv mf(r)$ . Moreover, $()_f$ has a left adjoint $f_!:Mod_R \to Mod_S$ given by $f_!(N) = N\otimes_R S$ , which can be viewed as an S-module by means of the module action $(n\otimes s)s' = n\otimes ss'$ . Clearly the unit $\eta_N : N \to N\otimes_R S$ and counit $\varepsilon_M : M\otimes_R S \to M$ of the adjunction are given by $\eta_N(n) = n\otimes 1$ and $\varepsilon_M(m\otimes s) = ms$ ; in particular, $\varepsilon_M = \mu_M$ . For $M \in Mod_S$ we shall write $\eta_M$ instead of $\eta_{(M)_f}$ , treating $M$ as an R-module.

Given $N \in Mod_R$ , the S-module $N\otimes_R S$ is said to be the S-module "induced by $N$ ". The descent problem we are interested in is characterizing those $M$ for which there is $N \in Mod_R$ such that $M \cong N\otimes_R S$ . We pursue this question by defining as in [5] a plausible category $Des(f)$ of S-modules for consideration.

First we observe that for $M \in Mod_S$ , $M\otimes_R S$ is endowed with two distinct module structures making it into an $S\otimes_R S$-module:

(1) $(m\otimes s)(s_1\otimes s_2) = ms_2\otimes ss_1$  (2) $(m\otimes s)(s_1\otimes s_2) = ms_1\otimes ss_2$ .

We shall adopt the convention of writing $S\otimes_R M$ ($\cong M\otimes_R S$) to mean the $S\otimes_R S$-module action given by the first module action: $(s\otimes m)(s_1\otimes s_2) = ss_1\otimes ms_2$ ; $M\otimes_R S$ will be assumed to have the second module action.

1.3 DEFINITION. The category Des(f) of descent data for f has as objects all pairs $(M,\varphi)$, where $M \in \mathrm{Mod}_S$ and where $\varphi : S\otimes_R M \to M\otimes_R S$ is an isomorphism in $\mathrm{Mod}_{S\otimes_R S}$ satisfying condition (P):

$$
\text{(P)}\qquad
\begin{array}{ccccc}
 & & S\otimes_R S\otimes_R M & & \\
 & \swarrow^{S\otimes \mathrm{twist}} & & \searrow^{S\otimes\varphi} & \\
S\otimes_R M\otimes_R S & & & & S\otimes_R M\otimes_R S \\
 & \searrow_{\varphi\otimes S} & & \swarrow_{\varphi\otimes S} & \\
 & M\otimes_R S\otimes_R S & \xrightarrow[M\otimes \mathrm{twist}]{} & M\otimes_R S\otimes_R S\ . &
\end{array}
$$

Precisely, (P) states that if $\varphi(1\otimes m) = \lim_k \sum_q m_{kq}\otimes s_{kq}$ and $\varphi(1\otimes m_{kq}) = \lim_n \sum_p m_{kqnp}\otimes s_{kqnp}$, then for all $s_1, s_2 \in S$, $\lim_k \sum_q m_{kq} s_1 \otimes s_2 \otimes s_{kq} = \lim_{kn} \sum_{qp} m_{kqnp} s_1 \otimes s_{kqnp} s_2 \otimes s_{kq}$ . A morphism $h:(M_1,\varphi_2) \to (M_2,\varphi_2)$ in Des(f) is a morphism $h:M_1 \to M_2$ in $\mathrm{Mod}_S$ such that $h\otimes S\circ\varphi_1 = \varphi_2\circ S\otimes h$ . (We shall have occasion later on to speak of $\mathrm{Des}(f)_1$ or $\mathrm{Des}(f)_\infty$ in particular.)

It is shown in [5] that in a natural way $N\otimes_R S$ comes equipped with an isomorphism $\varphi = \mathrm{twist}\otimes S : S\otimes_R N\otimes_R S \to N\otimes_R S\otimes_R S$ which makes it into descent data for f , i.e. that there is a functor $F : \mathrm{Mod}_R \to \mathrm{Des}(f)$ such that $F(N) = (N\otimes_R S,\varphi)$ . Thus, a method of solving our problem is to prove that under certain conditions F is an equivalence of categories, or at least a full and faithful embedding. We are aided in this task by the realization due to Beck (unpublished but cited in [5]) that Des(f) can be further described as the category of coalgebras arising from the comonad $(M \to M\otimes_R S,\ \mu,\ \eta_M\otimes S)$ on $\mathrm{Mod}_S$ of the adjunction $f_! \dashv (\ )_f$ . The coalgebras for this comonad are pairs $(M,\xi)$ , where $M \in \mathrm{Mod}_S$ and $\xi : M \to M\otimes_R S$ is a morphism in $\mathrm{Mod}_S$ satisfying the two standard coalgebra conditions (see [12]) . We denote this category of coalgebras with the obvious structure-compatible morphisms, Coalg .

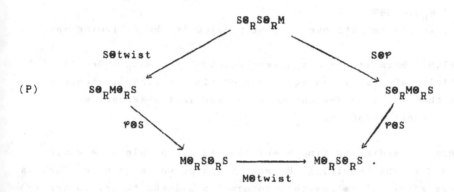

Explicitly, the equivalence of Des(f) and Coalg is given by the functor taking $(M,\varphi) \in \mathrm{Des}(f)$ to $(M,\bar\varphi)$ , where $\bar\varphi : M \to M\otimes_R S$ is defined by $\bar\varphi = \varphi\circ\mathrm{twist}\circ\eta_M$ . (The details of verifying that this correspondence respects all necessary commutativity relations are tedious and have been omitted from both [5] and our own presentation.) Viewing F then as a functor from $Mod_R$ to Coalg , it can be shown that F is exactly the usual comparison functor associated with the comonad $(-\otimes_R S, \mu, \eta\otimes S)$ .

Thus, we may restate our descent problem in the following way.

1.4. PROBLEM. When is the comparison functor $F : Mod_R \to Des(f)$ full and faithful? When is it an equivalence? In the first instance we shall say that f is a descent morphism and in the second, an effective descent morphism.

Before proceeding to find a solution to our problem, we would like to remark that requiring R and S to be unital Banach algebras is not necessary. It is possible to develop all the theory so far in the case that R and S are commutative Banach algebras with approximate units. The principal results concerning descent theory in this article remain true in this context, but the calculations are more involved in that some of the useful facts given in lemmas or propositions, e.g. 1.2.3 and 1.2.4, may not hold in the non-unital case and have to be circumvented.

## 2. Characterization of descent morphisms

Since the problem expressed above is a natural question in the context of comonad theory, namely when is the functor $f_! : Mod_R \to Mod_S$ comonadic, we may make use of the results due to Beck which answer this question in the general context. These results may be summarized as follows (e.g. see [12]):

2.1 THEOREM (Beck). Let F be the comparison functor $F : Mod_R \to$ Coalg . Then the following are equivalent:

(1) F is full and faithful,
(2) $-\otimes_R S$ reflects isomorphisms,
(3) $\eta_N : N \to N\otimes_R S$ is a regular monomorphism for all $N \in Mod_R$ .

Furthermore, F is an equivalence if and only if, in addition, $-\otimes_R S$ preserves equalizers of $-\otimes_R S$-split equalizer pairs.

The key concept necessary for $F$ to be full and faithful, that is to say, for $f$ to be a descent morphism, is that of a weak retract.

2.2 DEFINITION. A morphism $g:M \to N$ in $Mod_R$ is said to be a weak retract when $g'$ (the dual Banach space morphism) is a retraction in $Mod_R$, that is, when there exists $h:M' \to N'$ in $Mod_R$ such that $g' \circ h = M'$.

The following proposition characterizes weak retracts.

2.3 PROPOSITION. The morphism $g:M \to N$ is a weak retract in $Mod_R$ if and only if $P \otimes g : P \otimes_R M \to P \otimes_R N$ is a regular monomorphism in $Mod_R$ for every $P \in Mod_R$.

PROOF. Now suppose that $g$ is a weak retract and that $h:M' \to N'$ is such that $g' \circ h = M'$. This means that $L_R(P,g') \circ L_R(P,h) = L_R(P,M')$ and $L_R(P,h)' \circ L_R(P,g')' = L_R(P,M')'$. But $L_R(P,M')' \cong (P \otimes_R M)''$, so

$$L_R(P,g')' \cong (P \otimes g)'' : (P \otimes_R M)'' \to (P \otimes_R N)''$$

is a monomorphism in $Ban_\infty$ and in fact an isometric inclusion in $Ban_1$. Since $(P \otimes g)'' \circ i = i \circ (P \otimes g) : (P \otimes_R M) \to (P \otimes_R N)''$, where $i$ denotes the canonical isometric embedding of a space into its bidual, $P \otimes g$ is a monomorphism and an isometry as well if $(P \otimes g)''$ is. Thus, if we are working in $Ban_1$, $P \otimes g$ is a regular monomorphism.

If we are working in $Ban_\infty$, we must also verify that $P \otimes g$ has closed image. To this end, let $(x_n)$ be a sequence of elements in $P \otimes_R M$ such that $(P \otimes g)(x_n) \to y \in P \otimes N$. Clearly, $(P \otimes g)''(ix_n) \to iy \in (P \otimes N)''$. Letting $\hat{x}_n$, $\hat{y}$ denote the images of $i(x_n)$ and $i(y)$ in $L_R(P,M')'$ and $L_R(P,N')'$, respectively, we have

$$L_R(P,h)' \circ L_R(P,g')'(\hat{x}_n) = \hat{x}_n \to L_R(P,h)'(\hat{y}) .$$

But since $i(P \otimes_R M)$ is closed in $(P \otimes_R M)'' \cong L_R(P,M')'$, we can conclude that $L_R(P,h)'(\hat{y})$ is of the form $\hat{x}$ for some $x \in P \otimes_R M$, which means that $\hat{x}_n \to \hat{x}$, $x_n \to x$, and $y = (P \otimes g)(x)$, which completes this part of the proof.

Conversely, suppose that $P \otimes g$ is a regular monomorphism for every $P \in Mod_R$. We choose $P = M'$ to obtain a regular monomorphism

$M'\otimes g : M'\otimes_R M \to M'\otimes_R N$ . We are looking for a R-module morphism $h:M' \to N'$ such that $g'\circ h = M'$ . To this end we note the isomorphisms $L_R(M',N') \cong (M'\otimes N)'$ and $L_R(M',M') \cong (M'\otimes M)'$ , and realize that it is equivalent to find an element $k$ of $(M'\otimes N)'$ such that $k\circ M'\otimes g = r$ , where $r \in (M'\otimes M)'$ corresponds to $M' \in L_R(M',M')$ . In $Ban_1$ , since $M'\otimes g$ is an isometric inclusion, we can use the Hahn-Banach theorem to produce such a $k$ . In $Ban_\infty$ , since $M'\otimes g$ is merely a monomorphism with closed image, we need first to factor $M'\otimes g$ as $M'\otimes M \to im(M\otimes g) \to M'\otimes_R N$ , which is a $Ban_\infty$-isomorphism followed by an isometric inclusion. We extend along the isometric inclusion using the Hahn-Banach theorem as above to get our result. □

2.4 COROLLARY. If $f:R \to S$ is a weak retract in $Mod_R$ , then $f$ is a regular monomorphism.

PROOF. Taking $P = R$ in Proposition 2.3, we see that $R\otimes f:R\otimes_R R \to R\otimes_R S$ is a regular monomorphism. However, by 1.2.3, $R\otimes_R R \cong R$ , $R\otimes_R S \cong S$ , and $R\otimes f$ corresponds to $f$ . □

In view of 2.1, it is now easy to prove the theorem determining when $f$ is a descent morphism.

2.5 THEOREM. Let $f:R \to S$ be a morphism of commutative Banach algebras with unit. Then $f$ is a weak retract in $Mod_R$ if and only if $\eta_N:N \to N\otimes_R S$ is a regular monomorphism for every $N \in Mod_R$ .

PROOF. It follows from 2.3 that $f$ is a weak retract if and only if $N\otimes f:N\otimes_R R \to N\otimes_R S$ is a regular monomorphism for every $N \in Mod_R$ . But $N\otimes_R R \cong N$ and under this isomorphism $N\otimes f$ corresponds to $\eta_N$ since $\eta_N(n) = n\otimes 1 = N\otimes f(n\otimes 1)$ , which proves the theorem. □

2.6 COROLLARY. $F : Mod_R \to Des(f)$ is full and faithful, that is to say, $f$ is a descent morphism, if and only if $f:R \to S$ is a weak retract in $Mod_R$ .

We conclude this section by giving some examples of weak retracts in $Mod_R$ .

2.7 EXAMPLES. 1. In $Ban_1$ the canonical example of a weak retract which is not necessarily a retract (clearly, every retract is a weak retract) is the embedding $i_X:X \to X''$ for any $X \in Ban_1$ . In this case, $(i_X)'\circ i_{X'} = X'$ . Of course, $X$ and $X''$ can always be considered as X-modules, bringing this example into our context in a very trivial way. 2. A less trivial example is that of $c$, the Banach space of all real convergent sequences. It is well known that the dual of $c$ is $\ell_1$, and, hence, that $c'' = \ell_\infty$. Now $c$ and $\ell_\infty$ are also commutative Banach algebras with unit and modules over $c$. The embedding $i: c \to \ell_\infty$ is a weak retract in $Mod_R$ and is thus a descent morphism. 3. In general if $R$ is a commutative Banach algebra with approximate unit, then $R''$ can be endowed with a multiplication called the Arens multiplication making it into a Banach algebra with unit such that $i: R \to R''$ is an algebra homomorphism. $R''$ will not always be commutative, but it is commutative in various cases which include, for example, all commutative B*-algebras (see [2]). Now the R-module structure on $R''$ that derives from the homomorphism $i$ coincides with that deriving from the dual space structure, so $R''$ is unambiguously an R-module. Moreover, $i$ is a weak retract in $Mod_R$ and, hence, a descent morphism. 4. Not all weak retracts, of course, are of the form $i: M \to M''$. One example is the inclusion $C(S) \to B(S)$, where $S$ is a compact set and $C(S)$ and $B(S)$ denote, respectively, the continuous, bounded, real-valued functions on $S$. Clearly, $C(S)$ and $B(S)$ are commutative Banach algebras with unit. It is well known (for example, see [3]) that $C(S)' = rbaS$ and $B(S)' = baS$, where $(r)baS$ denotes the (regular) bounded additive set functions on the algebra generated by the closed subsets of $S$, so the inclusion $C(S)' \to B(S)'$ provides a $C(S)$-linear right inverse for $B(S)' \to C(S)'$. 5. Finally, an interesting example is provided by considering again the algebra $c$ and the map $f: c \to c \times c$ given by $y((x_n)) = ((x_{2n}),(x_{2n+1}))$. One sees that $f$ is a c-linear map when $c \times c$ is given the c-module structure:

$$((y_n),(z_n))(x_n) = (y_n x_{2n}, z_n x_{2n+1}) .$$

Since $c' = \ell_1$, it can be shown that $f': \ell_1 \amalg \ell_1 \to \ell_1$ is the map $f'((a_n),(b_n)) = (a_0+b_0, a_1, b_1, a_2, b_2, \ldots)$ and that it has a coretraction map $h: \ell_1 \to \ell_1 \amalg \ell_1$ mapping $(a_n)$ to $((b_n),(c_n))$, where $(b_n) = (a_0/2, a_1, a_3, \ldots, a_{2n+1}, \ldots)$, and $(c_n) = (a_0/2, a_2, a_4, \ldots, a_{2n}, \ldots)$. Moreover, it can be verified that $h$ is c-linear, so that $f$ is a descent morphism.

## 3. Characterization of effective descent morphisms

In view of Theorem 2.1 and Corollary 2.6, we see that the condition that $f$ be a weak retract in $Mod_R$ implies that $f$ is a descent morphism but not an effective descent morphism. We are still lacking the condition that $-\otimes_R S$ preserves equalizers of $-\otimes_R S$-split equalizer pairs. (We refer the reader to [12] or [5] for the definition of such equalizer pairs.) Our first result states that it is sufficient for $f$ to be a retract.

**3.1 THEOREM.** Let $f:R \to S$ be a retract in $Mod_R$. Then $F : Mod_R \to Des(f)$ is an equivalence of categories.

PROOF. Since retracts are also weak retracts, it remains to prove that $-\otimes_R S$ preserves equalizers of $-\otimes_R S$-split equalizer pairs. This has been done by Joyal and Tierney [5] by showing that $-\otimes_R S$-split equalizer pairs are already split when $f$ is a retract and, hence, are preserved by any functor, and their proof is equally applicable in our context. □

The notion and properties of retracts are, of course, well known. The following proposition gives a general instance in which a weak retract is always a retract.

**3.2 PROPOSITION.** Let $f:M \to N$ be a weak retract in $Mod_R$. Suppose that $M$ is the Banach space dual of some R-module $P$. Then $f$ is a retract.

PROOF. Let $h:M' \to N'$ be such that $f' \circ h = M'$. Then, as in the proof of 2.3, $L_R(Q,f') \circ L_R(Q,h) = L_R(Q,M')$ for every $Q \in Mod_R$. Therefore, $L_R(Q,f')$ is a surjection for all $Q \in Mod_R$. But $L_R(Q,N') \cong (Q \otimes_R N)' \cong L_R(N,Q')$ and $L_R(Q,M') \cong L_R(M,Q')$, with $L_R(Q,f')$ corresponding to $L_R(f,Q')$ under this isomorphism. Then, choosing $Q = P$, we have a surjection $L_R(f,M):L_R(N,M) \to L_R(M,M)$ and we may find $r \in L_R(N,M)$ such that $r \circ f = M$. □

We pause now to give a few examples of retracts.

**3.3 EXAMPLES.** 1. We know from example 2.7.3 that any canonical map $i:R \to R''$ of commutative Banach algebras, when $R''$ has the Arens multiplication, will be a retract if it happens that $R$ is already the dual space of a Banach module. One candidate for such an $R$ is the commutative Banach algebra $\ell_\infty$ of all bounded real sequences with the sup norm. It is well known that $\ell_1' = \ell_\infty$ and that the space $\ell_1$ is an $\ell_\infty$-module. Thus, any weak retract in $Mod_R$ defined with domain $\ell_\infty$ is a retract. 2. Another example of a retract in $Mod_R$, is given by the obvious map $f : B[-1,1] \longrightarrow B[-1,0] \times B[0,1]$. Clearly, the retraction map is provided by gluing functions together. 3. Of course, for any commutative Banach algebra $R$ with unit and $n \in \mathbb{N}$, $\prod_{i=1}^{n} R$ is a commutative Banach algebra with unit under pointwise multiplication and the sup norm. The obvious diagonal map $d : R \to \prod_{i=1}^{n} R$ is a retract in $Mod_R$ with any one of the projection maps as retraction. If we replace $\prod_{i=1}^{n} R$ by $\coprod_{i=1}^{n} R$ (of course, these are $Ban_\infty$-isomorphic), then $d$ is a retract in $Mod_R^{\infty}$ but not in $Mod_R^{1}$ since $d$ is no longer norm-decreasing in view of the fact that $\|(r)_i\|_{\amalg R} = \sum_{i=1}^{n} \|r\|$. 4. One can also alter the multiplication on $\coprod_{i=1}^{n} R$ and still obtain an R-module and commutative Banach algebra with unit of which $R$ is a retract in $Mod_R$. We define

$$(r_1,\ldots,r_n)(t_1,\ldots,t_n) = (r_1t_1, r_1t_2+r_2t_1+r_1r_2, \ldots, r_1t_n+r_nt_1+r_1r_n) ,$$

and call the resulting algebra $S$. $S$ has $(1,0,\ldots,0)$ as its unit. The map $R \to S$ which maps to $r$ to $(r,0,\ldots,0)$ is an isometric inclusion of Banach algebras, and the R-module structure on $S$ is given by $r(r_1,\ldots,r_n) = (rr_1,\ldots,rr_n)$. Clearly, $R$ is a retract of $S$ with the retraction being provided by the first projection.

We turn now to the problem of identifying particular instances of $S$ for which $-\otimes_R S$ preserves equalizers of $-\otimes_R S$-split equalizer pairs, or indeed all equalizers. The first thing we observe is that once we have found an $S$ such that $-\otimes_R S$ preserves equalizers in $Mod_R^{1}$, then we have found others with this property, too, as the next proposition and corollary describe.

3.4 PROPOSITION. If $-\otimes_R N : Mod_R^{1} \to Mod_R^{1}$ preserves equalizers, and if $g : M \to N$ is a weak retract in $Mod_R^{1}$, then $-\otimes_R M : Mod_R^{1} \to Mod_R^{1}$ also preserves equalizers.

PROOF. Let $K \xrightarrow{\;k\;} A \overset{\alpha}{\underset{\beta}{\rightrightarrows}} B$ be an equalizer in $Mod_R^{1}$. Then since $k\otimes N = eq(\alpha\otimes N, \beta\otimes N)$, we may consider the following diagram, where $q:Q \to A\otimes_R M$ is the equalizer of $(\alpha\otimes M, \beta\otimes M)$ and $m:K\otimes_R M \to Q$ is the unique map such that $q\circ m = k\otimes M$ :

$$\begin{array}{ccccc}
K\otimes_R N & \xrightarrow{k\otimes N} & A\otimes_R N & \overset{\alpha\otimes N}{\underset{\beta\otimes N}{\rightrightarrows}} & B\otimes_R N \\
\uparrow K\otimes g & \nwarrow h \quad Q \quad \searrow q & \uparrow A\otimes g & & \uparrow B\otimes g \\
 & \nearrow m & & & \\
K\otimes_R M & \xrightarrow{k\otimes M} & A\otimes_R M & \overset{\alpha\otimes M}{\underset{\beta\otimes M}{\rightrightarrows}} & B\otimes_R M .
\end{array}$$

Since

$$\alpha\otimes N\circ A\otimes g\circ q = B\otimes g\circ\alpha\otimes M\circ q = B\otimes g\circ\beta\otimes M\circ q = \beta\otimes N\circ A\otimes g\circ q ,$$

there is a unique map $h:Q \to K\otimes_R N$ such that $K\otimes N\circ h = A\otimes g\circ q$ . We note by 2.3 that $A\otimes g$ and $K\otimes g$ are isometric inclusions, which implies, since $k\otimes N$ is one, too, that $k\otimes M$ is isometric. It then follows that $m$ , $q$ , and $h$ are also isometric inclusions. Since

$$k\otimes N\circ h\circ m = A\otimes g\circ q\circ m = A\otimes g\circ k\otimes M = k\otimes N\circ K\otimes g ,$$

we also conclude that $h\circ m = K\otimes g$ .

Now in the bidual of the above diagram, we will have the addition of the maps $(s\_)':(-\otimes N)'' \longrightarrow (-\otimes M)''$ arising from the coretraction $s:M' \to N'$ of $g$ via the maps

$$(-\otimes_R N)' \cong L_R(-,N') \underset{L_R(-,s)}{\overset{L_R(-,g')}{\rightleftarrows}} L_R(-,M') \cong (-\otimes_R M)' ,$$

that is, $(s\_)'$ will denote $L_R(-,s)'$ under this isomorphism. Clearly, $s\_'\circ(-\otimes g)'' = (-\otimes_R M)''$ .

We claim that $(K\otimes_R M)'' = eq((\alpha\otimes M)'',(\alpha\otimes N)'')$ , that is, that $m''$ is an isomorphism. We see easily that

$$s_K'\circ h''\circ m'' = s_K'\circ(K\otimes g)'' = (K\otimes_R M)'' .$$

Moreover,

$$\begin{aligned} q''\circ m''\circ s_K'\circ h'' &= (k\otimes M)''\circ s_K'\circ h'' = s_A'\circ(k\otimes N)''\circ h'' \\ &= s_A'\circ(A\otimes g)''\circ q'' = q'' . \end{aligned}$$

The fact that $q''$ is a monomorphism implies that $m''\circ s_K'\circ h'' = N''$ .

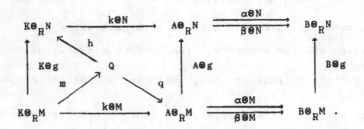

Finally, since $m$ is an isometric inclusion, the following diagram is a pullback

$$\begin{array}{ccc} (K\otimes_R M)'' & \xrightarrow{\ m''\ } & N'' \\ {\scriptstyle i}\uparrow & & \uparrow{\scriptstyle i} \\ (K\otimes_R M) & \xrightarrow[\ m\ ]{} & N\ , \end{array}$$

and the fact that $m''$ is an isomorphism implies that $m$ is, as well. Therefore, $K\otimes_R M$ is the equalizer of $(\alpha\otimes M, \beta\otimes M)$ . □

The usefulness of this proposition is not merely restricted to $\mathrm{Mod}_R^{\ 1}$ in view of the following obvious result.

3.5 LEMMA. If $N \in \mathrm{Mod}_R$ and $\otimes_R N : \mathrm{Mod}_R^{\ 1} \to \mathrm{Mod}_R^{\ 1}$ preserves equalizers, then $\otimes_R M : \mathrm{Mod}_R^{\ \infty} \to \mathrm{Mod}_R^{\ \infty}$ preserves equalizers when $M$ is any R-module such that $N \cong M$ in $\mathrm{Mod}_R^{\ \infty}$ .

PROOF. Given $\alpha, \beta : A \rightrightarrows B$ in $\mathrm{Mod}_R^{\ \infty}$ , we can renorm $B$ in such a way that $\|\alpha\|, \|\beta\| \le 1$ . This, of course, is equivalent to saying that there is a $C \in \mathrm{Mod}_R$ and a $\mathrm{Mod}_R^{\ \infty}$-isomorphism $\varphi : B \to C$ such that $\|\varphi\circ\alpha\| \le 1$ and $\|\varphi\circ\beta\| \le 1$ . Clearly, $eq(\varphi\circ\alpha, \varphi\circ\beta) = eq(\alpha,\beta)$ , call it $K$ . By hypothesis, then, $K\otimes_R N = eq(\varphi\circ\alpha\otimes N, \varphi\circ\beta\otimes N)$ in $\mathrm{Mod}_R^{\ 1}$ . Moreover, since $\varphi\circ\alpha\otimes N = \varphi\otimes N\circ\alpha\otimes N$ with $\varphi\otimes N$ a $\mathrm{Mod}_R^{\ \infty}$-isomorphism, $K\otimes_R N = eq(\alpha\otimes N, \beta\otimes N)$ in $\mathrm{Mod}_R^{\ \infty}$ . Since $-\otimes_R N \cong -\otimes_R M$ in $\mathrm{Mod}_R^{\ \infty}$ , the result follows. □

Thus, we now can turn our attention without loss of generality to finding R-modules $N$ such that $-\otimes_R N$ preserves equalizers in $\mathrm{Mod}_R^{\ 1}$ . The most obvious example is that of $\ell_1(I,R)$ , where $I$ is a set and

$$\ell_1(I,R) = \left\{ (r_i)_{i\in I} \;\middle|\; r_i \in R\ ,\ \sum_i \|r_i\| < \infty \right\}$$

with $\|(r_i)\| = \sum_i \|r_i\|$ . We note that in $\mathrm{Mod}_R^{\ 1}$ , $\ell_1(I,R) = \coprod_i R$ , the coproduct of $R$ over $I$ ; for finite $I$ , $\ell_1(I,R)$ is also the coproduct in $\mathrm{Mod}_R^{\ \infty}$ .

3.6 PROPOSITION. The functor $-\otimes_R \ell_1(I,R) : \mathrm{Mod}_R{}^1 \to \mathrm{Mod}_R{}^1$ preserves equalizers.

PROOF. From the relations $A\otimes_R \coprod_i R \cong \coprod_i (A\otimes_R R) \cong \coprod_i A$ we see that $-\otimes_R \ell_1(I,R)$ is just the copower functor, which clearly preserves isometric inclusions. It is easy from this fact to show that equalizers are then preserved. □

Putting together the results 3.6, 3.4 and 3.5, we obtain as a corollary the following theorem.

3.7 THEOREM. Let $f:R \to S$ be a weak retract in $\mathrm{Mod}_R$, where $S$ is a weak retract of $\ell_1(I,R)$ in $\mathrm{Mod}_R$, then $F : \mathrm{Mod}_R \to \mathrm{Des}(f)$ is an equivalence of categories.

PROOF. By 2.1 and 2.6, it suffices to prove that $-\otimes_R S$ preserves equalizers. If $f \in \mathrm{Mod}_R{}^1$, then by 3.7 and 3.5, $-\otimes_R S$ preserves equalizers in $\mathrm{Mod}_R{}^1$. If $f$ is not norm-decreasing, we can renorm $S$ so that $\|f\| \le 1$, that is, find a $\mathrm{Mod}_R{}^\infty$-isomorphism $\varphi:S\to S_0$ such that $\varphi\circ f:R \to S_0$ is a weak retract in $\mathrm{Mod}_R{}^1$.
Since $S_0$ is still a weak retract of $\ell_1(I,R)$, $-\otimes_R S$ preserves equalizers in $\mathrm{Mod}_R{}^\infty$, which gives us the result. □

Let us examine the type of $S$ that we have identified in the last theorem as yielding an effective descent morphism $f:R \to S$.

3.8 DEFINITION. Let $A \in \mathrm{Mod}_R{}^1$. Then $A$ is said to be projective when the functor $L_R(A,-) : \mathrm{Mod}_R{}^1 \to \mathrm{Mod}_R{}^1$ preserves regular epimorphisms. Clearly, this condition can be expressed by saying that if $p:M \to N$ is a regular epimorphism in $\mathrm{Mod}_R{}^1$, then for every $\mathrm{Mod}_R{}^1$-map $g:A \to N$ there exists a $\mathrm{Mod}_R{}^1$-map $h:A \to M$ such that $p\circ h = g$.

The proof of the following proposition may be found in [6].

3.9 PROPOSITION. $A \in \mathrm{Mod}_R{}^1$ is projective if and only if $A$ is a retract of $\ell_1(I,R)$, for some set I. (N.B.: The direction ($\Leftarrow$) does not hold when $A$ is not unital.)

Thus, we see by 3.7 and 3.9 that in particular $f:R \to S$ is an effective descent morphism in $Mod_R^{\,1}$ whenever $S$ is a projective object in $Mod_R^{\,1}$ or a weak retract of one. On the other hand, the $Mod_R^{\,\infty}$ case is somewhat different. For instance, for $I$ finite, $\ell_\infty(I,R)$, the space of all I-indexed families $(r_i)$ such that $\sup\|r_i\| < \infty$, is topologically isomorphic to $\ell_1(I,R)$. Thus, $f:R \to S$ is an effective descent morphism in $Mod_R^{\,\infty}$ whenever $S$ is a weak retract of $\ell_\infty(I,R)$.

## 4. A Morita theorem

As an application of our theory, we shall prove a Morita theorem giving conditions under which there is an equivalence between the categories of modules over $R$ and modules over $L_R(S,S)$. As before, $R$ and $S$ are understood to be commutative Banach algebras with unit connected by means of a homomorphism $f:R \to S$. $L_R(S,S)$ is, of course, a unital non-commutative Banach algebra with composition of endomorphisms as its multiplication. Our method for obtaining this result will be to utilize Theorem 3.7, which gives conditions to ensure the equivalence of $Mod_R^{\,\infty}$ and $Des(f)_\infty$, and to show that in certain instances $Des(f)_\infty$ is equivalent to $Mod_{L_R(S,S)}$, the category of left $L_R(S,S)$-modules in $Ban_\infty$. Our results will obtain only in the $Ban_\infty$-module setting.

We first notice that there are Banach algebras homomorphisms $R \xrightarrow{f} S \xrightarrow{\nu} L_R(S,S)$ connecting $R$, $S$, and $L_R(S,S)$, where $\nu(s):S \to S$ is the map $\nu(s)(s') = s's$. Therefore, each left $L_R(S,S)$-module can be viewed as an $R$- or an S-module. In particular, $L_R(S,S)$ has the S-module structure given by $(s\alpha)s' = \alpha(s')s$ for $\alpha \in L_R(S,S)$, $s,s' \in S$. Hence, a left $L_R(S,S)$-module $M$ may be presented as an S-module together with a bilinear map of Banach spaces $L_R(S,S) \times M \to M$ or with a map of R-modules $\mu:L_R(S,S)\otimes_R M \to M$ satisfying the usual module properties. We will write $\mu(\alpha\otimes m)$ as $\alpha m$ for simplicity. Clearly, the existence of $\mu$ is equivalent to the existence of a convenient R-linear Banach algebra homomorphism $h:L_R(S,S) \to L_R(M,M)$. Now both $L_R(S,S)$ and $L_R(M,M)$ have $S\otimes_R S$-module structures as well. In fact, for any S-modules $N$ and $P$, $L_R(N,P)$ carries a right $S\otimes_R S$-module structure

$L_R(N,P)\otimes_R S\otimes_R S \to L_R(N,P)$ defined by $\alpha(s_1\otimes s_2)(n) = s_1\alpha(ns_2)$ , for $\alpha \in L_R(N,P)$ , $s_1,s_2 \in S$ , $n \in N$ . It turns out that the morphism h obtained above is actually an $S\otimes_R S$-module homomorphism.

4.1 PROPOSITION. If $M \in \mathrm{Mod}_S$ , then a structure of a left $L_R(S,S)$-module is equivalent to the existence of a $\mathrm{Mod}_R^{\infty}$-homomorphism of Banach algebras $h: L_R(S,S) \to L_R(M,M)$ which is also $S\otimes_R S$-linear.

PROOF. One sees directly that h corresponds to a map of R-modules $\mu: L_R(S,S)\otimes_R M \to M$ satisfying the usual module properties. It remains to show that h is $S\otimes_R S$-linear. Let $\alpha \in L_R(S,S)$ and $s_1\otimes s_2 \in S\otimes_R S$ . Now for $m \in M$ , $h(\alpha)(m) = \alpha m \in M$ . Then

$$h(\alpha(s_1\otimes s_2))(m) = \alpha(s_1\otimes s_2)m = s_1(\alpha(-s_2))m$$

and

$$h(\alpha)(s_1\otimes s_2)(m) = s_1(h(\alpha)ms_2) = s_1(\alpha ms_2) .$$

But $s_1(\alpha(-s_2))m = s_1(\alpha(ms_2)) = s_1(\alpha ms_2)$ by the module action property $(\beta\alpha)m = \beta(\alpha m)$ . □

We would now like to show that there is a one-to-one correspondence between $S\otimes_R S$-linear maps from $S\otimes_R M$ to $M\otimes_R S$ and $S\otimes_R S$-linear maps from $L_R(S,S)$ to $L_R(M,M)$ for certain S-modules M . The map we have in mind is the following one:

$$(*) \qquad \lambda: L_{S\otimes_R S}(S\otimes_R M, M\otimes_R S) \to L_{S\otimes_R S}(L_R(S,S), L_R(M,M))$$

defined by $\lambda(\varphi)(\alpha)(m) = \lim_k \sum_q m_{kq}\alpha(s_{kq})$ , for where $\varphi \in L_{S\otimes_R S}(S\otimes_R M, M\otimes_R S)$ , $\alpha \in L_R(S,S)$ , $m \in M$ , and $\varphi(1\otimes m) = \lim_k \sum_q m_{kq}\otimes s_{kq}$ .

That $\lambda$ does indeed preserve $S\otimes_R S$ maps is shown by the following calculations. Let $\varphi: S\otimes_R M \to M\otimes_R S$ be $S\otimes_R S$-linear. Then

$$\begin{aligned}\lambda(\varphi)(\alpha(s_1\otimes s_2))(m) &= \lim_k \sum_q m_{kq}\alpha(s_1\otimes s_2)(s_{kq}) ,\\ &= \lim_k \sum_q m_{kq}s_1\alpha(s_2 s_{kq}) ,\end{aligned}$$

and $(\lambda(\varphi)(\alpha)(s_1\otimes s_2))(m) = s_1\lambda\varphi(\alpha)(ms_2)$ . But since $\varphi$ is $S\otimes_R S$-linear and by the definition of the $S\otimes_R S$-module structure on $L_R(M,M)$ ,

$$\begin{aligned}
\varphi(1\otimes ms_2) = \varphi(1\otimes m)(1\otimes s_2) &= (\lim_k \sum_q m_{kq}\otimes s_{kq})(1\otimes s_2)\,, \\
&= \lim_k \sum_q m_{kq}\otimes s_2 s_{kq}\,, \\
s_1\lambda\varphi(\alpha)(ms_2) &= s_1 \lim_k \sum_q m_{kq}\alpha(s_2 s_{kq}) \\
&= \lim_k \sum_q m_{kq} s_1 \alpha(s_2 s_{kq})\,,
\end{aligned}$$

which establishes that $\lambda(\varphi)$ is $S\otimes_R S$-linear.

In the case that $M \in \mathrm{Mod}_R^{\infty}$ is a retract of $\ell_1(I,R)$, for $I$ finite, we will be able to establish this correspondence and restrict it to get the desired result about $\mathrm{Des}(f)_{\infty}$. We note that in $\mathrm{Mod}_R^{\infty}$, $\ell_1(I,R)$ is both the coproduct and product of $R$ over $I$; hence, we may rephrase this condition on $M$ by saying that $M$ is a retract of a finite biproduct $\bigoplus_{i=1}^{n} R$. A series of lemmas will be needed to establish that $\lambda$ is an isomorphism in this case.

**4.2 LEMMA.** Let $M$ be an S-module which is a retract in $\mathrm{Mod}_R^{\infty}$ of $\bigoplus_{i=1}^{n} R$ for some $n \in \mathbb{N}$. For any S-module $N$, the morphism $\pi_{M,N}: N\otimes_R M \longrightarrow L_R(L_R(M,R),N)$ defined by $\pi_{M,N}(n\otimes m)(\alpha) = n(\alpha(m))$ is an isomorphism in $\mathrm{Mod}_{S\otimes_R S}^{\infty}$.

PROOF. It is easy to check that $\pi_{M,N}$ is a R-module map, $S\otimes_R S$-linear, and natural in both $M$ and $N$. Suppose that $M = \bigoplus_{i=1}^{n} R$. Then $N\otimes_R M$ and $L_R(L_R(\bigoplus_{i=1}^{n} R,R),N)$ are both isomorphic to $\bigoplus_{i=1}^{n} N$ with $\pi_{M,N} = \bigoplus_{i=1}^{n} \pi_{R,N}$. Since $\pi_{R,N}$ is clearly an isomorphism, so is $\pi_{M,N}$.

Now suppose that the result holds for $M$ and let $P$ be a retract of $M$ in $\mathrm{Mod}_R^{\infty}$ with maps $P \underset{i}{\overset{r}{\rightleftarrows}} M$ such that $r\circ i = P$. Then, by naturality of $\pi$, we obtain the following commutative diagram:

$$\begin{array}{ccc}
N\otimes_R M & \xrightarrow{\ \pi_{M,N}\ } & L_R(L_R(M,R),N) \\
N\otimes r \downarrow \uparrow N\otimes i & & L_R(L_R(r,R),N) \downarrow \uparrow L_R(L_R(i,R),N) \\
N\otimes_R P & \xrightarrow[\ \pi_{P,N}\ ]{} & L_R(L_R(P,R),N)\,.
\end{array}$$

One can easily check that $N\otimes r\circ\pi_{M,N}{}^{-1}\circ L_R(L_R(i,R),N) = \pi_{P,N}{}^{-1}$, establishing the result. □

4.3 LEMMA. Let $M$ be an S-module which is a retract in $\mathrm{Mod}_R^{\infty}$ of $\bigoplus_{i=1}^{n} R$ for some $n \in \mathbb{N}$. The morphism $\theta_M : M\otimes_R L_R(M,R) \longrightarrow L_R(M,M)$ defined by $\theta_M(m\otimes\alpha)(m') = m(\alpha(m'))$, is an isomorphism in $\mathrm{Mod}^{\infty}_{S\otimes_R S}$.

PROOF. The proof is similar to that of Lemma 4.2. □

4.4 LEMMA. For any S-modules $M$, $N$, $P$, $Q$, we have the following isomorphism:

$$\gamma_{M,N,P,Q} : L_{S\otimes S}(M\otimes_R N, L_R(P,Q)) \cong L_{S\otimes S}(M\otimes_R P, L_R(N,Q)) .$$

PROOF. The obvious isomorphism clearly holds when $L_{S\otimes S}$ is replaced by $L_R$. Moreover, it is easy to check that this isomorphism takes $S\otimes_R S$-linear morphisms to $S\otimes_R S$-linear morphisms. □

We now return to the map $\lambda$ defined by (*). An analysis of $\lambda$ shows that it fits into the following diagram.

$$\begin{array}{ccc}
L_{S\otimes_R S}(S\otimes_R M, M\otimes_R S) & \xrightarrow{\;L_{S\otimes_R S}(S\otimes_R M,\pi_{S,M})\;} & L_{S\otimes_R S}(S\otimes_R M, L_R(L_R(S,R),M)) \\
\Big\downarrow \lambda & & \Big\downarrow \gamma \\
L_{S\otimes_R S}(L_R(S,S), L_R(M,M)) & \xrightarrow[\;L_{S\otimes_R S}(\theta_S, L_R(M,M))\;]{} & L_{S\otimes_R S}(S\otimes_R L_R(S,R), L_R(M,M)),
\end{array}$$

which can be shown to be commutative by a standard calculation. This proves that $\lambda = L_{S\otimes_R S}(\theta_R^{-1}, L_R(M,M))\circ\gamma\circ L_{S\otimes_R S}(S\otimes_R M, \pi_{S,M})$ which gives us the following theorem.

4.4 THEOREM. If $M$ is an S-module which is a retract in $\mathrm{Mod}_R^{\infty}$ of $\bigoplus_{i=1}^{n} R$ for some $i \in \mathbb{N}$, then the map

$$\lambda : L_{S\otimes_R S}(S\otimes_R M, M\otimes_R S) \to L_{S\otimes_R S}(L_R(S,S), L_R(M,M))$$

is an isomorphism in $Mod_R^\infty$ .

PROOF. We put the comment preceding the statement of the theorem together with Lemmas 4.2-4.4. □

Upon further examination of $\lambda$ , we obtain the following result.

4.5 PROPOSITION. The isomorphism $\lambda$ retricts to a bijection between pairs $(M,\varphi)$ which are descent data and pairs $(M,\psi)$ , where $\psi$ provides an $L_R(S,S)$-module structure on the S-module M ; thus, it determines an equivalence of the categories $Des(f)_\infty$ and $Mod_{L_R(S,S)}^\infty$.

PROOF. The arguments used to establish this result are similar to those involved in establishing the equivalence of the categories Des(f) and Coalg . □

4.6 COROLLARY (Morita Theorem). Let $f:R \to S$ be a weak retract of Banach algebras in $Mod_R^\infty$ , where S is a retract in $Mod_R^\infty$ of $\bigoplus_{i=1}^n R$ , for some $n \in \mathbb{N}$ . Then the categories $Mod_R^\infty$ and $Mod_{L_R(S,S)}^\infty$ are equivalent.

PROOF. The result follows directly from Theorems 3.7 and 4.5. □

We close by recalling that some of the examples mentioned earlier fit into the context of the Morita theorem.

4.7 EXAMPLES. 1. As discussed in 3.4.3, for any commutative Banach algebra R with unit, the finite biproduct $S = \bigoplus_{i=1}^n R$ has R as a retract in $Mod_R^\infty$ under the diagonal map. Thus, $Mod_R^\infty \sim Mod_{L_R(S,S)}^\infty$ . 2. The module S described in 4.4.4, namely $\bigoplus_{i=1}^n R$ with a multiplication defined by

$$(r_1,\ldots,r_n)(t_1,\ldots t_n) = (r_1t_1, r_1t_2+r_2t_1+r_1r_2, \ldots r_1t_n+r_nt_1+r_1r_n) ,$$

also has $R$ as a $\mathrm{Mod}_R^\infty$-retract under the embedding $r \mapsto (r,0,\ldots,0)$. Therefore, $\mathrm{Mod}_R^\infty \sim \mathrm{Mod}_{L_R(S,S)}^\infty$.

## REFERENCES


1. Cigler, J., Losert, V., and Michor, P., Banach modules and functors on categories of Banach spaces, Lecture Notes in Pure Appl. Math. 46, Marcel Dekker, Inc., New York and Basel, 1979.

2. Civin, P., and Yood, B., The second conjugate space of a Banach algebra as an algebra, Pac. J. Math. 11 (1961), 847-870.

3. Dunford, N. and Schwartz, J., Linear Operators Part I, Interscience Publishers, Inc., New York, Fourth Printing, 1967.

4. Graven, A.W.M., Injective and Projective Banach Modules, Nederl. Akad. Wetensch. Indag. Math. 41 (1979), 253-272.

5. Joyal, A. and Tierney, M, An extension of the Galois theory of Grothendieck, Memoirs Amer. Math. Soc. 51 (1984).

6. Kaijser, S., On Banach Modules I, Math. Proc. Camb. Phil. Soc. 90 (1981), 423-444.

7. Khelemskiĭ, A. Ya., Flat Banach modules and amenable algebras, Trans. Moscow Math. Soc. (1985), 199-244.

8. Knus, M.-A., and Ojanguren, M., Théorie de la Descente et Algébres d'Azumaya, Lecture Notes in Math. 389, Springer-Verlag, Berlin, Heidelberg, New York, 1975.

9. Rieffel, M.A., Induced Banach representations of Banach algebras and locally compact groups, J. Fnl. Anal. 1 (1967), 443-491.

10. Rieffel, M.A., Induced representation of $C^*$-algebras, Adv. in Math. 13 (1974), 176-257.

11. Rieffel, M.A., Morita equivalence for $C^*$-algebras and $W^*$-algebras, J. Pure Appl. Alg. 5 (1974), 51-96.

12. Schubert, H., Categories, Springer-Verlag, New York, Heidelberg, Berlin, 1972.

**Pseudofunctors and non-abelian weak equivalences.**

**by Dominique BOURN.** (*)

The main tool for the development of a cohomology theory has tradionally been the notion of chain complex. Various attempts have been made to give a non abelian equivalent to this notion. The most usual equivalent is the notion of simplicial object, since, in any abelian category, the two notions coïncide via the Dold-Kan theorem. Duskin [9] and Glenn [10],for instance, used it in their works on cohomology. Another is the notion of crossed complex, extensively used by Brown and Higgins [7] in their generalization of the Seifert-Van Kampen theorem.

There is one more candidate in the concept of n-groupoïd , since, given an abelian category $\mathbb{A}$, there is, for each integer $n$, a natural (in $n$) equivalence between the category $C^n(\mathbb{A})$ of positive chain complexes of length $n$ (n-complexes for short) and the category n-Grd $\mathbb{A}$ of internal n-groupoïds in $\mathbb{A}$ [5].

Furthermore, while the usual Dold-Kan equivalence between the category $C^{\cdot}(\mathbb{A})$ of positive chain complexes and the category Simpl $\mathbb{A}$ of simplicial objects in $\mathbb{A}$ did not strictly exchange chain homotopies with simplicial homotopies, the above equivalences do strictly exchange chain homotopies with what is known as (higher) pseudonatural transformations [5]. In that sense, the two notions of n-complexes and n-groupoïds seem to be more closely connected.

It is therefore justified, starting from the notion of weak equivalence between chain complexes, to investigate the nature of its image in n-Grd $\mathbb{A}$. This is the aim of this paper.

When $n = 1$, the weak chain equivalences (between 1-complexes) are in one to one correspondance with the usual internal weak equivalences (i.e. internally fully faithful and essentially surjective functors).

At level 2, we are in the following situation :

$$\text{2-Grd } \mathbb{A} \xrightarrow{\ (\ )_1\ } \text{Grd } \mathbb{A}$$

The forgetful functor $(\ )_1$, has a fully faithful right adjoint Gr and consequently is a fibration [3]. It has also a left adjoint dis (the discrete functor) which admits itself a left adjoint $\pi_1$. The weak chain equivalences (between the 2-complexes) are in one to one correspondance with the internal 2-functors $f_2 : X_2 \to Y_2$ such that :

(*) This paper is in final form and will not be published elsewhere.

(1) $f_2$ is $(\ )_1$-cartesian ;

(2) $\pi_1(f_2)$ is a weak equivalence in Grd /A.

Such a 2-functor will be called a weak 2-equivalence.

Now at level n, we have the same situation :

$$n\text{-Grd}\ /\!A \xrightarrow{(\ )_{n-1}} (n-1)\ \text{Grd}\ /\!A$$

with analogous Gr, dis and $\pi_{n-1}$. The weak equivalences (between n-complexes) are in one to one correspondance with the internal n-functors $f_n : X_n \to Y_n$ such that :

(1) $f_n$ is $(\ )_{n-1}$-cartesian ;

(2) $\pi_{n-1}(f_n)$ is a weak (n - 1)-equivalence

such a n-functor being called a weak n-equivalence.

The ultimate equivalence between $C^{\cdot}(/\!A)$ and $\infty$-Grd /A exchanges the weak equivalences with the $\infty$-functors $f_\infty : X_\infty \to Y_\infty$ having their projections $f_n$ in n-Grd /A such that the $\pi_{n-1}(f_n)$ are weak (n - 1)-equivalences. Part 1 is devoted to establish these points.

Now, it is clear that the definition of a weak n-equivalence has a meaning not only in an abelian category A but also in any exact category E. In particular, the question may be asked of what these notions become in the set theoretical context. Such is the aim of Part 2.

At level 1, it is well known that, thanks to the axiom of choice, the notion of weak equivalence between ordinary groupoids coincides with that of ordinary equivalences (i.e. functors with an inverse up to isomorphisms). Here we shall only examine the highest commonly used categorical level, that is level 2. In the same way as, at level 1, the axiom of choice allows us to associate to each weak equivalence its inverse functor, it also allows us, at level 2, to associate to each weak 2-equivalence $f_2 : X_2 \to Y_2$, an inverse pseudofunctor $\varphi_2 : Y_2 \dashrightarrow X_2$ with appropriate pseudo-natural transformations $\varepsilon_2 : 1_{Y_2} \Rightarrow f_2 \cdot \varphi_2$ and $\gamma_2 : 1_{X_2} \Rightarrow \varphi_2 \cdot f_2$ and appropriate modification. Conversely a 2-functor $f_2 : X_2 \to Y_2$ having such a pseudo-inverse is a weak 2-equivalence. This observation is a new example of the close connexion between n-complexes and n-groupoids and of how these two notions are mutually enlightened.

On the other hand, the category 2-Grd of ordinary 2-groupoids being equivalent to the category of crossed modules over groupoids [7], the 2-weak equivalence are shown to be in one to one correspondance with what is expected to be a weak equivalence of crossed modules over groupoids.

Finally it is pointed out that, in the category 2-Grd, any pseudofunctor $\psi_2 : U_2 \dashrightarrow V_2$ has a canonical (up to isomorphism) representation :

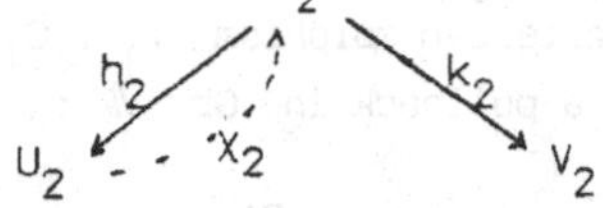


with $k_2$ and $h_2$ two 2-functors, the 2-functor $h_2$ having $\chi_2$ as a strict ($h_2 \cdot \chi_2) = 1$) pseudo-inverse (and being therefore a weak 2-equivalence) such that $k_2 \cdot \chi_2 = \psi_2$ . So a pseudo-functor appears as nothing but a "weakly representable 2-profunctor".

PART I. :

Let $\mathbb{A}$ be an abelian category.

1. The basic denormalization :

Let us recall that an internal category $C_1$ in $\mathbb{A}$ is an internal diagram :

$$C_0 \underset{d_1}{\overset{d_0}{\leftleftarrows}} \overset{d_0}{\rightarrow} m\,C_1 \underset{d_2}{\overset{d_0,\ d_1}{\Lleftarrow}} m_2\,C_1$$

where $C_0$ is called the object of object, $m\,C_1$ the objects of morphisms and where the object $m_2\,C_1$ of composable pairs of morphisms is the pull back of do along $d_1$ It must satisfy the usual conditions (see for instance [13] p. 47), briefly these of a simplicial object as far as level 3, when completed by the pullback of $d_0$ along $d_2$ . The internal functors are just natural transformations between such diagrams. An internal category is an internal groupoid, when furthermore, the following diagram is a pullback :

$$C_0 \xleftarrow{\ d_1\ } m\,C_1 \underset{d_2}{\overset{d_1}{\leftleftarrows}} m_2\,C_1$$

Now, the category $\mathbb{A}$ being abelian, an internal category is always an internal groupoid (see for instance [5], prop. 3). Let us denote by Grd $\mathbb{A}$ the category of the intenal groupoid in $\mathbb{A}$. It is not difficult to check that Grd $\mathbb{A}$ is again an abelian category.

We have a forgetful functor $(\ )_0$ : Grd $\mathbb{A} \to \mathbb{A}$, associating $C_0$ to $C_1$, which has a fully faithful right adjoint Gr, given for each object $C$ of $\mathbb{A}$ by the kernel equivalence associated to the terminal map $C \to 1$. This functor $(\ )_0$

has also a fully faithful left adjoint dis, defined for each object C by the groupoid whose any structural map is the identity on C. This functor dis has itself a left adjoint $\pi_0$, where $\pi_0(C)$ is the cokernel of $d_0$ and $d_1$.

The left exact functor $(\ )_0$, having a right adjoint, is actually a fibration. Let us recall that a $(\ )_0$-cartesian morphism $f_1 : C_1 \to C_1'$ is then a morphism such that the following square is a pullback in Grd /A :

$$\begin{array}{ccc} C_1 & \xrightarrow{f_1} & C_1' \\ \downarrow & & \downarrow \\ \mathrm{Gr}\, C_0 & \xrightarrow[\mathrm{Gr}\, f_0]{} & \mathrm{Gr}\, C_0' \end{array}$$

This condition is reduced here to the fact that the following square is a pullback in /A :

$$\begin{array}{ccc} m\, C_1 & \xrightarrow{m\, f_1} & m\, C_1' \\ {\scriptstyle \binom{d_0}{d_1}}\downarrow & & \downarrow{\scriptstyle \binom{d_0'}{d_1'}} \\ C_0 \times C_0 & \xrightarrow[f_0 \times f_0]{} & C_0' \times C_0' \end{array}$$

That is : $f_1$ is internally fully faithful.
A morphism $f_1$ will be said $(\ )_0$-invertible when its image by $(\ )_0$ is invertible.

Let us denote by $C^1(/A)$ the category of 1-complexes in /A, whose objects d are just the morphisms $d : D_1 \to D_0$ and morphisms $f : d \to d'$ the commutative squares :

$$\begin{array}{ccc} D_1 & \xrightarrow{f_1} & D_1' \\ d\downarrow & & \downarrow d' \\ D_0 & \xrightarrow[f_0]{} & D_0' \end{array}$$

We have a forgetful functor $(\ )_0 : C^1(/A) \to /A$ associating $D_0$ to d which has a fully faithful right adjoint Kr defined by $\mathrm{Kr}(D) = 1 : D \to D$. This functor $(\ )_0$ has also a fully faithful left adjoint t, defined by $t(D) = 0 \to D$, which has itself a left adjoint Cok, where Cok(d) is the cokernel of the morphism d.

Then this functor $(\ )_0$ is again a fibration. A morphism $f : d \to d'$ is $(\ )_0$-cartesian if and only if the previous square associated to f is a pullback.

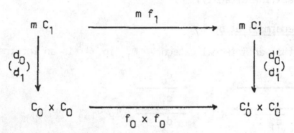

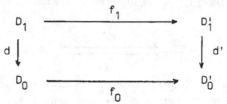

Now, the basic denormalization is well known :

$$\mathrm{Grd}\,/\!\mathcal{A} \underset{D}{\overset{N}{\rightleftarrows}} C^1(/\!\mathcal{A})$$

Where $N(C_1)$ is given by the composite : $\mathrm{Ker}\, d_1 \rightarrowtail m\, C_1 \xrightarrow{d_0} C_0$ , and, if we denote a morphism between $X_1 \times X_2 \times \ldots \times X_n$ and $Y_1 \times Y_2 \times \ldots Y_p$ by a $n \times p$ matrix of morphisms, $D(d)$ is given by the following diagram :

$$D_0 \;\underset{(0,\,1)}{\overset{(d,\,1)}{\leftleftarrows}}\!\!\rightarrow\; D_1 \times D_0 \;\overset{\begin{pmatrix}1&0&0\\0&d&1\end{pmatrix},\ \begin{pmatrix}1&1&0\\0&0&1\end{pmatrix},\ \begin{pmatrix}0&1&0\\0&0&1\end{pmatrix}}{\longleftarrow}\; D_1 \times D_1 \times D_0$$

It is clear that $(\ )_0 \cdot N = (\ )_0$ and $(\ )_0 \cdot D = (\ )_0$. Furthermore the three other pairs of functors (Gr, Kr), (dis, t) and ($\pi_0$, Cok) commute through this equivalence up to natural isomorphisms.

Therefore, these equivalences commuting with $(\ )_0$ , with Gr and Kr up to isomorphism and being left exact, do exchange $(\ )_0$-cartesian morphisms with $(\ )_0$-cartesian morphisms and trivially $(\ )_0$-invertible ones with $(\ )_0$-invertible ones.

2. <u>The weak equivalences at level 1</u> :

A 1-complex can be considered as a complex with no information beyond level 1 and so, as a complex with 0 beyond this stage. So $H^1(d)$ for any object d of $C^1(/\!\mathcal{A})$ is just the kernel of the morphism d and $H^0(d)$ its cokernel. Therefore a weak equivalence f between 1-complexes is such that its extensions Ker f and Cok f to the kernels and the cokernels are isomorphisms :

$$\begin{array}{ccc}
\mathrm{Ker}\, d & \overset{\mathrm{Ker}\, f}{\dashrightarrow} & \mathrm{Ker}\, d' \\
\downarrow & & \downarrow \\
D_1 & \xrightarrow{f_1} & D'_1 \\
d\downarrow & & \downarrow d' \\
D_0 & \xrightarrow[f_0]{} & D'_0 \\
\downarrow & & \downarrow \\
\mathrm{Cok}\, d & \underset{\mathrm{Cok}\, f}{\dashrightarrow} & \mathrm{Cok}\, d'
\end{array}$$

Such squares as $(d, f_1, f_0, d')$ are known as biexact squares (see for instance [12]). They satisfy any of the three following equivalent properties :

(i) to be a pushout and pullback.

(ii) to be such that the following sequence is exact :

$$0 \longrightarrow D_1 \xrightarrow[\binom{d}{f_1}]{} D_0 \times D_1' \xrightarrow[(f_0,\, d')]{} D_0' \longrightarrow 0$$

(iii) ($\alpha$) to be a pullback and ($\beta$) having Cok f an epi.

The third condition is equivalent to ($\alpha$) : f is $(\ )_0$-cartesian and ($\beta$) : Cok f is an epi. Now, according to the previous properties of N and D, f is a weak equivalence if and only if ($\alpha$) : D f is $(\ )_0$-cartesian and ($\beta$) : $\pi_0(D(f))$ is an epi. The condition ($\alpha$) is equivalent to : D(f) is internally fully faithful and the condition ($\beta$) to : D(f) is essentially surjective. In other words the condition ($\alpha$) and ($\beta$) are equivalent to : D(f) is an internal weak equivalence, in the usual sense.

We can sum up this result in the following proposition :

Proposition 1. : At level 1, the functors N and D exhange weak chain equivalences with internal weak equivalences.

3. Denormalization at level 2. :

An internal 2-groupoid [3] in $\mathbb{A}$ is an internal groupoid $C_2$ in Grd $\mathbb{A}$ :

$$C_1 \;\underset{d_1}{\overset{d_0}{\leftleftarrows}}\; m\, C_2 \;\underset{d_2}{\overset{d_0,\ d_1}{\Lleftarrow}}\; m_2\, C_2$$

such that its image by the functor $(\ )_0$ is a discrete groupoid in $\mathbb{A}$, in other words such that each structural map of $C_2$ is $(\ )_0$-invertible. Clearly it is sufficient that any structural map is $(\ )_0$-invertible. A 2-functor is a natural transformation of such diagrams in Grd $\mathbb{A}$. Let us denote by 2-Grd $\mathbb{A}$ the category of internal 2-groupoids in $\mathbb{A}$.

Again we have a forgetful functor $(\ )_1$ : 2-Grd $\mathbb{A} \to$ Grd $\mathbb{A}$ , associating $C_1$ to $C_2$ , which has a fully faithful right adjoint Gr , given, for each internal groupoid $C_1$ , by the kernel equivalence associated to $C_1 \to \mathrm{Gr}\, C_0$ :

$$\mathrm{Gr}\, C_0 \leftarrow \text{- - -}\; C_1 \;\underset{p_1}{\overset{p_0}{\leftleftarrows}}\; C_1 \times_0 C_1 \;\Lleftarrow\; C_1 \times_0 C_1 \times_0 C_1$$

where $(p_0, p_1)$ is the kernel pair associated to the dotted arrow, or equivalently where $C_1 \times_0 C_1$ is the product in the fiber above $C_0$ . This functor $(\ )_1$ is thus again a fibration. It has also a fully faithful left adjoint dis, defined as at level 1. This functor dis has itself a left adjoint $\pi_1$ , where $\pi_1(C_2)$ is the cokernel of $d_0$ and $d_1$.

Let us denote by $C^2(A)$ the category of 2-complexes in $\mathbb{A}$, whose objects $d$ are the sequences $D_2 \xrightarrow{d} D_1 \xrightarrow{d} D_0$ with $d^2 = 0$, and morphisms the chain transformations. We have a forgetful functor $(\ )_1 : C^2(\mathbb{A}) \to C^1(\mathbb{A})$, namely the truncation of the last element.

It has a fully faithful right adjoint $Kr$, namely the augmentation by the kernel :

$$Kr(d) = Ker\ d \rightarrowtail D_1 \xrightarrow{d} D_0 \ .$$

Consequently the functor $(\ )_1$ is a fibration whose $(\ )_1$-cartesian morphisms $f$ are the chain transformations such that : the following square is a pullback :

$$
\begin{array}{ccc}
D_2 & \xrightarrow{\ \ f_2\ \ } & D_2' \\
\downarrow & & \downarrow \\
Ker\ d & \xrightarrow[Ker(f)_1]{} & Ker\ d'
\end{array}
$$

This functor has also a fully faithful left adjoint $t$, namely the augmentation by $0$, which has itself a left adjoint $Cok_1$ where $Cok_1(d)$ is defined by the following dotted arrow :

$$
\begin{array}{ccc}
D_2 & \xrightarrow{\ d\ } & D_1 \xrightarrow{\quad} Cok\ d \\
 & {\scriptstyle d}\downarrow & \nearrow Cok_1(d) \\
 & D_0 &
\end{array}
$$

Now the denormalization at level 2 has the following properties (see [5]) :

$$
\begin{array}{ccc}
\text{2-Grd A} & \underset{D_2}{\overset{N_2}{\rightleftarrows}} & C^2(A) \\
{\scriptstyle (\ )_1}\downarrow & & \downarrow{\scriptstyle (\ )_1} \\
\text{Grd A} & \underset{D}{\overset{N}{\rightleftarrows}} & C^1(A)
\end{array}
$$

The two previous squares commute. Furthermore, the three other pairs of functors $(Gr, Kr)$, $(dis, t)$, $(\pi_1, Cok_1)$ commute through these equivalences up to isomorphism.

### 4. The weak equivalences at level 2. :

A chain transformation $f : d \to d'$ is a weak equivalence when its extensions to $H^0$, $H^1$, $H^2$ are isomorphisms. So, let us consider the following diagram :

$$
\begin{array}{ccccccc}
& & Ker\ d & & Cok\ d & & \\
& {\scriptstyle \delta}\nearrow & \downarrow{\scriptstyle i} & {\scriptstyle \rho}\nearrow & & \searrow{\scriptstyle Cok_1(d)} & \\
Ker\ d \rightarrowtail & D_2 & \xrightarrow{\ d\ } & D_1 & \xrightarrow{\ d\ } & D_0 & \twoheadrightarrow Cok\ d
\end{array}
$$

Let us denote by $\delta$ the factorization of $d : D_2 \to D_1$ through $\mathrm{Ker}\, d$, by $i$ the inclusion : $\mathrm{Ker}\, d \to D_1$ and by $\rho : D_1 \to \mathrm{Cok}\, d$ the cokernel.

The $H^0(d)$ is $\mathrm{Cok}\, d$, which is also ($\rho$ being an epi) the cokernel of $\mathrm{Cok}_1(d)$. The $H^2(d)$ is $\mathrm{Ker}\, d$ which is also ($i$ being a mono) the kernel of $\delta$. The $H^1(d)$ is the kernel of $\mathrm{Cok}_1(d)$ or equivalently the cokernel of $\delta$. Thus we have the two following diagrams where the columns are exact :

$$
\begin{array}{ccc}
H^1(d) & \xrightarrow{H^1(f)} & H^1(d') \\
\downarrow & & \downarrow \\
\mathrm{Cok}\, d & \xrightarrow{\mathrm{Cok}\, f_1} & \mathrm{Cok}\, d' \\
{\scriptstyle \mathrm{Cok}_1(d)} \downarrow & (1) & \downarrow {\scriptstyle \mathrm{Cok}_1(d')} \\
D_0 & \xrightarrow{f_0} & D'_0 \\
\downarrow & & \downarrow \\
H^0(d) & \xrightarrow{H^0(f)} & H^0(d')
\end{array}
\qquad ; \qquad
\begin{array}{ccc}
H^2(d) & \xrightarrow{H^2(f)} & H^2(d') \\
\downarrow & & \downarrow \\
D_2 & \xrightarrow{f_2} & D'_2 \\
{\scriptstyle \delta} \downarrow & (2) & \downarrow {\scriptstyle \delta} \\
\mathrm{Ker}\, d & \xrightarrow{\mathrm{Ker}\, f_1} & \mathrm{Ker}\, d' \\
\downarrow & & \downarrow \\
H^1(d) & \xrightarrow{H^1(f)} & H^1(d')
\end{array}
$$

whence the following proposition :

<u>Proposition 2.</u> : The chain transformation $f$ is a weak equivalence if and only if the squares (1) and (2) are biexact or equivalently (1) is biexact ($\alpha$) and (2) left exact (a pullback) ($\beta$).

<u>Remark.</u> : The condition ($\alpha$) means exactly that $\mathrm{Cok}_1(f)$ is a weak equivalence, the condition ($\beta$) that $f$ is $(\ )_1$-cartesian. Now, according to the properties of $N$, $D$, $N_2$, $D_2$, ($\alpha$) is equivalent to : $\pi_1(D_2(f))$ is a weak equivalence and ($\beta$) to : $D_2(f)$ is $(\ )_1$-cartesian.

Whence the following definition and proposition :

<u>Definition 1.</u> : A 2-functor $f_2 : X_2 \to Y_2$ is called a weak 2-equivalence when :

(1) $f_2$ is $(\ )_1$-cartesian ; (2) $\pi_1(f_2)$ is a weak equivalence.

<u>Proposition 3.</u> : The functors $N_2$ and $D_2$ exchange weak chain equivalences with weak 2-equivalences.

## 5. <u>The denormalization at level n.</u> :

We shall define n-groupoids by induction. Let us suppose already defined the categories $(n-2)$-Grd/$\mathbb{A}$ and $(n-1)$-Grd/$\mathbb{A}$ of respectively internal $(n-2)$-

groupoids and (n - 1)-groupoids in $\mathbb{A}$, with a forgetful functor :

$$(\ )_{n-2} : (n-1)\text{-}\mathrm{Grd}\,\mathbb{A} \longrightarrow (n-2)\text{-}\mathrm{Grd}\,\mathbb{A}$$

having a fully faithful right adjoint Gr.

<u>Definition 2.</u> : An internal n-groupoid $C_n$ in $\mathbb{A}$ is an internal groupoid in $(n-1)\text{-}\mathrm{Gr}\ \mathbb{A} : C_{n-1} \leftleftarrows m\,C_n \;\substack{\leftarrow\\ \leftarrow\\ \leftarrow}\; m_2\,C_n$ whose image by $(\ )_{n-2}$ is discrete.

We have a forgetful functor $(\ )_{n-1} : n\text{-}\mathrm{Gr}\,\mathbb{A} \rightarrow (n-1)\text{-}\mathrm{Grd}\,\mathbb{A}$ associating $C_{n-1}$ to $C_n$. It has a fully faithful right adjoint Gr associating to $C_{n-1}$ the groupoid associated to the kernel pair of $C_{n-1} \rightarrow \mathrm{Gr}\ C_{n-2}$. The functor $(\ )_{n-1}$ has again a fully faithful left adjoint dis which has a left adjoint $\pi_{n-1}$.

On the other hand, let us denote by $C^i(\mathbb{A})$ the category of i-complexes in $\mathbb{A}$, whose objects are the positive complexes of length i and morphisms the chain transformations.
Let us denote by $(\ )_{n-1} : C^n(\mathbb{A}) \rightarrow C^{n-1}(\mathbb{A})$ the truncation of the last element. This functor has a fully faithful right adjoint Kr, namely the augmentation by the kernel. It has also a fully faithful left adjoint t, namely the augmentation by 0, which has itself a left adjoint $\mathrm{Cok}_n$, defined in the same way as at level 2.

Now the denormalization at level n has the following properties (see [5]) :

$$\begin{array}{ccc}
n\text{-}\mathrm{Gr}\,\mathbb{A} & \underset{D_n}{\overset{N_n}{\rightleftarrows}} & C^n(\mathbb{A}) \\
{\scriptstyle (\ )_{n-1}}\downarrow & & \downarrow{\scriptstyle (\ )_{n-1}} \\
(n-1)\text{-}\mathrm{Grd}\,\mathbb{A} & \underset{D_{n-1}}{\overset{N_{n-1}}{\rightleftarrows}} & C^{n-1}(\mathbb{A})
\end{array}$$

In the same way as at level 2, the two previous squares commute. Furthermore the three other pairs (Gr, Kr), (dis, t) and $(\pi_{n-1}, \mathrm{Cok}_{n-1})$ commute through these equivalences up to isomorphism.

These equivalences, commuting with $(\ )_{n-1}$, with Gr and Kr up to isomorphism, and being left exact, do exchange $(\ )_{n-1}$-cartesian morphisms with $(\ )_{n-1}$ cartesian. The same is trivially true for the $(\ )_{n-1}$-invertible morphisms.

6. <u>The weak equivalence at level n.</u> :

By induction, let us define the notion of weak n-equivalence.

Definition 3. : An internal n-functor $f_n : X_n \to Y_n$ is a weak n-equivalence when :

(1) $f_n$ is $(\ )_{n-1}$-cartesian ; (2) $\pi_{n-1}(f_n)$ is a $(\ )_{n-1}$-equivalence.

Proposition 4. : At level n, the functors $N_n$ and $D_n$ exchange weak chain equivalences with weak n-equivalences.

Proof. : By induction. Then just mimicing the proof of propositions 2 and 3.

7. The ultimate equivalence :

Let us denote by $\infty$-Grd /A the projective limit of the tower defined by the categories n-Grd /A and the functors $(\ )_{n-1}$. Clearly $C^{\cdot}(/A)$ is the projective limit of the tower defined by the $C^n(/A)$ and the $(\ )_{n-1}$. Now the functors $N_n$ and $D_n$ being natural in n determine an equivalence (see [5]) :

$$\infty\text{ - Grd } /A \underset{D_\infty}{\overset{N_\infty}{\rightleftarrows}} C^{\cdot}\ /A$$

Definition 4. : A $\infty$-functor $f_\infty : X_\infty \to Y_\infty$ is called a weak $\infty$-equivalence, when, for each n, its projection $f_n$ in n-Grd /A is such that $\pi_{n-1}(f_n)$ is a weak (n - 1)-equivalence.

Proposition 5. : The functors $N_\infty$ and $D_\infty$ do exchange the weak chain equivalences with the weak $\infty$-equivalences.

Proof. : Let $f : d \to d'$ be a chain equivalence. We saw previously that $H^{n-1}(d)$ is the kernel of $Cok_{n-1}$ and so f is a chain equivalence if and only if, for each n , $Cok_{n-1}[(f)_n]$ is a weak equivalence, what is equivalent to : $\pi_{n-1}(D_\infty f)_n$ is a weak (n - 1)-equivalence.

PART II. : PSEUDOFUNCTORS AND NON ABELIAN WEAK 2-EQUIVALENCES.

It is clear that the notion of internal n-groupoid exists in any left exact category $\mathbb{E}$. If, further more, $\mathbb{E}$ is exact in the sense of Barr [1], each fibration :

$$(\ )_{n-1} : n - \text{Grd}\,\mathbb{E} \longrightarrow (n - 1)\ \text{Grd}\,\mathbb{E}$$

is Barr-exact (see [3]) : each fiber is Barr-exact and each change of base functor is Barr-exact. Therefore the functor $\pi_{n-1}$ does exist and so the notion of internal weak n-equivalence has again a meaning in $\mathbb{E}$.

In this part II, we shall study the case $\mathbb{E}$ = Set, for n = 1, 2, since the higher levels have not been yet really entered upon.

In the case $n = 1$, we recover the usual notion of a weak equivalence, that is of a functor $F_1 : X_1 \to Y_1$ between 2-groupoids which is : ($\alpha$) fully faithful ; ($\beta$) essentially surjective, that means : for any object $Y$ in $Y_1$ there exists an object $X$ in $X_1$ and an isomorphism : $\varepsilon_Y : Y \to F_1(X)$.

It is well known that a weak equivalence determines, thanks to the axiom of choice, an inverse equivalence $G_1$ in the following way : choose for $G_1(Y)$ any of those given by the essential surjectivity and extend it to the morphisms in a unique way by the fully faithfulness. The isomorphisms $\varepsilon_Y$ are then organized into a natural isomorphism $\varepsilon : 1 \Rightarrow F_1 . G_1$. There is also a (unique) natural isomorphism $\eta : 1 \Rightarrow G_1 . F_1$ such that $F_1 . \eta = \varepsilon . F_1$. At last it appears that $G_1 . \varepsilon = \eta . G_1$ and consequently $F_1$ is an ordinary equivalence of groupoids.

We shall study now the case $n = 2$. Let $X_2$ be a 2-groupoid. Then $X_1$ denotes its underlying groupoid of 1-morphisms.

1. <u>The ( $)_1$-cartesian 2-functors.</u> :

Let $F_2 : X_2 \to Y_2$ be a 2-functor. It is ( $)_1$-cartesian if and only if the following square is a pullback :

$$\begin{array}{ccc} m\,X_2 & \xrightarrow{\;m\,F_2\;} & m\,Y_2 \\ \downarrow{\scriptstyle (d_0,\, d_1)} & & \downarrow{\scriptstyle (d_0,\, d_1)} \\ X_1 \times_0 X_1 & \xrightarrow[F_1 \times_0 F_1]{} & Y_1 \times_0 Y_1 \end{array}$$

where $X_1 \times_0 X_1$ is the groupoid with the same objects as $X_1$ and whose morphisms : $X \to X'$ are the pairs of morphisms $f_0$ , $f_1 : X \to X'$ in $X_1$. Consequently $F_2$ is ( $)_1$-cartesian if and only if : given $f_0$, $f_1 : X \to X'$ and a 2-cell $\nu : F_2(f_0) \Rightarrow F_2(f_1)$ there exists a unique 2-cell $\bar{\nu} : f_0 \Rightarrow f_1$ such that $F_2(\bar{\nu}) = \nu$. In other words, $F_2$ is ( $)_1$-cartesian if and only if its restriction "hom by hom" :

$$F_{XX'} : X_2(X, X') \longrightarrow Y_2(F_2 X, F_2 X')$$

is ( $)_0$-cartesian, that is fully faithful.

2. <u>The functors</u> $\pi_1(F_2)$. :

The groupoid $\pi_1(X_2)$ has the same objects as $X_1$ and as morphisms : $X \dashrightarrow X'$, the classes, modulo the 2-cells, of 1-morphisms of $X_2$.

Thus the functor $\pi_1(F_2)$ is a weak equivalence if and only if it is : ($\alpha$) essentially surjective, which only means here : for any object $Y$ in $Y_2$, there is an object $X$ in $X_2$ and an isomorphism : $\varepsilon_Y : Y \to F_2 X$ (we shall say that $F_2$ is essentially surjective).

($\beta$) fully faithful, which means :

($\beta_1$) full : for any pair $X$, $X'$ of objects in $X_2$ and any 1-morphism $h : F_2 X \to F_2 X'$, there exists a 1-morphism $\bar{h} : X \to X'$ and a 2-cell $\psi_h : F_2(\bar{h}) \Rightarrow k$. (which is always invertible since we are in a 2-groupoid).

($\beta_2$) faithful : for any pair of 1-morphisms $f_0$, $f_1 : X \to X$ such that there exists a 2-cell $\varphi : F_2(f_0) \Rightarrow F_2(f_1)$, then there exists a 2-cell $\bar{\varphi} : f_0 \Rightarrow f_1$, with no further condition.

3. The weak 2-equivalences. :

Now if $F_2$ is $(\ )_1$-cartesian (*), we can forget ($\beta_2$). Then (*) + ($\beta_1$) means exactly that for any pair $(X, X')$ of objects of $X_2$, the restriction $F_{XX'} : X_2(X, X') \to Y_2(F_2 X, F_2 X')$ of $F_2$ is a weak equivalence. In other words that means that $F_2$ is "hom by hom" a weak equivalence.
Whence the following result :

Proposition 6. : A 2-functor $F_2 : X_2 \to Y_2$ between two ordinary 2-groupoids is a weak 2-equivalence if and only if :

(1) $F_2$ is essentially surjective ; (2) $F_2$ is "hom by hom" a weak equivalence.

Remark : The category 2-Grd of ordinary 2-groupoids is known [7] to be equivalent to the category $Cross_2$ of crossed modules over groupoids :

$$\mathcal{G}_2 : \quad G_0 \underset{d_1}{\overset{d_0}{\Longleftarrow}} m\, G_1 \overset{\partial}{\longleftarrow} C_2$$

where the left hand graph is underlying to a groupoid $G_1$ and $C_2$ is a family of groups $C_2(X)\}_{X \in G_0}$, with an action of $m\,G_1$ on $G_2$ satisfying some coherence conditions. When $G_2$ is a 2-groupoid, its associated crossed module over $G_1$ has for $C_2$ the family of groups whose objects are the 2-cells whose 1-domain is an identity map.

Let us denote by $\pi_0(\mathcal{G}_2)$ the coequalizer of $d_0$ and $d_1$, by $Eq(d_0, d_1)$ the equalizer of $d_0$ and $d_1$. This $Eq(d_0, d_1)$ is a family of groups, indexed by $G_0$. Furthermore the images of the $C_2(X)$ by $\partial$ are normal subgroups of $Eq(d_0, d_1)(X)$.

Let us denote by $H^1\mathcal{G}_2)$ the quotient $Eq(d_0, d_1) / \partial C_2$. Finally let us denote by $H^2(\mathcal{G}_2)$ the kernel of $\partial$. Then, as it is expected, a morphism $h : \mathcal{G}_2 \to \mathcal{G}_2'$ in $Cross_2$ has its associated 2-functor a weak 2-equivalence if and only if $\pi_0(h)$, $H^1(h)$ and $H^2(h)$ are isomorphisms.

4. The associated pseudoinverse. :

Let $F_2$ be a weak 2-equivalence. Is it possible now, as it is the case at levet 1,to construct, thanks to the axiom of choice, an inverse equivalence ? The answer, here, is weaker, the "inverse" is not a regular 2-functor but only a pseudofunctor (see for instance [2] and [11]),what means that the composition of 1-morphisms is respected by this construction only up to 2-isomorphisms.

Indeed let us start by choosing a $G_2\,Y$ and a $\varepsilon_Y : Y \to F_2 \,.\, G_2\,Y$ in the same way as at level 1. Given an $h : Y \to Y'$, there is (by $\beta_1$) a morphism $G_2\,h : G_2\,Y \to G_2\,Y'$ and a 2-cell $\psi_h : F_2\,G_2\,h \Rightarrow \varepsilon_{Y'} \,.\, h \,.\, \varepsilon_Y^{-1}$ . Let us denote by $\varepsilon_h$ the 2-cell $\psi_h \cdot \varepsilon_Y$ :

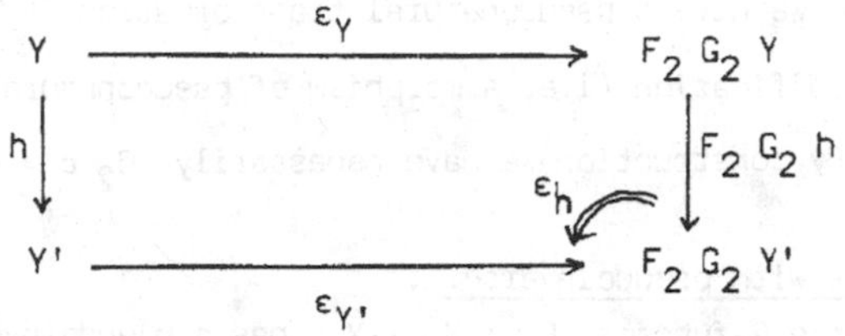


If $\nu : h \Rightarrow k$ is a 2-cell, there is obviously a unique 2-cell $\bar{\nu} : F_2\,G_2\,h \Rightarrow F_2\,G_2\,k$ such that $\varepsilon_k \bullet (\bar{\nu} \,.\, \varepsilon_Y) = (\varepsilon_{Y'} \,.\, \nu) \bullet \varepsilon_h$ . Let us denote by $G_2\,\nu$ the unique 2-cell : $G_2\,h \Rightarrow G_2\,k$ such that $F_2 \,.\, G_2\,\nu = \bar{\nu}$ . Now the fully faithfulness of $F_2$ implies that $G_{YY'} : Y_2(Y, Y') \to X_2(G_2\,Y, G_2\,Y')$ is a functor and $\varepsilon$ is a part of a pseudonatural transformation (called quasi-natural in [11]).

Let $h'$ be another 1-morphism : $Y' \to Y''$ . Let us compare now $G_2(h' \,.\, h)$ and $G_2\,h' \,.\, G_2\,h$ . For that let us consider the two following diagrams :

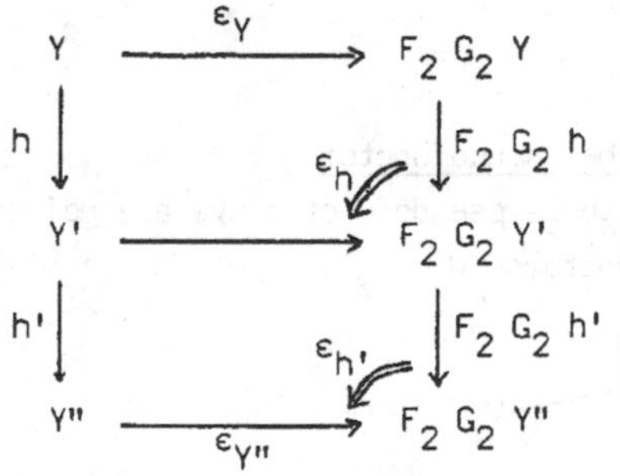


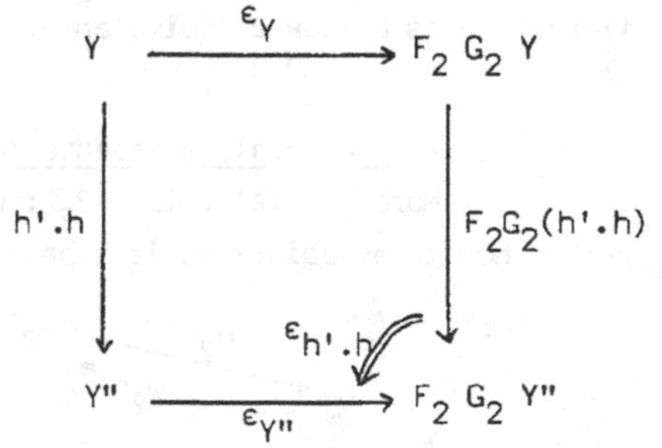


It is then clear that there is a unique 2-cell

$$\bar{\gamma}_{h'.h} : F_2\,G_2\,h' \,.\, F_2\,G_2\,h \Rightarrow F_2\,G_2(h' \,.\, h)$$

such that

$$(\varepsilon_{h'} \,.\, h) \bullet (F_2\,G_2\,h' \,.\, \varepsilon_h) = \varepsilon_{h'h} \bullet (\bar{\gamma}_{h'.h} \cdot \varepsilon_Y)$$

Let $\gamma_{h'.h}$ be the unique 2-cell : $G_2 h' . G_2 h \Rightarrow G_2(h' . h)$ such that $F_2(\gamma_{h'.h}) = \bar{\gamma}_{h'.h}$. Consequently $G_2$ is only compatible with the composition of 1-morphisms up to $\gamma_{h'.h}$.

Moreover, if $h = 1_Y$, then, by our choice, $G_2(1_Y)$ is a priori any 1-morphism : $G_2 Y \to G_2 Y$ such that $F_2[G_2(1_Y)] = 1_{F_2 G_2 Y}$. But already we have $F_2(1_{G_2 Y}) = 1_{F_2 G_2 Y}$ and therefore there exists a unique 2-cell $\gamma_Y : 1_{G_2 Y} \Rightarrow G_2(1_Y)$. These data $(G_2, \gamma_{h'h}, \gamma_Y)$ are the data of a pseudofunctor. The coherence conditions (see [2], [11], for details) hold, because of the uniqueness given by the faithfulness of $F_2$. Clearly the $\varepsilon$ determine a pseudonatural transformation $\varepsilon : 1_{Y_2} \Rightarrow F_2 . G_2$. Moreover we have a pseudonatural transformation $\eta : 1_{X_2} \Rightarrow G_2 . F_2$ with a modification (i.e. a morphism of pseudonatural transformations) $\theta : F_2 \eta \Rrightarrow \varepsilon F_2$. By construction we have necessarily $G_2 \varepsilon = \eta G_2$.

5. <u>The 2-functors with pseudoinverse.</u> :

We shall say that a 2-functor $F_2 : X_2 \to Y_2$ has a pseudoinverse when it is equipped with a pseudofunctor $G_2 : Y_2 \dashrightarrow X_2$, pseudonatural transformations $\varepsilon : 1_{Y_2} \Rightarrow F_2 . G_2$, $: 1_{X_2} \Rightarrow G_2 . F_2$ and a modification $\Theta : F_2 \eta \Rrightarrow \varepsilon F_2$. No more condition is required for $G_2 \varepsilon$ and $\eta G_2$, since the modification $\Theta$ determines always a modification $\psi : G_2 \varepsilon \Rrightarrow \eta G_2$. It is therefore a situation of i-quasi adjonction as defined in [11].

Then a careful but straitforward inspection allows to assert :

<u>Proposition 7.</u> : A 2-functor $F_2 : X_2 \to Y_2$ between 2-groupoids which admits a pseudo-inverse is a weak 2-equivalence.

6. <u>A general representation for the pseudofunctors.</u> :

More generally let $P_2 : U_2 \dashrightarrow V_2$ be a pseudofunctor. We are going to sketch how to associate to it a pair of 2-functors $H_2$, $K_2$ :

$$U_2 \xleftarrow{\;H_2\;} W_2 \xrightarrow{\;K_2\;} V_2, \qquad Q_2 : U_2 \dashrightarrow W_2$$

where $H_2$ has $Q_2$ as strict (i.e. $H_2 . Q_2 = 1_{U_2}$) pseudoinverse and satisfying $K_2 . Q_2 = P_2$.

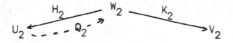

This construction is mimicing the usual construction of the comma object of $P_2$ (if it were a regular 2-functor) and $1_{V_2}$. It is also analogous to the mapping cylinder factorization in homotopy theory.

The objects of $W_2$ are the triples $(u, h, v)$ with $h : v \to P_2 u$, the morphisms : $(u, h, v) \to (u', h', v')$ are given by the triples $(f, \nu, g)$ satisfying the following condition :

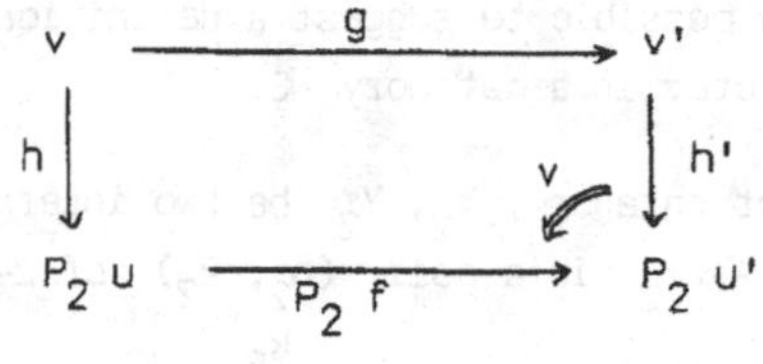


A 2-cell : $(f, \nu, g) \Rightarrow (f', \nu', g')$ is a pair of 2-cells $\lambda : f \Rightarrow f'$ and $\mu : g \Rightarrow g'$, coherent with $\nu$ and $\nu'$. The identity map associated to the object $(u, h, v)$ is given by the triple $(1, \gamma_U \cdot h, 1)$ :

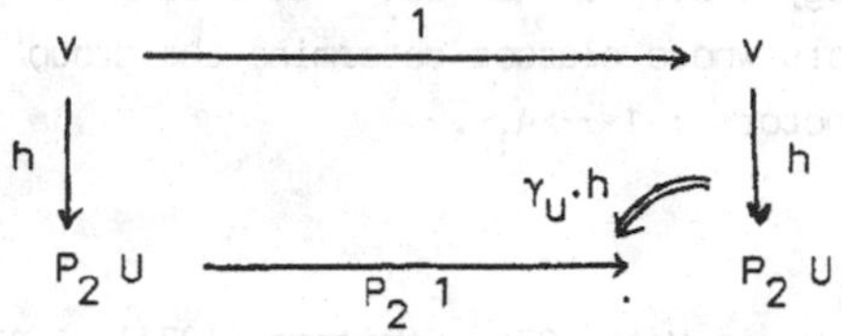


The composition is given by the following diagram :

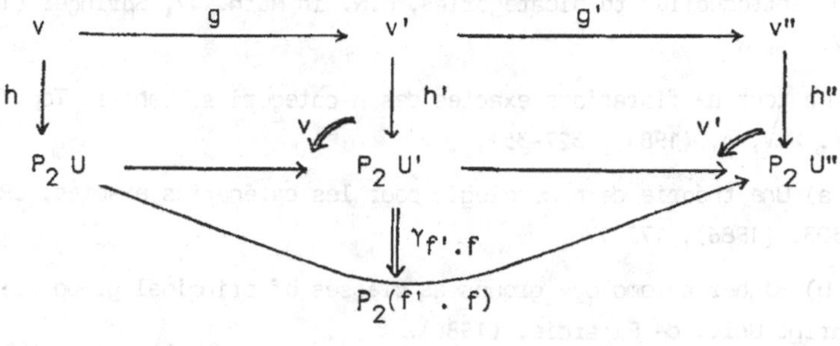


and the following formula :

$(f', \nu', g') \cdot (f, \nu, g) = (f' \cdot f, (\gamma_{f' . f} \cdot h) \bullet (P_2 f' \cdot \nu) \bullet (\nu' \cdot g), g' \cdot g)$.

The unitary and associativity conditions are satisfied because of the unitary and associativity coherence conditions of a pseudofunctor.

The 2-functor $H_2$ is defined by $H_2(u, h, v) = u$ and the 2-functor $K_2$ by $K_2(u, h, v) = v$. The 2-functor $H_2$ has a pseudo-inverse $Q_2$, defined by :

$$Q_2(u) = (u, 1, P_2 u),\ Q_2(f) = (f, 1, P_2 f),\ Q_2(\lambda) = (\lambda, P_2 \lambda).$$

The construction $Q_2$ is clearly not 2-functorial but pseudo-functorial. Furthermore we have $H_2 \cdot Q_2 = 1_{U_2}$ and $K_2 \cdot Q_2 = P_2$. Finally, we have a pseudonatural transformation $\eta : 1_{W_2} \Rightarrow Q_2 . H_2$

$$\eta(u, h, v) = (1, \gamma_U . h, h) : (u, h, v) \to (u, 1, P_2 u)$$

$$\eta(f, \nu, g) = (1, \nu^{-1})$$

So a pseudofunctor between 2-groupoids is nothing but a "weakly representable 2-profunctor". On the other hand, it is now possible to suggest a definition of what should be called an internal pseudofunctor in a category $\mathbb{E}$.

Definition 5. : Let $\mathbb{E}$ be a Barr-exact category, $X_2$, $Y_2$ be two internal 2-groupoids in $\mathbb{E}$, an internal pseudofunctor $X_2 \dashrightarrow Y_2$ is a pair $(h_2, k_2)$ of 2-functors :

$$X_2 \xleftarrow{h_2} Z_2 \xrightarrow{k_2} Y_2$$

with $h_2$ a weak 2-equivalence.

In the general cohomology theory of $\mathbb{E}$ with value is an internal abelian group $A$ in $E$ , [4], the 2-torsors whose classes determine the group $H^2(\mathbb{E}, A)$ are exactly the internal pseudofunctors : $1 \dashrightarrow A_1$ .

REFERENCES. :

[1] M. BARR, Exact categories, L.N. in Math. 236, Springer (1971), 1-120.

[2] J. BENABOU, Introduction to bicategories, L.N. in Math. 47, Springer (1967), 1-77.

[3] D. BOURN, La tour de fibrations exactes des n-categories, Cahiers Top. Geom. Diff. XXV, 4, (1984), 327-351.

[4] D. BOURN, a) Une théorie de cohomologie pour les catégories exactes, CRAS, T. 303, (1986), 173-176.

b) Higher cohomology groups as classes of principal group actions, Preprint Univ. de Picardie, (1985).

[5] D. BOURN, Another denormalisation theorem for the abelian chain complexes, (to appear).

[6] R. BROWN & M. GOLASINSKI, A model structure for the homotopy theory of crossed complexes, Preprint University of Wales, 87.12.

[7] R. BROWN & P.J. HIGGINS, Colimit theorem for relative homotopy groups, J. Pure Appl. Algebra, 22, (1981), 11-41.

[8] R. BROWN & P.J. HIGGINS, The equivalence of $\infty$-groupoids and crossed complexes, Cahiers Top. et Géom. Diff., 22, 4, (1981), 371-386.

[9] J. DUSKIN, Higher dimensional torsors and cohomology of topoï : the abelian theory, L.N. in Math. 753, Springer (1979).

[10] P. GLENN, Realization of cohomology classes in arbitrary exact categories, J. Pure Appl. ALgebra 25, (1982), 33-105 .

[11] J.W. GRAY, Formal category theory : adjointness for 2-categories, L.N. in Math. 391, Springer (1974).

[12] P. HILTON, Correspondances and exact squares, Proc. Conf. on Cat. Alg., La Jolla, Springer (1966).

[13] P.T. JOHNSTONE, Topos theory, Academic Press, (1977).

[14] G.H. KELLY & R. STREET, Review of the elements of 2-categories, L.N. in Math., 420, Springer (1974), 75-103.

Université de Picardie
**U.F.R. Maths & Informatique**
33, rue Saint Leu
80039 AMIENS Cedex - France.

# AN EXACT SEQUENCE IN THE FIRST VARIABLE FOR NON-ABELIAN COHOMOLOGY IN ALGEBRAIC CATEGORIES. A MAYER-VIETORIS SEQUENCE FOR NON-ABELIAN COHOMOLOGY OF GROUPS.

by

A.M. Cegarra, A.R. Garzon and P. Carrasco.

In many algebraic contexts such as commutative algebras, Lie algebras, etc., sequences similar to Hochschild-Serre's fundamental exact sequence in cohomology of groups

$$\text{(I)}\quad 0\longrightarrow Der(X,A)\longrightarrow Der(E,A)\longrightarrow Hom_{X\text{-}Mod}(N_{ab},A)\longrightarrow H^2(X,A)\longrightarrow H^2(E,A)$$

associated to an epimorphism $p:E\twoheadrightarrow X$ with kernel N and a X-module A, have been obtained. The purpose of the first paragraph of this paper is to get a 5-term exact sequence in the first variable for a general non-abelian cohomology, which provides exact sequences for the usual algebraic non-abelian cohomology theories. These sequences are the the natural versions of the sequences (I) in the case of non-abelian coefficients. Thus, for groups, we obtain in proposition (1.5), an exact sequence of sets with distinguished elements

$$\text{(II)}\quad *\longrightarrow Z^1_{\emptyset}(X,\mathfrak{F})\longrightarrow Z^1_{\emptyset p}(E,\mathfrak{F})\longrightarrow Hom_{XM}(N\hookrightarrow E,\mathfrak{F})\longrightarrow H^2(X,\mathfrak{F})\longrightarrow H^2(E,\mathfrak{F})$$

associated to an epimorphism $p:E\twoheadrightarrow X$ and a crossed module $\mathfrak{F}$, (an analogous sequence for Giraud's non-abelian cohomology of groups is described in [6]).

In the second paragraph the sequence (II) is used to obtain a 6-term Mayer-Vietoris exact sequence in non-abelian cohomology of groups. First we prove a coproduct theorem for Dedecker's non-abelian cohomology by showing there are natural bijections (proposition 2.1)

$$Z^1_{\emptyset}(G_1*G_2,\mathfrak{F}) \cong Z^1_{\emptyset_1}(G_1,\mathfrak{F}) \times Z^1_{\emptyset_2}(G_2,\mathfrak{F})$$

$$H^2(G_1*G_2,\mathfrak{F}) \cong H^2(G_1,\mathfrak{F}) \times H^2(G_2,\mathfrak{F}).$$

for $G_1*G_2$ the free product of groups $G_1$ and $G_2$. Then, if G is a group, which is the free product of subgroups $G_1$ and $G_2$ with amalgamated subgroup U, the sequence (II), applied to the epimorphism $p:G_1*G_2 \twoheadrightarrow G$ induced by the inclusions, gives an exact sequence

$$* \longrightarrow Z^1_{\emptyset}(G,\Phi) \longrightarrow Z^1_{\emptyset_1}(G_1,\Phi)\times Z^1_{\emptyset_2}(G_2,\Phi) \longrightarrow Der(U,G_1,G_2;\Phi) \longrightarrow H^2(G,\Phi) \longrightarrow H^2(G_1,\Phi)\times H^2(G_2,\Phi).$$

Moreover, as in the abelian case, the image I of the map $H^2(G,\Phi)\longrightarrow H^2(G_1,\Phi)\times H^2(G_2,\Phi)$ makes the diagram

$$\begin{array}{ccc} I & \longrightarrow & H^2(G_1,\Phi) \\ \downarrow & & \downarrow \\ H^2(G_2,\Phi) & \longrightarrow & H^2(U,\Phi) \end{array}$$

a pullback square.

**1.- A 5-term exact sequence in the first variable for non-abelian cohomology in algebraic categories.**

For the usual algebraic categories $\underline{C}$ as Groups, Associative algebras, etc., non-abelian cohomology theories were stablished by using adequate notions of 1 and 2-cocycles with coefficients in the corresponding crossed modules . Now, it is well known that these notions of crossed module correspond categorically to the concept of internal groupoid in $\underline{C}$ in such a way that there exists an equivalence of categories $\underset{\sim}{G}:XM(\underline{C})\longrightarrow GPD(\underline{C})$. Using this equivalence 1-cocycles of an object X of $\underline{C}$ with coefficients in a crossed module $\Phi$ correspond bijectively to morphisms in $\underline{C}$ from X to $(\underset{\sim}{G}(\Phi))_1$ and 2-cocycles to internal functors from the internal groupoid $\mathbb{G}X\times_X\mathbb{G}X \rightrightarrows \mathbb{G}X$ (where $\mathbb{G}X$ is the free object on the underlying set of X) to $\underset{\sim}{G}(\Phi)$, two 2-cocycles being equivalent if there exists a natural transformation between them. These considerations suggest a definition of 1 and 2-cocycles with coefficients in internal groupoids for any algebraic category $\underline{C}$; this resulting cohomology will permit us to obtain a general 5-term exact sequence associated to a surjective epimorphism in $\underline{C}$. This cohomology $H^2$ coincides (although not explicitely shown here) with the $H^1$ defined by Duskin in [10] where 1-cocycles are simplicial morphisms from the cotriple standar resolution to the nerve of the groupoid.

In what follows $\underline{C}$ will denote an algebraic category with cotriple $\mathbb{G}=(\mathbb{G},\rho,\varepsilon)$ associated to the monadic forgetful functor $\underline{C} \longrightarrow$ Sets.

For X an object of $\underline{C}$, $\eta_X: X \longrightarrow \mathbb{G}X$ is the canonical inclusion map.

Let us remember that an internal groupoid in $\underline{C}$, $\underset{\sim}{G}$, is an internal category in $\underline{C}$ in which all the morphisms are invertible, i.e., a diagram in $\underline{C}$:

$$\underset{\sim}{G}: \quad G_{1\,d_1}\!\times_{d_0} G_1 \xrightarrow{\ m\ } G_1 \overset{s_0}{\underset{d_1}{\overset{d_0}{\rightrightarrows}}} G_0$$

satisfying:

i) $d_0 s_0 = d_1 s_0 = id$.

ii) (Associativity) $m(m(x,y),z) = m(x,m(y,z))$.

iii) (Unities) $m(x,s_0d_1x) = x = m(s_0d_0x,x)$.

iv) (Inverses exist) for all $x \in G_1$ there exists a unique $x^{-1}$ of $G_1$ such that $m(x,x^{-1}) = s_0d_0x$ and $m(x^{-1},x) = s_0d_1x$.

The morphism m is called the multiplication of $\underset{\sim}{G}$ and $m(x,y)$ is usually denoted by xy.

A morphism of groupoids $\underset{\sim}{f} : \underset{\sim}{G} \longrightarrow \underset{\sim}{G}'$ is a commutative diagram:

$$\begin{array}{ccccccc}
\underset{\sim}{G}: & G_{1\,d_1}\!\times_{d_0} G_1 & \xrightarrow{m} & G_1 & \rightleftarrows & G_0 \\
\underset{\sim}{f}=(f_1,f_0)\ \downarrow & \downarrow f_1\times f_1 & & \downarrow f_1 & & \downarrow f_0 \\
\underset{\sim}{G}': & G'_{1\,d_1}\!\times_{d_0} G'_1 & \xrightarrow{m'} & G'_1 & \rightleftarrows & G'_0
\end{array}$$

The corresponding category $\underline{C}$ is denoted by GPD($\underline{C}$).

In categories like Groups, Algebras, etc., this concept of internal groupoid is equivalent to that of crossed module in the corresponding sense, [4], [12]. For example, in the case $\underline{C}$=Groups, given a diagram $G_1 \rightleftarrows G_0$, with $d_0s_0 = 1 = d_1s_0$, there is a unique candidate for the multiplication morphism m ; $m(x,y) = x-s_0d_1x+y$. Moreover, using that every element in $G_1$ can be expressed uniquely as a sum $h+s_0x$ with $h\in H = Ker(d_0)$ and $x\in G_0$, i.e., $G_1 = H \rtimes G_0$, the semidirect product, it is clear that m is a group morphism if and only if $h+h'-h = s_0d_1h+h'-s_0d_1h$, for all $h,h' \in H$.

So, if $\underset{\sim}{G}$ is a groupoid in Groups, $\Phi(\underset{\sim}{G}) = (H \xrightarrow{\delta} G_0, \mu)$ is a crossed module in the sense of Whitehead [18], where $H=Ker(d_0)$, $\delta=d_1/H$ and $\mu:G_0 \longrightarrow Aut(H)$ is the morphism given by $\mu(x)(h)={}^{x}h=s_0x+h-s_0x$. That is, $\delta$ is a morphism of $G_0$-groups (considering $G_0$ as $G_0$-group via conjugation) and ${}^{\delta(h)}h'=h+h'-h$, for all $h,h' \in H$.

Moreover, a morphism of groupoids $\underset{\sim}{f} : \underset{\sim}{G} \longrightarrow \underset{\sim}{G}'$ induces a morphism of crossed modules $\Phi(\underset{\sim}{f})=(f_1/Ker(d_0), f_0):\Phi(\underset{\sim}{G}) \longrightarrow \Phi(\underset{\sim}{G}')$, so we have a functor $\Phi(-)$ from GPD(Groups) to the category of crossed modules XM, which is an equivalence. A pseudo inverse for $\Phi(-)$ is defined by associating to any crossed module $\Phi=(H \xrightarrow{\delta} \pi, \mu)$ the groupoid

$$\underset{\sim}{G}(\Phi) = (H\times_{\mu}\pi)\,{}_{d_1}\times_{d_0}(H\times_{\mu}\pi) \xrightarrow{\ m\ } H\times_{\mu}\pi \underset{d_1}{\overset{d_0}{\rightrightarrows}} \pi \quad (s_0 : \pi \to H\times_{\mu}\pi),$$

where: $d_0(h,y)=y$, $d_1(h,y)=\delta(h)+y$, $s_0(y)=(0,y)$ and $m((h,y),(h',\delta(h)+y))= = (h+h',y)$, and to a crossed module morphism $\Gamma=(j,\tau):\Phi \longrightarrow \Phi'$ the groupoid morphism $\underset{\sim}{G}(\Gamma) = (f_1,\tau):\underset{\sim}{G}(\Phi) \longrightarrow \underset{\sim}{G}(\Phi')$ where $f_1(h,y)= =(j(h),\tau(y))$.

Thus, the category of crossed modules XM is equivalent to the category GPD(Groups).

Let us note that if N is a normal subgroup of a group E, then considering N as an E-group by conjugation, one has the crossed module $N \hookrightarrow E$ which corresponds, by the above equivalence, to the groupoid obtained by taking the kernel pair of the projection $E \twoheadrightarrow E/N$, i.e., $E\times_{E/N}E \rightrightarrows E$, where the multiplication morphism is given by $m((e_0,e_1)(e_1,e_2)) = (e_0,e_2)$.

In Dedecker's non-abelian cohomology [7], a 1-cocycle from a group X to a crossed module $\Phi= (H \xrightarrow{\delta} \pi, \mu)$ is a pair $(d,\Theta)$ where $\Theta:X \longrightarrow \pi$ is a homomorphism and $d:X \longrightarrow H$ is a $\Theta$-derivation, i.e., $d(x+y)=d(x)+{}^{\Theta(x)}d(y)$, for all $x,y \in X$. Taking $\emptyset:X \longrightarrow \pi$ a fixed homomorphism, $Z^1_{\emptyset}(X,\Phi)$ will denote the set of $\emptyset$-derivations $d:X \longrightarrow H$, endowed with the base point $(0,\emptyset)$. 1-cocycles from a group X to a crossed module $\underline{\Xi}$ have a natural translation in terms of the groupoid $\underset{\sim}{G}(\underline{\Xi})$; they correspond bijectively to the homomorphisms $f:X \longrightarrow (\underset{\sim}{G}(\underline{\Xi}))_1 = H \times_\mu \pi$. The elements of $Z^1_\emptyset(X,\underline{\Xi})$ correspond to those homomorphisms with $d_0 f = \emptyset$. Under this correspondence, the distinguished element of $Z^1_\emptyset(X,\underline{\Xi})$ goes to $s_0\emptyset$. In the same way, the following definition includes all the clasical non-abelian 1-cocycles in other algebraic contexts:

**Definition 1.1.-** Given a groupoid $\underset{\sim}{G}$ in $\underline{C}$ and a morphism $\emptyset:X \longrightarrow G_0$ in $\underline{C}$, $\Gamma_\emptyset(X,\underset{\sim}{G})$ is the set of all morphism $f:X \longrightarrow G_1$ such that $d_0 f=\emptyset$. This set is pointed by $s_0\emptyset$.

Now, if $p:E \twoheadrightarrow X$ is an epimorphism of groups with kernel N, and A is an X-module, the obstruction to an element of Der(E,A) being in the image of the morphism $p^*:\mathrm{Der}(X,A) \longrightarrow \mathrm{Der}(E,A)$ can be measured by the exactness of the partial Hochschild-Serre exact sequence :

$$(1)\quad 0 \longrightarrow \mathrm{Der}(X,A) \xrightarrow{p^*} \mathrm{Der}(E,A) \longrightarrow \mathrm{Hom}_{X\text{-}Mod}(N_{ab},A)$$

But $\mathrm{Hom}_{X\text{-}Mod}(N_{ab},A) \cong \mathrm{Hom}_{E\text{-}Groups}(N,A)$ and it is clear that morphisms of E-groups from N to A are identified with crossed module morphisms of the form:

$$\begin{array}{ccc} N & \hookrightarrow & E \\ \downarrow & & \downarrow p \\ A & \xrightarrow{0} & X \end{array}$$

This pattern can be carried over to the non-abelian case in order to measure the obstruction to an element of $Z^1_{\emptyset p}(E,\underline{\Xi})$ being in the image of $p^*:Z^1_\emptyset(X,\underline{\Xi}) \longrightarrow Z^1_{\emptyset p}(E,\underline{\Xi})$.

**Definition 1.2.-** Given a crossed module $\underline{\Xi}=(H \xrightarrow{\delta} \pi,\mu)$, in the set $\mathrm{Hom}_{XM}(N \hookrightarrow E,\underline{\Xi})$ of crossed module morphisms, we will call any element of the form $(0,f)$ a <u>neutral element</u>. ( Note that if $(0,f)$ is a neutral element then $f=\theta p$ for some homomorphism $\theta:X \longrightarrow \pi$, and we will say that $(0,f)$ is a $\theta$-neutral element).

**Proposition 1.3**.- Let $p:E \longrightarrow X$ be an epimorphism with kernel $N$, $\mathfrak{F}=(H \xrightarrow{\delta} \pi,\mu)$ a crossed module, and $\emptyset:X \longrightarrow \pi$ a homomorphism. Then there exists a sequence

$$(2) \quad * \longrightarrow Z^1_{\emptyset}(X,\mathfrak{F}) \xrightarrow{p^*} Z^1_{\emptyset p}(E,\mathfrak{F}) \xrightarrow{\delta^\circ} \mathrm{Hom}_{XM}(N \hookrightarrow E,\mathfrak{F}),$$

which is exact in the sense that $p^*$ is injective, and an element of $Z^1_{\emptyset p}(E,\mathfrak{F})$ is in the image of $p^*$ if and only if its image by $\delta^\circ$ is a neutral element.█

The above sequence is a particular case of a more general one which we will obtain for any algebraic category $\underline{C}$. Let us note that $\mathrm{Hom}_{XM}(N \hookrightarrow E,\mathfrak{F}) \cong \mathrm{Hom}_{GPD(Groups)}(Ex_X E \rightrightarrows E, \underset{\sim}{G}(\mathfrak{F}))$, $\Theta$-neutral elements corresponding to groupoid morphisms of the form $(s_0\Theta p d_0, \Theta p)$.

**Proposition 1.4**.- Let $\underset{\sim}{G}$ be an internal groupoid in $\underline{C}$, $p:E \twoheadrightarrow X$ a surjective epimorphism and $\emptyset:X \longrightarrow G_0$ a morphism. There exists a sequence

$$(3) \quad * \longrightarrow \Gamma_{\emptyset}(X,\underset{\sim}{G}) \xrightarrow{p^*} \Gamma_{\emptyset p}(E,\underset{\sim}{G}) \xrightarrow{\delta^\circ} \mathrm{Hom}_{Gpd(\underline{C})}(Ex_X E \rightrightarrows E,\underset{\sim}{G})$$

which is exact in the sense that $p^*$ is injective, and an element of $\Gamma_{\emptyset p}(E,\underset{\sim}{G})$ is in the image of $p^*$ if and only if it is mapped by $\delta^\circ$ to a neutral element; by a neutral element of $\mathrm{Hom}_{Gpd(\underline{C})}(Ex_X E \rightrightarrows E,\underset{\sim}{G})$ we understand a groupoid morphism of the form $(s_0\Theta p d_0, \Theta p)$ for any morphism $\Theta:X \longrightarrow G_0$.

**Proof**. First of all we define the connecting map $\delta^\circ$. For $g \in \Gamma_{\emptyset p}(E,\underset{\sim}{G})$ let $\delta^\circ(g)=(g_1,g_0)$ where $g_0=d_1 g:E \longrightarrow G_0$ and $g_1:Ex_X E \longrightarrow G_1$ is given by $g_1(z_0,z_1)=g(z_0)^{-1}g(z_1)$. If $g=fp$, $f \in \Gamma_{\emptyset}(X,\underset{\sim}{G})$, then $g_1(z_0,z_1)= =fp(z_0)^{-1}fp(z_1)=s_0 d_1 fp(z_0)$ and so $g_1=s_0 d_1 fpd_0$ and $\delta^\circ p^*(f)$ is always a neutral element.

Conversely, if $g \in \Gamma_{\emptyset p}(E,\underset{\sim}{G})$ is such that $\delta^\circ(g)=(g_1,g_0)$ is a neutral element, then $g_1(z_0,z_1)=g(z_0)^{-1}g(z_1)=s_0 d_1 g(z_0)$, which implies that $g(z_1)=g(z_0)s_0 d_1 g(z_0)=g(z_0)$ for all $(z_0,z_1) \in Ex_X E$, and so $g$ factors through $p$, i.e., there exists a morphism $f:X \longrightarrow G_1$ such that $fp=g$. $f$ belongs to $\Gamma_{\emptyset}(X,\underset{\sim}{G})$, since $d_0 fp=d_0 g=\emptyset p$ and so $d_0 f=\emptyset$. Therefore $g \in \mathrm{Img}(p^*)$.█

As Hochschild-Serre's sequence shows, the Eilenberg-MacLane cohomology group $H^2$ allows one to extend the sequence (1) to a five term exact sequence. We will show that in the non-abelian case, Dedecker's 2-cohomology can be used to obtain a five term exact sequence, whose three first terms are just those of the sequence (2).

Let us recall,[7], that a Dedecker 2-cocycle of a group X with coefficients in a crossed module $\mathfrak{H}=(H\xrightarrow{\delta}\pi,\mu)$ is a pair of maps $(f:X\times X \longrightarrow H, \sigma:X \longrightarrow \pi)$ satisfying: i) $\sigma(x)+\sigma(y) = \delta f(x,y)+\sigma(x+y)$ and ii) ${}^{\sigma(x)}f(y,z)+f(x,y+z) = f(x,y)+f(x+y,z)$, $x,y,z \in X$. Two 2-cocycles $(f,\sigma)$ and $(f',\sigma')$ are equivalent if there exists a map $\tau:X \longrightarrow H$ such that: i) $\sigma'(x)= \delta\tau(x)+\sigma(x)$ and ii) $f'(x,y)= \tau(x)+{}^{\sigma(x)}\tau(y)+f(x,y)-\tau(x+y)$, $x,y \in X$.

This establishes an equivalence relation in the set $Z^2(X,\mathfrak{H})$ of 2-cocycles of X with coefficients in $\mathfrak{H}$, whose quotient set $H^2(X,\mathfrak{H})$ is, by definition, the 2-cohomology set of X with coefficients in $\mathfrak{H}$. There is in $H^2(X,\mathfrak{H})$ a subset of distiguished elements, which are those classes of 2-cocycles of the form $(0,\sigma)$ with $\sigma:X\longrightarrow\pi$ a group morphism. $(0,\sigma)$ is called a <u>neutral class</u>, or $\sigma$-neutral class.

**Proposition 1.5.-** Let $\mathfrak{H}=(H\longrightarrow\pi,\mu)$ be a crossed module, $p:E\twoheadrightarrow X$ an epimorphism and $\emptyset:X \longrightarrow \pi$ a group morphism. There exists a sequence extending the exact sequence (2)

$$* \longrightarrow Z^1_{\emptyset}(X,\mathfrak{H}) \longrightarrow Z^1_{\emptyset p}(E,\mathfrak{H}) \xrightarrow{\delta^\circ} \mathrm{Hom}_{XM}(N\hookrightarrow E,\mathfrak{H}) \xrightarrow{@^\circ} H^2(X,\mathfrak{H}) \xrightarrow{p^*} H^2(E,\mathfrak{H}),$$

whose exactness at the two last places can be expressed as follows: An element of $\mathrm{Hom}_{XM}(N\hookrightarrow E,\mathfrak{H})$ is in the image of $\delta^\circ$ if and only if its image by $@^\circ$ is the $\emptyset$-neutral class; an element of $H^2(X,\mathfrak{H})$ is in the image of $@^\circ$ if and only if its image by $p^*$ is a neutral class.∎

Sequences such as the above can also be obtained in other algebraic contexts, all of them being examples of a more general one in an algebraic category, which we will establish.

If one considers the crossed module $\ker(\varepsilon_X) \hookrightarrow \mathbb{G}(X)$, using that

$Ker(\varepsilon_X:\mathbb{G}X \twoheadrightarrow X)$ is the free group on the set $\{\eta x+\eta y-\eta(x+y)/x,y \in X\}$, it is clear that Dedecker's 2-cocycles correspond bijectively to crossed module morphisms from $Ker(\varepsilon_X) \hookrightarrow \mathbb{G}X$ to $\underline{\Phi}$, so $Z^2(X,\underline{\Phi}) \overset{\beta}{\cong} \cong Hom_{GPD(Groups)}(\mathbb{G}X \times_X \mathbb{G}X \rightrightarrows \mathbb{G}X, \underset{\sim}{G}(\underline{\Phi}))$. Moreover, it is straightforward that giving a map $\tau:X \longrightarrow H$ which establishes an equivalence between 2-cocycles $(f,\sigma),(f',\sigma')$ is equivalent to giving a morphism $h:\mathbb{G}X \longrightarrow (\underset{\sim}{G}(\underline{\Phi}))_1 = Hx_\mu\pi$, which defines a natural transformation from $\beta(f,\sigma)$ to $\beta(f',\sigma')$, if one regards a groupoid morphism as an internal functor. Under this bijection, a 2-cocycle of the form $(0,\sigma)$, $\sigma:X \twoheadrightarrow \pi$, goes to the groupoid morphism $(s_0\sigma\varepsilon_X d_0,\sigma\varepsilon_X)$ .

This fact suggests the following definition, which includes also other definitions of non-abelian 2-cohomology sets [12], [8].

**Definition 1.6.-** Given an internal groupoid $\underset{\sim}{G}$ in $\underline{C}$, and an object X in $\underline{C}$, a 2-cocycle of X with coefficients in $\underset{\sim}{G}$ is an element of the set $Hom_{GPD(\underline{C})}(\mathbb{G}X x_X \mathbb{G}X \rightrightarrows \mathbb{G}X,\underset{\sim}{G})$. Two 2-cocycles $(f_1,f_0)$ and $(g_1,g_0)$ are equivalent if there exists a natural transformation from $(f_1,f_0)$ to $(g_1,g_0)$, i.e., if there exists a morphism $h:\mathbb{G}X \longrightarrow G_1$ satisfying:

i) $d_0h = f_0$, $d_1h = g_0$,

ii) $h(z_0)g_1(z_0,z_1) = f_1(z_0,z_1)h(z_1)$ for all $(z_0,z_1) \in \mathbb{G}X\ x_X \mathbb{G}X$.

This relation is an equivalence relation, and we define the 2-cohomology set of X with coefficients in $\underset{\sim}{G}$, $H^2(X,\underset{\sim}{G})$, to be the corresponding quotient set. There is in $H^2(X,\underset{\sim}{G})$ a subset of distinguished elements, called neutral elements, namely those classes of 2-cocycles of the form $(s_0\sigma\varepsilon_X d_0,\sigma\varepsilon_X)$ with $\sigma:X \longrightarrow G_0$ a morphism; the class containing the neutral 2-cocycle $(s_0\sigma\varepsilon_X d_0,\sigma\varepsilon_X)$ will be called the $\sigma$-neutral class.

This $H^2$ is functorial in both variables( contravariant in the first ).

Given the surjective epimorphism $p:E \twoheadrightarrow X$, there is a connecting map $\partial^*:Hom_{GPD(\underline{C})}(Ex_XE \rightrightarrows E,\underset{\sim}{G}) \longrightarrow H^2(X,\underset{\sim}{G})$ defined as follows:

Letting $t:\mathbb{G}X \longrightarrow E$ be any morphism such that $pt = \varepsilon_X$, $\mathbb{Q}°$ is the composition:

$$\mathrm{Hom}_{GPD(\underline{C})}(Ex_X E \overleftarrow{\rightrightarrows} E, \underset{\sim}{G}) \xrightarrow{(txt,t)^*} \mathrm{Hom}_{GPD(\underline{C})}(\mathbb{G}X\times\mathbb{G}X \overleftarrow{\rightrightarrows} \mathbb{G}X, \underset{\sim}{G}) \xrightarrow{pr} H^2(X,\underset{\sim}{G}),$$

which does not depend of the choice of $t$: If $t':\mathbb{G}X \longrightarrow E$ is another morphism with $pt' = \varepsilon_X$, the morphism $(t,t'):\mathbb{G}X \longrightarrow Ex_X E$ defines a natural transformation from $(txt,t)$ to $(t'xt',t')$.

**Theorem 1.7**.- Let $\underset{\sim}{G}$ be an internal groupoid in $\underline{C}$, $p:E \twoheadrightarrow X$ a surjective epimorphism and $\emptyset:X \longrightarrow G_0$ a morphism. There exists a sequence extending the exact sequence (3):

$$* \longrightarrow \Gamma_\emptyset(X,\underset{\sim}{G}) \xrightarrow{p^*} \Gamma_{\emptyset p}(E,\underset{\sim}{G}) \xrightarrow{\delta°} \mathrm{Hom}_{GPD(\underline{C})}(Ex_X E \overleftarrow{\rightrightarrows} E,\underset{\sim}{G}) \xrightarrow{\mathbb{Q}°} H^2(X,\underset{\sim}{G}) \xrightarrow{p^*} H^2(E,\underset{\sim}{G}),$$

whose exactness at the two last places can be expressed as follows: An element of $\mathrm{Hom}_{GPD(\underline{C})}(Ex_X E \overleftarrow{\rightrightarrows} E,\underset{\sim}{G})$ is in the image of $\delta°$ if and only if its image by $\mathbb{Q}°$ is the $\emptyset$-neutral class; an element of $H^2(X,\underset{\sim}{G})$ is in the image of $\mathbb{Q}°$ if and only if it is mapped by $p^*$ to a neutral element.

**Proof**. i) Exactness at $\mathrm{Hom}_{GPD(\underline{C})}(Ex_X E \overleftarrow{\rightrightarrows} E,\underset{\sim}{G})$:

Let $g\in\Gamma_{\emptyset p}(E,\underset{\sim}{G})$, then $\mathbb{Q}°\delta°(g)$ is the $\emptyset$-neutral class since if $t:\mathbb{G}X \longrightarrow E$ is any morphism such that $pt = \varepsilon_X$ then $h = gt$ defines a natural transformation from $(s_0\emptyset\varepsilon_X d_0, \emptyset\varepsilon_X)$ to $(txt,t)^*\delta°(g)$.

Conversely, if $(g_1,g_0) \in \mathrm{Hom}_{GPD(\underline{C})}(Ex_X E \overleftarrow{\rightrightarrows} E,\underset{\sim}{G})$ is such that $\mathbb{Q}°(g_1,g_0)$ is the $\emptyset$-neutral class, there is a natural transformation $h$ from $(g_1(txt),g_0 t)$ to $(s_0\emptyset\varepsilon_X d_0,\emptyset\varepsilon_X)$. Then for $e\in E$, choosing $x_e\in\mathbb{G}X$ such that $p(e) = \varepsilon_X(x_e)$, the morphism $f:E \longrightarrow G_1$ given by $f(e) = h(x_e)g_1(e,t(x_e))^{-1}$ is an element of $\Gamma_{\emptyset p}(E,\underset{\sim}{G})$ whose image by $\delta°$ is $(g_1,g_0)$.

ii) Exactness at $H^2(X,\underset{\sim}{G})$:

Let $(g_1,g_0) \in \mathrm{Hom}_{GPD(\underline{C})}(Ex_X E \overleftarrow{\rightrightarrows} E,\underset{\sim}{G})$, then $p^*\mathbb{Q}°(g_1,g_0) = [(g_1(txt)(\mathbb{G}p\times\mathbb{G}p), g_0 t\mathbb{G}p)]$ if $t:\mathbb{G}X \longrightarrow E$ satisfies $pt = \varepsilon_X$. Then $h = g_1(t\mathbb{G}p, \varepsilon_E)$ defines a natural transformation from

$(g_1(txt,t)(\mathbb{G}p\times\mathbb{G}p),g_0t\mathbb{G}p)$ to $(s_0g_0\varepsilon_E d_0,g_0\varepsilon_E)$. Therefore $p^*@^\circ(g_1,g_0)$ is a neutral element.

Conversely, let $[(f_1,f_0)] \in H^2(X,\underset{\sim}{G})$ be such that $p^*[f_1,f_0)]$ is the $\sigma$-neutral class for some $\sigma:E \longrightarrow G_0$, and let h be the natural transformation defining the equivalence. We consider the pair $(g_1,g_0)$ where $g_0=\sigma:E \longrightarrow G_0$ and $g_1:Ex_XE \longrightarrow G_1$ is given by $g_1(e,e') = h(z_e)^{-1}f_1(\mathbb{G}p\times\mathbb{G}p)(z_e,z_{e'})h(z_{e'})$, $z_e,z_{e'}\in\mathbb{G}E$ being any elements such that $\varepsilon_E(z_e) = e$ and $\varepsilon_E(z_{e'}) = e'$. (It is straightforward to see that $g_1$ does not depend on the choice of $z_e,z_{e'}$). Then $(g_1,g_0) \in \text{Hom}_{GPD(\underline{C})}(Ex_XE \rightleftarrows E,\underset{\sim}{G})$ and, in addition, taking $s:\mathbb{G}X \longrightarrow \mathbb{G}E$ a section of $\mathbb{G}p$ and $t = \varepsilon_E s$, $\bar{h} = hs$ defines a natural transformation from the 2-cocycle $(f_1,f_0)$ to $(txt,t)^*(g_1,g_0)$ so that $@^\circ(g_1,g_0) = [(f_1,f_0)]$. ∎

**2.- A 6-term Mayer-Vietoris sequence for non-abelian cohomology of groups.**

It is well known that the Mayer-Vietoris sequence provides a relationship between the cohomology of a group and the cohomology of certain subgroups of it when these subgroups "cover" it. That is, if G is a group which is the free product of groups $G_1$ and $G_2$ with amalgamated subgroup U, and A is a G-module, there is an exact sequence

$$(4)\quad 0\longrightarrow \text{Der}(G,A)\longrightarrow \text{Der}(G_1,A)\oplus\text{Der}(G_2,A)\longrightarrow\text{Der}(U,A)\longrightarrow H^2(G,A)\longrightarrow$$
$$\longrightarrow H^2(G_1,A)\oplus H^2(G_2,A)\longrightarrow\ldots\longrightarrow H^n(G,A)\longrightarrow H^n(G_1,A)\oplus H^n(G_2,A)\longrightarrow H^n(U,A)\ \ldots$$

Now, the Hochschild-Serre exact sequence associated to the short exact sequence $N\hookrightarrow G_1*G_2 \xrightarrow{p} G$, where $G_1*G_2$ is the free product of $G_1$ and $G_2$, p is the epimorphism induced by the inclusions, and A is a G-module:

$$0\longrightarrow\text{Der}(G,A)\longrightarrow\text{Der}(G_1*G_2,A)\longrightarrow\text{Hom}_{G\text{-mod}}(N_{ab},A)\longrightarrow H^2(G,A)\longrightarrow H^2(G_1*G_2,A),$$

is equivalent to the one consisting of the five first terms of the above sequence (4), which is a consequence of the coproduct theorem in group cohomology, [16], and the fact $\text{Hom}_{G\text{-mod}}(N_{ab},A)\cong\text{Der}(U,A)$, which is not difficult to prove.

We now will prove a coproduct theorem for non-abelian cohomology of groups, and, using the exact sequence of (1.5), obtain a Mayer-Vietoris exact sequence of six terms for this cohomology.

**Proposition 2.1**.- Let $G_1*G_2$ be the free product of the groups $G_1$ and $G_2$ and $\mathfrak{Q}=(H\xrightarrow{\delta}\pi,\mu)$ a crossed module.

i) For each homomorphism $\emptyset:G_1*G_2\longrightarrow\pi$ there exists a natural bijection of pointed sets,

$$Z^1_{\emptyset}(G_1*G_2,\mathfrak{Q}) \cong Z^1_{\emptyset_1}(G_1,\mathfrak{Q}) \times Z^1_{\emptyset_2}(G_2,\mathfrak{Q}) \quad ,$$

induced by the injections, where $\emptyset_j:G_j\longrightarrow\pi$ is the restriction of $\emptyset$, $j=1,2$.

ii) There is a natural bijection of sets with distinguished elements,

$$H^2(G_1*G_2,\mathfrak{Q}) \cong H^2(G_1,\mathfrak{Q}) \times H^2(G_2,\mathfrak{Q})$$

induced by the injections.

**Proof**. i) Given a $\emptyset_j$-derivation $d_j:G_j\longrightarrow H$ , $j=1,2$ , there exists a unique $\emptyset$-derivation $d_1*d_2:G_1*G_2\longrightarrow H$ which restricts to $d_j$ , $j=1,2$ . It is defined inductively on the length of the reduced words of $G_1*G_2$ by $(d_1*d_2)(w+c)=(d_1*d_2)(w)+{}^{\emptyset(w)}d_j(c)$ if $c\in G_j$.

ii) Given $(f_j,\sigma_j)$ , $j=1,2$, two 2-cocycles of $G_j$ with coefficients in $\mathfrak{Q}$, we define a 2-cocycle $(f_1*f_2,\sigma_1*\sigma_2)$ of $G_1*G_2$ with coefficients in $\mathfrak{Q}$, recursively on the length of the reduced words of $G_1*G_2$, as follows:

$$(\sigma_1*\sigma_2)(w+c)=(\sigma_1*\sigma_2)(w)+\sigma_j(c) \quad \text{if } c_j\in G_j$$

$$(f_1*f_2)(w+c,c'+w')=\begin{cases} {}^{(\sigma_1*\sigma_2)(w)}f_j(c,c') & \text{if } c,c'\in G_j \\ 0 & \text{if } c\in G_j,\ c'\in G_k \text{ and } j\neq k \end{cases} .$$

Its class in $H^2(G_1*G_2,\mathfrak{Q})$ only depends on the class of $(f_j,\sigma_j)$ in $H^2(G_j,\mathfrak{Q})$, $j=1,2$. In fact, let $\tau_j:G_j\longrightarrow H$ establish equivalences between $(f_j,\sigma_j)$ and $(f'_j,\sigma'_j)$, $j=1,2$ . Then the map $\tau:G_1*G_2\longrightarrow H$ defined by $\tau(w+c)=\tau(w)+{}^{(\sigma_1*\sigma_2)(w)}\tau_j(c)$ if $c\in G_j$, establishes an equivalence between $(f_1*f_2,\sigma_1*\sigma_2)$ and $(f'_1*f'_2,\sigma'_1*\sigma'_2)$. So $J([(f_1,\sigma_1)],[(f_2,\sigma_2)])=$ $=[(f_1*f_2,\sigma_1*\sigma_2)]$ defines a map $J:H^2(G_1,\mathfrak{Q})\times H^2(G_2,\mathfrak{Q})\longrightarrow H^2(G_1*G_2,\mathfrak{Q})$

which clearly satisfies $\Gamma J=1$. Moreover, for any 2-cocycle $(f,\sigma)$ of $G_1*G_2$ with coefficients in $\mathfrak{F}$ we have $J\Gamma[(f,\sigma)]=[(f,\sigma)]$, since the map $\tau:G_1*G_2\longrightarrow H$ given by $\tau(c)=0$ if $c\in G_j$, $j=1,2$, and $\tau(w+c)=\tau(w)+f(w,c)$, establishes an equivalence between $(f,\sigma)$ and $(f/G_1*f/G_2,\sigma/G_1*\sigma/G_2)$.∎

Now, let us suppose a group G is the free product of subgroups $G_1$ and $G_2$ with amalgamated subgroup U. Recall that if we select right coset systems $X_j$ for $G_j$modU, $j=1,2$, then any element g in G can be represented uniquely as $g=u+c_1+\ldots+c_q$ where $u\in U$, $c_k\in X_1\cup X_2$, $c_k\neq 1$ and $c_k,c_{k+1}$ are not both in $X_1$ or both in $X_2$.

$p:G_1*G_2\twoheadrightarrow G$ being the epimorphism induced by the inclusions, and $N=\mathrm{Ker}p$, one has that, for each G-module A, the set Der(U,A) can be identified with a certain set of crossed module morphisms from $N\hookrightarrow G_1*G_2$ to $A\xrightarrow{0}G$ ; to translate this fact to the non-abelian case we consider the set $\mathrm{Der}(U,G_1,G_2;\mathfrak{F})$ of all triples $(d,f_1,f_2)$ where $f_j:G_j\longrightarrow\pi$, $j=1,2$, are homomorphisms, $d:U\longrightarrow H$ is a $f_2/U$-derivation, i.e., $d(u+u')=d(u)+{}^{f_2(u)}d(u')$, and for all $u\in U$ the condition $f_1(u)-f_2(u)=\delta d(u)$ is satisfied. Note that the condition for d being a $f_2/U$-derivation is equivalent to the condition $d(u+u')={}^{f_1(u)}d(u')+d(u)$ for all $u\in U$.

**Lemma 2.2**.- There exists a natural bijection

$$\mathrm{Hom}_{XM}(N\hookrightarrow G_1*G_2,\mathfrak{F})\overset{\Omega}{\cong}\mathrm{Der}(U,G_1,G_2;\mathfrak{F})$$

which carries neutral elements of the set $\mathrm{Hom}_{XM}(N\hookrightarrow G_1*G_2,\mathfrak{F})$ to triples of $\mathrm{Der}(U,G_1,G_2;\mathfrak{F})$ of the form $(0,f/G_1,f/G_2)$ for any homomorphism $f:G\longrightarrow\pi$.

**Proof**. Using that N is the normal subgroup of $G_1*G_2$ generated by the elements $\alpha(u)=i_1(u)-i_2(u)$, $u\in U$, where $i_j:G_j\hookrightarrow G_1*G_2$, $j=1,2$, are the injections, the bijection is given by $\Omega(t,f)=(t\alpha,fi_1,fi_2)$.∎

Now, from (1.5),(2.1) and (2.2) we have

**Proposition 2.3.**- Let G be a group which is the free product of subgroups $G_1$ and $G_2$ with amalgamated subgroup U, $\Phi=(H\xrightarrow{\delta}\pi,\mu)$ a crossed module and $\emptyset:G\longrightarrow\pi$ a homomorphism. There exists an exact (in the sense of proposition (1.5)) sequence of sets with distinguished elements

$$*\longrightarrow Z^1_{\emptyset}(G,\Phi)\longrightarrow Z^1_{\emptyset_1}(G_1,\Phi)\times Z^1_{\emptyset_2}(G_2,\Phi)\longrightarrow Der(U,G_1,G_2;\Phi)$$
$$\longrightarrow H^2(G,\Phi)\longrightarrow H^2(G_1,\Phi)\times H^2(G_2,\Phi)$$

where $\emptyset_j=\emptyset/G_j$, j=1,2. ∎

As for the Mayer-Vietoris exact sequence in the abelian case, the image of the map $H^2(G,\Phi)\longrightarrow H^2(G_1,\Phi)\times H^2(G_2,\Phi)$ can be characterized using the cohomology set $H^2(U,\Phi)$, and so we extend the above sequence by one more place.

**Proposition 2.4.**- The image I of the map $H^2(G,\Phi)\longrightarrow H^2(G_1,\Phi)\times H^2(G_2,\Phi)$ makes the square

$$\begin{array}{ccc} I & \xrightarrow{pr} & H^2(G_1,\Phi) \\ {\scriptstyle pr}\downarrow & & \downarrow \\ H^2(G_2,\Phi) & \longrightarrow & H^2(U,\Phi) \end{array}$$

a pullback, where the maps $H^2(G_j,\Phi)\longrightarrow H^2(U,\Phi)$, j=1,2, are induced by the inclusions.

**Proof.** The commutativity of the square is clear. Let us suppose $(f_j,\sigma_j)$ are 2-cocycles of $G_j$ with coefficients in $\Phi$, j=1,2, such that their restrictions to U are equivalent, and let $\tau:U\longrightarrow H$ be a map defining the equivalence; we will prove the existence of a 2-cocycle $(f,\sigma)$ of G with coefficients in $\Phi$ such that $(f/G_j\times G_j,\sigma/G_j)$ is equivalent to $(f_j,\sigma_j)$, j=1,2. Let us note that we can reduce to the case in which the restrictions to U are equal, since if one considers the map $\bar{\tau}:G_1\longrightarrow H$ defined by $\bar{\tau}/U=\tau$ and $\bar{\tau}(x)=0$ if $x\notin U$, the 2-cocycle equivalent to $(f_1,\sigma_1)$ under $\bar{\tau}$, also denoted $(f_1,\sigma_1)$, satisfies this condition.

We define $\sigma:G\longrightarrow\pi$ by $\sigma(u+c_1+\ldots+c_q)=\sigma_{j_1}(uc_1)+\sigma_{j_2}(c_2)+\ldots+\sigma_{j_q}(c_q)$,

$c_k \in X_{j_k}$, $j_k=1,2$, and $f: G\times G \longrightarrow H$ recursively by

$$f(u,v+c_1+\dots+c_q)=f_{j_1}(u,v+c_1) \quad \text{if} \quad c_1 \in X_{j_1}$$

$$f(c,v+c_1+\dots+c_q)=\begin{cases} f_{j_1}(c,v+c_1) & \text{if } c,c_1\in X_{j_1} \\ -\left[{}^{\sigma_j(c)} f_{j_1}(v,c_1)\right] + f_{j_2}(c,v) & \text{if } c_1\in X_{j_1}, c\in X_{j_2}, j_1\neq j_2 \end{cases}$$

$$f(u+c_1+\dots+c_s,w)=\begin{cases} {}^{\sigma(u+c_1+\dots+c_{s-1})}f(c_s,w)+f(u+c_1+\dots+c_{s-1},c_s+w) & \text{if } s\geq 2 \\ -f_{j_1}(u,c_1)+{}^{\sigma(u)}f(c_1,w)+f(u,c_1+w) & \text{if } s=1 \text{ and } c_1\in X_{j_1}. \end{cases}$$

Thus, $(f,\sigma)$ is a 2-cocycle as required.∎

## References


[1] **AZNAR E.R.**- Cohomologia no abeliana en categorias de interes. Alxebra 33,1983

[2] **BARR,M.-BECK,J.**.- Homology and standard construction. L. N. in Math. 80, 357-375, Springer, 1969.

[3] **BULLEJOS,M.**.- Cohomologia no abeliana, la sucesion exacta larga. Tesis. Universidad de Granada. 1985.

[4] **CEGARRA,A.M.-AZNAR, E.R.**.- An exact sequence in the first variable for the torsor cohomology. The 2-dimensional theory of obstructions.J.P.A.A. 39,197-250, 1986.

[5] **CEGARRA,A.-GARZON,A.**.- La principalite dans la theorie de l'obstruction de dimension 3. Un allongement de la suite de cohomologie non-abelienne des groupes. C. R. Acad. Sc. Paris, T. 302, I, 15, 523-526, 1986.

[6] **CEGARRA,A.-CARRASCO,P.-GARZON,A.**.- La suite fondamentale de Hochschild-Serre pour la cohomologie non abelienne des groupes I. C. R. Acad. Sc. Paris, t.305, I,505-508,1987.

[7] **DEDECKER,P.**- Cohomologie non-abelienne, Mimeographie,Fac.Sc.Lille 1965.

[8] **DEDECKER,P.-LUE,A.S.T.**.- A non-abelian two-dimensional cohomology for associative algebras. Bull. Amer. Math. Soc .72, 1044-50, 1966.

[9] **DUSKIN,J.** - Simplicial methods and the interpretation of triple cohomology. Memoir A.M.S. vol 3, issue 2,163, 1975.

[10] **DUSKIN,J.**.- Non abelian monadic cohomology and low dimensional obstruction theory. Math. Forschung Ins. Oberwolfach, Tagunsbericht, 33,1976.

[11] **HIGGINGS,P.J.**.- Notes on Categories and groupoids. Van Nostrand, 32, 1972.

[12] **LAVENDHOMME,R.-ROISIN,J.R.**.-Cohomologie non abelienne de structures algebraiques. J. of Algebra 67, 385-414, 1980.

[13] **LODAY,J.L.**.- Cohomologie et groupe de Steinberg relatifs, J. Algebra 54, 178-202, 1978.

[14] **LUE,A.S.T.**.- Non-abelian cohomology of associative algebras. Quart. J. Math. Oxford (2), 19, 159-180. 1968.

[15] **MAC LANE, S.**.- Homology, Springer, Berlin, 1967.

[16] **STAMMBACH,U.**.- Homology in group theory. L.N. in Math. 359, 1973.

[17] **VAN OSDOL, D.H.**.- Long exact sequences in the first variable for algebraic cohomology theories, J.P.A.A. 23 (3), 271-309, 1982.

[18] **WHITEHEAD,J.H.C.**.- Combinatorial homotopy II, Bull. A.M.S. 55, 496-543, 1949.

This paper is in final form and will not be published elsewhere.

Departamento de Algebra.
Universidad de Granada.
Granada 18071.
España.

# LOCALLY HILBERT CATEGORIES

Yves DIERS

Département de Mathématiques.

Institut des Sciences - Université de Valenciennes

F - 59326 VALENCIENNES Cedex (France)

## 0. Introduction.

The notion of locally Hilbert category provides an axiomatic description of these categories of algebraic structures for which the geometrical interpretation of objects, given by the Hilbert Nullstellensatz in the case of commutative algebras, is sufficiently faithful, to enable one to reconstruct the category from it. Such a category $\mathbb{A}$ is, indeed, equivalent to the category of realizations in Set of the category $\Sigma\mathbb{A}$ of algebraic sets on $\mathbb{A}$, by associating to an object $A$, the left exact functor $P_A : \Sigma\mathbb{A} \to \text{Set}$ which assigns, to an algebraic set $\Sigma$, the set $P_A(\Sigma)$ of points of $\Sigma$ over $A$.

The construction of algebraic sets on $\mathbb{A}$ and their morphisms, which form the category $\Sigma\mathbb{A}$, seems easy, up to equivalences of categories and isomorphisms of functors, but their construction on the nose needs some care and involves several categories and functors, precise equivalences of categories and isomorphisms of functors. Our construction gives exactly the classical affine algebraic varieties over an algebraically closed field k, when it is performed in the category k-$\mathbb{A}$lgcRed of reduced k-algebras. It describes a sort of non-additive algebraic geometry.

In a locally Hilbert category, as in any category of commutative algebras, the notion of codisjunctors is needed in order to describe the "objects of fractions", and will be recalled. Simple objects are defined as non terminal objects without non trivial regular quotients. Any non terminal object has a simple quotient and the class of simple objects cogenerates the category. We prove that any simple object has an algebraically closed extension, using the notion of algebraically closed objects introduced by S. Fakir in any locally finitely presentable category [6]. The main feature of a locally Hilbert category is that any algebraically closed simple object $L$ is such that every finitely presentable object of it is a subobject of a power of $L$. It follows that, if $J : \mathbb{A}_0 \to \mathbb{A}$ denotes the inclusion functor of the full subcategory $\mathbb{A}_0$ of finitely presentable objects in $\mathbb{A}$, then the functor $\text{Hom}_{\mathbb{A}}(J(-),L) : \mathbb{A}_0^{op} \to \text{Set}$ is a embedding i.e. injective on objects and morphisms, and therefore it induces an isomorphism between the dual of $\mathbb{A}_o$ and a subcategory of Set. This subcategory of Set is proved to be isomorphic to the category $\Sigma\mathbb{A}$ of algebraic sets on $\mathbb{A}$, but not identical to it, and it is at this point that we examine closely the relations between algebra and geometry.

Locally Hilbert categories form a special class of locally Zariski categories [5] and we don't use here the full power of its axomatic.

1. Definition of locally Hilbert categories.

1.0. Notations. We are going to use the results of P. Gabriel and F. Ulmer on locally finitely presentable categories [7], the results of M. Barr and P.M. Grillet on regular categories [1] and [8], and the notion of codisjunctor introduced in [4] and recalled here.

Let us be in a complete and cocomplete category $\mathbb{A}$.

(1) A pair of parallel morphisms $(g,h) : C \rightrightarrows A$ is said to be codisjoint if any morphism $u : A \to X$ which satisfies $ug = uh$ necessarily has a terminal object as its codomain.

(2) One says that a morphism $f : A \to B$ codisjoints a pair of morphisms $(g,h) : C \rightrightarrows A$, if the pair $(fg,fh)$ is codisjoint.

(3) A codisjunctor of a pair of morphisms $(g,h) : C \rightrightarrows A$ is a morphism $f : A \to B$ which codisjoints the pair $(g,h)$ and is such that any morphism $u : A \to X$ which codisjoints $(g,h)$ factors in a unique way through f.

(4) A pair of morphisms is said to be codisjunctable if it admits a codisjunctor.

(5) An object A is said to be codisjunctable if the coproduct of A with itself exists and the pair of inductions $A \rightrightarrows A \coprod A$ is codisjunctable, or equivalently in a finitely cocomplete category, if and only if, any pair of morphisms $(g,h) : A \rightrightarrows C$ with domain A, is codisjunctable.

(6) A pair of morphisms $(g,h) : C \rightrightarrows A$ is conjoint if it has, as a codisjunctor, the morphism $A \to 1$ whose codomain is the terminal object.

Given an object A of $\mathbb{A}$, let us write $(p_1,p_2) : A\times A \rightrightarrows A$ for the product of A by itself. A relation on A is a subobject $r : R \rightrightarrows A\times A$ of $A\times A$ which can be identified with the pair of morphisms $(r_1,r_2) : R \rightrightarrows A$ defined by $r_1 = p_1 r$ and $r_2 = p_2 r$. Thus, the previous notions can be applied to relations on A. A congruence on A is a relation on A which is the kernel pair of some morphism $f : A \to B$ [8, §.5]. Thus, the previous notions can also be applied to congruences on A.

For any pair of morphisms $(f : A \to B,\ g : A \to C)$, the pushout of which is the pair $(h : B \to D,\ k : C \to D)$, the morphism h will be called : the pushout of g along f. By pushing out along a fixed morphism $f : A \to B$, one gets the pushout functor : $A/\mathbb{A} \to B/\mathbb{A}$ induced by f. The product $(p_1 : A\times B \to A,$ $p_2 : A\times B \to B)$ of two objects is said to be couniversal [9,II.4.5] if, for any morphism $f : A\times B \to C$, the pushouts $q_1,q_2$ of $p_1,p_2$ along f give rise to a product $(q_1,q_2)$. A morphism $f : A \to B$ is flat if the pushout functor $A/\!\mathbb{A} \to B/\!\mathbb{B}$ induced by $f$ preserves monomorphisms. An object $A$ is flatly codisjunctable if it is codisjunctable and the codisjunctor of $A \rightrightarrows A \coprod A$ is a flat morphism.

An object $A$ is simple if it has precisely two regular quotients which are $1_A : A \to A$ and $0_A : A \to 1$. A simple quotient of an object is a regular quotient whose codomain is simple. One says that the simple objects of $\mathbb{A}$ satisfy the amalgamation property, if, for any pair of monomorphisms $m : K \to M$, $n : K \to N$ between simple objects, there exists a pair of monomorphisms $p : M \to L$, $q : N \to L$, with $L$ simple, such that $pm = qn$.

For any object $A$, the lattice of congruences on $A$ is denoted by Cong(A) and the lattice of quotient objects of $A$ is denoted by Quot(A).

1.1. Definition. A category is called a locally Hilbert category if it satisfies the following axioms :

(1) It is cocomplete

(2) Products of pairs of objects are couniversal.

(3) It has a proper generating set whose objects are projective, finitely presentable, flatly codisjunctable, and finite coproducts of which are noetherian.

(4) Conjoint morphisms are equal.

(5) The initial object is simple.

(6) Its simple objects satisfy the amalgamation property.

(7) For any pair of codisjunctable congruences $r_1, r_2$ on an object $A$, having respective codisjunctors $d_1, d_2$ :

$$r_1 \vee r_2 = 1_{A \times A} \text{ in Cong(A)} \Rightarrow d_1 \vee d_2 = 1_A \text{ in Quot(A)}.$$

According to [1] and [7] such a category is locally finitely presentable, locally noetherian, and regular. Thus it is complete.

## 2. Examples of locally Hilbert categories.

All the algebras considered are unitary and commutative, and all the homomorphisms of algebras preserve units. They are reduced if they have no non zero nilpotent.

2.1. $\mathbb{A}$lgclRed(k) : Category of commutative reduced algebras over a field k.
As an algebraic category, it is cocomplete. As products of pairs of algebras are defined by idempotent elements, they are couniversal. The polynomial algebra $k[X]$ is a projective finitely presentable proper generator in $\mathbb{A}$lgclRed(k), and any finite copower of it is of the form $k[X_1,\dots,X_n]$, thus is a noetherian object. This generator is codisjunctable, as any parallel pair of morphisms $(g,h) : k[X] \rightrightarrows A$ has, for codisjunctor, the canonical morphism $\ell_a : A \to A[a^{-1}]$ with $a = g(X)-h(X)$ [4]. It is flatly codisjunctable because $\ell_a$ is flat as an A-linear map, and this fact implies that $\ell_a$ is a flat morphism in $\mathbb{A}$lgclRed(k). Let $(g,h) : B \rightrightarrows A$ be a conjoint pair of morphisms in $\mathbb{A}$lgclRed(k) i.e. such that the null morphism $A \to \{0\}$ is its codisjunctor. Let $x$ be any element in $B$ and $a = g(x) - h(x)$.

The canonical morphism $\ell_a : A \to A[a^{-1}]$ codisjoints the pair $(g,h)$, thus it factors through the morphism $A \to \{O\}$, hence $A[a^{-1}] = \{O\}$. Then $a$ is nilpotent. As $A$ is reduced, $a = O$, hence $g(x) = h(x)$. As a result, conjoint morphisms are equal. The initial object in /AlgclRed(k) is k. It is a simple object. The classical amalgamation property for commutative fields implies the amalgamation property for simple objects in /AlgclRed(k). Let $A$ be an object of /AlgclRed(k) and $r_1, r_2$ be a pair of codisjunctable congruences on $A$ such that $r_1 \vee r_2 = 1_{A\times A}$ in Cong(A). The congruences $r_1, r_2$ are the congruences modulo respective radical ideals $I_1$, $I_2$ of A, and the relation $r_1 \vee r_2 = 1_{A\times A}$ implies $\sqrt{I_1+I_2} = A$ which implies $I_1+I_2 = A$. Thus, there exists $a_1 \in I_1$ and $a_2 \in I_2$ such that $a_1+a_2 = 1$. It is a classical result, that the pair of canonical morphisms ($\ell_1 : A \to A[a_1^{-1}]$, $\ell_2 : A \to A[a_2^{-1}]$) is then an effective counion of quotient objects of A [cf. 2], hence is a counion. Let $d_1, d_2$ be the respective codisjunctors of $r_1, r_2$. As the morphism $\ell_1$ (resp. $\ell_2$) codisjoints $r_1$ (resp. $r_2$), it factors through $d_1$ (resp. $d_2$). It follows that $(d_1, d_2)$ is a counion i.e. $d_1 \vee d_2 = 1_A$ in Quot(A). ∎

2.2. <u>/AlgclReg(k)</u> : <u>Category of commutative regular algebras over a field k</u> (in the sense of von Neuman).

2.3. <u>/Algc/AlgIRed(k)</u> : <u>Category of commutative reduced algebraic algebras over a field k</u>.

This is the full subcategory of /AlgclRed(k) whose objects are the algebras whose elements are algebraic over k. It is a locally presentable category [3,2.13. 1.1.], whose objects of the form $k[X]/(P(X))$ with $P(X) \neq O$ make up an adequate generating set.

2.4. <u>/Algc/AlgIReg(k)</u> : <u>Category of commutative regular algebraic algebras over a field k</u> [3].

This is the full subcategory of /AlgclReg(k) whose objects are algebraic algebras.

2.5. <u>/Algc/Alg$ep(k)</u> : <u>Category of commutative separable algebraic algebras over a field k</u> [3].

This is the full subcategory of /Algc/AlgIReg(k) whose objects are algebras whose elements have separable minimal polynomial.

2.6. <u>p-/Algc(k)</u> : <u>Category of commutative p-algebras over a p-field k, with p a prime</u> [1O].

2.7. <u>IBool</u> : <u>Category of Boolean algebras.</u>

2.8. <u>$\mathbb{B}ool^{Cont(G)}$</u> : <u>Category of continuous representations of a profinite group G in Boolean algebras</u> [3, 2.14. 3.1.].

2.9. <u>Any locally preGalois category</u>. [3, 2.16] ; more generally any locally simple category with a simple initial object and satisfying the amalgamation property [3, 2.1.].

3. Simple objects.

In the sequel, one considers a locally Hilbert category $\mathbb{A}$, the initial object of which is denoted by $K$, and the terminal object, assumed to be unique, is denoted by $1$.

3.0. Proposition. The terminal object $1$ is strict, has no proper subobject, is finitely presentable, and is not a filtered colimit of non terminal objects.

Proof : The proper generating set mentionned in axiom (3) of locally Hilbert categories is not empty, otherwise $K$ would be isomorphic to $1$, and thus would have only one regular quotient. Let $A$ be an object of this generating set. The pair $(1_A, 1_A) : A \rightrightarrows A$ has a codisjunctor whose codomain must be a strict terminal object. Therefore $1$ is strict. Let $g : B \to 1$ be a monomorphism. Let $f : K \to B$ be the unique morphism. Because $K$ is simple, the morphism $gf : K \to 1$ is a regular epimorphism. It is the coequalizer of the pair of projections $(p_1, p_2) : K \times K \rightrightarrows K$. The relation $gfp_1 = gfp_2$ implies $fp_2 = fp_1$, thus $B \simeq 1$, and $g$ is an isomorphism. Therefore $1$ has no proper subobject. The object $1$ is finitely generated as being a regular quotient of the finitely presentable object $K$. Because the category is locally noetherian, $1$ is finitely presentable. Let us consider a filtered colimit $(\alpha_i : A_i \to 1)_{i \in \mathbb{I}}$. There exists an index $i \in \mathbb{I}$ and a morphism $\beta_i : 1 \to A_i$ such that the unit morphism $1_1 : 1 \to 1$ factors in the form $1_1 = \alpha_i \beta_i$. As $1$ is strict, $A_i$ is terminal. Consequently, $1$ cannot be a filtered colimit of non terminal objects. ▪

3.1. Proposition. Any non terminal object has a simple quotient.

Proof : Let $A$ be a non terminal object. Let $\mathrm{QuotReg}^*(A)$ be the ordered set of non terminal regular quotients of $A$ [11, p. 122 ]. It is not empty, as $1_A$ belongs to it. Let $C$ be a decreasing chain in $\mathrm{QuotReg}_*(A)$, and let $q : A \to Q$ be the cointersection of members of $C$. Then the object $Q$ is the filtered colimit of the codomains of members of $C$. According to proposition 3.0., the object $Q$ is not terminal, hence $q \in \mathrm{QuotReg}^*(A)$. It follows that $\mathrm{QuotReg}^*(A)$ is a decreasing inductive ordered set. By Zorn lemma, $\mathrm{QuotReg}^*(A)$ has a minimal element $s : A \to S$. It is a simple quotient of $A$ because any non terminal regular quotient $p : S \to P$ is such that $ps \simeq p$, thus $p$ is isomorphic. ▪

3.2. Theorem. The class of simple objects cogenerates the category.

Proof : Let $(f,g) : A \rightrightarrows B$ be a pair of distinct morphisms. According to axiom (3) of locally Hilbert categories, there exist a codisjunctable object $X$ and a morphism $u : X \to A$ such that $fu \neq gu$. Let $d : B \to D$ be the codisjunctor of $(fu, gu)$. According to axiom (4) of locally Hilbert categories, $D$ is not terminal.

According to proposition 3.1., $D$ has a simple quotient $s : D \to S$. Then the pair $(sdfu, sdgu)$ is codisjoint. As $S$ is not terminal, one has $sdfu \neq sdgu$, hence $sdf \neq sdg$. As a result, there exist a simple object $S$ and a morphism $h : B \to S$ such that $hf \neq hg$. Therefore, the class of simple objects cogenerates the category. ∎

3.3. Proposition. Any filtered colimit of simple objects is a simple object.

Proof : Let $(\alpha_i : L_i \to L)_{i \in \mathbb{I}}$ be a filtered colimit such that the objects $L_i$ are simple. According to proposition 3.0., the object $L$ is not terminal. Thus, it has a least two regular quotients $1_L : L \to L$ and $O_L : L \to 1$. Let $q : L \to Q$ be a proper regular quotient of $L$. Then $q$ is not monomorphic. According to the axiom (3) of locally Hilbert categories, there exist a finitely presentable object $X$ and a pair of distinct morphisms $(u,v) : X \rightrightarrows L$ such that $qu = qv$. There exists an index $i \in \mathbb{I}$ and a pair of morphisms $(u_i, v_i) : X \rightrightarrows L_i$ such that $(\alpha_i u_i, \alpha_i v_i) = (u,v)$. Then $u_i \neq v_i$. Let $q_i : L_i \to Q_i$ be the coequalizer of $(u_i, v_i)$. It is a proper regular quotient of $L_i$. As $L_i$ is simple, $Q_i$ must be terminal. Because $q$ is the pushout of $q_i$ along $\alpha_i$, $Q$ is terminal. As a result, $L$ has exactly two regular quotients which are $1_L : L \to L$ and $O_L : L \to 1$, that is to say, $L$ is simple. ∎

## 4. Algebraically closed simple objects.

### 4.0. Algebraically closed objects.

Following S. Fakir [6, definition 5.5], a monomorphism $m : M \to N$ is algebraically closed if, for any monomorphism $f : A \to B$ whose domain is finitely generated and codomain is finitely presentable, and any pair of morphisms $p : A \to M$, $q : B \to N$ such that $qf = mp$, there exists a morphism $d : B \to M$ such that $df = p$. Fakir has proved [ 6, proposition 5.3 ] that, in the categories of algebraic structures, this notion coincides with the notion of algebraically closed subalgebras defined by means of compatible systems of equations. Following [6, definition 6.1.], an object $A$ is algebraically closed if any monomorphism with domain $A$ is algebraically closed.

### 4.1. Algebraically closed extensions.

Let $H$ be a simple object.

An extension of $H$ is a pair $(L, \ell)$ of a simple object $L$ and a morphism $\ell : H \to L$, necessarily monomorphic. An extension $(L, \ell)$ of $H$ is algebraically closed if $L$ is an algebraically closed object in the sense of S. Fakir (4.0.). An extension $(L, \ell)$ of $H$ is finite if it is a finitely presentable object in the category $H/\mathbb{A}$ of objects of $\mathbb{A}$ under $H$.

4.2. Proposition. If $H$ is a simple object, $(M,m)$ a finite extension of $H$, and $(L, \ell)$ an algebraically closed extension of $H$, there exists a morphism $f : M \to L$ such that $fm = \ell$.

<u>Proof</u> : a) Let us prove, firstly, that there exists a morphism $g : X \to Y$ with finitely presentable domain and codomain, such that $m$ is the pushout of $g$ along some morphism $\beta : X \to H$. The category $H/\mathbb{A}$ of objects of $\mathbb{A}$ under $H$ is regular and locally finitely presentable, and the set of objects of the form $(H \amalg X, i_X)$ where $X$ runs over the finitely presentable objects of $\mathbb{A}$, and $i_X : H \to H \amalg X$ is the canonical induction, is a proper generating set in $H/\mathbb{A}$. Because the object $(M,m)$ is finitely presentable in $H/\mathbb{A}$, there exists a pair of finitely presentable objects $X, Y$ in $\mathbb{A}$, and a pair of morphisms $(u,v) : (H \amalg X, i_X) \rightrightarrows (H \amalg Y, i_Y)$ whose coequalizer in $H/\mathbb{A}$ is $q : (H \amalg Y, i_Y) \to (M,m)$. But the morphisms $u,v : H \amalg X \rightrightarrows H \amalg Y$ in $\mathbb{A}$ are of the form $u = \langle i_Y, p\rangle$, $v = \langle i_Y, n\rangle$ where $(p,n) : X \rightrightarrows H \amalg Y$ is a pair of morphisms whose coequalizer is $q : H \amalg Y \to M$. Because $\mathbb{A}$ is a locally finitely presentable category, the object $H$ is the filtered colimit of the finitely presentable objects above it ; i.e. $H = \varinjlim_{(H_o, \alpha_o) \in (\mathbb{A}_o, H)} H_o$, where $\mathbb{A}_o$ denotes the category of finitely presentable objects in $\mathbb{A}$. Then

$$H \amalg Y = \varinjlim_{(H_o, \alpha_o) \in (\mathbb{A}_o, H)} (H_o \amalg Y).$$

Because $X$ is finitely presentable, there exist $(H_o, \alpha_o) \in (\mathbb{A}_o, H)$ and a pair of morphisms $(m_o, n_o) : X \rightrightarrows H_o \amalg Y$, such that $p = (\alpha_o \amalg 1_Y)m_o$ and $n = (\alpha_o \amalg 1_Y)n_o$. Let $q_o : H_o \amalg Y \to M_o$ be the coequalizer of $(m_o, n_o)$. Then $q$ is the pushout of $q_o$ along $\alpha_o \amalg 1_Y$. Hence, if $i_o : H_o \to H_o \amalg Y$ denotes the canonical induction, $m = q\, i_Y$ is the pushout of $q_o i_o$ along $\alpha_o$. Moreover, the objects $H_o$ and $M_o$ are finitely presentable in $\mathbb{A}$. As a result, there exists a morphism $f_o : H_o \to M_o$ between finitely presentable objects, and morphisms $\alpha_o : H_o \to H$, $\beta : M_o \to M$ such that $(m, \beta)$ is the pushout of $(\alpha_o, f_o)$.

b) The morphism $f_o$ factors in the form $f_o = m_o g_o$ with $g_o : H_o \to N_o$ a regular epimorphism so that $N_o$ is finitely generated, and with $m_o : N_o \to M_o$ a monomorphism. There exists a unique morphism $\gamma : N_o \to H$ such that $\gamma g_o = \alpha_o$ and $m\gamma = \beta m_o$. Then $(m, \beta)$ is the pushout of $(\gamma, m_o)$. According to the amalgamation property, there exists a simple object $T$ and morphisms $s : M \to T$, $t : L \to T$ such that $sm = t\ell$. Then $s\beta m_o = sm\gamma = t\ell\gamma$. Because the object $L$ is algebraically closed there exists a morphism $h : M_o \to L$ such that $hm_o = \ell\gamma$. Hence there exists a morphism $f : M \to L$ such that $fm = \ell$.∎

4.3. Theorem. Any simple object has an algebraically closed extension.

Proof : Let H be a simple object.

a) Let $(f : H \to A,\ g : H \to B)$ be a pair of morphisms with non terminal codomains. According to proposition 3.1.., A and B have simple quotients $m : A \to M$ and $n : B \to N$. According to the amalgamation axiom (6), there exists a simple object L, and a pair of morphisms $p : M \to L$, $q : N \to L$ such that $pmf = qng$. Consequently, the pushout of $(f,g)$ has a non terminal codomain. By induction, one proves that the generalized pushout of finitely many morphisms $f_n : H \to A_n$ with non terminal codomain has a non terminal codomain. It follows that generalized pushouts of arbitrary families of morphisms $(f_i : H \to A_i)$ with non terminal codomains, have non terminal codomains, because there are filtered colimits of generalized finite pushouts (proposition 3.0.).

b) Let $F$ be the class of morphisms $f : H \to A$ such that A is not terminal, and $(A,f)$ is finitely presentable in $H/\mathbb{A}$. Up to isomorphisms, $F$ is a set. Let $g : H \to B$ be the generalized pushout of $F$. According to a), B is not terminal. According to proposition 3.1., B has a simple quotient $q : B \to L_1$. Let $\ell_1 = qg$. Then $(L_1, \ell_1)$ is an extension of H such that, for any morphism $f : H \to A$ in $F$, there exists a morphism $h : A \to L_1$ such that $hf = \ell_1$.

c) Let us build up a sequence of morphisms $\ell_n : L_{n-1} \to L_n$ by induction on $n \in \mathbb{N}^*$, in the following way. Let $L_o = H$ and $\ell_1 : L_o \to L_1$ as constructed in b). If $(\ell_n)_{n<p}$ is built up, the morphism $\ell_p : L_{p-1} \to L_p$ is the morphism built up in (b) but starting with $L_{p-1}$ instead of H. Let $(\alpha_n : L_n \to L)_{n \in \mathbb{N}}$ be the colimit of the diagram $(L_n)_{n\in\mathbb{N}}$. The object L is simple, as it is a filtered colimit of simple objects (proposition 3.3.).

d) Let us show that L is algebraically closed. Let $\alpha : X \to Y$ be a monomorphism with a finitely generated domain and a finitely presentable codomain, $t : L \to T$ be a monomorphism, and $\beta : X \to L$, $\gamma : Y \to T$ be such that $\gamma\alpha = t\beta$. Because $(\alpha_n : L_n \to L)_{n \in \mathbb{N}}$ is a monomorphic filtered colimit, there exist $n \in \mathbb{N}$ and a morphism $\beta_n : X \to L_n$ such that $\alpha_n \beta_n = \beta$. Let $(v : L_n \to V,\ w : Y \to V)$ be the pushout of $(\beta_n, \alpha)$. Because $(Y,\alpha)$ is finitely presentable in $X/\mathbb{A}$, $(V,v)$ is finitely presentable in $L_n/\mathbb{A}$. Because there exists a morphism $V \to T$, the object V is not terminal. According to the construction of the extension $\ell_{n+1} : L_n \to L_{n+1}$, there exists a morphism $u : V \to L_{n+1}$ such that $uv = \ell_{n+1}$. Then the morphism $\alpha_{n+1}\, u\, w : Y \to L$ is such that : $\alpha_{n+1}\, u\, w\, \alpha = \alpha_{n+1}\, u\, v\, \beta_n = \alpha_{n+1}\, \ell_{n+1}\, \beta_n = \alpha_n\, \beta_n = \beta$. As a result, L is algebraically closed. ■

4.4. Theorem. If L is any algebraically closed simple object, then every finitely presentable object is a subobject of a power of L.

Proof : Let A be a finitely presentable object. Let $(u,v) : X \rightrightarrows A$ be a pair of distinct morphisms. According to axiom (3) of locally Hilbert categories, there exists a codisjunctable object $C$ and a morphism $w : C \to X$ such that $uw \neq vw$. Let $(f,g) = (uw, vw)$ and $d : A \to D$ the codisjunctor of $(f,g)$. Let us prove that the object $D$ is finitely presentable. There exists a filtered diagram of finitely presentable objects $(D_i)_{i \in \mathbb{I}}$, the colimit of which is $(\delta_i : D_i \to D)_{i \in \mathbb{I}}$. There exist an index $i_o \in \mathbb{I}$ and a morphism $d_o : A \to D_{i_o}$ such that $\delta_{i_o} d_o = d$. One will assume that $i_o$ is the initial object in $\mathbb{I}$, otherwise replace $\mathbb{I}$ by $i_o/\mathbb{I}$. For any $i \in \mathbb{I}$, let $\alpha : i_o \to i$ be the unique morphism in $\mathbb{I}$, and let $d_i = D_\alpha d_o$: $A \to D_i$ and $(f_i, g_i) = (d_i f, d_i g)$. Let $q : A \to Q$ be the coequalizer of $(f,g)$ and, for any $i \in \mathbb{I}$, $q_i : D_i \to Q_i$ be the coequalizer of $(f_i, g_i)$. The pushout of $q$ along $d$ is, on one hand, the morphism $D \to 1$, and, one the other hand, the filtered colimit of the diagram of morphisms $(q_i)_{i \in \mathbb{I}}$. According to proposition 3.0., there exists an index $i \in \mathbb{I}$ such that $Q_i = 1$, so that $d_i$ codisjoints $(f,g)$. Therefore, $d_i$ factors through $d$ in a morphism $e : D \to D_i$. The relation $\delta_i e d = \delta_i d_i = d$ implies $\delta_i e = 1_D$. The object $D$ is thus finitely presentable as being a split quotient of the finitely presentable object $D_i$. According to the axiom (4) of locally Hilbert categories, the object $D$ is not terminal. According to proposition 3.1., $D$ has a simple quotient $s : D \to H$. Because the category is locally noetherian, the object $H$ is finitely presentable. It follows that, if one denotes by $h : K \to H$ the unique morphism, then $(H,h)$ is a finite extension of $K$. According to proposition 4.2., there exists a morphism $m : H \to L$. Then the morphism $m\,s\,d : A \to L$ codisjoints $(f,g)$. As $L$ is not terminal, it follows that $m\,s\,d\,f \neq m\,s\,d\,g$, hence $m\,s\,d\,u \neq m\,s\,d\,v$. As a result, for any pair of distinct morphisms $(u,v) : X \rightrightarrows A$, there exists a morphism $t : A \to L$ such that $tu \neq tv$. Consequently, if one denotes by $L^{\mathrm{Hom}_{\mathbb{A}}(A,L)}$ the product of $L$ by itself $\mathrm{Hom}_{\mathbb{A}}(A,L)$ times, with projections $p_t : L^{\mathrm{Hom}_{\mathbb{A}}(A,L)} \to L$, one gets a monomorphism $j : A \to L^{\mathrm{Hom}_{\mathbb{A}}(A,L)}$ defined by $p_t j = t$ for any $t : A \to L$. Therefore, $A$ is a subobject of $L^{\mathrm{Hom}_{\mathbb{A}}(A,L)}$. ∎

Let $\mathbb{A}_o$ denotes the full subcategory of $\mathbb{A}$ whose objects are the finitely presentable objects, and let $J : \mathbb{A}_o \to \mathbb{A}$ be the inclusion functor.

4.5. Corollary. For any algebraically closed simple object $L$, the functor $\mathrm{Hom}_{\mathbb{A}}(J(-),L) : \mathbb{A}_o^{op} \to \mathbf{Set}$ is an embedding, thus, it induces an isomorphism between the dual of the category $\mathbb{A}_o$ and a subcategory of $\mathbf{Set}$.

Proof : By theorem 4.4., an object $A$ in $\mathbb{A}_o$ is terminal if and only if $\mathrm{Hom}_{\mathbb{A}}(A,L) = \emptyset$. If $A$ is not terminal and $B$ is any object in $\mathbb{A}_o$ such that $\mathrm{Hom}_{\mathbb{A}}(A,L) = \mathrm{Hom}_{\mathbb{A}}(B,L)$, then, $\mathrm{Hom}_{\mathbb{A}}(A,L) \cap \mathrm{Hom}_{\mathbb{A}}(B,L) = \mathrm{Hom}_{\mathbb{A}}(A,L) \neq \emptyset$. Consequently $A = B$ [12, 1.1.1.]. It follows that the functor $\mathrm{Hom}_{\mathbb{A}}(J(-),L)$ is injective on objects. By theorem 4.4., the functor $Hom_{\mathbb{A}}(J(-),L)$ is faithful. As a result, this functor is an embedding, and thus, it induces an equivalence of categories between $\mathbb{A}_o^{op}$ and its image in $\mathbb{S}et$ [12, 4.1.3.].

Which is this subcategory of $\mathbb{S}et$ isomorphic to $\mathbb{A}_o^{op}$? It is the purpose of the following section to describe it.

5. Algebraic sets.

Let $(L,\ell)$ be a fixed algebraically closed simple extension of the initial object K.

5.0. The category of generators : $\mathbb{N}$.

The category $\mathbb{N}$ is the full subcategory of $\mathbb{A}$, the objects of which are representative of all finite coproducts of generators in $\mathbb{A}$ mentioned in axiom (3) of locally Hilbert categories. The objects of $\mathbb{N}$ are finitely presentable, noetherian and projective. Let us denote by $J_o : \mathbb{N} \to \mathbb{A}_o$, $J_1 : \mathbb{N} \to \mathbb{A}$ and $J : \mathbb{A}_o \to \mathbb{A}$, the inclusion functors.

5.1. The category of congruences : $\mathbb{C}ong\mathbb{N}$.

The category $\mathbb{C}ong\mathbb{N}$ of congruences on objects of $\mathbb{N}$ has, as its objects the pairs $(N,r)$ of an object N of $\mathbb{N}$ and a congruence r on N in the category $\mathbb{A}$, and, as its morphisms $(N,r) \to (M,s)$, the morphisms $f : N \to M$ in $\mathbb{N}$ such that the direct image of r by f is included in s i.e. $(f\times f)r$ factors through s. An object $(N,r)$ of $\mathbb{C}ong\mathbb{N}$ is said to be proper if the congruence r is proper i.e. distinct from $1_{N\times N}$, and it is said to be improper if not.

5.2. The quotient functor : $Q : \mathbb{C}ong\mathbb{N} \to \mathbb{A}_o$.

The quotient functor $Q : \mathbb{C}ong\mathbb{N} \to \mathbb{A}_o$ assigns, to an object $(N,r)$, the codomain of a chosen quotient $q_r : N \to N/r$ of N by r, and to a morphism $f : (N,r) \to (M,s)$, the morphism $Q(f) : Q(N,r) \to Q(M,s)$ defined by $Q(f)q_r = q_s f$.

5.2.0. Proposition. The quotient functor $Q : \mathbb{C}ong\mathbb{N} \to \mathbb{A}_o$ is full and essentially surjective.

Proof : Let $(N,r)$, $(M,s)$ be a pair of objects of $\mathbb{C}ong\mathbb{N}$ and let $f : Q(N,r) \to Q(M,s)$ be a morphism. Let $q_r : N \to Q(N,r)$, $q_s : M \to Q(M,s)$ be the chosen regular epimorphisms. Because the object N is projective, there exists a morphism $g : N \to M$ such that $q_s g = f q_r$. Let $r = (r_1,r_2) : R \rightrightarrows N$ and $s = (s_1,s_2) : S \rightrightarrows M$. The relations $q_s g r_1 = f q_r r_1 = f q_r r_2 = q_s g r_2$ imply that the pair of morphisms $(gr_1, gr_2)$ factors through the pair $(s_1,s_2)$. Consequently $g : (N,r) \to (M,s)$ is a morphism in $\mathbb{N}$, such that $Q(g) = f$. Let A be any object in $\mathbb{A}_o$. It is a regular quotient of an object N of $\mathbb{N}$, let $q : N \to A$. If r denotes the kernel pair of q, then $(N,r)$ is an object of $\mathbb{N}$ such that $Q(N,r)$ is isomorphic to A. ∎

5.3. <u>The algebraic set functor</u> $\Sigma : (\mathbb{C}\mathrm{ong}\mathbb{N})^{op} \to \mathbb{S}\mathrm{et}$.

The <u>algebraic set functor</u> $\Sigma : (\mathbb{C}\mathrm{ong}\mathbb{N})^{op} \to \mathbb{S}\mathrm{et}$ assigns, to any object $(N,r)$, the set $\Sigma(N,r)$ of morphisms $x : N \to L$ which coequalizes $r$, and to any morphism $f : (N,r) \to (M,s)$, the map $\Sigma(f) : \Sigma(M,s) \to \Sigma(N,r)$ defined by $(\Sigma(f))(y) = yf$.

5.3.0. <u>Proposition. The algebraic set functor</u> $\Sigma : (\mathbb{C}\mathrm{ong}\mathbb{N})^{op} \to \mathbb{S}\mathrm{et}$ <u>is isomorphic to the composite functor</u> $\mathrm{Hom}_{\mathbb{A}}(J(-),L) \circ Q^{op} = \mathrm{Hom}_{\mathbb{A}}(JQ(-),L)$, <u>and is injective on proper objects.</u>

<u>Proof</u> : One defines a natural transformation $\alpha : \mathrm{Hom}_{\mathbb{A}}(JQ(-),L) \to \Sigma$ whose value at $(N,r)$ is the map $\alpha_{(N,r)} : \mathrm{Hom}_{\mathbb{A}}(N/r,L) \to \Sigma(N,r)$ defined by $\alpha_{(N,r)}(y) = q_r y$ where $q_r : N \to N/r = Q(N,r)$ is the canonical morphism. It is isomorphic, by the universal property of quotient objects. Let $(N,r)$ be a proper object of $\mathbb{C}\mathrm{ong}\mathbb{N}$, and $(M,s)$ be any object of $\mathbb{C}\mathrm{ong}(\mathbb{N})$ such that $\Sigma(N,r) = \Sigma(M,s)$. Let $q_s : M \to M/s$ be the quotient of $M$ by $s$. The objects $N/r$ and $M/s$ are finitely presentable. According to theorem 4.4., the canonical morphisms

$$J_r : N/r \to L^{\mathrm{Hom}_{\mathbb{A}}(N/r,L)} \simeq L^{\Sigma(N,r)}, \text{ and } J_s : M/s \to L^{\mathrm{Hom}_{\mathbb{A}}(M/s,L)} \simeq L^{\Sigma(M,s)},$$

are monomorphic. Because $r$ is proper, $N/r$ is not terminal and $L^{\Sigma(N,r)}$ neither. Thence $\Sigma(N,r)$ is not empty. As $\Sigma(N,r) = \Sigma(M,s)$ is included in $\mathrm{Hom}_{\mathbb{A}}(N,L)$ as well as in $\mathrm{Hom}_{\mathbb{A}}(M,L)$, it follows that $\mathrm{Hom}_{\mathbb{A}}(N,L) \cap \mathrm{Hom}_{\mathbb{A}}(M,L)$ is not empty. Consequently, $N = M$. Then the relation $J_r q_r = J_s q_s$ implies $r = s$. As a result, $(N,r) = (M,s)$. ∎

5.3.1. <u>Corollary. The image of the functor</u> $\Sigma : (\mathbb{C}\mathrm{ong}\mathbb{N})^{op} \to \mathbb{S}\mathrm{et}$ <u>is a subcategory of</u> $\mathbb{S}\mathrm{et}$.

<u>Proof</u> : By noticing that an object $(N,r)$ is improper if and only if, its image $\Sigma(N,r)$ is empty, it is immediate to see that the sets $\Sigma(N,r)$ with $(N,r) \in \mathbb{C}\mathrm{ong}(\mathbb{N})$, and the maps $\Sigma(f)$ with $f$ in $\mathbb{C}\mathrm{ong}(\mathbb{N})$, are the respective objects and morphisms of a subcategory of $\mathbb{S}\mathrm{et}$. ∎

5.4. <u>The category of algebraic sets</u> : $\Sigma\mathbb{A}$.

<u>The category</u> $\Sigma\mathbb{A}$ <u>of algebraic sets on</u> $\mathbb{A}$ is the subcategory of $\mathbb{S}\mathrm{et}$, image of the algebraic set functor $\Sigma : (\mathbb{C}\mathrm{ong}\mathbb{N})^{op} \to \mathbb{S}\mathrm{et}$. The set $\Sigma(N,r)$ is called the <u>algebraic set defined by</u> $(N,r)$ and its elements are called <u>points</u>. For any object $N \in \mathbb{N}$, the algebraic set $\Sigma(N, \Delta_N) = \mathrm{Hom}_{\mathbb{A}}(N,L)$ is called the <u>algebraic N-space</u> and is denoted by $L^N$. The functor $(\mathbb{C}\mathrm{ong}\mathbb{N})^{op} \to \Sigma\mathbb{A}$ induced by the functor $\Sigma$ will still be denoted by $\Sigma$.

5.5. <u>The coordinate functor</u> : $A(-) : (\Sigma\mathbb{A})^{op} \to \mathbb{A}_o$.

5.5.0. <u>Proposition.</u> The functor $Q : \mathbb{C}\mathrm{ong}\mathbb{N} \to \mathbb{A}_o$ <u>factors uniquely through the functor</u> $\Sigma^{op} : \mathbb{C}\mathrm{ong}\mathbb{N} \to (\Sigma\mathbb{A})^{op}$ <u>in a functor</u> $A(-) : (\Sigma\mathbb{A})^{op} \to \mathbb{A}_o$.

Proof : As the functor $\Sigma^{op}$ is surjective, it is sufficient to prove that, for any pair of objects $(N,r)$, $(M,s)$ (resp. of morphisms $(f,g) : (N,r) \rightrightarrows (M,s)$) such that $\Sigma(N,r) = \Sigma(M,s)$ (resp. $\Sigma(f) = \Sigma(g)$), one has $Q(N,r) = Q(M,s)$ (resp. $Q(f) = Q(g)$). Let $(N,r)$, $(M,s)$ such that $\Sigma(N,r) = \Sigma(M,s)$. If $\Sigma(N,r) = \emptyset$ then $\Sigma(M,s) = \emptyset$, and $(N,r)$ and $(M,s)$ are improper objects. Thus $Q(N,r) = 1 = Q(M,s)$. If $\Sigma(N,r) \neq \emptyset$, then $(N,r) = (M,s)$ (proposition 5.3.O.), so $Q(N,r) = Q(M,s)$. Let $(f,g) : (N,r) \rightrightarrows (M,s)$ such that $\Sigma(f) = \Sigma(g)$. Let $q_r : N \to N/r$, $q_s : M \to M/s$ be the respective quotients of $N,M$ by $r,s$. For any morphism $h : M/s \to L$, one has $hq_sf = (\Sigma(f))(h) = (\Sigma(g))(h) = hq_sg$. As $M/s$ is a subobject of a power of $L$ (Theorem 4.4.), it follows that $q_sf = q_sg$. Then $Q(f)q_r = q_sf = q_sg = Q(g)q_r$, hence $Q(f) = Q(g)$. ■

5.5.1. The functor $A(-) : (\Sigma\mathbb{A})^{op} \to \mathbb{A}_o$ is called the coordinate functor and, for any algebraic set $\Sigma \in \Sigma\mathbb{A}$, the object $A(\Sigma)$ is called the coordinate object of $\Sigma$.

5.5.2. Proposition. The inclusion functor $\Sigma\mathbb{A} \to \mathbf{Set}$ is isomorphic to the composite functor $\mathrm{Hom}_{\mathbb{A}}(J(-),L) \circ A(-)^{op} = \mathrm{Hom}_{\mathbb{A}}(JA(-),L)$.

Proof : Let $I : \Sigma\mathbb{A} \to \mathbf{Set}$ be the inclusion functor and $\Sigma : (\mathbb{C}\mathrm{ong}\mathbb{N})^{op} \to \Sigma\mathbb{A}$. It is immediate to see that the natural isomorphism

$$\alpha : \mathrm{Hom}_{\mathbb{A}}(JQ(-),L) = \mathrm{Hom}_{\mathbb{A}}(JA(-),L) \circ \Sigma \longrightarrow I \circ \Sigma$$

determines a natural isomorphism $\beta : \mathrm{Hom}_{\mathbb{A}}(JA(-),L) \to I$. ■

5.5.3. Theorem. The coordinate functor $A(-) : (\Sigma\mathbb{A})^{op} \to \mathbb{A}_o$ is an equivalence of categories.

Proof : The functor $A(-)$ is faithful because the functor $\mathrm{Hom}_{\mathbb{A}}(JA(-),L)$ is so (proposition 5.5.2.). It is full because the functor $A(-)\circ\Sigma^{op} = Q$ is so (proposition 5.2.O.) and the functor $\Sigma : (\mathbb{C}\mathrm{ong}\mathbb{N})^{op} \to \Sigma\mathbb{A}$ is surjective on objects. The functor $A(-)$ is essentially surjective because the functor $A(-)\circ\Sigma^{op} = Q$ is so (proposition 5.2.O.). ■

5.5.4. Corollary. The category $\Sigma\mathbb{A}$ is equivalent to the image of the functor $\mathrm{Hom}_{\mathbb{A}}(J(-),L) : \mathbb{A}_o^{op} \to \mathbf{Set}$.

Proof : It follows from theorem 5.5.3. and corollary 4.5. ■

## 5.6. The Zariski topology.

5.6.O. Proposition. The set of algebraic subsets of an algebraic set is closed under arbitrary intersections and finite unions.

Proof : Let $\Sigma = \Sigma(N,r)$ be an algebraic set. Any algebraic subset of $\Sigma$ is of the form $\Sigma(N,s)$ with $r \leq s$. Let $(\Sigma(N,s_i))_{i \in I}$ be a family of algebraic subsets of $\Sigma$. For any $i \in I$, let $q_i : N \to N/s_i$ be the quotient of $n$ by $s_i$.

Let $q : N \to Q$ be the generalized pushout of the family of morphisms $(q_i)_{i\in I}$. It is a regular epimorphism, the kernel pair of which is denoted by $s$. Then $\Sigma(N,s) = \bigcap_{i\in I} \Sigma(N,s_i)$. It follows that the set of algebraic subsets of $\Sigma$ is closed under arbitrary intersections. Let $\Sigma(N,s)$, $\Sigma(N,t)$ be a pair of algebraic subsets of $\Sigma$. Let $s = (s_1,s_2) : S \rightrightarrows N$, $t = (t_1,t_2) : T \rightrightarrows N$, and $s\wedge t$ be the intersection of the relations $s$ and $t$. The relation $s\wedge t$ is a congruence on $N$. Let us prove that $\Sigma(N,s\wedge t) = \Sigma(N,s) \cup \Sigma(N,t)$. The relations $s\wedge t \leq s$ and $s\wedge t \leq t$ imply $\Sigma(N,s) \subset \Sigma(N,s\wedge t)$ and $\Sigma(N,t) \subset \Sigma(N,s\wedge t)$, thus $\Sigma(N,s) \cup \Sigma(N,t) \subset \Sigma(N,s\wedge t)$. Let $x : N \to L$ such that $x \notin \Sigma(N,s)$ and $x \notin \Sigma(N,t)$ i.e. such that $x$ does not coequalize $s$ nor $t$. According to axiom (3) of locally Hilbert categories, there exist codisjunctable objects $X$ and $Y$, and morphisms $u : X \to S$, $v : Y \to T$ such that $xs_1u \neq xs_2u$ and $xt_1v \neq xt_2v$. Then $x$ codisjoints $(s_1u,s_2u)$ and $(t_1v,t_2v)$. Let $d : N \to D$, $\delta : N \to \Delta$ be the respective codisjunctors of $(s_1u,s_2u)$, $(t_1v,t_2v)$. Then $d$ codisjoints $s$, and $\delta$ codisjoints $t$. Let $e : N \to E$ be the cointersection of $(d,\delta)$. Then $e$ codisjoints $s$ and $t$. The morphism $x$ factors through $d$ and $\delta$, thus through $e$. Let us prove that $e$ codisjoints $s\wedge t$. Let $q_s : N \to N/s$, $q_t : N \to N/t$, $q_{s\wedge t} : N \to N/s\wedge t$ be the respective quotients of $N$ by $s,t$, $s\wedge t$, and let $m : N/s\wedge t \to N/s$, $n : N/s\wedge t \to N/t$ be the morphisms which satisfy $mq_{s\wedge t} = q_s$ and $nq_{q\wedge t} = q_t$. The morphism $w = (n,m) : N/s\wedge t \to (N/s)\times(N/t)$ is monomorphic. Let $\bar{q}_s$, $\bar{q}_t$, $\bar{q}_{s\wedge t}$, $\bar{w}$ be the respective pushouts of $q_s, q_t, q_{s\wedge t}, w$ along $e$. As $e$ codisjoints $s$ and $t$, one has : $\bar{q}_s = \bar{q}_t = O_E : E \to 1$. Because finite products are couniversal, the codomain of $\bar{w}$ is $1$. Because the morphisms $d$ and $\Delta$ are flat (axiom (3) of locally Hilbert categories), the morphism $e$ is flat thus $\bar{w}$ is monomorphic. Because 1 has no proper object (proposition 3.O.), $\bar{w}$ is monomorphic. Therefore $\bar{q}_{s\wedge t}$ is the morphism $O_E : E \to 1$. As a result, the morphism $e$ codisjoints $s\wedge t$. It follows that $x$ codisjoints $s\wedge t$. Thus $x$ does not coequalize $s\wedge t$ i.e. $x \notin \Sigma(N,s\wedge t)$. As a result $\Sigma(N,s) \cup \Sigma(N,t) = \Sigma(N,s\wedge t)$. Thus the set of algebraic subsets of $\Sigma$ is closed under finite unions ∎

5.6.1. <u>Definition.</u> The <u>Zariski topology</u> on an algebraic set is the topology, the closed sets of which are its algebraic subsets.

5.6.2. <u>Proposition</u>. <u>The inclusion functor</u> : $\Sigma\mathbb{A} \to \mathbf{Set}$ <u>lifts to an inclusion functor</u> : $\Sigma\mathbb{A} \to \mathbf{Top}$.

<u>Proof</u> : Equipped with its Zariski topology, any algebraic set becomes a topological space, and it is immediate to see that morphisms of algebraic sets become continuous maps. ∎

## 6. <u>The structure theorem</u>.

Let $A$ be an arbitrary object in $\mathbb{A}$.

6.0. The *A-algebraic set functor* : $\Sigma_A : (\mathbb{C}\mathrm{ong}\mathbb{N})^{op} \to \mathbb{S}\mathrm{et}$.

The *A-algebraic set functor* $\Sigma_A : (\mathbb{C}\mathrm{ong}\mathbb{N})^{op} \to \mathbb{S}\mathrm{et}$ assigns to $(N,r)$ the set $\Sigma_A(N,r)$ of morphisms $x : N \to A$ which coequalize $r$, called the *set of solutions of* $(N,r)$ *over* $A$, and, to a morphism $f : (N,r) \to (M,s)$, the map $\Sigma_A(f) : \Sigma_A(M,s) \to \Sigma_A(N,r)$ defined by $\Sigma_A(f)(y) = yf$. For $A = L$, one gets the functor $\Sigma$ (5.3.).

6.1. *Proposition. The functor* $\Sigma_A : (\mathbb{C}\mathrm{ong}\mathbb{N})^{op} \to \mathbb{S}\mathrm{et}$ *factors through* $\Sigma : (\mathbb{C}\mathrm{ong}\mathbb{N})^{op} \to \Sigma\mathbb{A}$ *in a functor* $P_A : \Sigma\mathbb{A} \to \mathbb{S}\mathrm{et}$ *which is isomorphic to the functor* $\mathrm{Hom}_{\mathbb{A}}(JA(-),A)$.

*Proof* : By the universal property of quotient objects, the functor $\Sigma_A$ is isomorphic to the functor $\mathrm{Hom}_{\mathbb{A}}(JQ(-),A) = \mathrm{Hom}_{\mathbb{A}}(J(-),A) \circ Q^{op} = \mathrm{Hom}_{\mathbb{A}}(J(-),A) \circ A(-)^{op} \circ \Sigma = \mathrm{Hom}_{\mathbb{A}}(JA(-),A) \circ \Sigma$. According to [12, Proposition 16.6.6.], there exists a functor $P_A : \Sigma\mathbb{A} \to \mathbb{S}\mathrm{et}$ isomorphic to $\mathrm{Hom}_{\mathbb{A}}(JA(-),A)$ and such that $P_A \circ \Sigma = \Sigma_A$. As $\Sigma$ is surjective on objects and morphisms such, a functor $P_A$ is uniquely defined. ∎

6.2. *Definition.* The functor $P_A : \Sigma\mathbb{A} \to \mathbb{S}\mathrm{et}$ is the *realization of* $\Sigma\mathbb{A}$ *over* $A$.

The functor $P_A : \Sigma\mathbb{A} \to \mathbb{S}\mathrm{et}$ preserves finite limits because the functor $\mathrm{Hom}_{\mathbb{A}}(JA(-),A)$ does so, thus $P_A$ is a realization of $\Sigma\mathbb{A}$ according to the following.

6.3. *Definition*. A *realization* of $\Sigma\mathbb{A}$ is a finitely continuous functor $P : \Sigma\mathbb{A} \to \mathbb{S}\mathrm{et}$.

The *category* $\mathbb{R}\mathrm{eal}(\Sigma\mathbb{A})$ *of realizations of* $\Sigma\mathbb{A}$ is the full subcategory of the functor category $[\Sigma\mathbb{A}, \mathbb{S}\mathrm{et}]$ whose objects are the realizations.

6.4. *Proposition. Any realization of* $\Sigma\mathbb{A}$ *is isomorphic to the realization of* $\Sigma\mathbb{A}$ *over some object of* $\mathbb{A}$.

*Proof* : Let $P : \Sigma\mathbb{A} \to \mathbb{S}\mathrm{et}$ be a realization of $\Sigma\mathbb{A}$. As the coordinate functor $A(-) : (\Sigma\mathbb{A})^{op} \to \mathbb{A}_0$ is an equivalence of categories, there exists a functor $F : \mathbb{A}_0^{op} \to \mathbb{S}\mathrm{et}$ such that $F \circ A(-)^{op} \simeq P$. According to [7], the finitely continuous functor $F$ is isomorphic to a representable functor $\mathrm{Hom}_{\mathbb{A}}(J(-),A)$ with $A \in \mathbb{A}$. Then, $P \simeq \mathrm{Hom}_{\mathbb{A}}(J(-),A) \circ A(-)^{op} = \mathrm{Hom}_{\mathbb{A}}(JA(-),A) \simeq P_A$. ∎

6.5. *Theorem. Any locally Hilbert category* $\mathbb{A}$ *is equivalent to the category* $\mathbb{R}\mathrm{eal}(\Sigma\mathbb{A})$ *of realizations of its category of algebraic sets.*

*Proof* : According to [7], $\mathbb{A} \sim \mathrm{Cont}_{\aleph_0}[\mathbb{A}_0^{op}, \mathbb{S}\mathrm{et}]$, and according to theorem 5.5.3., $\mathrm{Cont}_{\aleph_0}[\mathbb{A}_0^{op}, \mathbb{S}\mathrm{et}] \sim \mathrm{Cont}_{\aleph_0}[\Sigma\mathbb{A}, \mathbb{S}\mathrm{et}] = \mathbb{R}\mathrm{eal}(\Sigma\mathbb{A})$. ∎

## REFERENCES

[1] M. BARR : Exact categories and categories of sheaves - Lecture Notes in Mathematics 236, pp. 1-120, Springer-Verlag, Berlin/Heidelberg/New-York, 1971.

[2] Y. DIERS : Sur les familles monomorphiques régulières de morphismes. Cahiers de Topologie et géométrie différentielle, Vol XXI-4, 1980, pp. 411-425.

[3] Y. DIERS : Categories of Boolean sheaves of simple algebras. Lecture Notes in Mathematics 1187, Springer-Verlag, Berlin/Heidelberg/New-York / Tokio, 1986.

[4] Y. DIERS : Codisjunctors and singular epimorphisms in the category of commutative rings, J. Pure. Appl. Algebra. (to appear).

[5] Y. DIERS : Locally Zariski categories. Publication I.R.M.A. de l'Université de Lille 1, 1987.

[6] S. FAKIR : Objects algébriquement clos et injectifs dans les catégories localement présentables. Bull. Soc. Math. France, Mémoire 42, 1975.

[7] P. GABRIEL and F. ULMER : Lokal präsentierbare Kategorien. Lecture Notes in Mathematics 221, Springer-Verlag , Berlin/Heidelberg/New-York, 1971.

[8] P.M. GRILLET : Exact categories and categories of sheaves. Lecture Notes in Mathematics 236, pp. 121-222, Springer-Verlag, Berlin/Heidelberg/New-York, 1971.

[9] A. GROTHENDIECK, M. ARTIN, J.L. VERDIER : Théorie des Topos et Cohomologie Etale des schemas, Lecture Notes in Mathematics 269, Springer-Verlag, Berlin/Heidelberg/New-York, 1972.

[10] N.H. MACCOY and D. MONTGOMERY : A representation of generalized Boolean rings - Duke Math. J 3 .1937, pp. 455-459.

[11] S. MAC LANE : Categories for the working Mathematician, Springer-Verlag, New-York/ Heidelberg/Berlin, 1971.

[12] H. SCHUBERT : Categories, Springer-Verlag, Berlin/Heidelberg/New-York, 1972.

This paper is in final form.

# SUR LA CONVERGENCE DE SUITES DE FONCTIONS SUIVANT DES FILTRES

Iole F. DRUCK et Gonzalo E. REYES

Résumé

On définit deux notions de convergence d'une suite de fonctions continues suivant un filtre sur N : convergence ponctuelle et convergence accessible, et on étudie leurs rapports. On montre l'existence d'une suite de polynômes qui est "universelle" au sens que toute fonction est la limite ponctuelle de cette suite suivant un filtre approprié. En termes non-standard, ceci veut dire que "toute fonction est infiniment près d'un plynôme". Des questions semblables pour les fonctions $C^p$ sont aussi abordées.

## Introduction.

Le but de cette note est de généraliser les notions de convergence de fonctions (ponctuelle et uniforme sur les compacts) en employant des filtres arbitraires sur N , plutôt que le filtre de Fréchet des ensembles cofinis (c'est-à-dire, le filtre des complémentaires des ensembles finis). Notre résultat principal est l'existence d'une suite *universelle* de polynômes qui converge ponctuellement à n'importe quelle fonction ensembliste suivant un filtre approprié. Nous avons des résultats analogues pour le cas de fonctions $C^p$. Ces résultats admettent une reformulation dans l'Analyse non-standard que nous donnons à la fin de cette note. Une autre reformulation, en termes de la Géométrie différentielle synthétique peut se trouver dans Druck [à paraître].

## 1. Convergence suivant un filtre.

Soient $f: R^m \to R$ une fonction quelconque, $(f_k)_{k\in N}$ une suite de fonctions continues et $\Phi$ un filtre sur N .

### 1.1 Définition.

(a) f est la *limite ponctuelle de* $(f_k)_{k\in N}$ *suivant* $\Phi$ ($f = \mathrm{limp}_\Phi f_k$) ssi $\forall \varepsilon > 0\ \forall t \in R^m$ $\exists F \in \Phi\ \ \forall n \in F\, |f_n(t) - f(t)| \leq \varepsilon$.

(b) f est la *limite accessible de* $(f_k)_{k\in N}$ *suivant* $\Phi$ ($f = \mathrm{lima}_\Phi f_k$) ssi $\forall \varepsilon > 0\ \ \forall$compact $K \subset R^m$ $\exists F \in \Phi\ \ \forall n \in F\ \ \forall t \in K\, |f_n(t) - f(t)| < \varepsilon$.

On remarquera que la limite en question en (a) et (b) est unique (quand elle existe) et que si $\Phi$ est le filtre de Fréchet des ensembles cofinis, alors (a) se réduit à la notion habituelle de convergence poncturelle et (b) à la notion de convergence uniforme sur les compacts.

**1.2 Proposition.** Si $f = \mathrm{lima}_\Phi f_k$ , alors $f = \mathrm{limp}_\Phi f_k$

**Preuve:** Ceci est une conséquence immédiate des définitions.

La réciproque de 1.2 n'est pas vraie, mais on peut donner des conditions nécessaires et suffisantes pour le passage de la convergence ponctuelle à la convergence accessible en termes de la notion d'équicontinuité.

**1.3 Définition.** Une suite $(f_k)_{k\in N}$ est *équicontinue par rapport* à $\Phi$ ssi $\forall \varepsilon > 0\ \forall$compact $K \subset R^m\ \ \exists F \in \Phi\ \ \exists \delta > 0\ \ \forall n \in F\ \ \forall t \in K\ \ \forall r \in R^m (|r| < \delta \to |f_n(t+r) - f_n(t)| < \varepsilon)$ .

**1.4 Proposition.** $f = \mathrm{lima}_\Phi f_k$ ssi $f = \mathrm{limp}_\Phi f_k$ , f est continue et la suite $(f_k)_{k\in N}$ est équicontinue par rapport à $\Phi$ .

**Preuve:** Même preuve que pour le cas classique où $\Phi$ est le filtre de Fréchet. Finalement, la preuve classique nous permet de conclure:

**1.5 Propositon.**

(a) Si $f = \lim a_\Phi f_k$, alors

$$\int_0^{x_i} f(x_1,\ldots,u,\ldots,x_n)du = \lim a_\Phi \int_0^{x_i} f_k(x_1,\ldots,u,\ldots,x_n)du .$$

(b) Si $f = \lim a_\Phi f_k$ et la suite $\partial f_k/\partial x_i$ possède une limite accessible, alors $\partial f/\partial x_i = \lim a_\Phi \partial f_k/\partial x_i$.

## 2. Une suite universelle

Dans cette section nous démontrons notre résultat principal:

**2.1 Théorème.** Pour tout $m \in N$ il existe une suite de polynômes à $m$ variables $(f_k)_{k\in N}$ qui est universelle au sens que pour toute fonction $f: R^m \to R$ il existe un filtre $\Phi_f$ sur $N$ tel que $f = \lim p_{\Phi_f} f_k$.

**Preuve:** Si $n \in N$, $a_1,\ldots,a_n$ sont $n$ points distincts de $R^m$ et $a = (a_1,\ldots,a_n)$, on définit

$$P_a : U_a \times R^n \times R^m \to R$$

au moyen de la formule $P_a(x,y,z) = \sum_{i=1}^{n} y_i \prod_{j=1}^{m} P_i^j(z_j)/P_i^j(x_{ij})$, où

$P_i^j(t) = \prod_{\ell \in L_i^j(a)} (t - x_{\ell j})$, $L_i^j(a) = \{\ell : a_{\ell j} \neq a_{ij}\}$ et

$$U_a = \{b \in (R^m)^n : \forall i = 1,\ldots,n \;\; \forall j = 1,\ldots,m \;\; P_i^j(b_{ij}) \neq 0\} .$$

On remaquera que $U_a$ est un voisinage ouvert de $a$ et que

$$P_a(a,y,a_{i1},\ldots,a_{im}) = y_i$$

pour tout $i = 1,\ldots,n$. En effet, si $i \neq i'$, alors $a_{ij_o} \neq a_{i'j_o}$ pour un certain $j_0$, i.e. $i' \in L_i^{j_o}(a)$ et ceci implique que $P_i^{j_o}(a_{i'j_o}) = 0$. Mais alors $\prod_{j=1}^{m} P_i^j(a_{i'j}) = 0$ d'où la conclusion. On peut remarquer aussi que $P_a(a,y,z)$ se réduit à la formule d'interpolation de Lagrange pour $m = 1$. Une dernière remarque: $P_a = P_b$ ssi pour tout $i \leq n$ et pour tout $j \leq m$ $L_i^j(a) = L_i^j(b)$ . Ceci montre qu'il n'existe qu'un nombre fini de $P_a$ pour $n$ fixe et donc qu'un nombre dénombrable de polynômes $P_a(p,q,X_1,\ldots,X_m)$ dans $m$ indéterminées $X_1,\ldots,X_m$, pour $n$ appartenant à $N$, $p$ appartenant à $(Q^m)^n$ et $q$ appartenant à $Q^n$. Soit $(f_k)_{k\in N}$ une énumération arbitraire de ces polynômes. Nous allons démontrer que cette suite est universelle.

Soit $f: R^m \to R$ une fonction arbitraire. Pour chaque $\varepsilon > 0$ et chaque $\xi \in R^m$, définissons

$$F^f_{\varepsilon,\xi} = \{n \in N : |f_n(\xi) - f(\xi)| < \varepsilon\} .$$

Nous allons démontrer que ces ensembles constituent une base de filtre; c'est-à-dire, si $\varepsilon > 0$ et si $a_1,\ldots,a_n$ sont $n$ points de $R^m$, alors $\bigcap_{i=1}^{n} F^f_{\varepsilon,a_i} \neq 0$.

Considérons la fonction $P_a$, où $a = (a_1,\ldots,a_n)$ et définissons $y_i = f(a_i)$. Par continuité on peut choisir $(p,q) \in U_a \times Q^n$ de façon que

$$| P_a(p,q,a_i) - P_a(a,y,a_i) | < \varepsilon .$$

Mais d'après les remarques suivant la définition de $P_a$, $P_a(a,y,a_i) = f(a_i)$ et $P_a(p,q,X_1,\ldots,X_m)$ est un des polynômes dans l'énumération $(f_k)_{k\in N}$, disons le $k^{ième}$. On a démontré donc que $| f_k(a_i)-f(a_i) | < \varepsilon$ pour $i = 1,2,\ldots,n$. Autrement dit, on a trouvé un nombre $k \in \bigcap_{i=1}^{n} F^f_{\varepsilon,a_i}$. Soit maintenant $\Phi_f$ le filtre engendré par les $F^f_{\varepsilon,a}$. Il est évident que $f = \lim p_{\Phi_f} f_k$ et ceci conclut la preuve du théorème.

Pour les fonctions $C^p$ on a un résultat analogue:

**2.2 Proposition.** Pour tout $m \in N$ il existe une suite de polynômes à $m$ variables $(f_k)_{k\in N}$ telle que pour toute fonction $f \in C^p(R^m)$, il existe un filtre $\Phi_f$ engendré par un nombre dénombrable d'ensembles tel que pour tout multi-index $\alpha$ avec $|\alpha| \leq p$ on a

$$\partial^\alpha f/\partial x^\alpha = \lim a_\Phi\, \partial^\alpha f_k/\partial x^\alpha .$$

**Preuve:** Soit $(f_k)_{k\in N}$ une énumération arbitraire de tous les polynômes à $m$ variables à coefficients rationnels et soit $f \in C^p(R^m)$. Définissons por $\varepsilon$ rationnel positif et $K$ un compact de $R^m$ de la forme $[-k,k]^m$

$$F^f_{\varepsilon,K} = \{k \in N : \forall\alpha < p \forall x \in K\, | \partial^\alpha f/\partial x^\alpha - \partial^\alpha f_k/\partial x^\alpha | < \varepsilon\}.$$

On montre, une fois encore, que ces ensembles constituent une base de filtre. Etant donné qu'une réunion finie de compacts est encore un compact, il suffit de montrer que ces ensembles ne sont pas vides. Mais ceci découle de Narasimhan [1968, page 33].

Nous finissons cette section en étudiant le cas d'une fonction à une variable et qui possède $p$ dérivées, non nécessairement continues.

**2.3 Proposition.** Pour tout $p \in N$ il existe une suite de polynômes $(f_k)_{k\in N}$ telle que pour toute fonction $f$ d'une variable réelle possédant $p$ dérivées il existe un filtre $\Phi_f$ de façon que $\forall\alpha \leq p$

$$d^\alpha f/dx^\alpha = \lim p_{\Phi_f}\, d^\alpha f_k/dx^\alpha .$$

**Preuve:** Nous avons besoin du résultat suivant dont la preuve est donnée dans Natanson [1965, vol. III, p. 15]:

**2.4 Lemme.** Etant donnés $n$ réels distincts $x_1,\ldots,x_n$ et pour chaque $i \leq n$ une suite de réels $(y_i, y_i^{(1)},\ldots,y_i^{(p)})$ il existe un unique polynôme $P(X)$ de degré $\leq n(p+1)-1$ tel que pour tout $\alpha \leq p$

$$d^\alpha P/dx^\alpha(x_i) = y_i^{(a)}$$

On remarquera que $P$ ("le polynôme d'interpolation d'Hermite") dépend des paramètres $x_i$ et $y_i^{(j)}$ d'une façon continue (par unicité). Soit $(f_k)_{k\in N}$ une énumération arbitraire de tous ces polynômes dont les paramètres sont rationnels. Nous allons montrer que cette suite a la propriété de 2.3. Soit, en effet, $f$ donnée satisfaisant l'hypothèse. Pour chaque $\varepsilon > 0$ et chaque $x \in R$, nous définissons

$$F^f_{\varepsilon,x} = \{n \in N : | d^\alpha f_n/dx^\alpha(x_i) - d^\alpha f/dx^\alpha(x_i) | < \varepsilon,\ \forall i \leq n,\ \forall\alpha \text{ avec } |\alpha| \leq p\}.$$

Il nous reste à montrer que ces ensembles constituent une base de filtre et on procède d'une façon semblable à 2.1: si $x_1,\ldots,x_n$ sont des réels distincts et $\varepsilon > 0$, on pose $y_i^{(\alpha)} = d^\alpha f/dx^\alpha(x_i)$ et on trouve (par continuité) des rationnels $q_i,\ldots,q_i^{(p)}$ pour chaque $i \leq n$ tels que

$$| d^\alpha P/dx^\alpha(x_i, y_i^{(1)},\ldots,y_i^{(p)}) - d^\alpha P/dx^\alpha(q_i, q_i^{(1)},\ldots,q_i^{(p)}) | \leq \varepsilon .$$

Mais $P(q_i,q_i^{(1)},\ldots,q_i^{(p)},X)$ est l'un des polynômes de l'énumération, disons $f_k$ et ceci montre que ce k appartient à l'intersection des $F^f_{\varepsilon,x_i}$.

## 3. Interprétation non-standard.

Dans cette dernière section, nous allons donner une reformulation de nos résultats de la section 2 en termes de l'Analyse non-standard (voir, par example, Robinson et Zakon [1969] pour les définitions et les notations suivantes).

Plaçons nous dans un "enlargement" $*\mathcal{R}$ de l'utrastructure habituelle $\mathcal{R}$ engendrée par l'ensemble N de nombres naturels. Nous écrivons *N, *R, etc. pour les extensions de N, R, etc. dans $*\mathcal{R}$. Nous employons les notations $a \simeq b$ pour exprimer que la différence entre a et b est infinitésimal et $*R_{fin}$ pour l'ensemble des hyperréels dont la différence avec un réel standard est infinitésimale.

**3.1 Proposition.** Pour tout $m \in N$ il existe une suite de polynômes non-standards $(p_k)_{k\in *N}$ telle que pour toute fonction standard $f: R^m \to R$ il existe $\omega \in *N$ tel que

$$\forall x \in R^m f(x) \simeq p_\omega(x).$$

**3.2 Proposition.** Pour tout $m \in N$ il existe une suite de polynômes non-standards $(p_k)_{k\in *N}$ telle que pour toute fonction $f \in C^p(R^m)$, il existe $\omega \in *N$ tel que pour tout multi-indice $\alpha$ tel que $|\alpha| \leq p$

$$\forall x \in *R^m_{fin} *(\partial^\alpha f/\partial x^\alpha)(x) \simeq \partial^\alpha p_\omega/\partial x^\alpha(x).$$

**3.3 Proposition.** Pour tout $m \in N$ il existe une suite de polynômes non-standards $(p_k)_{k\in *N}$ telle que pour toute fonction standard $f: R \to R$ ayant p dérivées il existe $\omega \in *N$ de façon que pour tour $\alpha \leq m$

$$\forall x \in R d^\alpha f/dx^\alpha(x) \simeq d^\alpha p_\omega/dx^\alpha(x).$$

**Preuve:** Les preuves de ces propositions sont semblables et nous ne donnerons que celle de 3.2.

La proposition 2.2 affirme l'existence d'une suite "universelle" de polynômes $(f_k)_{k\in N}$ à m variables. Ceci nous permet de définir une fonction $p: N \times R^m \to R$ au moyen de la formule $p(k,x) = f_k(x)$ dont l'extension $*p: *N \times *R^m \to *R$ à l'ultrastructure $*\mathcal{R}$ nous donne la suite cherchée: $p_k(x) = *p(k,x)$ pour tout $k \in *N$ et tout $x \in *R^m$. Soit en effet $f \in C^p(R^m)$ une fonction standard et soit $\omega$ n'importe quel élément de la monade du filtre $\Phi_f$ donné par 2.2. Il suffit de montrer que pour tout multi-indice tel que $|\alpha| \leq p$

$$\forall x \in R^m \ \forall i \simeq 0 \quad \partial^{\alpha *}f/\partial x^\alpha(x+i) \simeq \partial^\alpha p_\omega/\partial x^\alpha(x+i).$$

En effet, soient $\varepsilon > 0$, $x_0 \in R^m$ et $\alpha$ un multi-indice tel que $|\alpha| \leq p$. Alors $i \in B(0,\delta)$, où $B(0,\delta)$ est la boule de $R^m$ de centre 0 et rayon $\delta$. On remarquera que l'énoncé $\forall n \in F_{\varepsilon,B(x_0,\delta)} \forall t \in B(x_0,\delta)$ $|\partial^\alpha f_n/\partial x^\alpha(t) - \partial^\alpha f/\partial x^\alpha(t)| < \varepsilon$ est vrai dans $\mathcal{R}$, par la définition même de convergence accessible. Le principe du transfert nous permet de conclure que $|\partial^\alpha p_\omega/\partial x^\alpha(x_0+i) - \partial^{\alpha *}f/\partial x^\alpha(x_0+i)| < \varepsilon$ est vrai dans $*\mathcal{R}$, ce qui conclut la preuve de 3.2.

**Remerciements.** Les auteurs tiennent à remercier M. Làszlo Csirmaz pour avoir suggéré l'idée de base pour la preuve de 2.1. Le premier auteur voudrait remercier l'aide financière de l'"Université de São Paulo ainsi que de la Fundação de Amparo à Pesquisa do Estado de São Paulo, Brésil, tandis que le deuxième aimerait remercier l'appui financier du Conseil de recherches en sciences naturelles et en génie du Canada et du Ministère de l'Education du Gouvernement du Québec.

## BIBLIOGRAPHIE

I. F. Druck [ à paraître], Un modèle de filtres pour l'Analyse réelle lisse, Thèse de doctorat, Université de Montréal, 1986.

R. Narasimhan [1968], Analysis on real and complex manifolds, North Holland, Amsterdam.

I.P. Natanson [1965], Constructive Function Theory, traduit par J.R. Schulenberger, Frederick Ungar Publishing Co., New York.

A. Robinson et E. Zakon [1969], A set-theoretical characterization of enlargements, dans Applications of Model Theory to Algebra, Analysis, and Probability, W.A. Luxemburg (éditeur), Holt, Rinehart and Winston, New York.

Universidade de São Paulo.
Université de Montréal.

This paper is in final form and will not be published elsewhere.

# THE AZUMAYA COMPLEX OF A COMMUTATIVE RING

JOHN W. DUSKIN
DEPARTMENT OF MATHEMATICS
STATE UNIVERSITY OF NEW YORK AT BUFFALO

## INTRODUCTION.

This paper has as its aim the "geometric" description of three important groups classically associated with a commutative ring R as the first (and only) three non-trivial homotopy groups of a reduced (and hence connected) simplicial Kan-complex AZ(R) which we will call the Azumaya complex of the ring. This description, which places the Brauer group Br(R) of equivalence classes of Azumaya algebras over the ring as $\pi_1$ of the complex, the Picard group Pic(R) of isomorphism classes of invertible modules over it as $\pi_2$, and the multiplicative group $G_m(R)$ [$= R^\times$, the group of units] of the ring as $\pi_3$, is stable under arbitrary change of the commutative base ring and (among other things) allows the recovery of several well-known exact sequences linking these groups as instances of exact sequences of groups obtained in simplicial homotopy theory. This conference paper will be devoted to a direct description of the simplicial complex itself with the applications which involve the computation of a particular spectral sequence to appear elsewhere.

The complex we construct here is, in fact, one which is associated with a particular "3-category" $AZ_R$ which has a single object (i.e.,0-cell) the commutative ring R. Its 1-cells are Azumaya algebras over R,

$$R \xrightarrow{A} R \,,$$

the 2-cells are invertible (left) $A \otimes_R B^o$- modules $M: A \Rightarrow B$,

$$R \underset{B}{\overset{A}{\rightrightarrows}} R \quad (M \Downarrow) \,,$$

and finally, the 3-cells are isomorphisms of such modules $i: M \Rrightarrow N$,

$$R \underset{B}{\overset{A}{\rightrightarrows}} R \quad (M \overset{i}{\Rrightarrow} N)$$

Composition is defined using the appropriate tensor products and while not strictly associative (this is what Benabou would probably call a "tricategory") this fact seems to cause no real problem since, at the very least, the bicategory [ Benabou(1967) ] coming from the vertical composition $-\otimes_B- : AZ_R(A,B) \times AZ_R(B,C) \longrightarrow AZ_R(A,C)$ can be replaced with an equivalent (strictly associative) 2-category if one remembers that Morita Theory identifies the groupoid $AZ_R(A,B)$ of invertible left $A \otimes_R B^o$-modules and isomorphisms of such modules with the groupoid of R-functorial equivalences of the corresponding categories of right modules $-\otimes M : Mod(A) \longrightarrow Mod(B)$ and natural isomorphisms of such functors [ Bass(1968)].

The structure of $AZ_R$ is (quite independently of any question of associativity) "groupoid-like" in the sense that the 3-cells are genuine isomorphisms, the 2-cells are equivalences (in the categorical sense) and, in fact, even the 1-cells are "invertible" (i.e.,equivalences in the 2-categorical sense) as well, since to be an Azumaya algebra is equivalent to being an R-algebra A for which there exists an R-algebra $A^o$ together with a

Morita equivalence $M : A \otimes_R A^o \Rightarrow R$ (a 2-cell, by definition!) since this latter is equivalent to the algebra isomorphism $A \otimes_R A^o \xrightarrow{\sim} End(M)$ which occurs in the conventional definition of the notion [Knus-Ojanguren(1974) or Orzech-Small(1975) ].

The author,of course, is aware that knowledge of this structure, or at least fragments of it, has been around for quite a while now, known at the very least to Grothendieck[(1968)] and Zelinsky[(1976), (1977) ] along with the immediate observations that the relevant groups $G_m$, Pic, and Br are all "groups of (equivalence) classes of auto-equivalences" in this trigroupoid with

$$Br(R) \xrightarrow{\sim} \{ [A] : R \longrightarrow R \mid [A]=[B] \Leftrightarrow \exists\ M : A \Rightarrow B \},$$

(Morita equivalence of Azumaya R-algebras) and composition by (horizontal) tensor product of R-algebras,

$$Pic(R) \xrightarrow{\sim} \{ [M] : R \Rightarrow R \mid [M]=[N] \Leftrightarrow \exists\ i : M \Rrightarrow N \},$$

(i.e.,isomorphism classes of modules) and composition by (vertical) tensor product of modules, or if one prefers, quite simply natural isomorphism classes of categorical equivalences of Mod(R) with itself under composition of such equivalences. Finally,

$$G_m(R) \xrightarrow{\sim} \{ i : R \Rrightarrow R \} \xrightarrow{\sim} Aut( id : Mod(R) \longrightarrow Mod(R) )$$

with the equivalence relation discrete and composition given by composition of module isomorphisms.

Anteriorly,one can consider the "groupoid with multiplication" [Benabou(1967)] one obtains by taking isomorphism classes of invertible bimodules of R-algebras as morphisms of R-algebras and takes tensor product of R-algebras as a binary "product". $Aut(A) = Pic_R(A)$ and $Aut_R(R) = Pic(R)$ with the set of connected components of the subgroupoid of Azumaya algebras forming the Brauer group of the ring. The appropriate higher dimensional analog under consideration here would be the "bicategory with multiplication" which either uses the Benabou bicategory of Azumaya R-algebras and invertible bimodules ( or its 2-category " replacement" of categories of A-modules and R-functorial equivalences between them) and gives it a multiplication by using the tensor product. The result may be viewed as a "tricategory with a single object" just as Benabou observed that a "category with multiplication" could be viewed equivalently as a bicategory with a single object. The "2-groupoid" of auto-equivalences of R with itself is the replacement for the category PIC(R) of projective R-modules of rank 1, which under tensor product is the so-called Picard category of R. It is the motivating example for the notion of a "Picard category" [Deligne(1973)] viewed as a groupoid with a binary product which is associative and commutative up to coherent isomorphism.

The tricategory structure referred to above is not, however, the structure which we wish to study here, rather it is the simplicial complex which may be associated to such structures which will concern us .

The existence of this structure, which is a generalization of the "nerve of a 3-category" as R.Street and the author [Duskin-Street] used the term,or more properly in the particular case at hand, the "classifying complex of the double complex associated with a 2-groupoid with multiplication", has been hinted at before (most notably by Fröhlich and Wall [ Fröhlich and Wall (1971), Wall(1974)] but as far as the author has been able to determine, never fully developed or exploited.

At least one of the reasons for this will readily become apparent to readers of this paper : The method of "naive computation" which we have employed here to obtain a full combinatorial description of the n-simplices of the complex AZ(R) becomes increasingly and even dauntingly complicated as the dimension of the simplices increases.

Now, part of this is complication is inevitable if one really wishes to properly take into account the associativity isomorphisms as they are given ( and even here we make use of Morita theory to suppress a portion of them and note, in passing, that there appears to be a 2-dimensional Morita Theory which would allow the suppression of the remainder). But even if one has full associativity, the direct description of the simplicial object associated with an n-category following the pattern used here for the Azumaya complex is combinatorially quite complicated and almost physically impossible to present in complete detail in this fashion in dimensions higher than five. Fortunately for our purposes here, a complete description is unnecessary beyond dimension four since the complex here is 4-coskeletal and is thus completely determined by its truncation, and in any case, the point we primarily wish to make is only that the Kan-complex structure is really there (even without the presence of genuine associativity at the category level) so that the full armature of simplicial homotopy theory is applicable to these fascinating and much studied abelian groups. For readers having a more sophisticated knowledge of category theory than the minimal amount required to follow the approach used here an alternative approach using recent work of R. Street is outlined at the end of the paper.

DESCRIPTION OF THE AZUMAYA COMPLEX

The unique 0-simplex of AZ(R) is the commutative ring R. The 1-simplices are Azumaya algebras

$$R \xrightarrow{A} R$$

with o-face and 1-face the ring R. The degenerate 1-simplex $s_0(R)$ is the commutative ring R considered as an Azumaya algebra over itself.

2-simplices are triplets of Azumaya algebras $(A_0, A_1, A_2)$ together with an invertible $A_1 \otimes (A_2 \otimes A_0)^o$-module $M : A_1 \Rightarrow A_2 \otimes_R A_1$, denoted by

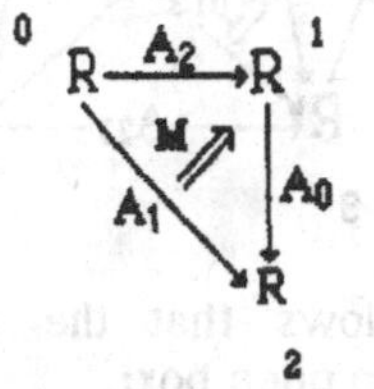


and with faces defined (as opposite the numbered vertex) by projection. Degeneracies $s_0(A)$ and $s_1(A)$ are defined using the bimodules which represent the isomorphisms $A \xrightarrow{\sim} R \otimes_R A$ and $A \xrightarrow{\sim} A \otimes_R R$, respectively.

The 3-simplices are tetrahedra of compatible 3-simplices defined as above:

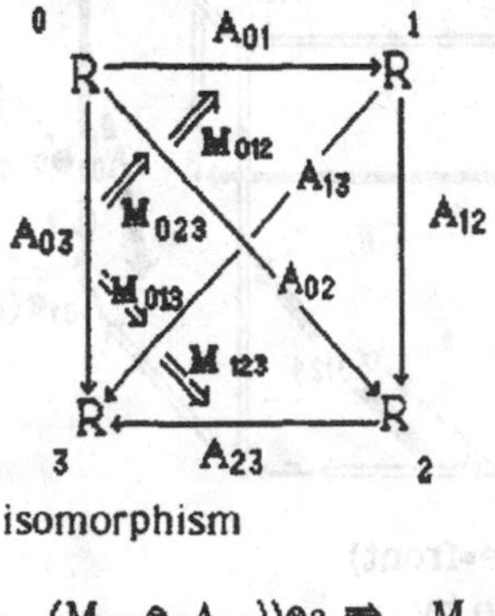


together with a bimodule isomorphism

$$\alpha_{0123} : (M_{023} \otimes_{A_{02} \otimes A_{23}} (M_{012} \otimes_R A_{23})) \otimes a \Rightarrow M_{013} \otimes_{A_{01} \otimes A_{13}} (A_{01} \otimes_R M_{123})$$

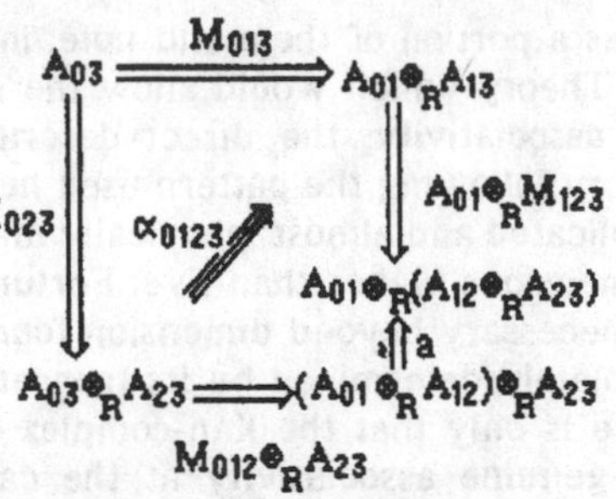


where a is the canonical associativity isomorphism (which shifts parentheses to the right) viewed as a bimodule.

The 4-simplices are constructed out of families of compatible 3-simplices:

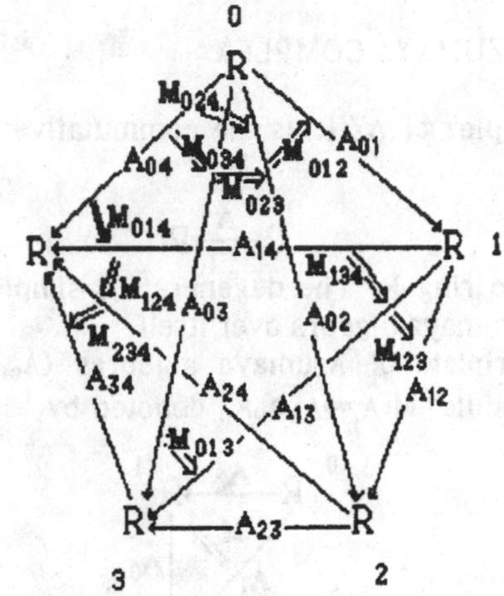


from whose compatibility it follows that the squares that specify the bimodule isomorphisms fit together to form an open box:

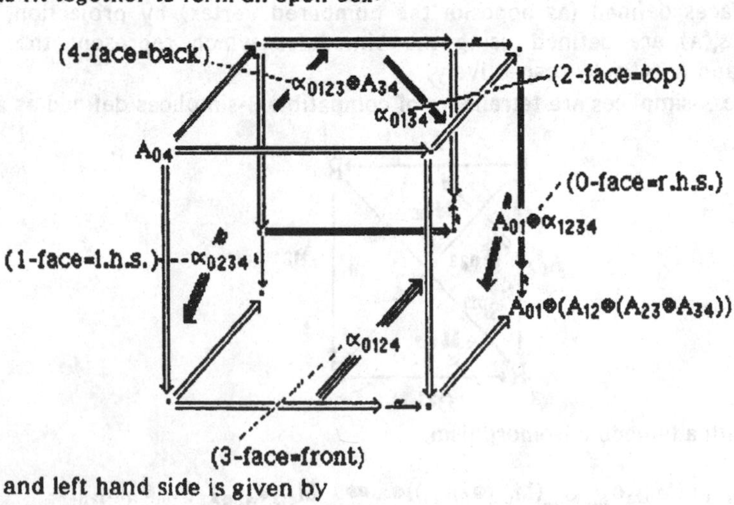


whose back and left hand side is given by

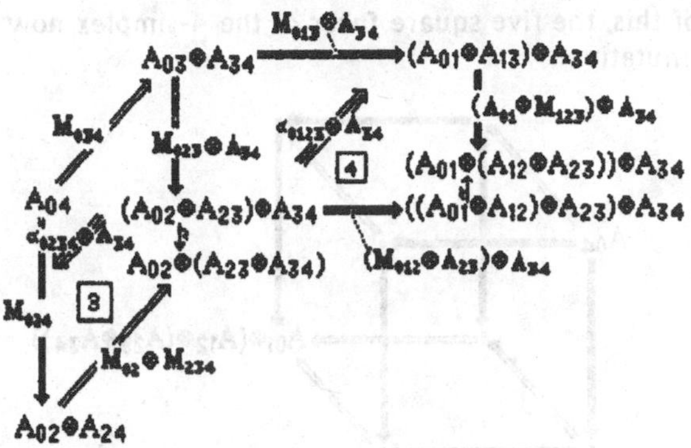

and whose front, top, and right hand side is given by

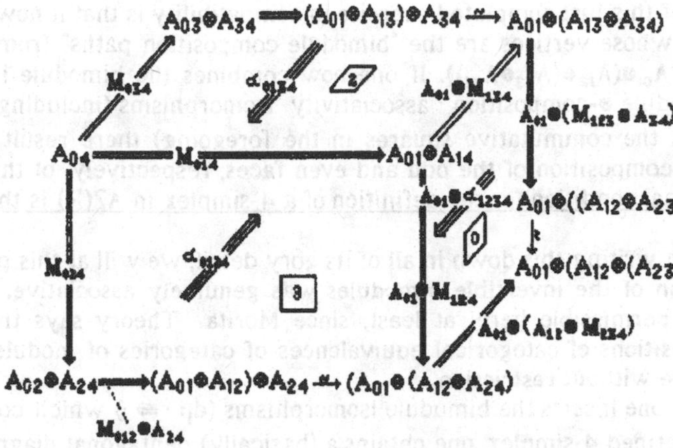

If one now uses the associativity isomorphisms for $\circledast_R$, then the back and bottom of the box may be filled out by a diagram, which by use of the pentagon axiom for $\circledast_R$ (as well as its naturality and 2-bifuntoriality over bimodule $\circledast_A$-composition), is readily seen to be commutative:

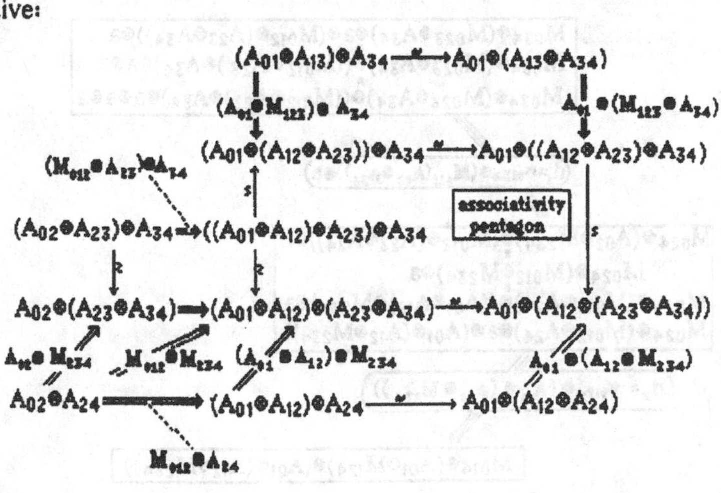

As a consequence of this, the five square faces of the 4-simplex now fit into a cube whose bottom face is commutative:

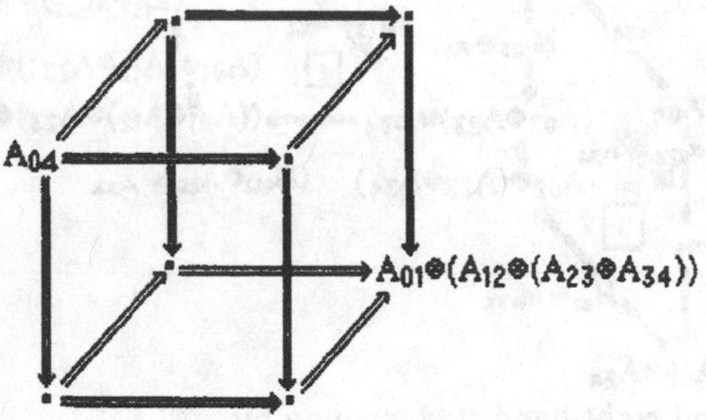


The point of this just completed exercise in compatibility is that it now allows us to form a new graph whose vertices are the "bimodule composition paths" from the algebra $A_{04}$ to the algebra $A_{01}\otimes(A_{12}\otimes(A_{23}\otimes A_{34}))$. If one now combines the bimodule isomorphisms $\alpha_{ijkl}$ with the bimodule $\otimes$-composition associativity isomorphisms (including the identity isomorphisms from the commutative squares in the foregoing) there result exactly two paths formed from composition of the odd and even faces, respectively, of the component 3-simplices. The final condition for the definition of a 4-simplex in AZ(R) is that these two paths be equal.

Rather than writing this down in all of its gory detail, we will at this point write as if the $\otimes$-composition of the invertible bimodules was genuinely associative. There is no doubt that this is permissible here, at least, since Morita Theory says that these are "really" just compositions of categorical equivalences of categories of modules which are certainly associative without restriction.

Thus when one inserts the bimodule isomorphisms ($d_i$: $\cdot \Rrightarrow \cdot$) which come from the faces $d_i$ of the so defined 4-simplex, one obtains a (basically) pentagonal diagram in which the promised two distinct paths corresponding to the next higher order of composition occur:

The left hand (odd faces) side of the pentagon is

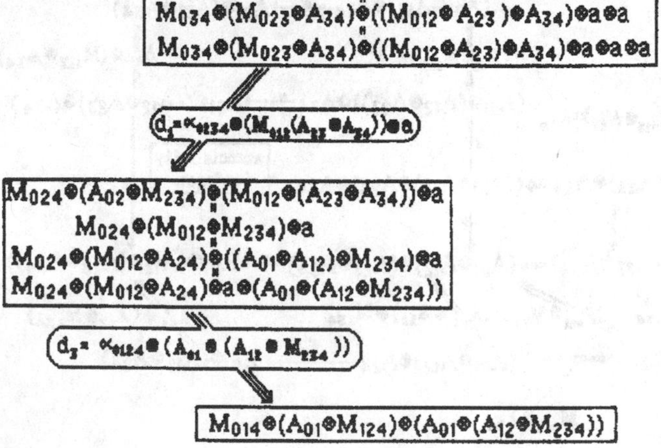


and the right hand (even faces) side is

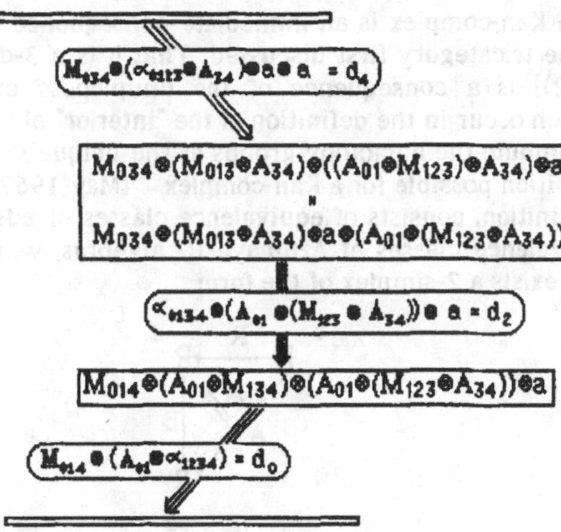


so that the final condition is that the composition of these two (just shown to be composable) sequences of isomorphisms be equal:

$$(d_3) \qquad\qquad \cdot \qquad\qquad (d_1) \qquad =$$

$$\{\alpha_{0124}\otimes(A_{01}\otimes(A_{12}\otimes M_{234}))\}\cdot\{\alpha_{0234}\otimes(M_{012}\otimes(A_{23}\otimes A_{34}))\otimes a\} =$$

$$\{M_{014}\otimes(A_{01}\otimes\alpha_{1234})\}\cdot\{\alpha_{0134}\otimes(A_{01}\otimes(M_{123}\otimes A_{34})\otimes a\}\cdot\{M_{034}\otimes(\alpha_{0123}\otimes A_{34})\otimes a\otimes a\}$$

$$= \qquad (d_0) \qquad\qquad \cdot \qquad\qquad (d_2) \qquad\qquad \cdot \qquad\qquad (d_4).$$

It is not difficult (although it *is* tedious) to verify that the obvious definition of the degeneracies does indeed work (using the appropriate ⊗-axioms ) and that the resulting truncated complex may be completed to a full simplicial complex AZ(R) by taking its coskeleton.

For our purposes what is most interesting about this complex is summarized in the following

PROPOSITION. For any commutative ring R, the simplicial complex AZ(R) has the following properties:

(a) AZ(R) is a reduced (and hence connected) Kan-complex;

(b) it has at most three non-trivial homotopy groups, viz.:

$$\pi_1(\, AZ(R)\,) = Br(R),$$

$$\pi_2(\, AZ(R)\,) = Pic(R), \text{ and}$$

$$\pi_3(\, AZ(R)\,) = G_m(R), \text{ the group of units of } R;$$

(c) it is a 3-dimensional hypergroupoid.

That it is a Kan-complex is an immediate consequence of the "quasi-invertibility" of the arrows in the tricategory first discussed. That it is a 3-dimensional hypergroupoid [Glenn(1977),(1982)] is a consequence of the uniqueness of the inverses (bimodule isomorphisms) which occur in the definition of the "interior" of the 3-simplices.

We now compute the homotopy groups at the unique vertex R following the usual combinatorial definition possible for a Kan-complex [May(1967) ]:

$\pi_1$ , by definition, consists of equivalence classes of edges of the form $A:R \longrightarrow R$ , i.e. here just equivalence classes of Azumaya R- algebras, where two edges A , B are equivalent if there exists a 2-simplex of the form

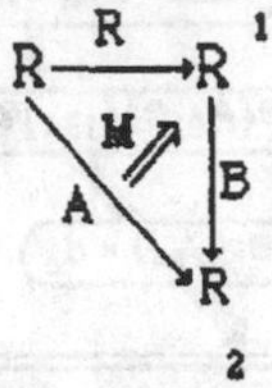


Since $R \otimes_R B \xrightarrow{\sim} B$, this is just Morita equivalence, as asserted. Furthermore the multiplication on $\pi_1$ is determined by any representative 2-simplex

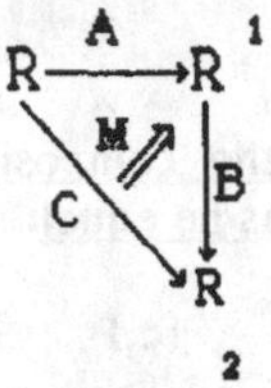


with $[C] = [A] \cdot [B]$, but by definition of the 2-simplices, $[C] = [A \otimes_R B]$ , so that the product is the usual one using the tensor product of the algebras. Thus the fundamental group of the complex is the Brauer group of the ring, as asserted.

$\pi_2$ , again by the standard combinatorial definition, consists of equivalence classes of 2-simplices of the form

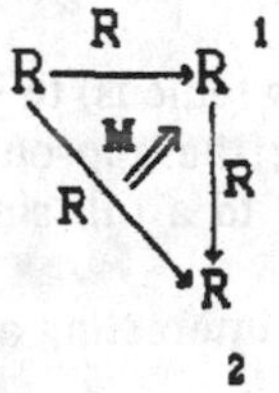


i.e.,just equivalence classes of invertible R-modules, where 2- simplices $M_0$ and $M_1$ are equivalent provided there exists a 3-simplex of the form

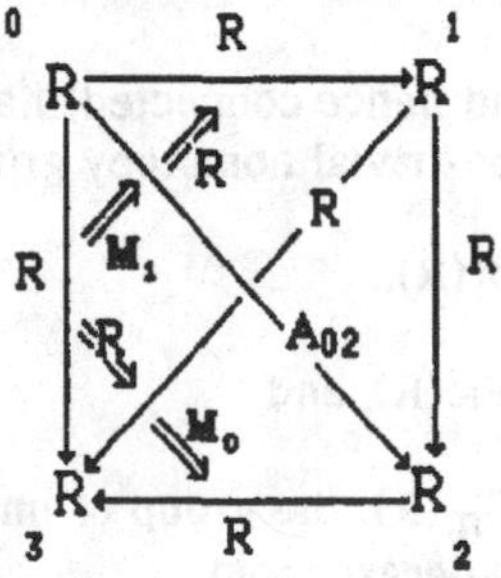

i.e.,a 3-simplex whose 0-face is $M_0$ and whose 1-face is $M_1$, but with all other faces degenerate. This means by our definition of a 3-simplex, just a square of the form

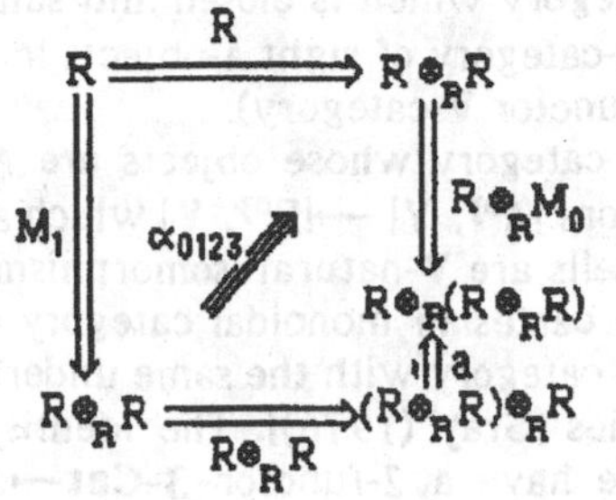


and thus equivalently, a square

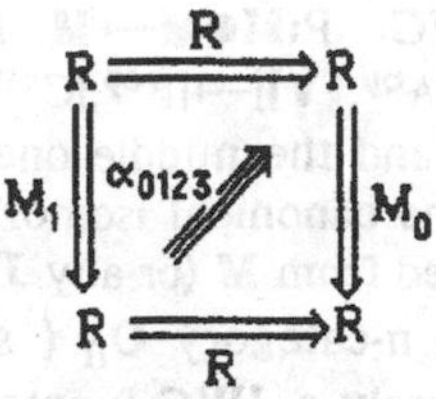


or, quite simply, an isomorphism of R-modules $\alpha\colon M_1 \xrightarrow{\sim} M_0$ as in the traditional definition.

Again, as before, the multiplication in $\pi_2$ is defined through its representatives using any 3-simplex whose 3-face (say) is degenerate

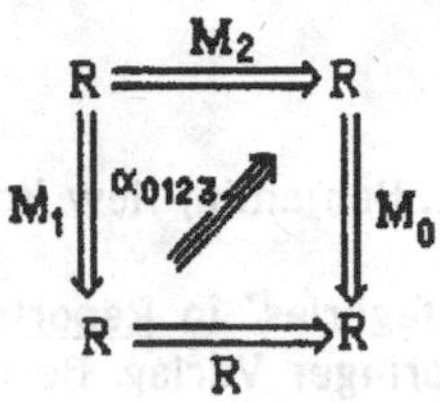


so that $[M_2 \otimes M_0] = [M_2]\cdot[M_0] = [M_1]$, and $\pi_2$ is the Picard group, as asserted.

Finally, $\pi_3$ consists of equivalence classes of 3-simplices, all of whose faces are similarly "at the base point", R:

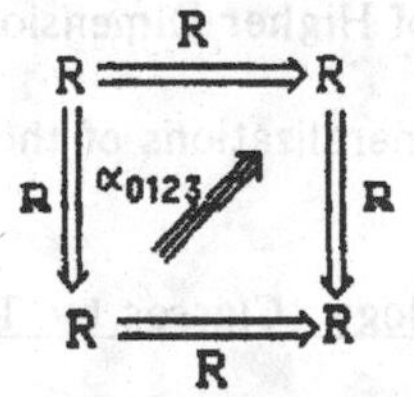


i.e.,an R-linear isomorphism of R with itself, or equivalently, a unit of R. The equivalence relation requires the existence of a 4-simplex with $d_0 = \alpha_0$ and $d_1 = \alpha_1$, with all other faces degenerate. This thus here means nothing more than the collapse of everything other than $\alpha_0 = \alpha_{1234}$ and $\alpha_1 = \alpha_{0234}$ and thus, by our definition which requires that the defining diagram be commutative, that $\alpha_0 = \alpha_1$. This also guarantees that the multiplication in $\pi_3$ is by composition of isomorphisms, and thus equivalently, by multiplication of units. Consequently, $\pi_3 = G_m(R)$, as asserted, and the full Proposition is established.

An alternative approach to obtaining a simplicial complex having properties equivalent to those of AZ(R) has been communicated in outline by Ross Street: "Let **V** be any symmetric monoidal category which is closed and suitably complete. For a monoid A in **V** write $[A^{op}, \mathbf{V}]$ for the **V**-category of right A-objects in **V** (that is, A is a one object **V**- category and this is the functor **V**-category).

Let **M** denote the 2-category whose objects are Azumaya monoids in **V**, whose arrows $M{:}A \longrightarrow B$ are **V**-functors $[A^{op}, \mathbf{V}] \longrightarrow [B^{op}, \mathbf{V}]$ which are equivalences of **V**-categories, and whose 2-cells are **V**-natural isomorphisms.

Write **2-Cat** for the cartesian monoidal category of 2-categories and 2-functors. Write **JWG** for the monoidal category with the same underlying category, but with Gray's tensor product of 2-categories [Gray (1976)]. The identity functor **2-Cat⟶JWG** has a monoidal enrichment, so we have a 2-functor **3-Cat⟶JWG-Cat** which allows us to regard 3-categories as **JWG** categories.

PROPOSITION: If **V** is strict monoidal then the following multiplication enriches **M** with a monoidal structure in **JWG**: $P{:}\mathbf{M}\bullet\mathbf{M} \longrightarrow \mathbf{M}$, $P(A,B)=A\bullet B$; for $M{:}A\longrightarrow C$, $P(M,B)$ is the composite $[(A\bullet B)^{op},V] \longrightarrow [B^{op},[A^{op}, \mathbf{V}]] \longrightarrow [B^{op},[C^{op}, \mathbf{V}]] \longrightarrow [(C\bullet B)^{op}, \mathbf{V}]$ where the first and last arrows are isomorphisms and the middle one is $[B^{op},M]$; $P(A,N)$ is similar, and $P(B,N)P(M,C) \longrightarrow P(M,D)P(A,N)$ is the canonical isomorphism.

A simplicial set **X** is obtained from **M** (or any **JWG** category) as follows. Let $\mathbf{T}_n$ be the underlying 3-category of the n-category $\mathbf{O}_n$ ( so that the two agree for $n<4$). An element of **X** of dimension n is merely a **JWG**-functor $x{:}\mathbf{T}_n \longrightarrow M$ which identifies the 3-source of each 4-cell in $\mathbf{O}_n$ with its 3-target. For the construction of the complex relevant to the one of this paper, any strict monoidal category **V** monoidally equivalent to Mod(R), e.g., the category of cocontinuous endofunctors of Mod(R), may be used."

BIBLIOGRAPHY

Bass,H.: Algebraic K-Theory ,W.A. Benjamin, New York (1968)

Benabou, J.: "Introduction to Bicategories" in Reports of the Midwest Category Seminar, Lec.Notes in Math.#47, pp.1-77,Springer Verlag, Berlin (1967)

Deligne,P.: "La Formule de Dualité Globale" in SGA-4, Lec.Notes in Math.#305,pp.481-587,Springer Verlag, Berlin (1973)

Duskin,J. and Street.R.: "Nerves of Higher Dimensional Categories", *In Preparation*

Fröhlich,A. and Wall, C.T.C.: "Generalizations of the Brauer Group", *Preprint* , Liverpool University (1971)

Glenn, P.: Realization of Cohomology Classes by Torsors under Hypergroupoids , Ph.D. Thesis, S.U.N.Y.,Buffalo(1977)

--------:"Realization of Cohomology Classes in Arbitrary Exact Categories", J.Pure Appl. Alg.,v.25, pp.33-105 (1982)

Gray,J.W.:"Coherence for the Tensor Product of 2-Categories,and Braid Groups" in Algebra,Topology and Category Theory (a Collection of Papers in Honor of Samuel Eilenberg),pp.106-147,Academic Press(1976)

Grothendieck,A.: "Le group de Brauer I, II, III" in Dix Exposés sur la Cohomologie des Schémas, pp.46-188,North Holland, Amsterdam (1968)

Knus, M.A. and Ojanguren,M. : Théorie de la Déscent et Algèbres de Azumaya, Lec.Notes in Math.#389, Springer Verlag, Berlin(1974)

May, J.P. : Simplicial Objects in Algebraic Topology , Van Nostrand Math. Studies #11, Van Nostrand, Princeton (1967)

Orzech, M. and Small, C.: The Brauer Group of Rings, Lec. Notes in Pure Appl.Math. #11,Marcel Dekker, New York (1975)

Street,R.: "The Algebra of Oriented Simplices". *Preprint* Macquarie University (1984) to appear J.Pure Appl.Alg.

Wall, C.T.C.: "Equivariant Algebraic K-Theory" in Segal,G.: New Developments in Topology. Lon. Math. Soc. Lec. Notes.#11, Cambridge Univ. Press., Cambridge (1974)

Zelinski,D.: "Long Exact Sequences and the Brauer Group", in Brauer Groups, Evanston Ill.1975, Lec.Notes in Math.#549, Springer Verlag, Berlin (1976)

------ "Brauer Groups", in Ring Theory II, Lec.Notes in Pure Appl.Math.#26, Marcel Dekker, New York (1977)

This paper is in final form and will not be published elsewhere.

# THE SPECTRUM LATTICE OF BAER RINGS AND POLYNOMIALS

by

L. ESPAÑOL

Departamento de Matemáticas

Colegio Universitario de La Rioja

26001 Logroño (España)

**Introduction.** This paper is concerned with (commutative unitary) rings and (bounded distributive) lattices. Models of these theories are in a topos **S** in which one can work as in set theory but without using either the axiom of choice or the law of the excluded middle. When **S**=Set, theorems in this paper are new versions or generalizations of well known results about prime ideals in commutative algebra, proved by elementary methods.

Recall [1], [2] that in Set there are two spectrum functors $Rg^{op} \to Sp$ (Rg is the category of rings) and $Lat^{op} \to Sp$ (Lat is the category of lattices) valued in the category Sp of spectral spaces and spectral maps. In each case points are prime ideals and the topology is the Zariski topology. The canonical base of Spec(A) is formed by the subsets $D(a)=\{p \in Spec(A) | a \notin p\}$, $a \in A$. The basic construction in this paper is the composition functor $D: Rg \to Sp^{op} \to Lat$ and the canonical maps $d_A: A \to D(A)$ given by $d_A(a)=D(a)$ (called the *reticulation* of A in [12]). I use Joyal's construction of D(A) [9], [3].

In Section 1 I set down some basic facts about D(A) related with baer and regular rings (in the sense of von Neumann). Section 2 contains a notion of content of polynomials [4] and a generalization to toposes of the relation between prime ideals in A and A[X]. Finally, in Section 3 we give elementary proofs of the transfer theorems: A domain (baer) implies A[X] domain (baer).

**Acknowledgements.** This work was supported by CAICYT, the spanish organization for the advancement of research. The final version of this paper was obtained during my stay in Milano (Italy, july 1987), where I profited from stimulating discussions with F.W. Lawvere and G.C. Meloni.

## 1. The spectrum lattice.

Let A be a ring in a topos **S**. D(A) is the set of radicals $[a_1,\dots,a_n]$ of finite generated ideals $(a_1,\dots,a_n)$ of A, ordered by inclusion. It is easy to check that D(A) is a lattice with the following operations:

$$[a_1,\dots,a_n]\wedge[b_1,\dots,b_m]=[a_ib_j \mid 1\le i\le n,\ 1\le j\le m]$$

$$[a_1,\dots,a_n]\vee[b_1,\dots,b_m]=[a_1,\dots,a_n,b_1,\dots,b_m]$$

In particular we have $[a_1,\dots,a_n]=[a_1]\vee\dots\vee[a_n]$ and

$$[a_1,\dots,a_n]=1 \Leftrightarrow 1=\Sigma t_ia_i \quad (t_i\in A)$$

$$[a_1,\dots,a_n]=0 \Leftrightarrow a_i\in N,\ 1\le i\le n$$

where N is the nilradical of A. We call D(A) the *spectrum lattice* of A.

A map $d:A\to L$ from a ring A into a lattice L is called a *support* of A if it satisfies the following conditions:

(i) $d(0)=0$ and $d(1)=1$

(ii) $d(ab)=d(a)\wedge d(b)$

(iii) $d(a+b)\le d(a)\vee d(b)$

It is easy to check that $d_A:A\to D(A)$, $d_A(a)=[a]$, is an universal support of A, i.e., $d_A$ is a support and for any support $d:A\to L$ there exist a unique lattice homomorphism $d^-:D(A)\to L$ such that $d^-d_A=d$. In fact we have $d^-[a_1,\dots,a_n]=d(a_1)\vee\dots\vee d(a_n)$. Given a support $d:A\to L$, the subset $d^{-1}(0)$ is an ideal of A, in particular $N=d_A^{-1}(0)$. Moreover, supports $A\to\{0,1\}$ are characteristic functions of of prime ideals of A. The spectrum lattice functor $D:Rg\to Lat$ is obtained by defining $D(f)([a_1,\dots,a_n])=[f(a_1),\dots,f(a_n)]$ for any ring homomorphism.

**Remarks.** Now we recall some well known properties of D(A). We omit the new proofs, which are elementary so that these results hold in an arbitrary topos.

1. If $f:A\to A/N$ is the canonical epimorphism, then D(f) is an isomorphism, $D(A/N)\cong D(A)$.

**2.** Given a lattice L, let C(L) be the boolean algebra of all complemented elements of L. Moreover the set E(A) of all idempotents of a ring A is a boolean algebra with $e\wedge e'=ee'$, $e\vee e'=e+e'-ee'$ and $\neg e=1-e$. For any support $d:A\to L$, the restriction of d to E(A) is a lattice homomorphism into C(L), i.e. $d:E(A)\to C(L)$.

**3.** For any ring A, the universal support $d_A$ induces an isomorphism $E(A)\cong CD(A)$. If $[a_1,\dots,a_n]\in CD(A)$ with $\neg[a_1,\dots,a_n]=[b_1,\dots,b_m]$ then

$$(a_ib_j)^{r_{ij}}=0\ ,\ 1=\Sigma t_ia_i^r+\Sigma s_jb_j^r\ \ (r=\Sigma r_{ij})$$

and $[a_1,\dots,a_n]=[e]$ where $e=\Sigma t_ia_i^r$ is an idempotent.

We include here two examples. Let us consider the topos **sh**(B) of sheaves over a boolean algebra B in **S**. A result in [5] is that a ring in this topos is a ring A in **S** with a lattice homomorphism $\alpha:B\to E(A)$, the spectrum lattice being D(A) with $\delta:B\to CD(A)$ the map given by $\alpha$ and Remark 3. As another example, let $\mathfrak{C}$ the classifying topos of rings and $\mathfrak{G}$ the subtopos of $\mathfrak{C}$ formed by finite products preserving functors (called Gaeta topos by Lawvere) which is the classifying topos for rings such that $x^2=x$ implies $x=0$ or $x=1$. The functor D is the spectrum lattice of the generic ring in $\mathfrak{C}$ and also in $\mathfrak{G}$, so that D satisfies in $\mathfrak{G}$ the formula: x complemented implies $x=0$ or $x=1$.

The next remark needs some definitions. A lattice L is *stonian* if for any $x\in L$ there exists $x^*\in C(D)$ such that (i)$x\wedge x^*=0$, and (ii)$x\wedge y=0$ implies $y\leq x^*$. A lattice L is stonian if and only if there exists a lattice homomorphism $k:L\to C(L)$ left adjoint to the inclusion map $i:C(L)\to L$, that is: $k(x)\leq y$ (in C(L)) iff $x\leq y$ (in L). Actually, $k(x)=x^{**}=\neg x^*$. Boolean lattices are stonian, and an stonian lattice L is boolean if and only if for all $x\in L$, $x^*=0$ implies $x=1$. A ring A is said *regular* if for any $a\in A$ there exits $b\in A$ such that $a^2b=a$; then the element $a'=b^2a$ is uniquely determined by the conditions $a^2a'=a$ , $aa'^2=a'$. Moreover $a^*=aa'$ is an idempotent such that $aa^*=a$, $a'a^*=a'$. A ring A is *baer* if for any $a\in A$, $Ann(a)=(a^*)$ where $a^*$ is an idempotent. Regular rings are baer, and a baer ring A is regular if and only if for all $a\in A$, $a^*=1$ implies a is unit. As far as the lattice D(A) is concerned, after Remark 1 we can suppose A is *nilpotent-free*, i.e.: $a^2=0$ implies $a=0$. Baer rings are nilpotent-free.

**4.** Let A be a ring. Then D(A) is stonian (resp. boolean) if and only if A/N is a baer (resp. regular) ring. The universal support of a regular ring A is the map $(.)^*:A\to E(A)$. If A is a baer ring, then this map is the universal *-support of A, where a *-support [4] is a support $d:A\to L$ such that $d(a)=0$ implies $d(a^*)=0$.

## 2.The radical content of a polynomial.

This section's goal is to define in a topos $\mathcal{S}$ the well-known adjunction in Set $i^*\dashv(.)^e:Spec(A)\to Spec(A[X])$ given by $P\cap A\le p$ if and only if $P\le p[X]$, where $i:A\to A[X]$ is the inclusion, P is a prime ideal of A[X], p is a prime ideal of A and p[X] is the set of all polynomials with coefficients in p.

So, we are looking for a left adjoint $\gamma$ to $D(i):D(A)\to D(A[X])$ such that $\gamma D(i)=id$. This adjunction means an equivalence between $\gamma([f_1,\dots,f_p])\le[b_1,\dots,b_m]$ in D(A) and $[f_1,\dots,f_p]\le[b_1,\dots,b_m]$ in D(A[X]).

**2.1.Lemma.** If $f=a_0+a_1X+\dots+a_nX^n\in A[X]$ and $b_1,\dots,b_m\in A$. The following statements are equivalent:

(i) $[f]\le[b_1,\dots,b_m]$ in D(A[X])

(ii) $[a_0,\dots,a_n]\le[b_1,\dots,b_m]$ in D(A)

Proof: We shall check the case m=1 because the general case is similar. The lemma is trivial when $f\in A$. Suppose (i), that is $f^r=gb$, $r\ge1$, $g\in A[X]$. If $f=pX+a_0$ and $g=g_rX^r+\dots+t_0$, with $p,g_r\in A[X]$, $t_0\in A$, then $a_0{}^r=t_0b$, that is $[a_0]\le[b]$, and $p^r=g_rb$. But $deg(p)=deg(f)-1$, so that we obtain by induction $a_i{}^{r_i}=t_ib$, i.e. $[a_i]\le[b]$, $1\le i\le n$. Hence it results (ii).

Conversely, if $a_i{}^{r_i}=t_ib$, $0\le i\le n$, then we can find $g_i\in A[X]$ such that for $r=r_0+\dots+r_n$, $f^r=(a_0+a_1X+\dots+a_nX^n)^r=a_0{}^{r_0}g_0+\dots+a_n{}^{r_n}g_n$

hence $f^r=(t_0g_0+\dots+t_ng_n)b$, i.e. $[f]\le[b]$ in D(A[X]). □

After Lemma 2.1 it is clear that to find $\gamma\dashv D(i)$ is equivalent to show that $c:A[X]\to D(A)$, $c(f)=[a_0,\dots,a_n]$, is a support. We shall say that c(f) is the *radical content* of the polynomial f. Note that the *Gauss content* of f is the ideal $(a_0,\dots,a_n)$ of A, so that the radical content is the radical of the Gauss content.

**2.2.Theorem.** For any ring A we have

(i) The radical content $c:A[X]\to D(A)$ is a support

(ii) The lattice homomorphism $\gamma=c^{-}:D(A[X])\to D(A)$ is a left adjoint to $D(i)$ such that $\gamma D(i)=id$.

Proof. (i)We give here a sketch of the proof in [4]. The nontrivial part is to show the *Gauss-Joyal equality*

$$c(fg) = c(f)\wedge c(g)$$

for rings in toposes. If R is the universal regular ring over A, then [3] D(R) is the boolean algebra freely generated by D(A), so that we only need to prove the Gauss-Joyal equality for a regular ring R. The basic fact is to show that $fg=0$ implies $a_ib_j=0$ for all coefficients $a_i$ of f and $b_j$ of g, the proof being by induction over the two indices. Then we prove that R[X] is baer with $f^*=a_0{}^*\vee\ldots\vee a_n{}^*=c(f)$, where the join is taken in the boolean algebra $E(R)=E(R[X])$ which is the spectrum lattice of R, so that the equality follows easily.

(ii) is an easy consequence of (i) and Lemma 2.1. □

Note that $\gamma\dashv D(i)$ is just the adjunction $k\dashv i$ given in Section 1 when A is a baer ring and $L=D(A[X])$.

Given a lattice L, let Supp(A,L) be the set of all supports from A to L, which is a poset with the order induced by D. The *extension* of a support $d:A\to L$ is the support $d^e:A[X]\to L$ given by $d^e=d^{-}c$, that is $d^e(f)=d(a_0)\vee\ldots\vee d(a_n)$. For instance, c is the extension of $d_A$. The *contraction* of a support $d:A[X]\to L$ is the support $d^c=di:A\to L$. So that we have monotone maps $(.)^e$ and $(.)^c$ between Supp(A[X],L) and Supp(A,L).

**2.3.Corollary.** There is an adjunction $(.)^c\dashv(.)^e$ such that $(.)^{ec}=id$. □

To give in Set a support $d:A\to\{0,1\}$ is equivalent to give a prime ideal $p=d^{-1}(0)$ in A, and $f\in(d^e)^{-1}(0)$ if and only if $a_0,\ldots,a_n\in p$, hence $(d^e)^{-1}(0)=p[X]$. Similarly, given a support $d:A[X]\to\{0,1\}$ with $P=d^{-1}(0)$, its contraction $d^c:A\to\{0,1\}$ is the support defined by $(d^c)^{-1}(0)=P\cap A$.

## 3.Transfer theorems

In the proof of Theorem 2.2(i) we have used that R regular implies R[X] baer. It is also true that K field implies K[X] domain, and this statement is equivalent to the Gauss-Joyal equality for the field K. Now we prove transfer theorems for domains and baer rings.

A *domain* is a non trivial ring such that ab=0 implies a=0 or b=0, and a lattice is said *domain* if it is non trivial and $x \wedge y=0$ implies x=0 or y=0. Let us note that the above definition of domain is not equivalent in toposes to other forms of the definition which are classically equivalent (for instance: ab=0 and a≠0 implies b=0). The same note is valid for fields. In this paper, a *field* is a non trivial ring such that a=0 or a is unit. A complete account of the diferent notions of fields and domains in toposes can be found in [7].

**3.1.Lemma.** Let A be a ring. Then:

(i) A/N is a domain ⇔ D(A) is domain

(ii) A/N is a field ⇔ D(A)={0,1}

Proof: Use that [S]=0 iff S⊂N. □

**3.2.Theorem.** If A is a domain then A[X] is a domain.

Proof: fg=0 implies $c(f) \wedge c(g)=0$ because c is a support; hence c(f)=0 or c(g)=0 by Lemma 3.1(i). But in any ring c(f)=0 means $a_0,\dots,a_n \in N$, that is f is nilpotent; but domains are nilpotent-free, so that c(f)=0 implies f=0. □

The transfer theorem for baer rings was proved classically in [11] and in [4] for rings in topos. Now we finish this paper giving a new proof based on the Pierce representation as considered in [5]: given a ring A, let $A^{\sim}$ be the sheaf over B=E(A) given by $A^{\sim}(e)=Ae$.

Remember from [7] that fields are *decidable* (i.e.: x=0 or ¬(x=0)) but not domains in general, and a domain is decidable iff it satisfies x=0 or y=0 or ¬(xy=0), condition introduced in [10], where the following is proved , given here in an elementary form.

**3.3.Theorem.** $A^{\sim}$ is a decidable domain iff A is baer.

Proof: We use boolean sheaf semantics. Suppose $A^{\sim}$ decidable domain in **sh**(B) and a∈A. Because $A^{\sim}$ is decidable there exists a partition 1=p+q in B such that ap=0 and for all e≤q ae=0 implies e=0. Hence, if $a^*=1-p=q$ we have $aa^*=a$. Moreover, taking ab=0 in the level $a^*$, since $A^{\sim}$ is domain there is a partition $a^*=u+v$ such that au=0 or bv=0; but au=0 with $u \le a^*=q$ implies u=0 and so $a^*b=vb=0$. Conversely, take elements a,b∈A and the induced partition 1=u+v+w where $u=1-a^*$, $v=a^*(1-b^*)$ and $w=a^*b^*$. In each level of this partition we have respectively: au=0, av=0 and ¬(ab=0) because if e≤w then abe=0 implies $e=ew=ea^*b^*=0$. □

**3.4.Theorem.** If A is baer then A[X] is baer.

Proof: After Theorems 3.2 and 3.3 we only need to prove that, in **sh**(B), $A^{\sim}$ decidable implies $A^{\sim}[X]$ decidable, but it is trivial to show this result in any topos. □

## References.


[1]N.Bourbaki, **Algèbre commutative**, Hermann, Paris, 1961.

[2]A.Brezuleanu et R.Diaconescu, Sur la duale de la catégorie des treillis, **Rev. Roum. Math. Pures et Appl. XIV,3**, (1969), 311-323.

[3]L.Español, Le spectre d'un anneau dans l'algèbre constructive et applications à la dimension, **Cahiers de Topo. et Géo. Diff. XXIV,2**,(1983),133-144.

[4]L.Español, Aspectos de primer orden de los *-anillos y versión constructiva del lema de Gauss, **Rev. Acad. Ciencias Zaragoza 38,** (1983), 11-14.

[5]L.Español, Dimension of boolean valued lattices nd rings, **J. Pure and Appl. Algebra 42,** (1986), 223-236.

[6]P.T.Johnstone, **Topos Theory,** Academic Press, New York, 1977.

[7]P.T.Johnstone, Rings, fields and spectra, **J. of Algebra 49,** (1977), 238-260.

[8]A.Joyal, Spectral spaces and distributive lattices, **Notices A.M.S. 18,** (1971), 393.

[9]A.Joyal, Les théorèmes de Chevalley-Tarski et remarques sur l'algèbre constructive, **Cahiers de Topo. et Géo. Diff. XVI, 3,** (1975), 256-258

[10]J.F.Kennison, Integral domain type representations in sheaves and other topoi, **Math. Z. 151,** (1976), 35-56.

[11]D.Saracino and V.Weispfening, On algebraic curves over commutative regular rings, in **Model theory and algebra: A memorial tribute to Abraham Robinson,** D.Saracino and V.Weispfening eds., Springer L.N.M. 498, (1975), 307-383.

[12]H.Simmons, Reticulated rings, **J. of Algebra 66,** (1980), 169-192.


This paper is in final form and will not be published elsewhere.

# Morse germs in S.D.G.

Felipe Gago

Depto. de Alxebra. Universidade de Santiago de Compostela
Santiago de Compostela, Galicia, Spain

In this paper we present some preliminary work on the study of singularities in the context of S.D.G.. We concentrate on germs of functions to $R$ and use the version of Mather's Theorem proved in [B/G] to give some examples of stable germs. We find the normal form of a Morse germ and use it to prove that they are stable.

On Morse germs.

We recall some definitions and results from [B] and [B/G].

If $x_o \in R^n$, a germ at $x_o$ is a function $f \in R^X$, where $X = \neg\neg\{x_o\}$, easily seen to be equal to $x_o + \neg\neg\{0\}$.

A point $x \in X$ is is called a singularity (or a critical point) of $f$ if $f$ is constant on $x + D(n)$. In the terminology of [B], $x \in X$ is is a singularity of $f$ if $\pi \circ j^1{}_x f = \pi \circ J^1 f(x) = 0$ where $J^1 f : X \to J^1{}_0(R^n) = R^{D(n)} \approx R \times R^n$ is defined by

$$J^1 f(x)(d) = f(x+d) = f(x) + \partial f/\partial x\,(x).d \equiv$$
$$\equiv (f(x), \partial f/\partial x_1(x), \ldots, \partial f/\partial x_n(x)) = (f(x), \partial f/\partial x\,(x)),$$

and could be described by the law $[x \in X \mapsto$ restriction of $f$ to $x + D(n)]$.

We let $S_1 \hookrightarrow R^{D(n)}$ be the subobject defined by $\pi^{-1}\{0\}$, where $\pi: R \times R^n \to R^n$ is the projection.

$S_1 = [[g \in R^{D(n)} \mid \forall d \in D(n)\;\; g(d) = g(0)]$, $S_1$ is a submanifold of $R^{D(n)}$ of codimension n, for $\pi$ is a submersion [B].

It is clear that if $x \in R^n$ is a singularity of $f$ then $J^1 f(x) \in S_1$.

By definition, a singularity is non-degenerate if $J^1 f \pitchfork_x S_1$, i.e., $T_{J^1 f(x)} R^{D(n)} = \mathrm{Im}[d(J^1 f)_x] + T_{J^1 f(x)} S_1$

DEFINITION

Call $f \in R^X$ a Morse germ if $\forall x \in X$ [$x$ singularity of $f \Rightarrow x$ non-degenerate].

Next proposition gives a characterization of non-degenerate singularities.

PROPOSITION

If $x \in X$ is a non-degenerate singularity, then $\pi \circ J^1 f$ is a submersion at $x$.

Proof.

$\pi$ is a projection, then so is $\pi^D$ and we have

$$\forall \tau \in T_{J^1 f(x)}(R \times R^n)\ (\tau \in T_{J^1 f(x)} S_1 \leftrightarrow \pi^D(\tau) = 0)$$

Therefore, for any non-degenerate singularity $x$ of $f$, $T_{J^1 f(x)} R^{D(n)} = \mathrm{Im}[d(J^1 f)_x] + \mathrm{Ker}[d(\pi)_{J^1 f(x)}]$ (1)

Now, after 2.3 in [B], it suffices to show that $d(\pi \circ J^1 f)_x$ is locally surjective, and this is easily seen using that $\pi^D$ is a submersion ( hence $d(\pi)_{J^1 f(x)}$ locally surjective,) and (1)

We can prove in this context the key result of Morse theory, namely that singularities are isolated

COROLLARY

A Morse germ has, at most, one singularity.

Proof.-

If $x \in X$ is a singularity, then it must be non–degenerate and, by proposition above, $\pi \circ J^1 f$ is a submersion at $x$. Preimage theorem ([B] Corollary 2.5) says precisely that the subobject $[[x \in X \mid x \text{ singularity of } f]] = (\pi \circ J^1 f)^{-1}\{0\} \subset X$ is a submanifold of codimension $n$, hence of dimension 0

The second important result of the theory is an easy consequence of Thom's Transversallity Theorem for germs ([B/G], Corollary 2.6)

PROPOSITION

The object of Morse germs is dense in $R^X$, with respect to the Weak topology introduced in [B/G].

On the stability of Morse germs.

This is the central part of our work. We will show that all Morse germs are stable.

We recall from [B/G] that $f \in R^X$ is called stable if $\gamma_f$ is locally surjective at $(id_R, id_R)$, with the notations as in [B] and [B/G]. Recall also that the notions of (right) equivalence are defined in a similar manner to the classical one.

We now begin with an easy exercise.

EXERCISE

Any germ $f \in R^X$ of the form $[(t_1,\ldots,t_n) \mapsto c + u_1 t_1^2 + \cdots + u_n t_n^2]$ with the $u_i$'s invertible is stable.

Proof.

Using Mather's Theorem ([B/G], Theorem 4.4) we have to check for infinitesimal stability, i.e.,

$$\forall \omega \in \mathrm{Vect}(f)\ \exists \sigma \in \mathrm{Vect}(R^n)\ \exists \tau \in \mathrm{Vect}(R)\ [\omega = \alpha_f(\sigma) \oplus \beta_f(\tau)].$$

So, let $\omega$ be any given vector field along $f$, $\omega(x) = (f(x), \underline{\omega}(x))$

We may assume that the principal part at 0, $\underline{\omega}(0)$, equals 0. Otherwise we can consider any vector field $\tau$ on $R$, such that $\tau(f(0)) = \underline{\omega}(0)$and then take $\omega - \tau \circ f$.

Now $\underline{\omega} : \Delta(n) \to R$ and $\underline{\omega}(0) = 0$.Therefore, there are functions $h_1,\ldots,h_n : \Delta(n) \to R$ such that $\underline{\omega}(\underline{t}) = \sum_{i=1}^{n} h_i(\underline{t}) t_i$, provided that the ring $R$ is a Fermat ring ([Pe], pag.47).

The required vector field $\sigma$ on $R^n$ is the one whose principal part $\underline{\sigma}(\underline{t}) = ((1/2)u_1^{-1}h_1(\underline{t}),\ldots,(1/2)u_n^{-1}h_n(\underline{t}))$.

Our next goal is to show that also in this context is possible to find a "normal" form for every Morse germ.

THEOREM

Every Morse germ with a critical point is right equivalent to the sum of a quadratic form and a constant.

REMARKS

(1) After a change of coordinates (isomorphism) we may assume that the critical point is the $0 \in R^n$, and that $g(0) = 0 \in R$ (otherwise take $g(x)-g(0)$ .)

(2) We may assume (see Lemma below) that the 2-jet of $g$ is of the form $a_1x_1^2+\cdots+a_nx_n^2$, where all the $a_i$'s are invertible in $R$.

(3) with the above reductions, we have that $g = f + \phi$, where $f = a_1x_1^2+\cdots+a_nx_n^2$, and $\phi$ is a germ vanishing on $D_2(n)$. Notice that $\phi$ has a zero of order three at 0.

Proof of the theorem.

We will prove that $f$ is equivalent to $g = f + \phi$. For this purpose we introduce in this context a new method to test for stability, namely the homotopic method (cf. [G], and [T/L] for its classical counterpart):

First, join $f$ to $g$ by the path $f + t\,\phi$, with $t \in [0,1]$

Then show that it is possible to find a one-parameter family of local diffeomorphisms

$$x \in \Delta(n) \mapsto \Phi(t,x) \in \Delta(n)$$

such that

$$(f + t\,\phi)\Phi(t,x) = f(x) \qquad \forall x \in \Delta(n)\ \forall t \in [0,1] \qquad (2)$$

$$\Phi(0,x) = x \qquad \forall x \in \Delta(n)$$

$$\Phi(t,0) = 0 \qquad \forall t \in [0,1]$$

Then $\Phi(1,-)$ will do the job.

We will obtain these $\Phi_t$'s as the integral curves of suitable vector fields $\delta_t$, or rather a time-dependent vector field $\delta$ (see [B/G] Proposition 4.1) which happen to be compactly supported inthe sense of [B/D]

$$d\Phi/dt(x,t) = \delta(\Phi(x,t),t).$$

The equations for $\delta$ can be obtained by taking derivatives in (2) with respect to the parameter $t$.

$$\phi(\,\Phi(x,t)\,) + d(f + t\,\phi)/dt\Phi(x,t).\ d\Phi/dt(x,t) \equiv 0 \text{ for each } t\,.$$

So, if the principal part of $\delta$ is expressed by the functions $(\delta_{t1},..,\delta_{tn})$, we have the equation

$$\phi_{|\Phi(x,t)} \equiv -\sum_{i=1}^{n} \delta_{ti} y_i{}_{|\Phi(x,t)}\,, \text{ where } y_i = 2a_ix_i + t\,\phi_{x_i}\,.$$

so $-\phi \equiv \sum_{i=1}^{n} \delta_{ti} y_i$ , both sides seen as functions on $(x,t)$.

Now, for each $t \in [0,1]$

$$|\partial y/\partial x_{|(0,t)}| = \begin{vmatrix} 2a_1 & & 0 \\ & \ddots & \\ 0 & & 2a_n \end{vmatrix} \# 0$$

since $\phi_{x_i}$ vanishes on $D(n)$, as $\phi$ did on $D_2(n)$.

By the theorem of infinitesimal inversion [Pe], for each $t$, $y$ is a bijection and $(y,t)$ defines a new system of coordinates.

In this new system $\phi$ takes the form $\phi(y,t) = \sum_{i=1}^{n} \psi_i(y,t) y_i$ as $\phi$ has no component in $t$ and $\psi_i(0,t)=0$, $\phi(0)=0$

Therefore $\delta_{ti} = -\psi_i$ as functions on $y$ , and the integral curves give the wanted solution.

To finish the theorem we have to prove the claim we made in remark (2); this is the following

LEMMA

If $f(x_1,\ldots,x_1)$ has a second–order Taylor polynomial of the form $\sum_{i\,j=1}^{n} a_{ij} x_i x_j$ and the singularity at zero is non-degenerate, then it is possible to find coordinates $z_i$ such that the new Taylor polynomial is $a_1 z_1^2 + \cdots + a_n z_n^2$.

Proof.

We are saying that there exists a linear isomorphism $\phi$ represented by a non singular matrix A such that $f \circ \phi$ has the desired Taylor polynomial.

$$\partial/\partial x (f \circ \phi)|_a = \partial f/\partial x\, |_{\phi(a)}\, \phi'_a = \partial f/\partial x\, |_{\phi(a)}\, A$$

$$\partial^2 (f \circ \phi)/\partial x^2 = A^t\, \partial^2 f/\partial x^2\, A$$

Therefore, the result will be proved if we show that there exists a matrix A which diagonalizes the symmetric bilinear form associated to $\partial^2 f/\partial x^2$, the Hessian of $f$, and one can see that the classical proof is intuitionistically valid for any local ring.

COROLLARY

Morse germs are stable.

Proof.

Immediate after EXERCISE and THEOREM above.

References


[B] M. BUNGE, Synthetic aspects of $C^\infty$–mappings. *Journal of Pure and Applied Algebra* **28** (1983) 41-63

[B/D] M. BUNGE and E.J. DUBUC, Local concepts in S.D.G. and germ representability, in E.G.K. López-Escobar & C. Smith (editors), *Mathematical Logic and Theoretical Computer Science*, pp. 39-158, M. Dekker, Inc., 1987.

[B/G] M.BUNGE and F.GAGO, Synthetic aspects of $C^\infty$–mappings,II: Mather's theorem for infinitesimally represented germs. (to appear in the *Journal of Pure and Applied Algebra*)

[G] F. GAGO, *Internal Weak Opens, internal stability and Morse Theory for synthetic germs*, (manuscripts to be presented as tesis at McGill University).

[Pe] J. PENON. *De l'infinitesimal au local*, Thèse de Doctorat d'Etat, Université Paris VII, 1985.

[T/L] R. THOM and H.I. LEVINE, *Singularities of Differentiable Mappings, I, Bonn 1959*, reprinted in C.T.C.WALL (ed.) *Proceedings of Liverpool Singularities-Symposium* Springer Lecture Notes in Mathematics **192** (1972), 1-89.


This paper is in final form and will not be published elsewhere.

# Modal and Tense Predicate Logic: Models in Presheaves and Categorical Conceptualization(*)

S. Ghilardi - G. C. Meloni
Dipartimento di Matematica "Federigo Enriques"
Milano

The aim of this paper is to study *first order logic* of two *tense operators*:

$\Diamond$ "generated subpresheaf" and

$\Box$ "maximum contained (or co-generated) subpresheaf"

(for the propositional case see, for example, [3]). Such operators satisfy the Lemmon ($K_t$) axioms for tense logic (see [12]) and the Lewis ($S4$) axioms for modal systems (see [7]) (but all the results and techniques presented in this paper also apply to many common normal modal systems and to the "co-intuitionistic" logic of the subtraction (see [11]); furthermore, as summarized in [4], these techniques are fundamental in the study of our *distributive linear logic* (see also [5])).

While the mathematical meaning of such operators is quite clear we point out their *logical interest* due for example to the fact that they falsify the following well-known formulae

$$\Diamond\forall\xi A(\xi) \vdash \forall\xi\Diamond A(\xi)$$
$$\Diamond(y_1 \neq y_2) \vdash y_1 \neq y_2$$
$$\forall\xi\Box A(\xi) \vdash \Box\forall\xi A(\xi)$$
$$y_1 \neq y_2 \vdash \Box(y_1 \neq y_2).$$

In spite of the *fact* that their informal truth is extremely dubious, the first two formulae are provable in any normal modal logic (see [7]) and the last two in usual tense logic (see [12]). The relevance of presheaves in mathematics and the naturality of the generated subpresheaf operator have been a *guide* in realizing that the *general reason* for such contradictions (between semantic falsity and syntactic provability) is *noncommutativity of substitution with respect to modal operators* (see [1]). However, in usual logical languages the commutativity between substitution and modal operators is a consequence of the traditional definitions of formula and substitution. Thus a certain number of problems and paradoxes which have arisen in first order modal logic can be solved by a simple and completely natural *modification of the logical language* which is usually too closely related to the ordinary informal or natural language to express the above noncommutativity. This modification can be done because *language is not a primary datum* but has its *foundation* in its relationship to *things* (and in general to natural semantic frameworks) whose features it must take into account.

In the context of categorical logic, by an analogous modification of the language (i.e. the indication of *implicit variables*), the problem of (partially) empty domains (e.g. inclusive logic) has already been conceptually solved. However, in modal and tense logic, it should be expressed more, that is *at which moment in the constuction of a formula implicit variables are introduced.* Besides, more generally, we need *substitution as a primitive operator*, like connectives and quantifiers. We

---

(*) Lavoro eseguito con il contributo del CNR, Contratto n. 87.01000.01.

will not enter into the details of a presentation of such a language because we think that the meaning of this work is better understood in the context of *invariant conceptualization* of theories (see [10]). In this case the notion of tense (or modal) theory is directly better expressed by *doctrines* (see [9]) because attributes cannot be viewed as objects of a logical category (since the doctrinal comprehension schema is not valid in these modal contexts).

The main technical result in the paper is the *completeness* theorem for tense theories with respect to models in presheaves. The *general* axioms of such theories can be strengthened or weakened to have completeness with respect to presheaves with *injective* or *surjective* transitions, or again with respect to structures which are more general than presheaves because they allow, as transitions, *relations* instead of functions (a table of these results is given afterwards). The technique used in the proof seems partially new. While usually completeness is demonstrated by means of Henkin theories (that is maximal, consistent and *rich* extensions of the given theory) here we cannot follow this suggestion because there are formulae that are provable in any Henkin theory but not valid in the presheaf semantics. It results, from an analysis of this fact, that *we cannot treat individuals of a given world as constants of a larger language.* The method we have found to overcome these difficulties is in our opinion interesting (although not necessary) also in the classic and intuitionistic cases because its main features consist of *keeping the initial language fixed* during the proof and "*working more on the semantic side*".

Finally we underline that our main motivation with respect to this research lies in the general project of *isolating the logical aspects arising from mathematical practice*, mainly in order to *utilize them in developing and understanding mathematics itself.* However, we argue that the analysis here presented appears *natural and significant* not only in mathematical contexts but also with respect to *linguistic and philosophic problems.* The general reason for this duality lies in the fact that (*spatial and temporal*) *variability* is, in a certain sense, a *common object* of study of both areas. Consequently it is not surprising, for example, that the *mathematical generated subpresheaf operation*, expressed in terms of condition of truth, gives rise to an *interesting definition* of a *possibility operator* which completely *avoids problems* concerning truth of statements *about non existing individuals.* Moreover the use of *substitution as a primitive operator* allows us to *maintain* both *standard modal propositional logic* and *standard rules for quantification*, thus giving a (and *maybe* the) solution for some traditional modal problems.

In this way the (partially) common object of study (i.e. variability) becomes a tool for understanding the *relations between mathematic and philosophic contributions*, relations not completely manifest with a superficial analysis. In such a *comparative* perspective it could perhaps be that some acquired results will appear of a *greater importance* with respect to the past, while others will reveal all their *extreme particularism.*

**1.** Let $\mathbf{C}$ be a small *category.* The objects $\sigma, \tau, \varepsilon, \varrho, \ldots$ of $\mathbf{C}$ can be considered (according to the different interpretations) bodies or spaces, states, worlds, instants of time. The arrows $\iota, \kappa, \varphi, \pi, \ldots$ will then represent transformations, processes, ways of accessibility and temporal developments. From the point of view of the completeness theorem $\mathbf{C}$ can be reduced, if you prefer, to a preorder.

Consider now, as a $\mathbf{C}$-universe, a *presheaf* (or *left* $\mathbf{C}$-*set*) $U$ (better seen here as a functor $\mathbf{C} \longrightarrow \mathbf{Set}^{op}$). If $\sigma \xrightarrow{\iota} \tau$ is an arrow of $\mathbf{C}$, $s \subset U_\sigma$ and $t \subset U_\tau$ then, when $s = U_\iota(t)$, we write $s = \iota t$ and read (for example)

" $s$ is *the* restriction (along $\iota$ ) of $t$ "

" $t$ is *an* extension (along $\iota$ ) of $s$ "

" $s$ is *the* past counterpart or *the ancestor* (with respect to the development $\iota$ ) of $t$ "

" $t$ is *a* future counterpart or *a descendant* (with respect to the development $\iota$ ) of $s$ ".

By (*semantic*) *attribute* of type the given presheaf $U$ we mean, by definition, a family $A = \{A_\sigma\}$ such that, for every object $\sigma$ of $\mathbf{C}$, $A_\sigma$ is a subset of $U_\sigma$. If $A = \{A_\sigma\}$ and $A' = \{A'_\sigma\}$ are attributes of type $U$ then we use the logic notation $A \vdash A'$ to denote the inclusion of $A$

in $A'$, i.e. the fact that $A_\sigma \subseteq A'_\sigma$ for every object $\sigma$ of $\mathbf{C}$. We also use the usual validity notation $s \models^U_\sigma A$, or more simply $s \models_\sigma A$ or $s \models A$, to mean (for $U$ C-universe, $A$ attribute of type $U$, $\sigma$ object of $\mathbf{C}$ and $s$ element of $U_\sigma$ ) that $s \in A_\sigma$. In this context *tense operators* are operations on attibutes in the sense that $\Diamond A$ is the minimum subpresheaf containing $A$ (or subpresheaf generated by $A$ ) and $\Box A$ is the maximum subpresheaf contained in $A$ (or subpresheaf co-generated by $A$ ). Equivalently this means that

($\Diamond$) $\quad s \models_\sigma \Diamond A \iff$ there exist an object $\varepsilon$, an arrow $\varphi : \sigma \longrightarrow \varepsilon$ and an element $e \in U_\varepsilon$ such that $s = \varphi e$ and $e \models_\varepsilon A$

and

($\Box$) $\quad s \models_\sigma \Box A \iff$ for every object $\varrho$ and every arrow $\pi : \varrho \longrightarrow \sigma$, $\pi s \models_\varrho A$.

These operators satisfy, in a very general setting (i.e. with $\mathbf{C}$ graph and $U_\iota$ relations "without composition"), the following functoriality and adjointness conditions (which are equivalent to $(K_t)$ axioms)

$$(F\Diamond) \qquad A_1 \vdash A_2 \implies \Diamond A_1 \vdash \Diamond A_2$$

$$(F\Box) \qquad A_1 \vdash A_2 \implies \Box A_1 \vdash \Box A_2$$

$$(ADJ) \qquad \Diamond A_1 \vdash A_2 \iff A_1 \vdash \Box A_2.$$

When $\mathbf{C}$ is a category and $U$ is a presheaf (or more generally a "relational presheaf", see next point **2.**) then $(S4)$-conditions hold, namely

$$(S4) \qquad A \vdash \Diamond A \quad \text{and} \quad \Diamond\Diamond A \vdash \Diamond A$$

or equivalently

$$\Box A \vdash A \quad \text{and} \quad \Box A \vdash \Box\Box A.$$

Furthermore, if we write $Bf$ for the inverse image of an attribute $B$ of type $V$ along a *natural transformation* between presheaves $f : U \longrightarrow V$, we have that

$$s \models^U_\sigma Bf \iff sf \models^V_\sigma B$$

(where $s \in U_\sigma$ and $sf$ ($\in V_\sigma$) denotes $f_\sigma(s)$ ). The following condition of commutativity between substitution and co-generated subpresheaf operator is then satisfied

$$(C) \qquad (\Box B)f = \Box(Bf),$$

while for the generated subpresheaf operator we only have

$$(SC) \qquad \Diamond(Bf) \vdash (\Diamond B)f.$$

In fact if $U$ and $V$ are presheaves, $B$ is an attribute of type $V$ and $y_1 : V \times U \longrightarrow V$ is the first projection, then(*) $(\Diamond B)y_1 \vdash \Diamond(By_1)$ is not valid in general because

$\langle s_1, s_2\rangle \models^{V\times U}_\sigma (\Diamond B)y_1 \iff$ there are an object $\varepsilon$, an arrow $\varphi : \sigma \longrightarrow \varepsilon$ and an element $e_1 \in V_\varepsilon$ such that $s_1 = \varphi e_1$ and $e_1 \models^V_\varepsilon B$

and

$\langle s_1, s_2\rangle \models^{V\times U}_\sigma \Diamond(By_1) \iff$ there are an object $\varepsilon$, an arrow $\varphi : \sigma \longrightarrow \varepsilon$

(*) To avoid misunderstanding we point out that the notations $(\Diamond B)y_1$ and $\Diamond(By_1)$ have been used (e.g. in [7]) in a completely different meaning.

and elements $e_1 \in V_\varepsilon$ and $e_2 \in U_\varepsilon$ such that

$$s_1 = \varphi e_1, \quad s_2 = \varphi e_2 \quad \text{and} \quad e_1 \models^V_\varepsilon B.$$

Similarly if $U$ is a presheaf, $\langle x,x\rangle : U \longrightarrow U\times U$ is the diagonal map and $B$ is an attribute of type $U\times U$, then $(\Diamond B)\langle x,x\rangle \vdash \Diamond(B\langle x,x\rangle)$ is not valid in general because

$$s \models^U_\sigma (\Diamond B)\langle x,x\rangle \iff \begin{array}{l}\text{there are an object } \varepsilon,\ \text{an arrow } \varphi:\sigma\longrightarrow\varepsilon\\ \text{and elements } e_1,e_2\in U_\varepsilon \text{ such that}\\ s=\varphi e_1=\varphi e_2 \text{ and } \langle e_1,e_2\rangle \models^{U\times U}_\varepsilon B\end{array}$$

and

$$s \models^U_\sigma \Diamond(B\langle x,x\rangle) \iff \begin{array}{l}\text{there are an object } \varepsilon,\ \text{an arrow } \varphi:\sigma\longrightarrow\varepsilon\\ \text{and an element } e\in U_\varepsilon \text{ such that}\\ s=\varphi e \text{ and } \langle e,e\rangle \models^{U\times U}_\varepsilon B.\end{array}$$

The above observations lead to the following results.

**Proposition.** *If* $U$ *is a presheaf then the following two conditions are equivalent:*
- *for every arrow* $\iota$, $U_\iota$ *is injective,*
- *for every attribute* $B$ *of type* $U\times U$, $(\Diamond B)\langle x,x\rangle \vdash \Diamond(B\langle x,x\rangle)$ *holds.*

*Furthermore, also the following two conditions are equivalent:*
- *for every arrow* $\iota$, $U_\iota$ *is surjective,*
- *for every presheaf* $V$ *and for every attribute* $B$ *of type* $V$, $(\Diamond B)y_1 \vdash \Diamond(By_1)$ *holds.*

Note that in the last condition we can omit $V$ (in the sense that we can put $V=\mathbf{1}$, the terminal object) when $\mathbf{C}$ is a preorder.

Now we discuss the *syntactic* counterpart of the universes just introduced. A *tense predicative theory* is defined, in its invariant (i.e. indipendent of any presentation) form, as a *tense doctrine* $\mathcal{D}$, that is as a pair $\langle\mathcal{T},\mathcal{A}\rangle$ such that:
- $\mathcal{T}$ is a small category with finite products, whose objects are the *types* of the theory and whose arrows are the *terms*;
- $\mathcal{A}$ is a contravariant functor (the functor "*(syntactic) attributes*") from $\mathcal{T}$ to the category whose objects are the boolean algebras equipped with two modal operators $\Diamond$ and $\Box$ satisfying $(F\Diamond)$, $(F\Box)$, $(ADJ)$ and $(S4)$ (where $\vdash$ is now the order relation of the boolean algebras, i.e. the syntactic notion of provability) and whose arrows are the boolean morphisms satisfying $(C)$ (note that the left to right inclusion of $(C)$ is equivalent to $(SC)$ );
- for every term $t: X\longrightarrow Y$, the *substitution operator* $\mathcal{A}X \xleftarrow{\mathcal{A}t} \mathcal{A}Y$ has left adjoint $\exists_t$ (the *existential quantifier* (along $t$ )) satisfying Frobenius and Beck conditions (the latter only with respect to certain specified pullbacks existing in any category with products (see [9])); note that, for an attribute $B\in\mathcal{A}Y$, we write $Bt$ for $(\mathcal{A}t)B$.

The fundamental features of this conceptualization of the logical syntax lie in the general fact that attributes (i.e. formulae in the usual presentations) live on types and consequently it is possible to form conjunctions and disjunctions only after such attributes have been pulled back on to the same type; furthermore such operation of pulling back (or substitution of a term in an attribute) do not completely commute with respect to $\Diamond$.

In a *purely syntactic* context, and therefore in particular in the categories of presheaves, we have the following

**Proposition.** *In every tense predicative theory the following axiom schemata are equivalent*

$$\Diamond\exists\xi A(\xi,\beta) \vdash \exists\xi\Diamond A(\xi,\beta)$$
$$\forall\xi\Box A(\xi,\beta) \vdash \Box\forall\xi A(\xi,\beta)$$
$$\exists\xi\Box A(\xi,\beta) \vdash \Box\exists\xi A(\xi,\beta)$$
$$\Diamond\forall\xi A(\xi,\beta) \vdash \forall\xi\Diamond A(\xi,\beta)$$
$$(\Diamond B)y_1 \vdash \Diamond(By_1)$$
$$\Box(By_1) \vdash (\Box B)y_1$$

*and also the following axiom schemata are equivalent*

$$\Diamond(y_1 \neq y_2) \vdash y_1 \neq y_2$$
$$y_1 \neq y_2 \vdash \Box(y_1 \neq y_2)$$
$$\Diamond(y_1 = y_2) \vdash y_1 = y_2$$
$$y_1 = y_2 \vdash \Box(y_1 = y_2)$$
$$(\Diamond B)\langle x,x\rangle \vdash \Diamond(B\langle x,x\rangle)$$
$$\Box(B\langle x,x\rangle) \vdash (\Box B)\langle x,x\rangle,$$

*where* $\Box = \neg\Diamond\neg$ *and* $\Diamond = \neg\Box\neg$.

Now it is clear that in traditional logical systems for modal and tense logic certain unpleasant principles are provable because using the usual inductive definition of substitution we *cannot express,* for example, the difference between $(\Diamond B)y_1$ and $\Diamond(By_1)$ or between $(\Diamond B)\langle x,x\rangle$ and $\Diamond(B\langle x,x\rangle)$. However, these *differences* are crucial because assuming their equality is equivalent, as explained, to assuming the same incriminated principles. Traditional proofs, if analyzed carefully, contain in fact exchanges between substitutions and modal operators. So we really *do not need to modify classical rules for identity and quantification* to have a modal logic that agrees with our expectations.

After universes and theories we discuss now the notion of model. Usually we shall interpret types as presheaves, terms as natural transformations and syntactic attributes as semantic attributes. Thus a *model* will consist of the following data:

1) a small category $\mathbf{C}$;
2) a finite product preserving functor

$$\hat{(-)}: \mathcal{T} \longrightarrow \left((\mathbf{Set}^{op})^{\mathbf{C}}\right)^{op};$$

3) a natural transformation $[\![-]\!]$ between the functors $\mathcal{A}$ and $\hat{(-)}\mathcal{P}$ (where $\mathcal{P}$ is the "semantic attributes functor") whose components $[\![-]\!]^X$ preserve the modal operators $\Diamond$ and $\Box$ and also commute with respect to existential quantifiers.

As usual, instead of $[\![-]\!]$ we can equivalently assign a validity relation

$$\models^X_\sigma \;\subseteq\; \hat{X}_\sigma \times \mathcal{A}(X)$$

(depending on $\sigma$ and $X$ ) connected to the natural transformation $[\![-]\!]$ by the relation

$$s \in [\![A]\!]^X_\sigma \iff s \models^X_\sigma A$$

and hence satisfying the following conditions (where $t: X \longrightarrow Y$ is a term, $1$ is the true and $0$ the false, $\sigma$ is an object, $s \in \hat{X}_\sigma$, $s' \in \hat{Y}_\sigma$, $A, A_1, A_2 \in \mathcal{A}(X)$ and $B \in \mathcal{A}(Y)$, where $st$ denotes $\hat{t}_\sigma(s)$ and where we omit $X$ and $\sigma$ in $\models^X_\sigma$ when they do not change along a condition)

(v) $\quad s \models A_1$ and $A_1 \vdash A_2 \implies s \models A_2$

| | | | |
|---|---|---|---|
| (1) | $s \models 1$ | | |
| $(\wedge)$ | $s \models A_1 \wedge A_2$ | $\iff$ | $s \models A_1$ and $s \models A_2$ |
| (0) | $s \not\models 0$ | | |
| $(\vee)$ | $s \models A_1 \vee A_2$ | $\iff$ | $s \models A_1$ or $s \models A_2$ |
| $(\neg)$ | $s \models \neg A$ | $\iff$ | $s \not\models A$ |
| $(t)$ | $s \models^X Bt$ | $\iff$ | $st \models^Y B$ |
| $(\exists)$ | $s' \models^Y \exists_t A$ | $\iff$ | there exists $s$ such that $st = s'$ and $s \models^X A$ |
| $(\Diamond)$ | $s \models_\sigma \Diamond A$ | $\iff$ | there exist $\varepsilon$, $\varphi : \sigma \longrightarrow \varepsilon$ and $e \in \hat{X}_\varepsilon$ such that $s = \varphi e$ and $e \models_\varepsilon A$ |
| $(\Box)$ | $s \models_\sigma \Box A$ | $\iff$ | for every $\varrho$ and every $\pi : \varrho \longrightarrow \sigma$, $\pi s \models_\varrho A$ |

and consequently

| | | | |
|---|---|---|---|
| $(\Box)$ | $s \models_\sigma \Box A$ | $\iff$ | for every $\varepsilon$, $\varphi : \sigma \longrightarrow \varepsilon$ and $e \in \hat{X}_\varepsilon$ if $s = \varphi e$ then $e \models_\varepsilon A$ |
| $(\Diamond)$ | $s \models_\sigma \Diamond A$ | $\iff$ | there exist $\varrho$ and $\pi : \varrho \longrightarrow \sigma$ such that $\pi s \models_\varrho A$. |

The truth conditions for the modal operators obtained from the mathematical notion of subpresheaf reflect a *current use in informal language*. Statements of the kind "it will always be true that $s$ will satisfy a given property" must mean something like "no matter how you consider future instants of time, no matter how you consider future counterparts of $s$, those counterparts will satisfy the property in question". This definition, although very natural, seems to meet with *difficulties* (truth value gaps, failure of usual modal and quantificational laws, and so on) within traditional methods *using infinitary assignments for untyped formulae*. On the contrary it is just this definition which is suggested by mathematics and which we use in our notion of model, but for this we need *finitary assignments, of variable lenght* according to the type.

**2.** The choice of presheaves as universes for tense predicate logic is, in our opinion, an interesting and natural generalization of the usual Kripke models. To explain better this fact we briefly discuss *one possible meaning* of such structures.

In this informal interpretation an object of the category is considered as a *possible world* and an arrow $\sigma \xrightarrow{\iota} \tau$ as a *possible temporal development* from the world $\sigma$, seen as the past, to the world $\tau$, seen as the future. But notice that, given $\sigma$ and $\tau$, there may be many, one or none at all of such developments. We can also have other developments $\tau \xrightarrow{\kappa} \sigma$ in the opposite direction and with respect to these $\tau$ plays the role of past and $\sigma$ of future. No assumption about the cyclicity of time is of course involved because $\sigma$ and $\tau$ are considered, in fact, as *abstract possible* worlds in the sense that they are *indipendent* (or *outside*) *of time.*

Thinking presheaves as functors $\mathbf{C} \longrightarrow \mathbf{Set}^{op}$ we want to underline that the direction of the time is the direction of the arrows of $\mathbf{C}$ and is opposite to the direction of set theoretical functions.

When we assign a universe $U$ for modal predicate logic, we associate to each world $\sigma$ a set $U_\sigma$ of elements living in $\sigma$ and to each temporal development $\sigma \xrightarrow{\iota} \tau$ some form of connection or identification between the individuals of $U_\sigma$ and $U_\tau$. In the case of presheaves such a connection has the form of a function

$$U_\sigma \xleftarrow{U_\iota} U_\tau$$

in the opposite direction with respect to the temporal development, but such a function must be seen itself as a temporal development from $U_\sigma$, the "past", to $U_\tau$, the "future". This is a very simple kind of development that can be found, in nature, for example in the living organisms reproducing by subdivision, where along a temporal development each individual living in the future has precisely one "ancestor" in the past and each individual living in the past has possibly many (or one, or none in case of death) "descendants" in the future. These observations hold also when $\mathbf{C}$ is just a graph, i.e. when we omit conditions $(S4)$. In these contexts it is in general better to assume the sets $U_\sigma$ as disjoint. In this way the only identifications between elements are those explicitly given by the connections $U_\iota$, without interferences with any identity relation.

Besides, if we are interested in more general kinds of development, we can consider the case in which

$$U_\iota : U_\sigma \longrightarrow\!\!\!\!\!\!\!\!\;\;+ \; U_\tau$$

is an arbitrary relation (with particular regard to the intermediate cases of partial functions and totally defined relations, in either directions). In this context, when we consider $(S4)$-systems, it is also better (although not necessary for completeness) to interpret the functoriality condition in the following less strict form

$$\mathbf{1}_{U_\sigma} \subseteq U_{\mathbf{1}_\sigma} \qquad \text{and} \qquad U_\iota U_\kappa \subseteq U_{\iota\kappa}$$

(where the composition of relations is of course the usual set-theoretical one, using existential quantifier). We shall call such a structure a *relational presheaf.* Relational presheaves form a category (with left limits) if as arrows we consider the families of set functions $\{f_\sigma : U_\sigma \longrightarrow V_\sigma\}_\sigma$ preserving the given relations

$$\begin{array}{ccccc} \sigma & & U_\sigma & \xrightarrow{\;f_\sigma\;} & V_\sigma \\ \downarrow \iota & & U_\iota \big\downarrow & \Rightarrow & \big\downarrow V_\iota \\ \tau & & U_\tau & \xrightarrow[\;f_\tau\;]{} & V_\tau \end{array}$$

i.e. such that for every $\sigma \xrightarrow{\iota} \tau$, $s \in U_\sigma$ and $t \in U_\tau$ (writing $s\,\iota\,t$ for $\langle s,t\rangle \in U_\iota$ and $sf\;\iota\;tf$ for $\langle sf, tf\rangle \in V_\iota$ )

$$(P) \qquad\qquad s\,\iota\,t \quad\Longrightarrow\quad sf\;\iota\;tf.$$

But condition $(C)$ is not valid in these relational contexts and we must only assume its left to right inclusion (which, as we know, is equivalent to $(SC)$ ). Furthermore, condition $(SC)$ is, in fact, equivalent to $(P)$. The definitions of theory and model can then be restated with the obvious modifications (e.g. $s \models^X_\sigma \Box A \iff$ for every $\varrho$, $\pi : \varrho \longrightarrow \sigma$ and $r \in \hat{X}_\varrho$, if $r\,\pi\,s$ then $r \models^X_\varrho A$ ).

Many equivalences between syntactic and semantic conditions can be proved in the same way discussed previously. Here is a partial table of such correspondences. For each case we have a completeness result (limited, till now, to countable doctrines when total relations are involved).

**Theorem.** *Every first order tense theory, based on classical logic (with substitution as a primitive operator) and satisfying* $(K_t)$, $(SC)$, $(S4)$ *and a group of equivalent axioms in the following first column, is complete with respect to the corresponding semantic condition, written in the second column.*

| | *Syntactic (equivalent) axioms* | *Corresponding semantic condition* |
|---|---|---|
| 0) | | relational presheaves |

| | | |
|---|---|---|
| 1) | $y_1 = y_2 \vdash \Box(y_1 = y_2)$<br>$y_1 \neq y_2 \vdash \Box(y_1 \neq y_2)$<br>$\Diamond(y_1 = y_2) \vdash y_1 = y_2$<br>$\Diamond(y_1 \neq y_2) \vdash y_1 \neq y_2$<br>$\Box(B\langle x,x\rangle) \vdash (\Box B)\langle x,x\rangle$<br>$(\Diamond B)\langle x,x\rangle \vdash \Diamond(B\langle x,x\rangle)$ | $U_\iota^{op}$ partial functions |
| 2) | $\Diamond\exists\xi A(\xi,\beta) \vdash \exists\xi\Diamond A(\xi,\beta)$<br>$\forall\xi\Box A(\xi,\beta) \vdash \Box\forall\xi A(\xi,\beta)$<br>$\exists\xi\Box A(\xi,\beta) \vdash \Box\exists\xi A(\xi,\beta)$<br>$\Diamond\forall\xi A(\xi,\beta) \vdash \forall\xi\Diamond A(\xi,\beta)$<br>$(\Diamond B)y_1 \vdash \Diamond(By_1)$<br>$\Box(By_1) \vdash (\Box B)y_1$ | $U_\iota^{op}$ totally defined |
| | 1) + 2) | presheaves |
| 3) | the dual axioms of 1) | $U_\iota$ partial functions |
| 4) | the dual axioms of 2) | $U_\iota$ totally defined |
| | 1) + 2) + 3) | presheaves with injective transitions |
| | 1) + 2) + 4) | presheaves with surjective transitions |
| | 1) + 2) + 3) + 4) | presheaves with bijective transitions . |

***Furthermore, without (S4)-conditions we have completness with respect to graphs instead of categories and without (SC)-conditions for some constant, or term different from n-tuples of projections, we have completeness with respect to non rigid designators, i.e. families $\{f_\sigma\}$ without condition (P). Finally we have completeness also without tense operators but with modal operators along only one direction of time, i.e. without $(K_t)$-axioms but in normal systems.***

We cannot here compare our systems with the most common ones (for a survey of these and for the terminology see, for example, [2]). In any case some simple observations are indispensable.

1) The converse of Barcan formula (contrarily to Kripke [8]) is provable in our sistems. However, it does not imply any nested condition on domains (totally defined relations, in our terminology). This last condition is expressed by the schema $\exists\xi\Box A(\xi,\beta) \vdash \Box\exists\xi A(\xi,\beta)$, not provable in our weakest systems.
2) Our interpretation of variables and quantifiers is strictly objectual. However, we can for instance allow some individual constants (to be treated in this case as *non rigid* designators) to represent arbitrary intensions and completeness follows *very easily*, from the proof given below, simply restricting condition (*A2*) to rigid terms. Notice that for an individual constant c, to have the converse of (*SC*) ensured, we have in general to require both (*SC*) and the "necessity of the identity".
3) Substitutivity of the identity (in the general conceptual form used, for example, in doctrines) is unlimited, in agreement with the fact that individuals which are the same must have the same properties. Thus in particular, if $P \in A(X)$ and $y_i : X\times X \longrightarrow X$ $(i = 1,2)$ are the

projections, then

$$(y_1 = y_2) \wedge (\Box P)y_1 \ \vdash \ (\Box P)y_2$$

holds. Starting from this we can prove

$$(y_1 = y_2) \wedge \Box(Py_1) \ \vdash \ \Box(Py_2)$$

while neither

$$y_1 = y_2 \ \vdash \ \Box(y_1 = y_2)$$

nor

(1) $$(c_1 = c_2) \wedge \Box(Pc_1) \ \vdash \ \Box(Pc_2)$$

(e.g. with $c_1, c_2 : \mathbf{1} \longrightarrow X$, i.e. individual constants) are provable. In fact we can only prove

(2) $$(c_1 = c_2) \wedge (\Box P)c_1 \ \vdash \ (\Box P)c_2.$$

Condition (1) is the formalization, for example, of Quine's famous paradox on the number of planets. Condition (2) has no counterintuitive meanings since $(\Box P)c$ is clearly a *de re* modality (see the semantic conditions). Many attempts have been made in the literature in order to distinguish the *de re* and the *de dicto* interpretations of $\Box P(c)$ and in particular the approach of Stalnaker and Thomason [13], using abstraction, seems to have some similarities with ours (using simply $(\Box P)t$ and $\Box(Pt)$ ) as far as the treatment of individual constant is concerned.

4) Similarly to the previous case, if $X \xrightarrow{u} \mathbf{1}$ and $P \in A(X)$, we have

$$\Box P \ \vdash \ (\exists_u \Box P)u$$

since rules for existential quantifier are standard. On the other way

(3) $$\Box(Pc) \ \vdash \ \exists_u \Box P$$

is not true and we can only prove $(\Box P)c \vdash \exists_u \Box P$. Condition (3) is clearly false when individuals split (and when $c$ is not rigid). Furthermore the correction(*)

$$\Box(Pc) \wedge \exists\xi\,\Box(\xi = c) \ \vdash \ \exists\xi\,\Box P(\xi)$$

suggested by Hintikka for $(S4)$-systems (see [6]) is provable in our context.

5) Finally, following [1], we have that partially denoting terms can be easily and naturally treated by considering many-sorted doctrines. For example, the partially denoting individual constants $c$ are the terms defined on subobjects of the terminal object, i.e. on types satisfying $\forall\xi_1\forall\xi_2(\xi_1 = \xi_2)$.

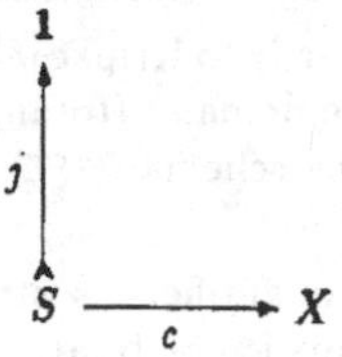


Note that we do not need to change anything since the given definitions of doctrine and model also work for free logics. For example, you can not even write $P(c) \vdash \exists\xi P(\xi)$ because of incompatibility of types. However, $\exists_j(Pc) \vdash \exists\xi P(\xi)$ is provable and correct because, also if $Pc$ is provable, $\exists_j(Pc)$ is not a theorem if $c$ is not a totally defined constant.

(*) Note that $\exists\xi\,\Box P(\xi)$ and $\exists_u \Box P$ are the same attribute, written in linguistic and in conceptual notation.

**3.** We prove now the completeness theorem for *one-sorted* tense doctrines. So $\mathcal{T}$ will have $\mathbf{1} = X^0$, $X = X^1$, $X^2, \ldots, X^n, \ldots$ as types. The non commutativity between substitutions and modal operators leads, as expected, to technical problems. In fact, sequents of the form

$$(4) \qquad (\Diamond A)u \vdash \Diamond(Au),$$

for $X \xrightarrow{u} \mathbf{1}$ and $A \in \mathcal{A}(\mathbf{1})$, are false even in the presheaf semantics. However, in any Henkin theory $H$ we cannot have $H \nvdash \forall_u((\Diamond A)u \to \Diamond(Au))$ (otherwise, being $H$ complete and rich, there would exist a constant $c$ such that $H \vdash \Diamond A$ and $H \vdash \neg((\Diamond(Au))c)$ so, applying $(SC)$, we get a contradiction). To overcome this difficulty we could drop $(SC)$ for the new constants, but the resulting proof looks unclear and rather complicated. To falsify $(4)$ we need individuals that "disappear" in a related world and for such individuals $c$ expressions like $\Diamond(Pc)$ or $\Box(Pc)$ cause confusion because, in general, they are neither true nor false, but simply meaningless. We solve the problem radically, making such expressions impossible to write. This is in fact *one meaning* of our dealing with *weak models* instead of theories.

A *classical model* $\sigma$ is, by definition, a model such that its category $\mathbf{C}$ is reduced to a singleton and without any condition requested for the validity of modal operators. To construct a generic tense model we consider the category $\mathbf{C}$ having as objects the classical models (thus $\mathbf{C}$ is not small, but this problem can be solved easily) and having as arrows between a pair of classical models $\sigma$ and $\epsilon$ all the relations $\varphi \subseteq \hat{X}_\sigma \times \hat{X}_\epsilon$ that, for every $n \geq 0$, $s \in \hat{X}^n_\sigma$ and $e \in \hat{X}^n_\epsilon$, with $s_k \varphi e_k$ (for $k = 1, \ldots, n$), satisfy the following conditions

$$(A1) \qquad \text{for every } A \in \mathcal{A}(X^n), \quad e \models^{X^n}_\epsilon A \implies s \models^{X^n}_\sigma \Diamond A$$

and

$$(A2) \qquad \text{for every term } t : X^n \longrightarrow X \qquad st\, \varphi\, et.$$

Note that the relations $\varphi$ will be substituted by totally defined, functional, ... relations, according to the various cases.

Now, the relational presheaf on $\mathbf{C}$, the interpretation of terms in it and the validity relation are given in the obvious way, simply by "putting together" the classical models. To see that the structure just defined is a generic tense model, the only non trivial things to prove are the following:

$$(G) \qquad \text{if } \nvdash A \text{ then there is a classical model falsifying } A,$$

$$(TM) \qquad \text{conditions } (\Diamond) \text{ and } (\Box) \text{ hold.}$$

But condition $(G)$ is equivalent to the usual completeness property for classical first order theories, so we prove $(TM)$ for which we need some new basic notions. If $\mathbf{S}_0$ is the skeleton of the category of finite sets then there is an obvious functor

$$\mathbf{S}_0^{op} \longrightarrow \mathcal{T}$$

associating the type $X^n$ to a finite set $\mathbf{n}$ and the $m$-ple of projections

$$p = \langle \pi^n_{P_1}, \ldots, \pi^n_{P_m} \rangle : X^n \longrightarrow X^m$$

to a function $P : \mathbf{m} \longrightarrow \mathbf{n}$.

We "approximate" classical models with *cartesian models*, that is classical models without conditions $(0)$, $(\vee)$, $(\neg)$ and $(\exists)$, and with condition $(t)$ restricted to $m$-ples of projections $p : X^n \longrightarrow X^m$. This means that the validity relation must satisfy the following properties of multifunctoriality and of commutation with $m$-ples of projections

$$(MF) \qquad A_1 \ldots A_n \vdash A \text{ and } s \models A_i \text{ (for every } i = 1, \ldots, n) \implies s \models A,$$

$$(MP) \qquad s \models Ap \iff sp \models A.$$

The basic result on cartesian models is the following

**Lemma.** *Let* $\hat{X}$ *be a set and* $\{\models_{\bullet}^{X^n}\}_{n\geq 0}$ *be a family of arbitrary relations such that* $\models_{\bullet}^{X^n} \subseteq \hat{X}^n \times A(X^n)$. *We can then consider the least (in the obvious sense) cartesian model* $\langle \hat{X}, \{\models^{X^n}\}_{n\geq 0}\rangle$ *containing the relations* $\{\models_{\bullet}^{X^n}\}_{n\geq 0}$, *i.e. the cartesian model generated by such relations. Then, for* $n \geq 0$, $s \in \hat{X}^n$ *and* $A \in A(X^n)$, *we have that* $s \models^{X^n} A$ *if and only if there exist* $m \geq 0$, $P : \mathbf{n} \longrightarrow \mathbf{m}$, $s' \in \hat{X}^m$, $k \geq 0$ *and, for every* $i = 1, \ldots, k$, $n_i \geq 0$, $P_i : \mathbf{n}_i \longrightarrow \mathbf{m}$ *and* $A_i \in A(X^{n_i})$

$$\begin{array}{ccccc} A_i & & & & A \\ X^{n_i} & \xleftarrow{\;p_i\;} & X^m & \xrightarrow{\;p\;} & X^n \\ & & s' & & s \end{array}$$

*such that* $A_1p_1 \ldots A_kp_k \vdash Ap$, $s'p = s$ *and, for every* $i = 1, \ldots, k$, $s'p_i \models_{\bullet}^{X^{n_i}} A_i$.

Then, using again and again the previous lemma, we can prove $(TM)$ (and also $(G)$, if you like it) by constructing the classical models we need as sums (divided by "identity") on chains of cartesian models. As an example we outline the proof of $(\Diamond)$ in the presheaf case.

Suppose you have $s \models_{\sigma}^{X^n} \Diamond A$, for $s \in \hat{X}_{\sigma}^n$ and $A \in A(X^n)$. We want to build a classical model $\epsilon$ and a function $\hat{X}_{\sigma} \xleftarrow{\varphi} \hat{X}_{\epsilon}$, satisfying $(A1)$ and $(A2)$, for which there exists $e \in \hat{X}_{\epsilon}^n$ such that $s = \varphi e$ and $e \models_{\epsilon}^{X^n} A$. We need the additional hypothesis of the doctrine being countable and we consider at most countable classical models. Fix an arbitrary countable set $S$ and construct two enumerations of pairs like $\langle c, \exists_u B\rangle$ and of pairs like $\langle c, C_1 \vee C_2\rangle$, respectively (where $m \geq 0$, $c \in S^m$, $B \in A(X^{m+1})$ and $C_1, C_2 \in A(X^m)$ are arbitrary and $p: X^{m+1} \longrightarrow X^m$ is the obvious projection). We are going to build a countable chain (for $\kappa \geq 0$ ) of cartesian models $\langle \hat{X}_{\kappa}, \{\models_{\kappa}^{X^h}\}_h\rangle$ and of functions $\hat{X}_{\sigma} \xleftarrow{\varphi_{\kappa}} \hat{X}_{\kappa}$ satisfying the following conditions:

(i) $\hat{X}_{\kappa}$ is a finite subset of $S$, whose elements are assigned in a given order $\langle c_{\kappa 1}, \ldots, c_{\kappa n_{\kappa}}\rangle = c_{\kappa}$,

(ii) $\{\models_{\kappa}^{X^h}\}_h$ is generated by a single clause of the form $c_{\kappa} \models_{\kappa} C_{\kappa}$, for a suitable $C_{\kappa}$,

(iii) $$\varphi_{\kappa} c_{\kappa} \models_{\sigma} \Diamond C_{\kappa}.$$

Let $\hat{X}_0$ be the set of elements of an arbitrary $n$-ple $c_0$ of distinct elements of $S$, let $C_0 = A$ and let's define $\varphi_0$ by putting $\varphi_0(c_{0i}) = s_i$, for every $i = 1, \ldots, n$. We consider now the case $\kappa$ odd. Take the first pair (if any) $\langle c, \exists_u B\rangle$, not yet considered in the first enumeration, such that $c \in \hat{X}_{\kappa}^m$ and $c \models_{\kappa} \exists_u B$. By (a slight improvement of) the basic lemma on cartesian models we have that there exists $l \geq 0$, $P : \mathbf{m} \longrightarrow \mathbf{l}$, $Q : \mathbf{n}_{\kappa} \longrightarrow \mathbf{l}$ and $c' \in \hat{X}_{\kappa}^l$

$$\begin{array}{ccccc} C_k & & & & \exists_u B \\ X^{n_k} & \xleftarrow{\;q\;} & X^l & \xrightarrow{\;p\;} & X^m \\ c_k & & c' & & c \end{array}$$

such that $C_\kappa q \vdash (\exists_u B)p$, $c'p = c$ and $c'q = c_\kappa$. Rewriting things in the following form

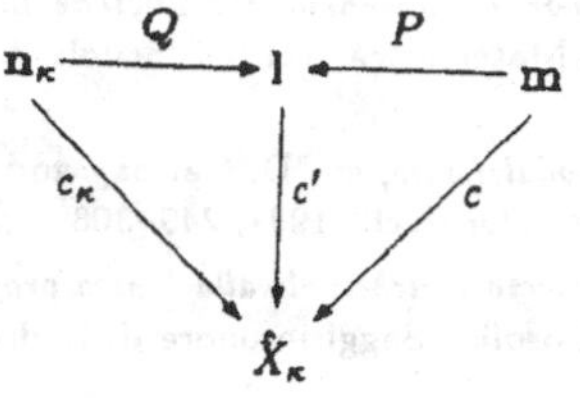


we have that, remembering that $c_\kappa$ is a bijection and putting $R = Pc'c_\kappa^{-1}$,

$$C_\kappa \vdash (\exists_u B)r \wedge C_\kappa \quad \text{and} \quad c_\kappa r = c.$$

By Beck and Frobenius conditions and by Barcan formula for $\Diamond$ (holding in presheaves) we get

$$\Diamond C_\kappa \vdash \exists_{u'} \Diamond((B(r\times 1)) \wedge (C_\kappa u')),$$

where $u'$ is defined by means of the following cartesian square

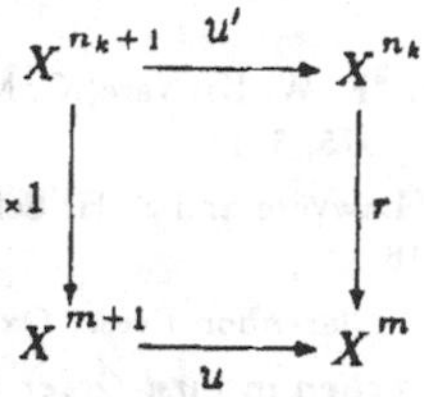


So by (iii) we can find an $a \in \hat{X}_\sigma$ such that

$$\langle \varphi_\kappa c_\kappa, a\rangle \models_\sigma^{X^{n_\kappa+1}} \Diamond((B(r\times 1)) \wedge (C_\kappa u')).$$

We put $\hat{X}_{\kappa+1} = \hat{X}_\kappa + \{d\}$ (where $d$ is a new element of $S$), $c_{\kappa+1} = \langle c_\kappa, d\rangle$, $C_{\kappa+1} = (B(r\times 1)) \wedge (C_\kappa u')$ and $\varphi_{\kappa+1} d = a$. Conditions (i), (ii) and (iii) are then verified and, moreover, $\langle c, d\rangle \models_{\kappa+1} B$. The reader has certainly guessed what we have to do for $\kappa$ even ($\neq 0$ ); so we simply observe that, applying $(SC)$ and the above argument, condition (iii) holds not only for the generating clause $c_\kappa \models_\kappa C_\kappa$, but also for every pair $\langle d, D\rangle$ such that $d \models_\kappa D$. Obviously, the union of the chain is a cartesian model; moreover it satisfies $(\vee)$, $(0)$, $(\neg)$ and, limited to quantification along projections, $(\exists)$. Note that not necessarily all the elements of $S$ have been utilized; in the extreme case we do not use any of them if $\hat{X}_e$ must be empty, taking into account in this way of inclusive logic. Now, if we divide by identity (i.e. if we take the quotient with respect to the congruence relation $[\![ y_1 = y_2 ]\!]$ ), we obtain, interpreting (via graphs) the terms as relations and using the existence and uniqueness properties provable in the syntax, a classical model $\varepsilon$. The requested $\varphi$ is then defined by the the universal properties of sums on chains and quotients, using $\Diamond(y_1 = y_2) \vdash y_1 = y_2$. Condition $(A1)$ holds by construction and a final short calculation shows it implies $(A2)$ in the present case.

## REFERENCES


1. C. Berta, G. C. Meloni, *Dottrine con Comprensione e Logiche Libere*, in "R. Ferro and A. Zanardo (eds.), Atti degli Incontri di Logica Matematica, vol. 3," Scuola di Specializzazione in Logica Matematica, Siena 1987, 177–185.
2. J. W. Garson, *Quantification in Modal Logic*, in "D. Gabbay and F. Guenthner (eds.), Handbook of Philosophical Logic, vol. II," Reidel, Dordrecht 1984, 249–308.
3. S. Ghilardi, G. C. Meloni, *Un approccio concettuale alla logica proposizionale cartesiana e distributiva*, in "C. Mangione (ed.), Scienza e filosofia. Saggi in onore di Ludovico Geymonat," Garzanti, Milano 1985, 446–467.
4. S. Ghilardi, G. C. Meloni, *Un approccio categoriale alla logica predicativa modale e lineare distributiva*, to appear in "Atti degli incontri di Logica Matematica, X incontro, Siena, Maggio 1987".
5. S. Ghilardi, G. C. Meloni, *Semantic Analysis of Distributive Linear Logic*, Preprint.
6. J. Hintikka, *Existential Presuppositions and Uniqueness Presuppositions*, in "K. Lambert (ed.), Philosophical Problems in Logic," Reidel, Dordrecht 1970, 20–55.
7. G. E. Hughes, M. J. Cresswell, "An Introduction to Modal Logic," Methuen, London 1968.
8. S. Kripke, *Semantical Considerations in Modal Logic*, Acta Philosophica Fennica **16** (1963), 83–94.
9. F. W. Lawvere, *Equality in Hyperdoctrines and Comprehension Schema as an Adjoint Functor*, in "A. Heller (ed.), Applications of Categorical Algebra," Proc. of Symp. in Pure Math., XVII, 1970, 1–14.
10. F. W. Lawvere, *Introduction to Part I*, in "F. W. Lawvere, C. Maurer and G. C. Wraith (eds.), Model Theory and Topoi," LNM 445, Springer 1975, 3–14.
11. F. W. Lawvere, *Introduction*, in "F. W. Lawvere and S. H. Schanuel (eds.), Categories in continuum Physics," LNM 1174, Springer 1986, 1–16.
12. A. N. Prior, "Past, Present and Future," Clarendon Press, Oxford 1967.
13. R. C. Stalnaker, R. H. Thomason, *Abstraction in First-Order Modal Logic*, Theoria **14**, 3, 1968, 203–207.



**Address of the authors:**

*Dipartimento di Matematica*
*Università degli Studi*
*via Cesare Saldini, 50*
*20133 MILANO (Italy)*


This paper is in final form and will not be published elsewhere.

# On the representability of partial morphisms in Top and in related constructs

Horst Herrlich

## Introduction

In a category, a partial morphism from A to B is a morphism from a "subobject" of A to B. Partial morphisms can be dealt with effectively provided they can be represented (in a specified manner) by ordinary morphisms. This is well known to be the case in each topos (with respect to monomorphic embeddings [5,9,1o]) or, more generally, in each quasitopos (with respect to strong monomorphic embeddings [13,18]), in fact it can be taken as part of the definition of a (quasi)topos. In these and in more abstract settings close relations between the representability of partial morphisms, the existence of subobject-classifiers, local cartesian closedness, the existence of partial products resp. of pullback-complements, and the exponentiability of embeddings have been established [3,13,17,18, in particular 4, Corollary 4.6]. Unfortunately in such constructs as Top and Unif partial morphisms (with respect to ordinary topological resp. uniform embeddings) fail to be representable [2,11,12]. In this paper we investigate the question whether Top and related constructs can be improved (with regard to the representability of partial morphisms) by passing to suitable subconstructs or superconstructs. It will be shown that the answer with respect to subconstructs is a clear no, with respect to superconstructs as an equally clear yes.

AMS Subj. Class.: 54 B 3o, 18 B 3o, 18 B 25, 18 B 15.

Key words: partial morphism, representation of partial morphisms; stable, pleasant stable; construct, hereditary construct, pleasant construct, topological construct; pleasant topological hull.

This paper is in final form and will not be published elsewhere.

## § 1 Representability of partial morphisms in abstract categories

The following concept is due to Rosolini [14], who however used the term dominion instead of stable.

1.1 Definition: In a category $\underline{A}$, a class M of $\underline{A}$-morphisms is called a stable provided it satisfies the following conditions:

(S1) Iso $\underline{A} \subset M \subset$ Mono $\underline{A}$,

(S2) M is composition-closed, i.e., $M \circ M \subset M$;

(S3) M is pullback-closed, i.e., for any pair $\bullet \xrightarrow{m} \bullet \xleftarrow{f} \bullet$ of morphisms with $m \in M$ there exists a pullback

$$\begin{array}{ccc} \bullet & \xrightarrow{\bar{m}} & \bullet \\ {\scriptstyle \bar{f}}\downarrow & & \downarrow{\scriptstyle f} \\ \bullet & \xrightarrow[m]{} & \bullet \end{array}$$

with $\bar{m} \in M$.

1.2 Definitions: Let M be a stable in $\underline{A}$.

(1) An M-partial morphism from A into B is a pair of morphisms $A \xleftarrow{m} \bullet \xrightarrow{f} B$ with $m \in M$.

(2) $B \xrightarrow{m_B} B^*$ represents M-partial morphisms into B, provided $m_B \in M$ and for each M-partial morphism $A \xleftarrow{m} \bullet \xrightarrow{f} B$ into B there exists a unique morphism $A \longrightarrow B^*$ such that

$$\begin{array}{ccc} \bullet & \xrightarrow{m} & A \\ {\scriptstyle f}\downarrow & & \downarrow \\ B & \xrightarrow[m_B]{} & B^* \end{array}$$

is a pullback diagram.

(3) A stable M in $\underline{A}$ is called pleasant provided for every $\underline{A}$-object B there exists $B \xrightarrow{m_B} B^*$ representing M-partial morphisms into B.

1.3 Examples of pleasant stables M in $\underline{A}$:

(1) $\underline{A}$ arbitrary; M = Iso $\underline{A}$. This stable is called improper, all others are called proper.

(2) $\underline{A}$ a topos; M = Mono $\underline{A}$ [5,9,1o]. In particular, in Set partial maps are representable by one-point-extensions.

(3) $\underline{A}$ a quasitopos; M = Strong Mono $\underline{A}$ [13,18].

(4) $\underline{A} = \underline{\mathrm{Set}}_p$, the construct of pointed sets; M = Mono $\underline{A}$. This cate-

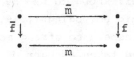

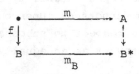

gory is not cartesian closed, hence not a quasitopos.

(5) $\underline{A}$ = Top; $M_c$ (all closed embeddings) and $M_o$ (all open embeddings) [2].

1.4 Proposition (cf. [14]) If M is a pleasant stable in $\underline{A}$ and A is an $\underline{A}$-object, then the following conditions are equivalent:

(1) A is M-injective,

(2) if $A \xrightarrow{m_A} A^*$ represents M-partial morphisms into A, then $m_A$ is a section,

(3) A is a retract of some object of the form B*.

## § 2 Representability of partial morphisms in hereditary constructs

2.1 Definitions:

(1) A construct $\underline{A}$ is a fibre-small concrete category over Set. To avoid pathologies we require further that $\underline{A}$ has precisely one object with empty underlying set.

(2) A construct $\underline{A}$ is called hereditary*, provided $\underline{A}$ has subspaces, i.e., every injective map $X \xrightarrow{f} |A|$ has a unique initial lift $B \xrightarrow{f} A$. Initial injective maps are called embeddings. The class of $\underline{A}$-embeddings is denoted by Emb $\underline{A}$.

(3) A construct $\underline{A}$ is called topological, provided every source $(X \xrightarrow{f_i} |A_i|)_{i\in I}$ has a unique initial lift $(A \xrightarrow{f_i} A_i)_{i\in I}$.

2.2 Proposition: If M is a pleasant stable in a hereditary construct $\underline{A}$, which has a concrete terminal object T, then $M \subset$ Emb $\underline{A}$.

Proof: Let $T \xrightarrow{m_T} T^*$ represent M-partial morphisms into T. Then M consists precisely of the pullbacks of $m_T$. Since $m_T$ is an embedding and (by heredity of $\underline{A}$) the class of embeddings is closed under the formation of pullbacks, $M \subset$ Emb $\underline{A}$.

2.3 Proposition

Let M be a proper pleasant stable in a hereditary construct $\underline{A}$ with concrete terminal object T. If $\underline{A}$ has discrete objects or if T is dis-

*Observe that the name "hereditary" has been used in [7,8,15,16] for a different concept.

crete, then M-representations $A \xrightarrow{m_A} A^*$ are preserved by the forgetful functor $\underline{A} \to \underline{Set}$, i.e., are one-point-extensions.

Proof: (1) Denote by $\emptyset$ the $\underline{A}$-object with empty underlying set, and by $T_o$ the discrete object with the same underlying set as T. By 2.2, $M \subset \text{Emb}\,\underline{A}$. If $m_T$ would be surjective, it would be an isomorphism and hence M, consisting of the pullbacks of $m_T$, would be improper. Hence $m_T$ is not surjective.

(2) No $m_A$ is surjective:

Case 1: $A = \emptyset$. Let $f: T_o \to T^*$ be a morphism, which as function is different from $m_T$, and let $A \xrightarrow{m} T_o$ be the pullback of $m_T$ along f. Then $A = \emptyset$. There exists a unique morphism $T_o \to \emptyset^*$ making

$$\begin{array}{ccc} \emptyset & \xrightarrow{m} & T_o \\ {\scriptstyle id}\downarrow & & \downarrow \\ \emptyset & \xrightarrow[m_\emptyset]{} & \emptyset^* \end{array}$$

a pullback diagram. Consequently $m_\emptyset$ is not surjective.

Case 2: $A \neq \emptyset$ and $T = T_o$. Then there exists a morphism $f: T \to A$. Hence there exists a morphism $T^* \to A^*$ making

$$\begin{array}{ccc} T & \xrightarrow{m_T} & T^* \\ {\scriptstyle f}\downarrow & & \downarrow \\ A & \xrightarrow[m_A]{} & A^* \end{array}$$

a pullback diagram. If $m_A$ would be surjective, then so would be $m_T$.

Case 3: $A \neq \emptyset$ and there exists a discrete object B with the same underlying set X as $T^*$. Let $m: C \to B$ be a pullback of $m_T$ along $id_X: B \to T^*$, and let $f: B \to A$ be an arbitrary morphism. Then there exist morphisms $B \to B^*$ and $B^* \to A^*$ such that

$$\begin{array}{ccc} C & \xrightarrow{m} & B \\ {\scriptstyle m}\downarrow & & \downarrow \\ B & \xrightarrow[m_B]{} & B^* \end{array} \qquad \text{and} \qquad \begin{array}{ccc} B & \xrightarrow{m_B} & B^* \\ {\scriptstyle f}\downarrow & & \downarrow \\ A & \xrightarrow[m_A]{} & A^* \end{array}$$

are pullback diagrams. If $m_A$ would be surjective, then so would be $m_B$, hence m, hence $m_T$.

(3) $A \xrightarrow{m_A} A^*$ is a one-point extension.

Let $a_1$ and $a_2$ be elements of $A^*$, which do not belong to the image of

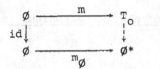

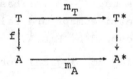

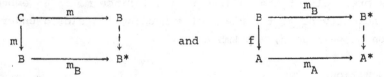

$m_A$. For i = 1,2 let $f_i$: $T_o \to A^*$ be the morphism with value $a_i$. Then

$$\begin{array}{ccc} \emptyset & \longrightarrow & T_o \\ \downarrow & & \downarrow f_i \\ A & \xrightarrow[m_A]{} & A^* \end{array}$$

is a pullback for i = 1 and for i = 2. Hence, by the uniqueness requirement, $f_1 = f_2$, i.e., $a_1 = a_2$.

2.4 Corollaries

(1) Top has precisely 3 proper pleasant stables, consisting of all closed, all open resp. all clopen embeddings.

(2) Unif has precisely one proper pleasant stable, consisting of all embeddings of subspaces, which are far from their complement.

(3) Met, the construct of metric spaces and non-expansive maps, has no proper pleasant stable.

(4) Pos, the construct of posets and order-preserving maps, has precisely 2 proper pleasant stables, consisting of all embeddings of lower sets resp. of upper sets.

2.5 Example: Let I be an arbitrary set. Consider the hereditary construct $\underline{A}$, whose objects are pairs $(X,(X_i)_{i\in I})$, consisting of a set X and a partition $(X_i)_{i\in I}$ of X, and whose morphisms $(X,(X_i)_{i\in I}) \xrightarrow{f} (Y,(Y_i)_{i\in I})$ are maps $X \xrightarrow{f} Y$ such that $f[X_i] \subset Y_i$ for each $i \in I$. Then the class M of embeddings is a pleasant stable in $\underline{A}$, and M-representations are Card(I)-point-extensions.

2.6 Remark: If $\underline{A}$ and M are as in 2.3 and $A \xrightarrow{m_A} A^*$ represents M-partial morphisms into A, then $A \xrightarrow{m_A} A^*$ is the "weakest" 1-point-M-extension of A. Vice versa, if $\underline{A}$ is as in 2.3 and M is a stable in $\underline{A}$ such that for each object A there exists a weakest 1-point-M-extension, then M need not be pleasant; the class M of embeddings in $\underline{Top}_1$ is a counterexample.

2.7 Definition: A hereditary construct $\underline{A}$ is called pleasant provided the class of embeddings is a pleasant stable in $\underline{A}$.

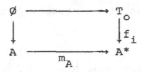

The following propositions 2.8, 2.1o and 2.12 have been demonstrated in [8] under slightly stronger assumptions. For the proof of the present 2.12 we need the additional fact that in a construct every subobject of a discrete object is discrete (here the assumption is needed that ∅ carries only one structure).

2.8 Proposition: For pleasant constructs $\underline{A}$ the following conditions are equivalent:

(1) A is injective in $\underline{A}$,

(2) A is of the form B* for some B.

2.9 Remark: For pleasant stables in a hereditary construct the analoque of 2.8 need not hold:

(1) If M is the class of open embedddings in Top, then
  (a) A is M-injective iff there exists some point of A, whose only neighbourhood is A,
  (b) A is of the form B* iff there exists some closed point of A, whose only neighbourhood is A.

(2) If M is the class of closed embeddings in Top, then
  (a) A is M-injective iff there exists some dense point in A,
  (b) A is of the form B* iff there exists some isolated, dense point in A.

2.1o Proposition: In pleasant constructs every object has an injective hull.
An extension A → B of a non-injective object A is an injective hull of A iff it represents partial morphisms into A.

2.11 Remark: For pleasant stables in a hereditary construct the analogue of 2.1o need not hold:
The class M of open (resp. closed) embeddings is a pleasant stable in Top; but no non-M-injective topological space has an M-injective hull, since the homeomorphisms are the only M-essential embeddings.

2.12 Proposition: For topological constructs $\underline{A}$ the following conditions are equivalent:

(1) $\underline{A}$ is pleasant,

(2) in $\underline{A}$, final sinks are hereditary, i.e., preserved under pullbacks along embeddings,

(3) in $\underline{A}$, coproducts and quotients are hereditary.

2.13 Proposition (F. Schwarz): A mono-topological construct is topological iff subobjects are representable.

## § 3 Pleasant subconctructs of Top and of related constructs

**3.1 Definition:** An object A of a construct is called *heretic*, provided partial morphisms into A (with respect to embeddings) are representable.

**3.2 Proposition**

(1) In any of the constructs Top, Unif and PMet (= pseudometric spaces and non-expansive maps) the equality

heretic = indiscrete

holds.

(2) In any of the constructs Met and Pos only the object with empty underlying set is heretic.

**Proof:** (1) Let $\underline{A}$ be any of the constructs Top, Unif or PMet. If an $\underline{A}$-object A is indiscrete, then an indiscrete one-point extension of A represents partial morphisms into A. Vice versa let $m_A\colon A \to A^*$ represent partial morphisms into A. Since $\underline{A}$ has discrete objects, $m_A$ has to be a one-point extension. Call the added point $\infty$. Let a be a point of A, let $f\colon T \to A$ be a morphism from a terminal object T into A with value a, and let $m\colon T \to B$ be an embeddding of T into an indiscrete object with two points. Then there exists a unique morphism $B \to A^*$ making

$$\begin{array}{ccc} T & \xrightarrow{\ m\ } & B \\ {\scriptstyle f}\downarrow & & \downarrow \\ A & \xrightarrow[\ m_A\ ]{} & A^* \end{array}$$

a pullback diagram. Hence the subspace of A* with underlying set $\{m_A(a), \infty\}$ is indiscrete. Since this is the case for each $a \in A$, the space A* and hence the space A have to be indiscrete.

(2) As in (1), replacing $T \xrightarrow{m} B$ by all possible embeddings of T into 2-point spaces.

**3.3 Proposition:** Let $\underline{A}$ be a pleasant full subconstruct of $\underline{B}$, which is closed under the formation of subspaces in B.

(1) If B = Top, then $\underline{A}$ consists of discrete spaces or of indiscrete spaces only.

(2) If $\underline{B}$ = Unif, then $\underline{A}$ consists of indiscrete spaces only or each member of $\underline{A}$ has a discrete underlying proximity.

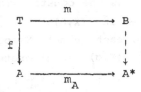

(3) If B = PMet, then there exists $r \in [0,\infty]$ such that for any pair $(x,y)$ of points of any member $(X,d)$ of A the inequalities $r \le d(x,y) \le 2r$ hold.

Proof: Since the proofs of the above results follow the same pattern, we restrict our attention to the case B = Top. If A contains a non-indiscrete space, then A contains a 2-element space $A = (\{0,1\},\tau)$ with $\{0\} \in \tau$. Let $m_A\colon A \to A^* = (\{0,1,2\}, \tau_A)$ with $m_A(0) = 0$ and $m_A(1) = 1$ represent partial morphisms into A. Then $\{0\} \in \tau_A$ or $\{0,2\} \in \tau_A$. Let $(X,\sigma)$ be an arbitrary A-object, let Y be a subset of X, let $\sigma_Y$ be the topology induced on Y by $\sigma$, let $f\colon (Y,\sigma_Y) \longrightarrow T$ be a continuous map and let $g\colon (X,\sigma) \to A^*$ be the unique continuous map making

$$\begin{array}{ccc} (Y,\sigma_Y) & \hookrightarrow & (X,\sigma) \\ f \downarrow & & \downarrow g \\ A & \xrightarrow[m_A]{} & A^* \end{array}$$

a pullback diagram.

Case 1: $\{0\} \in \tau$. Chose f as the constant map with value 0. Then $g^{-1}(0) = Y$ is open in $(X,\sigma)$.

Case 2: $\{0,2\} \in \tau$. Chose f as the constant map with value. Then $g^{-1}(1) = Y$ is closed in $(X,\sigma)$.
Hence in case 1 (resp. 2), each subset Y of X is open (resp. closed) in $(X,\sigma)$. Hence $(X,\sigma)$ is discrete.

## § 4 Pleasant superconstructs of hereditary constructs

4.1 Definition: Let A be a full subconstruct of a construct B. Then B is called a pleasant topological hull of A, provided the following conditions are satisfied:

(1) B is a pleasant topological construct,
(2) A is closed under the formation of subspaces in B,
(3) A is finally dense in B, i.e., for each B-object B the sink $(A_i \xrightarrow{f_i} B)_{i\in I}$, consisting of all morphisms with range B and with domain in A, is final,
(4) A is sub-initially dense in B, i.e. for each B-object B the source $(B \xrightarrow{f_i} A_i^*)_{i\in I}$, consisting of all morphisms with domain B and range in $\{A^* \mid A \in \mathrm{Ob}\,A\}$, is initial.

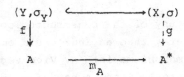

<u>4.2 Theorem</u>: Every hereditary construct has a (up to concrete isomorphism: unique) pleasant topological hull.

<u>Proof</u>: Up to the following 2 restrictions this result has been demonstrated in [8]. First, in [8] constructs are assumed to have the property that constant maps between objects are morphisms. This restriction is not essential. The constructions given [8] can immediately be carried over to the case considered here. Second, the "hereditary topological hull", constructed in [8], has not been shown to be fibre-small (and hence legitimate). Here follows the missing part: Let $\underline{A}$ and $\underline{B}$ be as in 4.1 with the exception that $\underline{B}$ is not required to be fibre-small. We will show that $\underline{B}$ has to be fibre-small. Let $X$ be a set. For a subset $Y$ of $X$, denote by $m_Y: Y \longrightarrow X$ the natural injection. For an equivalence relation $\rho$ on $Y$, denote by $\nu_\rho: Y \longrightarrow Y_\rho$ the natural surjection of $Y$ onto the set $Y_\rho$ of equivalence classes with respect to $\rho$. For an object $B$ and a subset $Y$ of the underlying set of $B$, let $B_Y$ be the subspace of $B$ with underlying set $Y$. Hence $m_Y: B_Y \longrightarrow B$ is an embedding. Let $\tau = \tau(X)$ be the collection of all triples $(Y,\rho,A)$, consisting of a subset $Y$ of $X$, an equivalence relation $\rho$ on $Y$, and an $\underline{A}$-object $A$ with underlying set $Y_\rho$. Since $\underline{A}$ is fibre-small, $\tau$ is a set. Hence the collection $P$ of all subsets of $\tau$ is a set as well. Let $\mathcal{B}$ be the collection of all $\underline{B}$-objects with underlying set $X$. It suffices to show that the map $H: \mathcal{B} \longrightarrow P$, defined by

$$H(B) = \{(Y,\rho,A) \in \tau \mid \nu_\rho : B_Y \longrightarrow A \in \mathrm{Mor}\ \underline{B}\},$$

is injective. Let $B$ and $B'$ be elements of $\mathcal{B}$ with $H(B) = H(B')$. To show $B = B'$ it suffices, by amnesticity (and symmetry), to demonstrate that $id_X: B' \longrightarrow B$ is a morphism. Hence, by sub-initial denseness, it suffices to show that a map $f: B' \longrightarrow A^*$ is a morphism, whenever $f: B \longrightarrow A^*$ is one. So assume $f: B \longrightarrow A^*$ to be a morphism. Let $m_A: A \longrightarrow A^*$ represent partial morphisms in to $A$, let $Y = f^{-1}[m_A[A]]$, and let $g: Y \longrightarrow A$ be the corresponding restriction of $f$. Then the diagram

$$\begin{array}{ccc} B_Y & \xrightarrow{\ m_Y\ } & B \\ {\scriptstyle g}\downarrow & & \downarrow{\scriptstyle f} \\ A & \xrightarrow[\ m_A\ ]{} & A^* \end{array}$$

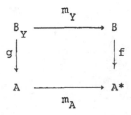

is a pullback diagram. Let $\rho$ be the equivalence relation induced on $Y$ by $g$ and let $Y_\rho \xrightarrow{m} |A|$ be the unique map with $g = m \circ \nu_\rho$. If $C \xrightarrow{m} A$ is an initial lift of $Y_\rho \xrightarrow{m} |A|$, then $C$, as a subspace of $A$, belongs to $\underline{A}$ and $\nu_\rho \colon B_Y \longrightarrow C$ is a morphism. So $(Y,g,C) \in H(B)$. Hence $(Y,\rho,C) \in H(B')$. In particular $\nu_\rho \colon B'_Y \longrightarrow C$ is a morphism. Hence so is $g = m \circ \nu_\rho \colon B'_Y \longrightarrow A$. Since $f\colon B' \longrightarrow A^*$ is the unique map, which could possibly make

$$\begin{array}{ccc} B'_Y & \xrightarrow{\;m_Y\;} & B' \\ {\scriptstyle g}\downarrow & & \downarrow{\scriptstyle f} \\ A & \xrightarrow[\;m_A\;]{} & A^* \end{array}$$

a pullback diagram, it follows that $f\colon B' \longrightarrow A^*$ is a morphism. This completes the proof.

<u>4.3 Remarks:</u>

(1) The fact that a pleasant topological hull exists for <u>all</u> hereditary constructs is highly surprising, since similar hulls (e.g., topological hulls = MacNeille completions, universal initial hulls, cartesian closed topological hulls = Antoine completions, concrete quasitopos hulls = Wyler completions) exist only under suitable smallness assumptions (stronger than fibre-smallness). For details and further references see [1,7].

(2) If the fibre-smallness assumption is dropped, a pleasant topological hull may still be constructed as before. I expect that in this setting it may fail to be legitimate.

(3) External characterizations and examples of pleasant topological hulls have been provided in [8] under the name "hereditary topological hulls". In particular, the construct of pretopological spaces is a pleasant topological hull of <u>Top</u>.

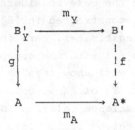

REFERENCES

[1] J. Adámek, H. Herrlich and G. E. Strecker, Least and largest initial completions I, II, Comment. Math. Univ. Carolinae 20 (1979) 43-77.

[2] P. I. Booth and R. Brown, Spaces of partial maps, fibred mapping spaces and the compact-open topology, Topol. Appl. 8 (1978) 181-195.

[3] B. Day, A reflection theorem for closed categories. J. Pure Appl. Algebra 2 (1972) 1-11

[4] R. Dyckhoff and W. Tholen, Exponentiable morphisms, partial products and pullback complements, J. Pure Appl. Algebra 49 (1987) 103-113.

[5] P. Freyd, Aspects of topoi, Bull. Austral. Math. Soc. 7 (1972)1-76.

[6] H. Herrlich, Are there convenient subcategories of Top? Topol. Appl. 15 (1983) 263-271.

[7] H. Herrlich, Topological improvements of categories of structured sets, Topol. Appl. 27 (1987) 145-155.

[8] H. Herrlich, Hereditary topological constructs, General Topology and its Relations to Modern Analysis and Algebra VI, Proc. Sixth Prague Topological Symposium 1986 (Heldermann Verlag 1988)249-262.

[9] P. Johnstone, Topos Theory, Academic Press 1977.

[10] A. Kock and G. C. Wraith, Elementary Toposes, Aarhus Lecture Note Series 30 (1971).

[11] L. G. Lewis, Jr., Open maps, colimits, and a convenient category of fibre spaces, Topol. Appl. 19 (1985) 75-89.

[12] S. Niefield, Cartesianness: topological spaces, uniform spaces and affine schemes, J. Pure Appl. Algebra 23 (1982) 147-167.

[13] J. Penon, Sur les quasi-topos, Cahiers Topol. Geom. Diff. 18 (1977) 181-218.

[14] G. Rosolini, Continuity and effectiveness in topoi, Thesis Carnegie Mellon Univ. 1986.

[15] F. Schwarz, Hereditary topological categories and topological universes, Quaestiones Math. 10 (1986) 197-216.

[16] F. Schwarz and R.-D. Brandt, Hereditary topological categories and applications to classes of convergence spaces. Preprint.

[17] W. Tholen, Exponentiable monomorphisms, Quaestiones Math. 9 (1986) 443-458.

[18] O. Wyler, Are there topoi in topology? Springer Lecture Notes Math. 540 (1976) 697-719.

Horst Herrlich
Feldhäuser Str. 69
2804 Lilienthal

Fed. Rep. Germany

# REMARKS ON LOCALIC GROUPS

J. Isbell*

State University of New York at Buffalo, Buffalo, New York, USA

I. Kříž

Charles University, Prague, Czechoslovakia

A. Pultr

Charles University, Prague, Czechoslovakia

J. Rosický

J. E. Purkyně University, Brno, Czechoslovakia

## INTRODUCTION

Locales are essentially a generalization of topological spaces — certainly a generalization of Hausdorff spaces. However, localic groups are not a generalization (nor 'essentially' a generalization) of topological groups. Reasons are mounting for thinking that localic groups are a considerable improvement on topological groups. This could be a mistake; proofs in locales must be written 'in a mirror', that is, in the dual category of frames, and enough basic questions remain unsettled to make i foolhardy to assert flatly that localic groups are better. Yet.

Hausdorff spaces are locales, by a full embedding, i.e. having the same morphisms. They do not in general have the same product spaces $A \times A$. The topologica product is a dense part of the locale product, containing all the points. It is the whole of the locale product if $A$ is Čech-complete (that is, a $G_\delta$ in a compact Hausdorff space). Otherwise, if $A$ has a continuous group structure, the multiplication on topological $A \times A$ may be inextensible over the locale product.

One of our main results is that every localic subgroup of a localic group is closed. Most of the rest of our results concern the relations between localic and topological groups, including the relation "containing no"; that is, the localic groups which have only one point. One simple *ad hoc* construction gives arbitrarily large localic groups, or vector spaces, or indeed $\mathcal{T}$-algebras for any finitary theory $\mathcal{T}$, whos points form a trivial $\mathcal{T}$-algebra (every morphism to it is epic. In groups, of course, that is a singleton.) This is to take an inverse mapping system of surjections of nonempty sets $S_\alpha$ such that the limit set is empty and every countable set of indice $\alpha$ is bounded; pass to free $\mathcal{T}$-algebras $A_\alpha$, considered as discrete localic $\mathcal{T}$-algebras, and form the inverse limit. It is nontrivial and maps epimorphically to all $A_\alpha$, but

* Supported by the U. S. National Science Foundation.

its algebra of points is trivial.

In abelian groups, we can embed every topological one in a localic one; in fact, there is a categorical equivalence between complete topological abelian groups A and localic abelian groups B whose points are dense (B being $A^{-}$). The topological groups G that are ipso facto localic groups, $G \times G$ being the same in locales as in spaces, include not only Čech-complete groups but also free topological groups generated by compact Hausdorff spaces.

## 1. PRELIMINARIES

1.1. A *frame* is a complete lattice A satisfying the distributivity rule $\left(\bigvee_I a_i\right) \wedge b = \bigvee_I (a_i \wedge b)$; a frame morphism $f: A \to B$ is a mapping preserving general joins and finite meets. The category obtained will be denoted by Frm. Its dual will be denoted by Loc and referred to as the category of *locales*. Although our subject is (in) locales, we will prefer the more convenient frame notation (i. e., the arrows will be written as in Frm so that, e.g., we may refer to a binary operation on a locale A but write it, nevertheless, as $m: A \to A \oplus A$).

We write

$$x \lhd y$$

if there is $r \in A$ such that $r \wedge x = 0$ and $r \vee y = 1$. A locale (frame) A is said to be *regular* if for all a in A

$$a = \bigvee \{x \mid x \lhd a\}.$$

We write

$$x \lhd\lhd y$$

if there exist $x_d$ for dyadic rationals d in the unit interval such that $x_0 = x$, $x_1 = y$, and for $d < d'$, $x_d \lhd x_{d'}$. A locale A is said to be *completely regular* if for all $a \in A$,

$$a = \bigvee \{x \mid x \lhd\lhd a\}.$$

1.2. A cover of a locale A is a subset $U \subseteq A$ such that $\bigvee U = 1$. For a cover U and an element $a \in A$ we put $Ua = \bigvee \{x \mid x \in U \text{ and } x \wedge a \neq 0\}$. For covers U, V write $U < V$ if $(\forall u \in U)(\exists v \in V) u \leq v$. We put

$$U^* = \left\{ \bigvee X \mid X \subseteq U \text{ and } (\forall x,y \in X)\ x \wedge y \neq 0 \right\}.$$

A system $\mathcal{U}$ of covers is a *subbasis for a uniformity* (cf. [27]) provided

$$(\forall U \in \mathcal{U})(\exists V \in \mathcal{U})\ V^* < U.$$

If $\mathcal{U}$ is a system of covers we write

$$x \lhd_{\mathcal{U}} y$$

if there is a $U \in \mathcal{U}$ such that $Ux < y$ (note that obviously $x \lhd_{\mathcal{U}} y \Rightarrow x \lhd y$) and put $A_{\mathcal{U}} = \left\{ a \in A \mid a = \bigvee \{x \mid x \lhd_{\mathcal{U}} a\} \right\}$. It can be proved (see [27]) that

- A is regular iff $A = A_{\mathcal{U}}$ for some system of covers $\mathcal{U}$.
- A is completely regular iff $A = A_{\mathcal{U}}$ for some subbasis for a uniformity $\mathcal{U}$.

A *uniformity* $\mathcal{U}$ on A is a subbasis for a uniformity such that $A = A_{\mathcal{U}}$ and, moreover,

$$U \in \mathfrak{U} \ \& \ U \prec V \Rightarrow V \in \mathfrak{U}.$$

$$U, V \in \mathfrak{U} \Rightarrow \{u \wedge v \mid u \in U, v \in V\} \in \mathfrak{U}. \quad \text{(Cf. [10, 27])}.$$

1.3. The trivial frame {0, 1} will be denoted by 2 , the unique morphism $2 \to A$ by $\sigma_A$. The coproduct of frames $A_i$ $(i \in J)$ will be denoted by $\oplus_J A_i$ (or by $A_1 \oplus A_2$, $A_1 \oplus A_2 \oplus A_3$ etc.). The coproduct of J copies of the same frame A will be denoted by $\oplus^J A$. We obviously have $A \oplus 2 = 2 \oplus A = A$ (consider the coproduct diagram

$$2 \xrightarrow{\ \sigma\ } A \xleftarrow{\ 1\ } A).$$

1.4. Let $\left(q_i: A_i \to A = \underset{J}{\oplus} A_j\right)_{i \in J}$ be a coproduct diagram. The element $\bigwedge_{k=1}^{n} q_{i_k}(a_{i_k})$ will be written as

$$(\ddagger) \qquad a_{i_1} \oplus \ldots \oplus a_{i_n}.$$

If $f_i: A_i \to B_i$ are frame morphisms, we have the obvious formula

$$(\oplus f_i)\left(a_{i_1} \oplus \ldots \oplus a_{i_n}\right) = f_{i_1}(a_{i_1}) \oplus \ldots \oplus f_{i_n}(a_{i_n}).$$

We will need some well known facts on coproducts of frames (see e.g. [6]), namely:

1.4.1. Proposition. $\oplus A_i$ is join-generated by the elements of the form (‡).

1.4.2. Proposition. $a_{i_1} \oplus \ldots \oplus a_{i_n} = 0$ iff $a_{i_k} = 0$ for some k.

1.4.3. Proposition. If $0 \neq a_{i_1} \oplus \ldots \oplus a_{i_n} \leq b_{i_1} \oplus \ldots \oplus b_{i_n}$ then $a_{i_k} \leq b_{i_k}$ for all k.

1.4.4. Proposition. The coproduct of (completely) regular frames is (completely) regular.

1.4.5. Proposition. Let $X \subsetneq A \times B$ be such that

(i) $(a, b) \in X,\ a' \leq a,\ b' \leq b \Rightarrow (a', b') \in X$.

(ii) $(a, b_i) \in X$ for $i \in J \Rightarrow (a, \bigvee_J b_i) \in X$.

(iii) $(a_i, b) \in X$ for $i \in J \Rightarrow (\bigvee_J a_i, b) \in X$.

Then we have the implication

$$x, y \neq 0 \text{ and } x \oplus y \leq \bigvee\{a \oplus b \mid (a, b) \in X\} \Rightarrow (x, y) \in X.$$

1.5. A frame morphism $f: A \to B$ is said to be *dense* if $f(a) = 0 \Rightarrow a = 0$. From 1.4.2: the coproduct of dense morphisms is dense.

1.6. Let $f: A \to B$ be a frame morphism. Put

$$n(f) = \bigvee\{x \mid f(x) = 0\}$$

(note that $f(n(f)) = 0$) and define $x \sim y$ to mean $x \vee n(f) = y \vee n(f)$. The relation ~ is obviously a congruence and we obtain a commutative diagram in Frm

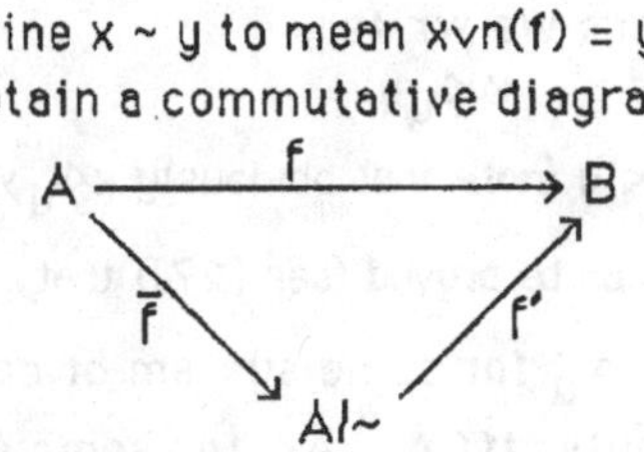


with $\bar{f}$ given by $\bar{f}(x) = [x]$ (the congruence class containing x), f' given by $f'([x]) = f(x)$.

The frame $A/\sim\ = c(f)$ will be referred to as the *closure* of f, and if f' is a monomorphism, f is said to be *closed*. If f is onto and f' an isomorphism, B is said to be a *closed sublocale* and the frame morphism f is called a *closed surjection*. The corresponding morphism in Loc is a *closed embedding*.

1.6.1. Lemma. The morphism f' is dense. Thus any morphism of locales decomposes uniquely (up to isomorphism) as a composition of a dense one and a closed embedding.

Proof. f' is dense by construction, and $\bar{f}$ a closed surjection. As for uniqueness, inspection shows that if g is a dense frame morphism and h a closed surjection, $(gh)' = g$.

1.6.2. Lemma. Let the diagram

$$\begin{array}{ccccc} A & \xrightarrow{\quad \bar{f} \quad} & c(f) & \xrightarrow{\quad f' \quad} & B \\ \Big\downarrow \alpha & & & & \Big\downarrow \beta \\ A_1 & \xrightarrow{\quad \bar{g} \quad} & c(g) & \xrightarrow{\quad g' \quad} & B_1 \end{array}$$

commute. Then there is exactly one $\gamma: c(f) \to c(g)$ such that the resulting squares commute.

Proof. We have $g(\alpha(n(f))) = 0$, hence $\alpha(n(f)) \le n(g)$. Thus we can put $\gamma([x]) = [\alpha(x)]$.

1.6.3. From 1.6.2 and 1.5, if $f_i: A_i \to B_i$ are closed frame morphisms, so is $f = \bigoplus_J f_i: \bigoplus_J A_i \to \bigoplus_J B_i$. For general $f_i: A_i \to B_i$, the factorization of $\bigoplus_J f_i$ after 1.6.1 is through the coproduct $\bigoplus_J c(f_i)$.

1.7. For a topological space X let $\Omega(X)$ denote the frame of open sets of X. If $f: X \to Y$ is a continuous mapping, $\Omega(f): \Omega(Y) \to \Omega(X)$ is the frame morphism given by $\Omega(f)(U) = f^{-1}(U)$. Let $(p_i: \times_J X_j \to X_i)_{i \in J}$ be a product of spaces. There is a uniquely determined morphism of frames

$$\pi: \bigoplus_J \Omega(X_j) \longrightarrow \Omega(\times_J X_j)$$

given by $\pi \circ q_i = \Omega(p_i)$ (where $q_i$ are the coproduct injections).

1.7.1. Observation: $\pi\left(U_{i_1} \oplus \dots \oplus U_{i_n}\right) = \times_J U_j$ where, for $j \neq i_k$, $U_j = X_j$. (Indeed, $\times U_j = \bigcap_k p_{i_k}^{-1}(U_{i_k})$.)

1.7.2. Corollary. The morphism $\pi$ is dense.

1.7.3. In the sequel we will also use the well known fact that $\Omega(X)$ is (completely) regular in the sense of 1.1 if (and only if) X is (completely) regular.

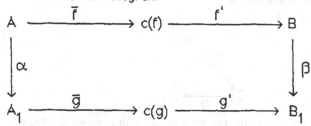

1.7.4. For each frame A there exist a topological space Pt(A) and a morphism $f_A: A \to \Omega(Pt(A))$ making $Pt: \mathbf{Loc} \to \mathbf{Sp}$ (**Sp** is the category of topological spaces) right adjoint to $\Omega: \mathbf{Sp} \to \mathbf{Loc}$ [16].

1.7.5. One sees easily that $\Omega(g): \Omega(X) \to \Omega(Y)$ is a closed surjection iff $g$ is a closed embedding.

## 2. REGULAR LOCALES AND ALGEBRAIC STRUCTURES

2.1. For a frame morphism $f: A \to B$, $f_*: B \to A$ is the right adjoint map, i.e. the uniquely determined monotone mapping such that $f(a) \leq b$ iff $a \leq f_*(b)$. Recall that $ff_*(b) \leq b$ and $a \leq f_*f(a)$, and that $f_*$ preserves meets.

2.2. Lemma. Let

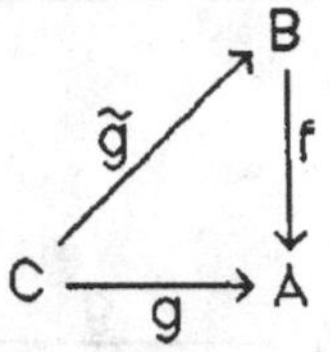


be a commutative triangle in **Frm**, let $f$ be dense and $C$ regular. Then

$$\tilde{g}(c) = \bigvee\{f_*g(x) \mid x \lhd c\}.$$

Proof. We have

(1) $\quad f_*g(z) = f_*f\tilde{g}(z) \geq \tilde{g}(z)$

so that

$$\tilde{g}(c) = \tilde{g}\left(\bigvee\{x \mid x \lhd c\}\right) = \bigvee\{\tilde{g}(x) \mid x \lhd c\} \leq \bigvee\{f_*g(x) \mid x \lhd c\}.$$

On the other hand, let $x \lhd c$. Then there is a $y$ such that $y \wedge x = 0$ and $y \vee c = 1$. Hence by (1),

(2) $\quad f_*g(y) \vee \tilde{g}(c) = 1.$

Further, $g(y) \wedge g(x) = 0$ so that $0 = ff_*(g(y) \wedge g(x)) = f(f_*g(y) \wedge f_*g(x))$. By density, $f_*g(y) \wedge f_*g(x) = 0$ and hence, using also (2), we obtain

$$f_*g(x) \lhd \tilde{g}(c).$$

Thus, $\tilde{g}(c) \geq \bigvee\{f_*g(x) \mid x \lhd c\}$.

2.3. Proposition. In **RegFrm**, the category of regular frames, the monomorphisms are exactly the dense morphisms.

Proof. By 2.2, dense morphisms are monomorphisms. Now, let $f: A \to B$ be not dense. Thus we have an $a \in A$, $a \neq 0$, such that $f(a) = 0$. Consider

$$C = \left\{(x, y) \in A \times A \mid x \vee a = y \vee a\right\}.$$

$C$ is a subframe of $A \times A$. It is regular, as follows. Let $u \lhd x$, $v \lhd y$ and $x \vee a = y \vee a$. Put

$$(uv)_1 = u \wedge (v \vee a), \quad (uv)_2 = v \wedge (u \vee a).$$

Then $\left((uv)_1, (uv)_2\right) \lhd (x, y)$ and $(uv)_1 \vee a = (uv)_2 \vee a$. We have $\bigvee\{(uv)_1 \mid u \lhd x, v \lhd y\} = \bigvee_{v \lhd y}\left(\bigvee_{u \lhd x} u \wedge (v \vee a)\right) = \bigvee_{v \lhd y}\left((v \vee a) \wedge \bigvee_{u \lhd x} u\right) = x \wedge \bigvee_{v \lhd y}(v \vee a) = x \wedge (y \vee a) = x$ and similarly for $(uv)_2$ and $y$ so that

$$\bigvee\left\{\left((uv)_1, (uv)_2\right) \mid u \lhd x, v \lhd y\right\} = (x, y).$$

Define $g_1, g_2: C \to A$ by putting $g_i(x_1, x_2) = x_i$. We have $g_1(a, 0) \neq g_2(a, 0)$ and

$g_i(x_1, x_2) = f(x_i) = f(x_i) \vee f(a) = f(x_i \vee a) = f(x_1 \vee a)$.

2.3.1. **Proposition.** Regular epimorphisms in **RegFrm** are exactly the closed injections. Consequently, the decomposition in 1.6 is the (regular epi, mono) factorization in this category.

Proof. By [16, p. 83], each regular epimorphism in **RegFrm** is a closed surjection. Now let $f: A \to B$ be a closed surjection and put $a = n(f)$. Take the $g_1$, $g_2: C \to A$ from the proof of 2.3. It is easy to see that f is the coequalizer of $g_1$ and $g_2$.

2.4. **Proposition.** The category **RegFrm** is well-powered.

Proof. Let $f: A \to B$ be a monomorphism. Then it is dense. For $a \in A$ put $B(a) = \{f(x) \mid x \lhd a\} \subset B$. By 2.2 (applied for $g = f$ and $\tilde{g} = 1_A$), $a = \bigvee\{f_*(b) \mid b \in B(a)\}$ for each $a \in A$ so that A has at most as many distinct elements as B has subsets.

2.5. Let M be a set. An *M-ary operation* on a locale A is a frame morphism $\alpha: A \to \oplus^M A$. A *type* $\tau$ is a system $(M_i)_{i \in J}$ of sets; an *algebraic structure* of the type $\tau$ on A is a system $(\alpha_i)_{i \in J}$ where $\alpha_i$ are $M_i$-ary operations. An *algebraic theory* $\mathcal{T}$ is, as usual, a type $\tau$ together with a system of equations $\mathfrak{E}$ required of some couples of derived operations. A *localic algebra*, briefly an *L-algebra*, of the theory $\mathcal{T} = (\tau, \mathfrak{E})$ is a locale A together with an algebraic structure of the type $\tau$ satisfying $\mathfrak{E}$. When there is no danger of confusion, we refer to A alone as the L-algebra.

The algebras of a theory $\mathcal{T}$ on topological spaces defined in the usual way will be called T-*algebras* (*topological* $\mathcal{T}$-*algebras*, if necessary for clarity).

Let $\hat{A} = (A, (\alpha_i))$, $\hat{B} = (B, (\beta_i))$ be L-algebras of the same type. A *homomorphism* from $\hat{A}$ to $\hat{B}$ is a frame morphism $f: B \to A$ such that, for all i, the diagram

$$\begin{array}{ccc} B & \xrightarrow{\quad f \quad} & A \\ \Big\downarrow{\scriptstyle \beta_i} & & \Big\downarrow{\scriptstyle \alpha_i} \\ \oplus^{M_i} B & \xrightarrow{\quad \oplus^{M_i} f \quad} & \oplus^{M_i} A \end{array}$$

commutes. If f is a surjection, we refer to it (or to the algebra A) as a *subalgebra* of B.

2.6. Recalling 1.6.3 and 1.6.2 we see that each homomorphism $f: B \to A$ of L-algebras can be decomposed as

$$B \xrightarrow{\quad \bar{f} \quad} c(f) \xrightarrow{\quad f' \quad} A$$

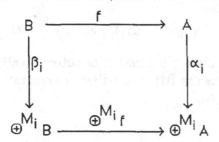

with $\bar{f}$ surjective and $f'$ dense, where c(f) is endowed (in a unique way) with an algebraic structure of the same type as $\hat{A}$ and $\hat{B}$ so that $\bar{f}$ and $f'$ are homomorphisms. If the algebras A, B are regular and of a theory $\mathcal{T}$, we see (taking into account that $f'$ and all the $\oplus^M f'$ are dense) that $c(f)$ is of the same theory.

In particular, considering surjective $f$, we see that the closures of subalgebras are again subalgebras.

2.7. Theorem. The category of regular L-algebras of a given theory $\mathcal{T}$ and all their homomorphisms is cocomplete and has equalizers. The forgetful functor from L-algebras to frames preserves colimits.

Proof. The existence (and preservation) of coproducts and coequalizers is standard (and does not require regularity). Given M-ary operations $\alpha^k: A_k \to \oplus^M A_k$ one obtains an M-ary operation $a: A = \bigoplus_{k\in K} A_k \to \oplus^M A$ by composing $\oplus_K \alpha^k$ with the obvious isomorphism $\bigoplus_{k\in K} \oplus^M A_k \to \oplus^M \bigoplus_{k\in K} A_k$. Given homomorphisms $f, g: A \to B$ with respect to operations $a: A \to \oplus^M A$, $b: B \to \oplus^M B$, consider the coequalizer $h: B \to C$ in RegFrm and for the operation on C the unique morphism $g: C \to \oplus^M C$ such that $g \circ h = \oplus^M h \circ b$.

Equalizers: Let $f, g: A \to B$ be homomorphisms with respect to the algebraic structures $(\alpha_i)_{i\in J}$, $(\beta_i)_{i\in J}$. Consider a system of representatives of all equivalence classes of dense homomorphisms $h_k: (E_k, (\varepsilon^k{}_i)) \to (A, (\alpha_i))$ (recall 2.4) such that $f \circ h_k = g \circ h_k$ and $(E_k, (\varepsilon^k{}_i))$ are of the given theory. Take the coproduct of (co-) algebras $q_k: E_k \to \bigoplus_{j\in K} E_j$, consider the homomorphism $h: \oplus E_j \to A$ satisfying $h \circ q_k = h_k$ and finally take $h': c(h) \to A$ (see 2.6). Clearly $h' = \mathrm{Equ}(f, g)$.

2.7.1. We do not know if (all of) 2.7 is true without "regular"; as noted, cocompleteness and cocontinuity are so. We do not know about products. Note, the Special Adjoint Functor Theorem cannot be applied freely, since Loc is not well-powered [16, II.2.10].

2.8. Take a T-algebra $(X, (\omega_i)_i)$ of a type $\tau = (M_i)_i$ (i.e., $\omega_i: X^{M_i} \to X$). Since we do not generally have $\Omega(X^{M_i}) = \oplus^{M_i} \Omega(X)$, the algebra cannot be automatically viewed as an L-algebra. In case the operations can be lifted to fit into commutative diagrams

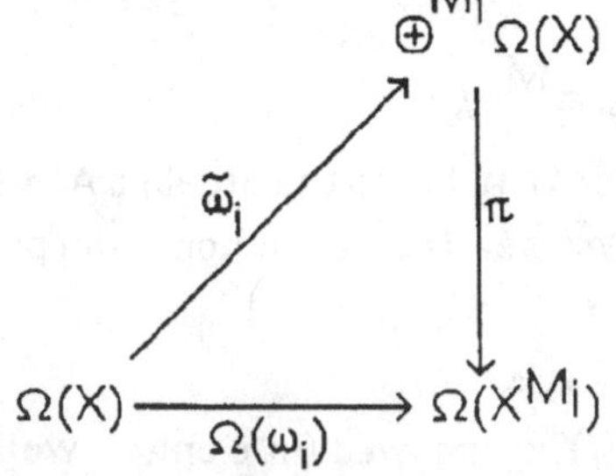


(recall 1.7) we speak of an LT-*algebra*. We will see later that even in some simple cases such a lifting may be impossible.

From 1.7, 2.2 and 2.3 we easily obtain by chasing diagrams the following consequences:

2.8.1. Proposition. In a regular LT-algebra the operations $\tilde{\omega}_i$ are uniquely determined by the $\omega_i$.

2.8.2. Proposition. If a regular LT-algebra is of a theory $\mathcal{T}$, the corresponding L-algebra is of the same theory.

2.8.3. Proposition. If $f:(X, (\omega_i)) \to (X', (\omega_i'))$ is a homomorphism between regular LT-algebras, $\Omega(f)$ is a homomorphism between the corresponding L-algebras

2.8.4. Since Pt is a right adjoint, it induces functors from L-algebras of a given theory to the corresponding T-algebras (operations take points to points).

The last result extends to:

Proposition. Let A be an L-algebra, X a regular LT-algebra and $h: A \to \Omega(X)$ a frame morphism. Then h is a homomorphism iff Pt(h) is a homomorphism.

Proof. The diagram (recall 1.7.4)

$$\begin{array}{ccc} \oplus^M A & \xrightarrow{\ \oplus^M \phi_A\ } & \oplus^M \Omega(Pt(A)) \\ & \searrow^{\phi} \qquad \nearrow_{\pi} & \\ & \Omega(Pt(\oplus^M A) = \Omega(Pt(A)^M) & \end{array}$$

commutes (if $q_i$, $\overline{q_i}$ are the coproduct injections, we have $\phi\circ q_i = \Omega(Pt(q_i))\circ\phi_A = \pi\overline{q_i}\phi_A = \pi\circ\oplus^M\phi_A\circ q_i$).

Let Pt(h) be a homomorphism; consider operations $\alpha_A: A \to \oplus^M A$, $\alpha_X: X^M \to X$. We have $\pi\tilde{\alpha}_X h = \Omega(\alpha_X)h = \Omega(\alpha_X)\Omega(Pt(h))\phi_A = \Omega(Pt(h)^M)\Omega(Pt(\alpha_A)\phi_A = \Omega(Pt(\alpha_A))\phi\alpha_A =$ $\Omega(Pt(h)^M)\circ\pi_{Pt(A)}\circ\oplus^M\phi_A\circ\alpha_A = \pi\circ\oplus^M\Omega(Pt(h))\circ\oplus^M\phi_A\circ\alpha_A = \pi\circ\oplus^M h\circ\alpha_A$. Thus, $\tilde{\alpha}_X\circ h = \oplus^M h\circ\alpha_A$.

Remark. Pt(A) is not necessarily an LT-algebra, as we will see later.

## 3. LOCALIC GROUPS

3.1. A *localic group* or L-*group* $(A, \mu, \iota, \varepsilon)$ consists of a frame A and frame morphisms $\mu: A \to A\oplus A$, $\iota: A \to A$, and $\varepsilon: A \to 2$, satisfying the equations

$$(\mu\oplus 1_A)\circ\mu = (1_A\oplus\mu)\circ\mu,$$
$$(\varepsilon\oplus 1_A)\circ\mu = 1_A = (1_A\oplus\varepsilon)\circ\mu,$$
$$\nabla\circ(\iota\oplus 1_A)\circ\mu = \nabla\circ(1_A\oplus\iota)\circ\mu = \sigma_A\circ\varepsilon;$$

$\nabla$ here is the codiagonal $A\oplus A \to A$, $\sigma_A: 2 \to A$.

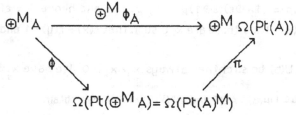

3.1.1. L-groups are just cogroups in the category of frames. Cogroups in any category satisfy $(\mu\oplus\mu)\circ\mu = (1\oplus\mu\oplus 1)\circ(\mu\oplus 1)\circ\mu$ and $\iota\circ\iota = 1$. Note also

$$\nabla(x\oplus y) = x\wedge y,$$

which is easily verified.

3.2. For each $a \in A$ such that $\varepsilon(A) = 1$ ('neighborhood of e') put

$U(a) = \{x\in A \mid x\oplus\iota(x) \leq \mu(a)\}$ (which is also $\{x\wedge y \mid x\oplus y \leq (1\oplus\iota)\mu(a)\}$);

$V(a) = \{x \mid \iota(x)\oplus x \leq \mu(a)\}$ $(= \{x\wedge y \mid x\oplus y \leq (\iota\oplus 1)\mu(a)\})$.

Proposition. $\mathcal{U} = \{U(a) \mid \varepsilon(a)=1\}$ and $\mathcal{V} = \{V(a) \mid \varepsilon(a) = 1\}$ are subbases for uniformities.

Proof. I. Each $U(a)$ is a cover. Indeed, $\bigvee U(a) = \bigvee\{x\wedge y \mid x\oplus y \leq (1\oplus\iota)\mu(a)\} = \bigvee\{\nabla(x\oplus y) \mid x\oplus y \leq (1\oplus\iota)\mu(a)\} = \nabla(\bigvee\{x\oplus y\} \mid x\oplus y \leq (1\oplus\iota)\mu(a)\})$. Thus, by 1.4.1, $\bigvee U(a) = (\nabla\circ(1\oplus\iota)\circ\mu)(a) = \sigma_A\varepsilon(a) = 1$.

II. For each $a$ such that $\varepsilon(a)=1$ there is a $b$ with $\varepsilon(b)=1$ and $b\oplus b \leq \mu(a)$; indeed, we have $(\varepsilon\oplus\varepsilon)\mu = (1_2\oplus\varepsilon)(\varepsilon\oplus 1)\mu = 1_2\oplus\varepsilon = \varepsilon$ and hence $1 = \varepsilon(a) = \bigvee\{\varepsilon(x)\oplus\varepsilon(y) \mid x\oplus y \leq \mu(a)$. Thus there are $x, y$ such that $\varepsilon(x) = \varepsilon(y) = 1$ and $x\oplus y \leq \mu(a)$. Put $b = x\wedge y$.

Now let $\{x_j\}_j \subseteq U(b)$ be suchthat always $x_j\wedge x_k \neq 0$. We have $x_j\oplus\iota(x_j)\oplus x_k\oplus\iota(x_k) \leq (\mu\oplus\mu)\mu(a) = (1\oplus\mu\oplus 1)(\mu\oplus 1)\mu(a)$. Applying $1\oplus(\nabla\circ(\iota\oplus 1))\oplus 1$ we obtain

$$x_j\oplus(x_j\wedge x_k)\oplus\iota(x_k) \leq (1\oplus\sigma\oplus 1)\mu(a).$$

Since $1\oplus\sigma\oplus 1$ sends $\bigvee y_i\oplus y'_i$ to $\bigvee y_i\oplus 1\oplus y'_i$ and since $x_j\wedge x_k \neq 0$ we see by 1.4.3 that $x_j\oplus\iota(x_k)\leq\mu(a)$. Hence $\bigvee x_j\oplus\iota(\bigvee x_j) = \bigvee_{j,k}(x_j\oplus\iota(x_k)) \leq \mu(a)$. Thus $U(b)^* \prec U(a)$.

3.3. Lemma. Let $(A, \mu, \iota, \varepsilon)$ be a localic group, let $B$ be a frame and let $\varepsilon(y) = 1$. If

$$x\oplus y\oplus z \leq (1_B\oplus\mu)(u) \quad \text{(respectively, } x\oplus y\oplus z \leq (\mu\oplus 1_B)(u))$$

then

$$x\oplus U(y)z \leq u \qquad \text{(respectively, } V(y)x\oplus z \leq u).$$

Proof. Let $v\in U(y)$, i.e. $v\oplus\iota(v) \leq \mu(y)$. We have

$$x\oplus v\oplus\iota(v)\oplus z \leq (1\oplus\mu\oplus 1)(1\oplus\mu)(u) = (1\oplus((\mu\oplus 1)\mu)(u) = (1\oplus((1\oplus\mu)\mu)(u) = (1\oplus 1\oplus\mu)(1\oplus\mu)(u).$$

Applying $1\oplus 1\oplus(\nabla\circ(\iota\oplus 1))$ to both sides we obtain

$$x\oplus U(y)z = x\oplus\bigvee\{v \mid v\in U(y),\ v\wedge z \neq 0\} \leq u.$$

3.4. Lemma. Put $M(x) = \{v \mid (\exists u)\ \varepsilon(u)=1,\ u\oplus v \leq \mu(x)\}$. Then $x = \bigvee M(x)$.

Proof. Since $x = (\varepsilon\oplus 1)\mu(x)$, we have $x = \bigvee\{\varepsilon(u)\oplus v \mid u\oplus v \leq \mu(x)\}$.

3.5. Theorem. If $A$ is a localic group then $A = A_{\mathcal{U}} = A_{\mathcal{V}}$. Hence each localic group is uniformizable and consequently completely regular.

Proof. Use 3.4. By 3.3, for each $v\in M(x)$ we have $v \lhd_{\mathcal{U}} x$.

3.6. A *free* L-algebra of a theory $\mathcal{T}$ over a frame $A$ is an L-algebra $F(A)$ together with a frame morphism $\eta_A: F(A) \to A$ such that for any L-algebra B of the theory and for any frame morphism $f: B \to A$ there is a unique homomorphism $\tilde{f}: B \to F(A)$ such that $\eta_A \circ \tilde{f} = f$.

We are now heading for the existence of free L-groups.

3.6.1. Proposition. For any frame there is a free L-monoid.

Proof. By [22; p. 168] it suffices to make certain that

$$C \oplus \prod_{i \in J} C_i \cong \prod_{i \in J} (C \oplus C_i)$$

for arbitrary frames $C, C_i$. It is easy to verify that the formula

$$f(x \oplus (x_i)_{i \in J}) = (x \oplus x_i)_{i \in J}$$

defines the required isomorphism.

3.6.2. A free L-monoid $M(A)$ over A can be explicitly described as follows:

$M(A) = \prod_{n=0}^{\infty} \oplus^n(A)$, $\varepsilon: M(A) \to 2$ is the projection $p_0: M(A) \to \oplus^0(A) = 2$,

$\mu: M(A) \to M(A) \oplus M(A) \cong \prod_{m,n=0}^{\infty} \oplus^{m+n} A$ is given by $p_{m,n} \circ \mu = p_{m+n}$

(so $\mu((a_n)) = (a_{m+n})$) and $\eta_A: M(A) \to A$ is $p_1: M(A) \to \oplus^1 A = A$.

From this, a free L-monoid over a (completely) regular frame is (completely) regular. Further,

$$Pt(M(A)) = \sum_{n=0}^{\infty} Pt(A)^n$$

and $Pt(M(A))$ is the free T-monoid over $Pt(A)$. Thus a free monoid over a frame without points has a unique point $\varepsilon$.

For any L-monoid A there is a unique homomorphism $\gamma_A: A \to M(A)$ such that $\eta_A \gamma_A = 1$. Consequently, $\gamma_A$ embeds A as a subframe into $M(A)$ with the (co-) multiplication preserved. This applies in particular for L-groups A.

3.6.3. Theorem. For any frame A, there is a free L-group over A.

Proof. The proof follows well known patterns (see e.g. [14, p. 43]). An emi-algebra $(A, \mu, \iota, \varepsilon)$ consists of an L-monoid $(A, \mu, \varepsilon)$ and a unary operation $\iota: A \to A$ such that

$$\varepsilon \circ \iota = \varepsilon$$

$$\iota \circ \iota = 1_A$$

$$\mu \circ \iota = \tau \circ (\iota \oplus \iota) \circ \mu$$

where $\tau: A \oplus A \to A \oplus A$ interchanges the injections.

Let A be a frame and $(M(A \times A), \mu', \varepsilon')$ the free monoid over $A \times A$. Let $p'_0, p'_1: A \times A \to A$, $p_n: M(A \times A) \to \oplus^n(A \times A)$ be the projections, $q^n_i: A \times A \to \oplus^n(A \times A)$ the in-

injections. Let $\tau^n: \oplus^n(A\times A) \to \oplus^n(A\times A)$, $\rho: A\times A \to A\times A$ satisfy the equations

$$\tau_n q_i = q_{n-i},\ p'_i\rho = p'_{1-i}.$$

There is a unique frame morphism $\iota': M(A\times A) \to M(A\times A)$ such that

$$p\circ\iota' = \oplus^n\rho\circ\tau_n\circ p_n.$$

By a straightforward calculation one can show that $(M(A\times A), \mu', \iota', \varepsilon')$ is a free emi-algebra over A with the universal morphism

$$\eta'_A: M(A\times A) \xrightarrow{p_1} A\times A \xrightarrow{p_0} A$$

Put

$$h_1: M(A\times A) \xrightarrow{m} M(A\times A)\oplus M(A\times A) \xrightarrow{1\oplus\iota'} M(A\times A)\oplus M(A\times A) \xrightarrow{\nabla} M(A\times A)$$

$$h_2: M(A\times A) \xrightarrow{\varepsilon'} 2 \xrightarrow{\sigma} M(A\times A)$$

Let $\tilde{h}_i: M(A\times A) \longrightarrow M(M(A\times A))$ be the induced homomorphisms of emi-algebras,

i.e., $\eta'_{M(A\times A)}\tilde{h}_i = h_i$.

Since L-groups are (completely) regular and regular frames are coreflective in **Frm** it suffices to construct free L-groups over regular frames. If A is regular, so are $M(A\times A)$ and $M(M(A\times A))$ and (recall 2.7) there exists the equalizer

$$F(A) \xrightarrow{h} M(A\times A) \underset{\tilde{h}_2}{\overset{\tilde{h}_1}{\rightrightarrows}} M(M(A\times A)).$$

It is easy to verify that F(A) with $\eta_A = \eta'_A\circ h$ is a free L-group over A.

3.6.4. From 2.8.4 we obtain

Corollary. If the free topological group G(X) over a topological space X is an LT-group, then $\Omega(G(X))$ is the free localic group over $\Omega(X)$.

3.7. Theorem. The category of localic groups is complete.

Proof. It is an immediate consequence of 3.6.3, 2.7 and Linton's theorem (see [21]).

## 4. CLOSEDNESS OF SUBGROUPS

4.1. The (pseudo) complement of an element a in a frame A, i.e.

$$\bigvee\{x \mid x\wedge a = 0\},$$

will be denoted by a'. Thus a' is the largest x such that $x\wedge a = 0$.

One sees easily that $a \le a''$ and that

$$a \le b'' \Rightarrow a'' \le b''.$$

Also note that $a \triangleleft b$ iff $a\vee b' = 1$.

4.2. Observation. If $a \triangleleft b$ then $a'' \triangleleft b$. Consequently, $a'' \triangleleft Ua$ for any cover U.

4.3. Lemma. Let $f: A \to B$ be dense. Then

$$f(x) = f(y) \Rightarrow x'' = y''.$$

Proof. $f(x\wedge y') = f(x)\wedge f(y') = f(y)\wedge f(y') = 0$, so $x\wedge y' = 0$; therefore $x \leq y''$, and $x'' \leq y''$. Likewise $y'' \leq x''$.

4.4. Lemma. Let $\mu: A \to A\oplus A$ be the binary operation in an L-group. Then

$$x\oplus y \leq \mu(z) \Rightarrow x''\oplus y'' \leq \mu(z).$$

Proof. Recall 3.4 and take a $v$ in $M(y)$. Then, for some $u$ with $\varepsilon(u) = 1$,

$$x\oplus u\oplus v \leq (1\oplus\mu)(x\oplus y) \leq (1\oplus\mu)(\mu(z)) = (\mu\oplus 1)(\mu(z))$$

and therefore, by 3.3 and 4.2, $x''\oplus v \leq V(u)(x)\oplus v \leq \mu(z)$. Thus $x''\oplus y = x''\oplus \bigvee M(y) \leq \mu(z)$. Similarly, $x''\oplus y'' \leq \mu(z)$.

4.5. Lemma. Let $(A, \mu, \iota, \varepsilon)$ be an L-group, B a frame and $f: A \to B$ a dense surjective morphism. Let $0 \neq y_1\oplus y_2 \leq (f\oplus f)\mu(u)$. Then there are $x_i$ such that $y_i = f(x_i)$ and $x_1\oplus x_2 \leq \mu(u)$.

Proof. Recall 1.4.5. Put $X = \{(f(x_1), f(x_2)) \mid x_1\oplus x_2 \leq \mu(u)\}$. If $u_i \leq f(x_i)$, choose first $v_i$ with $f(v_i) = u_i$ and then put $z_i = v_i\wedge x_i$. We have $f(z_i) = u_i\wedge f(x_i) = u_i$.

Further let $(y_{1j}, y_2)\in A$ for $j\in J$. Thus there are $x_{1j}$ and $x_{2j}$ such that $y_{1j} = f(x_{1j})$, $y_2 = f(x_{2j})$ and $x_{1j}\oplus x_{2j} \leq \mu(u)$. By 4.3, $x''_{2j} = x''_{2k}$ for any j, k, so that if we choose a fixed k and put $x''_2 = x''_{2k}$, we obtain from 4.4 that $x_{1j}\oplus x_2 \leq x_{1j}\oplus x''_2 \leq \mu(u)$. Hence $x_{1j}\oplus x_2 \leq \mu(u)$ and $(y_{1j}, y_2) = (f(x_{1j}), f(x_2)) \in X$. Now since $y_1\oplus y_2 \leq (f\oplus f)\mu(u) = \{z_1\oplus z_2 \mid (z_1, z_2)\in X\}$, we obtain by 1.4.5 that $(y_1, y_2)\in X$.

4.6. Theorem. If a homomorphism of frame cogroups $f: A \to A_1$ is onto and dense, it is an isomorphism.

Proof. We have to prove that it is one-one. Suppose not. Consider $x, y\in A$ with $f(x) = f(y)$, but not $y \leq x$. By the regularity of A (and 5.1) there is u such that $a = u'\vee x \neq 1$, but $u'\vee y = 1$. Then $f(a) = f(u')\vee f(x) = f(u')\vee f(y) = 1$, and therefore $(f\oplus f)\mu(a) = \mu_1 f(a) = 1 = 1\oplus 1$. Thus by 4.5 there are $x_1, x_2$ such that $f(x_i) = 1$ and $x_1\oplus x_2 \leq \mu(a)$. Since $\varepsilon_1(x_1) = \varepsilon f(x_1) = 1$, we can apply 3.3 to obtain $x_2 \triangleleft a$, and hence $x'_2\vee a = 1$. Since $a \neq 1$, $x'_2 \neq 0$ while $f(x'_2) = f(x'_2)\wedge f(x_2) = f(x'_2\wedge x_2) = 0$, in contradiction with density.

4.7. From 4.6 and 2.6 we have

Corollary. Each L-subgroup of an L-group is closed.

Since a closed sublocale of a spatial locale is spatial, we have also

Corollary. Each L-subgroup of an LT-group is spatial.

## 5. L-GROUPS WITHOUT POINTS

5.1. There are many localic groups which are not topological groups, for instance an uncountable power of the real line. (This was noted by Kirwan [19].) It is not a T-group, i.e. the product locale is not spatial, because every product of paracompact locales is paracompact ([10], see also [6]) while the product *space* is not normal [30]. At the same time, a spatial uncountable power of the line is a complete topological group that is not localic, by 4.7.

These two examples are not very far apart; the points of the locale product are dense, by 1.7.2. In this section we will show that there exist arbitrarily large localic groups with no point except the necessary one given by the unit operation e.

5.2. There are many examples of inverse mapping systems S of surjections of nonempty sets $S_\alpha$ with empty limit set. For instance, let T be any infinite set, U a larger infinite set, and I a directed set of subsets of U no larger than T but together covering U. For $\alpha \in I$, let $S_\alpha$ be the set of injections $f: \alpha \to T$ with $T - f(\alpha)$ as large as T. For $\alpha < \beta$ (which means $\alpha \subset \beta$), map $S_\beta$ to $S_\alpha$ by restriction. Since each $f \in S_\alpha$ omits enough values, it has extensions $g \in S_\beta$. But an element of the inverse limit set would define an injection of U into T.

Many of these systems S are $\omega_1$*-filtered,* i.e. every countable subset is bounded.

5.3. From any inverse mapping system of sets as in 5.2 (and $\omega_1$-filtered) we can construct large $\mathcal{T}$-algebras, for any finitary algebraic theory $\mathcal{T}$, which have no points except pseudoconstants. Pseudoconstants (like constants) are in the first place derived operations of $\mathcal{T}$. Recall, *constants* are those derived operations $w(x_1, \ldots, x_n)$ which are generated by the empty set of variables. *Pseudoconstant* means generated by each nonempty set of variables. (For instance, in the theory of semigroups with $x^2 = y^2$, $x^2$ is a pseudoconstant.) Then, the elements of an algebra which are values of some constant, or pseudoconstant, operation are called *constant elements*, or *pseudoconstant elements*, respectively.

If an algebra consists of pseudoconstants, every map to it is epic. (When $w(x_1)$ is pseudoconstant, a nonempty algebra has a unique element x satisfying $x = w(x)$, and morphisms preserve these elements.) The converse is false, witness the ring Q of rationals.

5.4. Proposition. For any $\omega_1$-filtered inverse mapping system $\Sigma$ of surjections of nonempty sets $S_\alpha$ whose limit is empty, and any finitary algebraic theory $\mathcal{T}$, the free $\mathcal{T}$-algebras on $S_\alpha$ considered as discrete spaces form an inverse system of epimorphisms of L-algebras $A_\alpha$ whose limit algebra projects epimorphically to all $A_\alpha$ but each point is pseudoconstant.

Proof. Since $\mathcal{T}$ is finitary, the discrete free algebras $FS_\alpha$ are topological algebras, and the $\Omega FS_\alpha$, L-algebras. The induced homomorphisms $FS_\beta \to FS_\alpha$ are surjective, so in Loc they are epimorphic. The locale underlying the limit L-algebra $A_\infty$ is the inverse limit locale by 2.7, and its projections to all $A_\alpha$ are epimorphic [12].

The points of the limit locale are the points of the inverse limit set, since Pt is a coreflector and thus preserves limits. Such a point p is given by its coordinates $p_\alpha \in FS_\alpha$. In free $FS_\alpha$, $p_\alpha$ is $w(x_1, \ldots, x_n)$ for some derived operation w of $\mathcal{T}$ and some $x_1, \ldots, x_n$ in $S_\alpha$. If w is not pseudoconstant, there is a unique smallest set $N_\alpha$ of generators $\{x_1, \ldots, x_n\}$ which generates $p_\alpha$ [13, proof of Theorem 2]. $N_\alpha$ is not empty, since constants are pseudoconstant. The connecting maps $c_{\beta\alpha}: S_\beta \to S_\alpha$ need not take $N_\beta$ into $N_\alpha$, but for each $x \in N_\alpha$ there is some $y \in N_\beta$ mapping to x — otherwise, since $N_\beta$ generates $p_\beta$, $c_{\beta\alpha}(N_\beta)$ and thus $N_\alpha - \{x\}$ would generate $p_\alpha$. Therefore $N_\beta$ has at least as many elements as $N_\alpha$. Then the number of elements is bounded; if it were unbounded, it would be unbounded on a countable set of indices α, which has a common successor since Σ is $\omega_1$-filtered, which is absurd. But being bounded and nondecreasing, it is finally constant. Then, since each $x \in N_\alpha$ is $c_{\beta\alpha}(y)$ for some $y \in N_\beta$, the connecting maps map $N_\beta$ to $N_\alpha$ bijectively. Then there is a point in the limit set of Σ, a contradiction.

In [24], Moerdijk constructs nontrivial localic vector spaces with no nonzero point in much the same way.

6. REMARKS ON LT-GROUPS

6.1. A completely regular frame is said to be Čech-complete if it is $G_\delta$ in its Čech-Stone compactification. Any Čech-complete frame is spatial and thus its (Čech-complete) square is also spatial [11]. Special cases are complete metric spaces and locally compact spaces. Thus we have

Proposition. Any Čech-complete topological group is an LT-group.

6.2. Consider a metric abelian group. If it is complete, it is an LT-group. If it is not complete, it is not closed in its completion and hence, by 4.7, it cannot be an LT-group.

More generally, a T-group X is called a SIN-group (for Small Invariant Neighborhoods) if the left and right uniformities coincide. The observation just made generalizes to

Proposition. A metrizable SIN-group (in particular, a metrizable abelian group) is an LT- group iff it is complete in the natural uniformity.

(A trivial case: all discrete groups are LT.)

6.3. The problem naturally arises to characterize the LT-groups among the T-groups. The observations on metric groups and the fact of natural uniformizability of T-groups suggest attacking the problem via completion (to use the more exact term, supercompletion). A supercompletion technique for locales has been, first, developed in [10], an alternative one in [20]. We need not go into details here. Let us just mention that

– the supercomplete uniform locales coincide with the absolutely closed uniform locales,

– the supercompletion is a reflection from the category of all uniform locales into that of supercomplete uniform locales (coreflection in the frame notation) $\rho_A: S(A) \to A$, all the morphisms $\rho_A$ being dense,

– the functor S preserves uniform embeddings and sums of frames,

– a locale carries a supercomplete uniformity iff it is paracompact.

Recently, one of the authors presented a characterization of LT-groups in terms of a relation between the supercompletions of the right and left uniformitie ([20]). Here we are going to present some more evident special criteria (which, however, include a general characterization for the commutative case).

6.3.1. Following [28], the infimum $\mathcal{L} \wedge \mathcal{R}$ of the right and left uniformities ($\mathcal{R}$, $\mathcal{L}$ respectively) on a T-group will be called the *lower uniformity*.

Proposition. Each T-group supercomplete in its lower uniformity is an LT-group.

Proof. Denote by $X_{\mathcal{L}}$, $X_{\mathcal{R}}$, $X_{\mathcal{L}\wedge\mathcal{R}}$ the corresponding uniform spaces. Evidently $\pi: \Omega(X_{\mathcal{L}}) \oplus \Omega(X_{\mathcal{R}}) \to \Omega(X_{\mathcal{L}} \times X_{\mathcal{R}})$ is a dense uniform embedding. Consider the followin diagram (the multiplication $m: X_{\mathcal{L}} \times X_{\mathcal{R}} \to X_{\mathcal{L}\wedge\mathcal{R}}$ is uniform – see [28]).

$$\begin{array}{ccc}
\Omega(X_{\mathcal{L}}) \oplus \Omega(X_{\mathcal{R}}) & \xrightarrow{\ \pi\ } & \Omega(X_{\mathcal{L}} \times X_{\mathcal{R}}) \\
\uparrow {\scriptstyle \rho_{\Omega(X_{\mathcal{L}})} \oplus \rho_{\Omega(X_{\mathcal{R}})}} & & \uparrow {\scriptstyle \rho_{\Omega(X_{\mathcal{L}} \times X_{\mathcal{R}})}} \\
S(\Omega(X_{\mathcal{L}}) \oplus \Omega(X_{\mathcal{R}})) & \xrightarrow{\ S(\pi)\ } & S(\Omega(X_{\mathcal{L}} \times X_{\mathcal{R}})) \\
 & & \uparrow {\scriptstyle S(\Omega(m))} \\
 & & \Omega(X_{\mathcal{L}\wedge\mathcal{R}}) \cong S(\Omega(X_{\mathcal{L}\wedge\mathcal{R}}))
\end{array}$$

Since $S(\pi)$ is a dense uniform embedding (in complete uniform locales), it is an isomorphism. Thus $\Omega(m)$ lifts to $(\rho\oplus\rho)\circ S(\pi)^{-1}\circ S\Omega(m)$.

6.3.2. Proposition. A SIN-group is an LT-group iff it is supercomplete.

Proof. The sufficiency has been proved in 6.3.1. Now, let $(X, m, \ldots)$ be a SIN group which is LT. Since m is uniform, it follows from [10; p. 31] that the lifted $\tilde{m}: \Omega(X) \to \Omega(x) \oplus \Omega(X)$ is uniform, too. Since S preserves sums, $\Omega(X)$ is a dense sub-group of an L-group S(X). Thus it is supercomplete by 4.7.

Remark. We cannot prove in this way that commutative L-groups are supercomplete. We do not know whether the operation $\mu$ must be uniformly continuous.

6.3.3. From 6.3.2, any Čech-complete SIN-group is supercomplete and hence complete and paracompact. This is not new, and in fact is true for all Čech-complete LT-groups [3, 4]. We obtain the following generalization of 6.2:

Proposition. Let X be a SIN-group with a compact subgroup K such that X/K is metrizable. Then the following conditions are equivalent:

(i) X is an LT-group,

(ii) X is complete,

(iii) X is Čech-complete.

Proof. (i) ⇒ (ii) by 6.3.2, (ii) ⇒ (iii) is proved in [3] and (iii) ⇒ (i) is contained in 6.1.

6.4. A space X is said to be a $k_\omega$-*space* if it is a k-space in virtue of a countable cover by compact (Hausdorff) sets $X_i$, i.e. a subset of X whose intersection with each $X_i$ is closed is a closed set. For our purposes, $k_\omega$ groups are just like Čech-complete groups.

6.4.1. Proposition. Each $k_\omega$-space is the union of an expanding sequence of compact sets $S_i$ such that for any sequence of open sets $V_i$, each $V_i$ containing $S_i$, there is an open covering $(U_i)$ such that each $U_i$ is contained in all $V_j$ with $j > i$.

Proof. We take, of course, compact sets $X_i$ such that sets meeting all of them in closed sets are closed; and put $S_i = \bigcup\{X_j \mid 1 \le j \le i\}$. If $V_i$ contains $S_i$ for each i, $V_i$ open, observe that $V_1 \cap S_2$ contains a closed $S_2$-neighborhood $U_{1,2}$ of $S_1$. $V_2$ contains $S_2$ and thus $U_{1,2}$, so it contains a closed $S_3$-neighborhood $U_{1,3}$ of $U_{1,2}$. Having gotten to $U_{1,j} \subset S_j \cap V_2 \cap \ldots \cap V_{j-1}$, $\bigcap\{V_i \mid 2 \le i \le j\}$ contains a closed $S_{j+1}$-neighborhood $U_{1,j+1}$ of $U_{1,j}$. Let $U_1 = \bigcup U_{1,j}$. Now $U_1 \cap S_2$ is open in $S_2$, for each of its points is in some $U_{1,j}$, and the next $U_{1,j+1}$ is a neighborhood in $S_{j+1} \supset S_2$. Similarly each $U_1 \cap S_j$ is open in $S_j$, so $U_1$ is open. Also $U_1 \supset S_1$. The same construction can be started with $V_i \cap S_{i+1}$ for each i, giving open $U_i$ containing $S_i$ and contained in all later $V_j$, as stated.

6.4.2. Theorem. If X and Y are σ-compact regular spaces and X satisfies the condition of 6.4.1, then $\Omega(X) \oplus \Omega(Y)$ is spatial.

Proof. It suffices to show that no open proper sublocale contains all the points, for $\Omega(X) \oplus \Omega(Y)$ is regular, so every sublocale is an intersection of open sublocales [10]. So consider $u \in \Omega(X) \oplus \Omega(Y)$ such that each point (x, y) is in some $b \times c$, $b \in \Omega(X)$ and $c \in \Omega(Y)$, where $b \oplus c \le u$. By standard compactness arguments, each compact $S \times T$, $S \subset X$ and $T \subset Y$, is contained in such a $b \times c$. In particular, covering X with $(S_i)$ as in 6.4.1 and covering Y with compact sets $T_i$, we get such open rectangles $b_i \times c_i \supset S_i \times T_i$. By the hypothesis on X, there is an open cover $(U_i)$ such that whenever $j > i$, $b_j \supset U_i$. Therefore $U_i \oplus c_{i+1}$, $U_i \oplus c_{i+2}$, and so on, are all $\leq u$. Then so is their join $U_i \oplus \bigvee(c_j \mid j > i) = U_i \oplus 1$; and so is *their* join 1.

We do not know whether 6.4.1 characterizes $k_\omega$-spaces.

6.5. Proposition. Any $k_\omega$ topological group is an LT-group.

This follows from 6.4.1, 6.4.2.

The $k_\omega$-spaces are less familiar than Čech-complete spaces; none are first countable except those which are locally compact (and thus Čech-complete). However, precisely in connection with topological groups, $k_\omega$-spaces frequently occur, for the free topological group on a compact space is $k_\omega$ [26]. More, the free T-group on any $k_\omega$-space is $k_\omega$ [26]. Thus

6.5.1. Proposition. The free T-group on a compact, or a $k_\omega$, space G is an LT-group and is the free L-group on G.

We do not know whether there are any LT-groups except the sort we have been seeing, where the space product is the locale product.

6.6. Finally, let us add some remarks on L-groups which are not LT.

6.6.1. Proposition. Let X be a SIN-group. Then $S\Omega(X)$ is an L-group and $Pt(\rho_{\Omega(X)})\colon X \to PtS\Omega(X)$ is a homomorphism of T-groups. Consequently, any SIN-group is a dense T-subgroup of an L-group.

Proof. Let $(X, \mu, \iota, \varepsilon)$ be a SIN-group. Put

$$\mu\colon S\Omega(X) \xrightarrow{S\Omega(m)} S\Omega(X \times X) \xrightarrow{\cong} S(\Omega(X) \oplus \Omega(X)) \xrightarrow{\cong} S\Omega(X) \oplus S\Omega(X).$$

It is easy to check that $(S\Omega(X), \mu, S\Omega(\iota), S\Omega(\varepsilon))$ is an L-group and that $Pt(\rho_{\Omega(X)})\colon X \to PtS\Omega(X)$ is a homomorphism.

6.6.2. Proposition. Each abelian L-group in which points are dense is super complete. Moreover, the category of these L-groups is isomorphic to the category of complete abelian T-groups.

Proof. Let $C_1$, $C_2$ be the indicated categories. It is easy to see that $S\Omega(f)\colon S\Omega(Y) \to S\Omega(X)$ is an L-group homomorphism for any T-group homomorphism $f\colon X \to Y$. Hence $S\Omega\colon C_2 \to C_1$ is a functor.

Let $(A, \mu)$ be an abelian L-group in which the points are dense. Since $\phi_A\colon A \to \Omega Pt(A)$ is a dense uniform embedding, $S(\phi_A)$ is an isomorphism. The composition

$$S\Omega Pt(A) \xrightarrow{S(\phi_A)^{-1}} S(A) \xrightarrow{\rho_A} A$$

is an L-group homomorphism (use the naturality of $\phi$ and $\rho$, 2.8.4, and the evident fact that

$$\rho_{A\oplus A}\colon S(A\oplus A) \cong S(A)\oplus S(A) \xrightarrow{\rho_A \oplus \rho_A} A\oplus A\,.)$$

Therefore, by 4.7, the L-groups A and SΩPt(A) are isomorphic. Thus A is supercomplete in the natural uniformity.

By [10], Pt(A) is complete in its natural uniformity. Thus Pt induces a functor $C_1 \to C_2$ and we obtain PtSΩ = id from [10; 3.5].

Acknowledgement. We are indebted to B. Banaschewski, A. H. Mekler and Evelyn Nelson for valuable comments and advice.

REFERENCES

1. A. Archangelskij, "O metrizacii topologičeskich prostranstv", *Bull. Acad. Polon. Sci., Ser. Math.* 8(1960), 589-595.
2. M. Artin and B. Mazur, *Etale Homotopy*, Lecture Notes in Math. 100, Springer-Verlag, 1969.
3. L. G. Brown, "Topologically complete groups", *Proc. Amer. Math. Soc.* 15(1972), 593-600.
4. M. M. Čoban, "O popolnenii topologičeskich grupp", *Vest. Mosk. Univ.* 25(1970), 33-38.
5. H. H. Corson, "The determination of paracompactness by uniformities", *Amer. J. Math.* 80(1958), 185-190.
6. C. H. Dowker and D. Strauss, "Sums in the category of frames", *Houston J. Math.* 3 (1976), 17-32.
7. M. I. Graev, "Svobodnyje topologičeskich gruppy", *Izv. Akad. Nauk SSSR* 12(1948), 279-324.
8. J. Isbell, "Supercomplete spaces", *Pacific J. Math.* 12(1962), 287-290.
9. J. Isbell, *Uniform Spaces*, Amer. Math. Soc., 1964.
10. J. Isbell, "Atomless parts of spaces", *Math. Scand.* 31(1972), 5-32.
11. J. Isbell, "Function spaces and adjoints", *Math. Scand.* 36(1975), 317-339.
12. J. Isbell, "Direct limits of meet-continuous lattices", *J. Pure Appl. Alg.* 23 (1982),33-35.
13. J. Isbell, "Generic algebras", *Trans. Amer. Math. Soc.* 275(1983), 497-510.
14. P. T. Johnstone, *Topos Theory*, Academic Press, 1977.
15. P. T. Johnstone, "Tychonoff's theorem without the axiom of choice", *Fund. Math.* 113(1981), 21-35.
16. P. T. Johnstone, *Stone Spaces*, Cambridge, 1982.
17. C. Joiner, "Free topological groups and dimension", *Trans. Amer. Math. Soc.* 220 (1976), 401-418.
18. A. Joyal and M. Tierney, "An extension of the Galois theory of Grothendieck", *Mem. Amer. Math. Soc.* 309, 1984.
19. F. C. Kirwan, "Uniform locales", Dissertation, Cambridge Univ., 1981.
20. I. Kříž, "A direct description of uniform completion in locales and a characteriza-

tion of LT-groups", *Cahiers Top. et Géom. Diff.* 27(1986), 19-34.

21. F. E. J. Linton, "Coequalizers in categories of algebras", Lect. Notes in Math. 80, Springer-Verlag, 1969; pp. 79-90.
22. S. MacLane, *Categories for the Working Mathematician*, Springer-Verlag, 1971.
23. E. Michael, "Bi-quotient maps and cartesian products of quotient maps", *Ann. Inst. Fourier* 18(1968), 287-302.
24. I. Moerdijk, "Prodiscrete groups and Galois toposes", to appear.
25. I. Moerdijk, "The classifying topos of a continuous groupoid", *Trans. Amer. Math. Soc.*, to appear.
26. E. T. Ordman, "Free k-groups and free topological groups", *Gen. Top. Appl.* 5 (1975), 205-219.
27. A. Pultr, "Pointless uniformities I. Complete regularity", *Comment. Math. Univ. Carol.* 25(1984), 91-104.
28. W. Roelcke and S. Dieroff, *Uniform Structures on Topological Groups and Their Quotients*, McGraw-Hill, 1981.
29. H. Simmons, "A framework for topology", *Proc. Wroclaw Logic Conf.* 1977, North-Holland, 1978, 239-251.
30. A. H. Stone, "Paracompactness and product spaces", *Bull. Amer. Math. Soc.* 54 (1948), 977-982.
31. G. C. Wraith, "Localic groups", *Cahiers Top. et Géom. Diff.* 22(1981), 61-66.

This paper is in final form.

Remark added 26 January 1988. P. T. Johnstone has written a paper giving "A simple proof that localic subgroups are closed". His proof is more conceptual than ours, i.e. it seems to answer the question we have been asked: 'Why are all localic subgroups closed?'.

# WEAK PRODUCTS AND HAUSDORFF LOCALES

By Peter Johnstone

Department of Pure Mathematics, University of Cambridge, England

and Sun Shu-Hao

Mathematics Institute, Sichuan University, Chengdu, P.R. China

## Introduction

It is well known (cf. [6], section 2) that one of the salient differences between the category of locales and that of spaces is that the notions of product in the two categories do not coincide: that is, if X and Y are spaces, with open-set lattices $\Omega(X)$ and $\Omega(Y)$, the locale product $\Omega(X) \times_1 \Omega(Y)$ is in general different from $\Omega(X \times Y)$. (Note: here and throughout the paper, our notation relating to locales is taken from [5].) In this paper, we introduce a construction which goes some way towards "reconciling" this difference: specifically, we show that the assignment $(\Omega(X),\Omega(Y)) \mapsto \Omega(X \times Y)$ is the restriction to spatial locales of a symmetric monoidal structure defined on the whole category of locales (which we denote by $(A,B) \mapsto A \otimes B$, and call the <u>weak product</u> structure), such that $A \otimes B$ is (naturally) a dense sublocale of $A \times_1 B$ for any A and B. (There is also an infinitary version of $\otimes$, which we shall define although we shall not investigate its properties in any great detail.) Consideration of the extent to which $\otimes$ differs from the categorical product leads us to introduce a new class of "weakly spatial" locales, which forms the largest (coreflective) subcategory of <u>Loc</u> on which the restriction of $\otimes$ yields the categorical product; thanks to a recent example of Kříž and Pultr, we know that it is strictly larger than the class of spatial locales.

In the second half of the paper, we use the weak product structure to introduce a new "conservative" Hausdorff axiom for locales (that is, a condition on locales which for spatial locales is equivalent to the Hausdorff condition on the corresponding space; it is well known that the "natural" Hausdorff axiom for locales, called the strong Hausdorff axiom in [4] and [5], is not conservative in this sense). Several other conservative Hausdorff axioms have been proposed by various authors [2,9,12,13]; we do not claim that our proposal represents the ultimate answer to the problem of finding a conservative Hausdorff axiom for locales (or even that such an ultimate answer exists), but we show that it has a number of desirable properties and that it compares well with the other axioms which have been proposed.

Before proceeding further, we should mention for the benefit of those readers who have absorbed the message of [6] that our arguments in this paper are <u>not</u> constructively valid; we shall freely use the Law of Excluded Middle when we need it (and indeed it seems unlikely that any of our main results could be proved without it). Also, to avoid needless pathology, we shall assume that all the topological spaces we mention satisfy at least the $T_0$ separation axiom.

<u>Acknowledgements</u>. The second author's research is supported by the Science Fund of the Chinese Academy of Sciences. He would like to thank Professor Liu Ying-Ming for his helpful suggestions. In addition, the first author is grateful to Aleš Pultr, Jan Paseka and Bohumil Šmarda for conversations and correspondence about the work of the Czech school of locale-theorists; and both authors are indebted to the referee of an earlier version of this paper, for his trenchant comments which have materially improved the present version.

## 1. Weak Products

We may think of an open set W in the product of two topological spaces X and Y as being specified by the family of "open rectangles" $U \times V$ which it contains (since W is the union of all such rectangles); and this point of view may be used (as in [6]) to motivate the definition of the locale product $A \times_1 B$ as a frame whose elements are certain lower (i.e. downward-closed) sets in the cartesian product $A \times B$. But we may also specify W by listing the <u>closed</u> rectangles which are disjoint from it; for the closed rectangles in $X \times Y$ include all closures of singletons, and a point $(x,y)$ belongs to W iff the closure of $\{(x,y)\}$ meets W. If we represent closed sets in X and Y by their open complements, we arrive at the idea that an element of $\Omega(X \times Y)$ may be represented by a certain upper set in $\Omega(X) \times \Omega(Y)$. Determining the relationship between this upper set and the lower set which we considered previously leads to the following definition (although it should be mentioned that it was also arrived at, from purely lattice-theoretic considerations and for entirely different purposes, by G.N. Raney [11]):

<u>Definition 1.1</u>. Let $(A_\gamma \mid \gamma \in \Gamma)$ be a family of partially ordered sets, and S a subset of the cartesian product $\prod_{\gamma\in\Gamma} A_\gamma$. We define

$$S^+ = \{\underset{\sim}{a} \in \prod A_\gamma \mid (\forall \underset{\sim}{x} \in S)(\exists \gamma \in \Gamma)(x_\gamma \leq a_\gamma)\}$$

and
$$S^- = \{\underset{\sim}{a} \in \prod A_\gamma \mid (\forall \underset{\sim}{x} \in S)(\exists \gamma \in \Gamma)(x_\gamma \geq a_\gamma)\}$$

where $x_\gamma$ denotes the $\gamma$th entry of the $\Gamma$-tuple $\underset{\sim}{x}$.

It is immediate that the maps $S \mapsto S^+$ and $S \mapsto S^-$ form a Galois connection from the power-set of $\prod A_\gamma$ to itself (i.e. that they are order-reversing, and that $S \subseteq T^-$ iff $T \subseteq S^+$); hence in particular the map $S \mapsto S^{+-}$ is a closure operation. Further, it is clear that $S^+$ is an upper set for any S, and $S^-$ is a lower set. (The connection with the ideas of the first paragraph is that, if W is an open subset of $X \times Y$, the lower set

$$I = \{(U,V) \mid U \times V \subseteq W\}$$

and the upper set

$$J = \{(U,V) \mid (X - U) \times (Y - V) \cap W = \emptyset\}$$

are related by $I = J^-$ and $J = I^+$.)

For our first three lemmas, we shall suppose that each $A_\gamma$ is a frame; adopting the notation of ([5], II 2.12), we shall write B for the product $\prod_{\gamma\in\Gamma} A_\gamma$, and A for the subset

$$\{\underset{\sim}{a} \in B \mid a_\gamma = 1 \text{ for all but finitely many } \gamma \in \Gamma\}.$$

<u>Lemma 1.2</u>. For any $S \subseteq B$, the set $S^- \cap A$ is a C-ideal, where C is the coverage defining the locale product $\prod_1 A_\gamma$ (cf. [5], II 2.12).

<u>Proof</u>. We have already observed that $S^-$ is a lower set in B, and so $S^- \cap A$ is a lower set in A. Thus we have to show that if $\underset{\sim}{a} \in A$, $a_{\gamma_0} = \bigvee\{b_i \mid i \in I\}$ for some $\gamma_0$, and $\underset{\sim}{b}_i \in S^-$ for each i, where

$$(\underset{\sim}{b}_i)_\gamma = b_i \quad \text{if } \gamma = \gamma_0, \quad = a_\gamma \quad \text{otherwise},$$

then $\underset{\sim}{a} \in S^-$. But by assumption, for each $x \in S$ and for each $i \in I$ we have either $(b_i \leq x_{\gamma_0})$ or $(\exists \gamma \neq \gamma_0)(a_\gamma \leq x_\gamma)$. So we either have $b_i \leq x_{\gamma_0}$ for all $i \in I$, whence $a_{\gamma_0} \leq x_{\gamma_0}$, or

$$(\exists i \in I)(\exists \gamma \neq \gamma_0)(a_\gamma \leq x_\gamma);$$

and in either case we deduce $(\exists \gamma \in \Gamma)(a_\gamma \leq x_\gamma)$. So $\underset{\sim}{a} \in S^-$. □

<u>Lemma 1.3</u>. The map $f: \prod_1 A_\gamma \to \prod_1 A_\gamma$ defined by $f(I) = I^{+-} \cap A$ is a nucleus on $\prod_1 A_\gamma$.

<u>Proof</u>. Lemma 1.2 assures us that this map is well-defined, and it follows easily from the remarks after Definition 1.1 that f is a closure operation on $\prod_1 A_\gamma$. So we need only verify that f preserves

finite intersections. Let $I_1$ and $I_2$ be C-ideals in A, and suppose $\underset{\sim}{b} \in (I_1 \cap I_2)^+$. Then for any $\underset{\sim}{x} \in I_1$ and $\underset{\sim}{y} \in I_2$ we have $x_\gamma \wedge y_\gamma \leq b_\gamma$ for some $\gamma$, or equivalently $x_\gamma \leq (y_\gamma \to b_\gamma)$; thus $(\underset{\sim}{y} \to \underset{\sim}{b}) \in I_1{}^+$, where the arrow denotes implication in the Heyting algebra B. Thus for any $\underset{\sim}{x} \in f(I_1) \subseteq I_1{}^{+-}$ we have $x_\gamma \leq (y_\gamma \to b_\gamma)$ for some $\gamma$, or equivalently $x_\gamma \wedge y_\gamma \leq b_\gamma$; thus we have shown that $\underset{\sim}{b} \in (f(I_1) \cap I_2)^+$. A similar argument now shows that $\underset{\sim}{b} \in (f(I_1) \cap f(I_2))^+$; so we have shown that $(I_1 \cap I_2)^+ \subseteq (f(I_1) \cap f(I_2))^+$. But the reverse of this inclusion is trivial since $I_j \subseteq f(I_j)$; so we actually have equality here, and hence $f(I_1 \cap I_2) = f(f(I_1) \cap f(I_2))$. Finally, the right-hand side of this equation equals $f(I_1) \cap f(I_2)$, since an intersection of f-closed C-ideals is f-closed. □

We denote the sublocale of $\prod_1 A_\gamma$ consisting of the fixed points of f by $\bigotimes_{\gamma\in\Gamma} A_\gamma$, and call it the <u>weak product</u> of the $A_\gamma$. Our next lemma provides us with a plentiful supply of elements of $\bigotimes A_\gamma$, and we shall use it repeatedly. Given $\underset{\sim}{b} \in B$, let us write $L(\underset{\sim}{b})$ for the set

$$\{\underset{\sim}{a} \in A \mid (\exists\, \gamma \in \Gamma)(a_\gamma \leq b_\gamma)\}.$$

<u>Lemma 1.4</u>. For any $\underset{\sim}{b} \in B$, $L(\underset{\sim}{b})$ is an f-closed C-ideal.

<u>Proof</u>. By definition, $L(\underset{\sim}{b}) = \{\underset{\sim}{b}\}^- \cap A$, from which the result follows easily. □

<u>Corollary 1.5</u>. $\bigotimes A_\gamma$ is a dense sublocale of $\prod_1 A_\gamma$.

<u>Proof</u>. The zero element of $\prod_1 A_\gamma$ is easily seen to be the C-ideal $L(\underset{\sim}{0})$, where $\underset{\sim}{0}$ is the $\Gamma$-tuple with all entries 0; so it belongs to $\bigotimes A_\gamma$. □

From now on, we shall restrict our attention to finite weak products (in fact to weak products of two factors — which we shall of course denote by $A_1 \otimes A_2$); this will enable us to simplify our notation, since we shall no longer have to distinguish between the product B and its subset A. However, it is worth noting here that most of the results of this section remain true for infinite weak products; we leave to the reader the task of generalizing the proofs which we give in the finite case.

We shall write $\Omega$ for the terminal object of <u>Loc</u>, i.e. the open-set lattice of the one-point space. (Of course, this is just the two-element frame $\{0,1\}$.)

<u>Corollary 1.6</u>. For any A, we have $A \otimes \Omega = A \times_1 \Omega$ $(\cong A)$.

<u>Proof</u>. It is easily seen that the C-ideals in $A \times \Omega$ are all of the form $L(a,0)$ for some $a \in A$; so they are all f-closed by 1.4, and the

map $a \mapsto L(a,0)$ gives the required isomorphism. □

<u>Proposition 1.7</u>. The assignment $(A,B) \mapsto A \otimes B$ defines a subfunctor of the product functor $(A,B) \mapsto A \times_1 B$.

<u>Proof</u>. We have to show that if $f: A \to A'$ and $g: B \to B'$ are locale maps, then $f \times_1 g: A \times_1 B \to A' \times_1 B'$ maps the sublocale $A \otimes B$ into $A' \otimes B'$; equivalently, that if $I$ is a C-ideal in $A' \times B'$, then

$$((f \times_1 g)^*(I))^{+-} = ((f \times_1 g)^*(I^{+-}))^{+-}.$$

Now $(f \times_1 g)^*(I)$ is the C-ideal in $A \times B$ generated by the set ($J$, say) of all pairs $(f^*(a), g^*(b))$ with $(a,b) \in I$; in general, it is not simply the downward-closure of $J$, but it certainly lies between $J$ and $J^{+-}$, since the latter is a C-ideal by 1.2. So

$$\begin{aligned}((f \times_1 g)^*(I))^+ = J^+ &= \{(a,b) \mid (\forall (x,y) \in I)(a \geq f^*(x) \text{ or } b \geq g^*(y)\} \\ &= \{(a,b) \mid (f_*(a), g_*(b)) \in I^+\}\end{aligned}$$

since $a \geq f^*(x)$ iff $f_*(a) \geq x$ and $b \geq g^*(y)$ iff $g_*(b) \geq y$. Since $I^+ = I^{+-+}$, we deduce that $((f \times_1 g)^*(I))^+ = ((f \times_1 g)^*(I^{+-}))^+$, from which the required equality follows. □

<u>Theorem 1.8</u>. The bifunctor $\otimes$ defines a symmetric monoidal structure on the category **Loc**, with unit $\Omega$.

<u>Proof</u>. It is straightforward to verify that the twist isomorphism $A \times_1 B \to B \times_1 A$, and the associativity isomorphism

$$A \times_1 (B \times_1 C) \to (A \times_1 B) \times_1 C,$$

restrict to isomorphisms between the corresponding weak products. The fact that these isomorphisms (and the isomorphism $A \otimes \Omega \to A$ of Corollary 1.6) are natural in $A$, $B$ and $C$, and that they satisfy the coherence conditions for a symmetric monoidal structure, follows from the fact that they are defined by restricting the canonical isomorphisms for the categorical product structure on **Loc**, which automatically have these properties. □

Having thus established the fundamental properties of the weak product structure on the category of all locales, we now turn our attention to how it relates to points and to spatial locales.

<u>Proposition 1.9</u>. For any $A$ and $B$, each point of $A \times_1 B$ factors through the inclusion $A \otimes B \to A \times_1 B$.

<u>Proof</u>. We note that this follows from 1.6 and 1.7, since we can regard a point $(p,q): \Omega \to A \times_1 B$ as a morphism

$$\Omega \xrightarrow{\sim} \Omega \times_1 \Omega \xrightarrow{p \times_1 q} A \times_1 B .$$

However, we may also give a direct proof, as follows. Let p and q be points of A and B, corresponding to prime elements $\bar{p}$ and $\bar{q}$ respectively. Then, for a C-ideal $I \subseteq A \times B$, we have

$$(p,q)^*(I) = 0 \iff (\forall (x,y) \in I)(p^*(x) \wedge q^*(y) = 0)$$
$$\iff (\forall (x,y) \in I)(x \leqslant \bar{p} \text{ or } y \leqslant \bar{q})$$
$$\iff (\bar{p},\bar{q}) \in I^+ .$$

Since $I^+ = I^{+-+}$, we deduce that $(p,q)^*(I) = (p,q)^*(I^{+-})$ for all I, which implies that (p,q) factors through the sublocale $A \otimes B$. □

Corollary 1.10. Let A and B be arbitrary locales, and C any spatial locale. Then any locale map $f: C \to A \times_1 B$ factors through $A \otimes B$.

Proof. As before, we have to show that $f^*(I) = f^*(I^{+-})$ for every C-ideal I. But if p is any point of C, then $p^*f^*(I) = p^*f^*(I^{+-})$ by 1.9, and since C is spatial the family of all such functions $p^*$ is jointly injective. □

It follows from 1.10 that $A \otimes B$ always contains the spatial part $\Omega(pt(A \times_1 B))$ of $A \times_1 B$. If A and B are themselve spatial, we can say more:

Proposition 1.11. If A and B are spatial locales, then $A \otimes B$ is spatial.

Proof. Let I and J be elements of $A \otimes B$ with $I \nsubseteq J$; then we have to find a point (p,q) of $A \otimes B$ with $(p,q)^*(I) = 1$ and $(p,q)^*(J) = 0$. Let (a,b) be any element of I - J. Then since $J = J^{+-}$ we have $(a,b) \notin J^{+-}$; i.e. there exists $(x,y) \in J^+$ with $a \nleqslant x$ and $b \nleqslant y$. Since A and B are spatial we can find points p and q with $p^*(a) = 1$, $p^*(x) = 0$, $q^*(b) = 1$ and $q^*(y) = 0$. Now since $(a,b) \in I$ we have

$$(p,q)^*(I) \geqslant p^*(a) \wedge q^*(b) = 1,$$

and since $(x,y) \in J^+$ we deduce $(\bar{p},\bar{q}) \in J^+$, whence $(p,q)^*(J) = 0$ by the argument in the proof of Proposition 1.9. □

Corollary 1.12. For any two spaces X and Y, we have

$$\Omega(X) \otimes \Omega(Y) \cong \Omega(X \times Y).$$

Proof. By [5], II 2.13, we know that $\Omega(X \times Y)$ is (isomorphic to) the spatial part of $\Omega(X) \times_1 \Omega(Y)$. (Actually, [5] proves this only in the case when X and Y are sober; but since the soberification functor preserves finite products it is true in general.) And 1.10 and 1.11 together tell us that this spatial part is exactly $\Omega(X) \otimes \Omega(Y)$. □

It is clear that, for a general locale A, the diagonal map

$\Delta_A : A \to A \times_1 A$ cannot always factor through $A \otimes A$; for if it did, the factorization would be the unit of an adjunction between $\otimes$ and the diagonal functor Loc $\to$ Loc $\times$ Loc, which would force $\otimes$ to be naturally isomorphic to $\times_1$. More specifically, if $A = B \times_1 C$ is any locale product which does not coincide with the weak product $B \otimes C$, then the composite

$$A \xrightarrow{\Delta_A} A \times_1 A \xrightarrow{p \times_1 q} B \times_1 C = A$$

is the identity (where p, q are the projections) and so does not factor through $B \otimes C$; hence by 1.7 $\Delta_A$ does not factor through $A \otimes A$.

We shall call a locale A <u>weakly spatial</u> if $\Delta_A$ factors through $A \otimes A$. By Corollary 1.10, every spatial locale is weakly spatial. But in fact we have a more general result: let us call a complete lattice <u>f-distributive</u> if it satisfies the particular case of the complete distributive law

$$\bigvee_{i \in I} \bigwedge_{j \in J_i} a_{i,j} = \bigwedge_{f \in F} \bigvee_{i \in I} a_{i,f(i)}$$

(where $F = \prod_{i \in I} J_i$ is the set of choice functions for the family of sets $(J_i \mid i \in I)$) in which the sets $J_i$, $i \in I$, are all finite. (By taking each $J_i$ to be a doubleton $\{0,1\}$, and $a_{i,0}$ to be independent of i, it is easy to show that an f-distributive lattice is necessarily a frame.) A strengthening of (the dual of) f-distributivity was studied by S. Papert in [8]; using Papert's techniques (or those used in the proof of Theorem I 2.3 in [3]), it is easy to show that a complete lattice A is f-distributive iff every $a \in A$ satisfies

$$a = \bigwedge\{b \in A \mid a \prec b\},$$

where $a \prec b$ holds iff, whenever $S \subseteq A$ is finite and $\bigwedge S \leqslant a$, there exists $s \in S$ with $s \leqslant b$. Clearly $a \prec b$ holds if there is a prime element p with $a \leqslant p \leqslant b$, from which it is easy to see that spatial frames are f-distributive. Recently, I. Křiž and A. Pultr [10] have constructed an example of an f-distributive frame which is not spatial; however, we have

<u>Lemma 1.13</u>. f-distributive frames are weakly spatial.

<u>Proof</u>. Let us recall that the diagonal map $\Delta_A$ corresponds to the frame homomorphism

$$\Delta^*: A \times_1 A \to A;\ I \mapsto \bigvee\{a \wedge b \mid (a,b) \in I\},$$

whose right adjoint $\Delta_*$ is given by $a \mapsto \{(b,c) \mid b \wedge c \leqslant a\}$. We have to prove that if A is f-distributive then every $\Delta_*(a)$ belongs to $A \otimes A$, i.e. satisfies $\Delta_*(a) = \Delta_*(a)^{+-}$. But if $a \prec b$ then $(b,b) \in \Delta_*(a)^+$;

hence if $(x,y) \in \Delta_*(a)^{+-}$ then for all b with $a \prec b$ we have either $x \leqslant b$ or $y \leqslant b$, so that $x \wedge y \leqslant \bigwedge\{b \in A \mid a \prec b\} = a$. □

We do not know whether the converse of Lemma 1.13 holds: it seems unlikely, but in the next section we shall see that for our notion of Hausdorff locales the properties "spatial", "weakly spatial" and "f-distributive" all coincide.

Lemma 1.14. The full subcategory WSLoc of weakly spatial locales is coreflective in Loc.

Proof. For any locale A, define $\check{A}$ to be the pullback

$$\begin{array}{ccc} \check{A} & \longrightarrow & A \otimes A \\ \downarrow & & \downarrow \\ A & \xrightarrow{\ \Delta\ } & A \times_1 A \end{array}$$

considered as a sublocale of A. Now define a transfinite descending sequence of sublocales $(A_\alpha)$ of A by $A_0 = A$, $A_{\alpha+1} = (A_\alpha)^\vee$, and $A_\alpha = \bigcap\{A_\beta \mid \beta < \alpha\}$ if $\alpha$ is a limit ordinal. Clearly, there must exist an ordinal $\gamma$ such that $A_\gamma = A_{\gamma+1}$, i.e. such that $A_\gamma$ is weakly spatial. But by 1.7 any locale map $f\colon B \to A$ restricts to a map $\check{B} \to \check{A}$ and hence to a map $B_\alpha \to A_\alpha$ for each ordinal $\alpha$; in particular, if B is weakly spatial (so that $B_\alpha = B$ for all $\alpha$) f itself factors (uniquely) through $A_\gamma \to A$. So $A_\gamma$ is the coreflection of A in WSLoc. □

We do not know of any example in which the sequence $(A_\alpha)$ in the proof of 1.14 requires more than a single step to converge; once again, we shall see in the next section that no such example can be a Hausdorff locale. We may also note that if $A = B \times_1 C$ is a locale product of weakly spatial locales, then the sequence converges in (at most) a single step; for an easy diagram-chase shows that in this instance $\check{A} = B \otimes C$, and we have

Lemma 1.15. (a) A weak product of weakly spatial locales is weakly spatial.

(b) $\otimes$ is the categorical product in the category WSLoc.

Proof. (a) If A and B are weakly spatial, then we have a map

$$A \otimes B \xrightarrow{\ \delta_A \otimes \delta_B\ } (A \otimes A) \otimes (B \otimes B) \xrightarrow{\ \varphi\ } (A \otimes B) \otimes (A \otimes B)$$

where $\delta_A$, $\delta_B$ are the factorizations of $\Delta_A$, $\Delta_B$ through the respective weak products, and $\varphi$ is the "middle four interchange" arising from the associativity and symmetry isomorphisms for $\otimes$ (cf. 1.8). A straightforward diagram-chase shows that this composite is the required factorization of $\Delta_{A \otimes B}$ through the weak product.

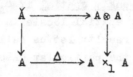

(b) is immediate from (a) and the definition of weakly spatial locales. ∎

It is clear that Lemma 1.15 characterizes the category WSLoc, in that it is the unique largest full subcategory of Loc with these two properties.

From 1.7, we have a commutative square

$$\begin{array}{ccc} A' \otimes B' & \xrightarrow{f \otimes g} & A \otimes B \\ \downarrow & & \downarrow \\ A' \times_1 B' & \xrightarrow{f \times_1 g} & A \times_1 B \end{array}$$

whenever we are given locale maps $f\colon A' \to A$ and $g\colon B' \to B$; it is of interest to know when this square is a pullback. Once again, this cannot be true in general, even when f and g are inclusions; for every locale is embeddable in a locally compact (spatial) locale (namely a power of the Sierpiński locale, cf. [4]), and for such locales the functors $\otimes$ and $\times_1$ coincide, by 1.12 and [5], Proposition II 2.13. However, there is an important special case when the result is true. Recall that a <u>locally closed</u> sublocale is one which is expressible as the intersection of an open and a closed sublocale; equivalently, a locally closed inclusion is one expressible as a composite (in either order) of an open and a closed inclusion.

<u>Lemma 1.16</u>. The square above is a pullback whenever f and g are locally closed inclusions.

<u>Proof</u>. Since a composite of pullback squares is a pullback, it suffices to prove the result in the two cases when one of the maps (g, say) is an identity and the other is either open or closed. Let us write j for the nucleus on A corresponding to f, and identify A' with the sublocale $A_j$. Now suppose we are given a C-ideal $I \subseteq A_j \times B$ such that $(f \times_1 id)_*(I)$ belongs to $A \otimes B$; we must show that $I \in A_j \otimes B$. An easy calculation shows that

$$(f \times_1 id)_*(I) = \{(a,b) \mid (j(a),b) \in I\},$$

from which it follows that if $(x,y) \in ((f \times_1 id)_*(I))^+$ then $(j(x),y) \in I^+$. The argument now splits into two cases:

If j is open and we are given $(a,b) \in I^{+-}$, then $(j_!(a),b)$ belongs to $((f \times_1 id)_*(I))^{+-} = (f \times_1 id)_*(I)$, where $j_!$ denotes the left adjoint of j. But $j(j_!(a)) = a$, so we deduce that $(a,b) \in I$.

If j is closed, let c denote the element $j(0)$ of A; then $(c,1) \in (f \times_1 id)_*(I)$, so that every element $(x,y)$ of $((f \times_1 id)_*(I))^+$

satisfies either $x \geq c$ or $y = 1$. Now suppose, once again, that $(a,b) \in I^{+-}$. Then if $(x,y) \in ((f \times_1 \mathrm{id})_*(I))^+$, we have either $j(x) \geq a$ or $y \geq b$ by the first paragraph; but since $j(x) = x \vee c$, we also have $j(x) = x$ or $y = 1$, whence we obtain $x \geq a$ or $y \geq b$. So $(a,b) \in ((f \times_1 \mathrm{id})_*(I))^{+-} = (f \times_1 \mathrm{id})_*(I)$, whence $(a,b) \in I$. □

As mentioned previously, we know from 1.12 and [5], Proposition II 2.13 that, if A is a locally compact (spatial) locale and B is any spatial locale, then $A \otimes B = A \times_1 B$. It is tempting to ask whether we can drop the spatial condition on B from this result; but we shall see shortly that this is not possible. Nevertheless, we can say a good deal about the class of locales A such that $A \otimes B = A \times_1 B$ for all locales B; we shall call such locales <u>distributed</u>. (Note that a distributed locale is necessarily weakly spatial; there does not seem to be any <u>a priori</u> reason why it should be spatial.)

<u>Proposition 1.17</u>. (a) Distributedness is inherited by locally closed sublocales and by retracts.

(b) Finite products of distributed locales are distributed.

(c) If A has an open covering by distributed sublocales, then A is distributed.

(d) If A has a distributed open sublocale whose closed complement is also distributed, then A is distributed.

(e) Injective locales (cf. [5], VII 4.9) are distributed.

(f) If A is locally compact and $A \otimes B = A \times_1 B$ for all Boolean locales B, then A is distributed.

<u>Proof</u>. (a) The first assertion is immediate from 1.16, and the second is a categorical triviality.

(b) This is also a categorical triviality, given the associativity of both $\otimes$ and $\times_1$.

(c) Suppose A has an open covering by good sublocales $A_j$, $j \in J$. Then for any B the family of maps $A_j \otimes B = A_j \times_1 B \to A \times_1 B$, $j \in J$, is jointly epimorphic; but they all factor through $A \otimes B$ by 1.7.

(d) This is similar to (c), the point being that the pair of inclusions $A_{u(a)} \to A$, $A_{c(a)} \to A$ is not just epimorphic but universally epimorphic.

(e) Since injective locales are retracts of powers of the Sierpiński locale S (i.e. the three-element frame $\{0,m,1\}$), it suffices by (a) to prove that powers of S are distributed. Consider a product of locales $A_\gamma$, $\gamma \in \Gamma$, where one factor $A_\delta$ is arbitrary and the others are all copies of S. Given $a \in A_\delta$ and a finite subset F of

$\Gamma - \{\delta\}$, let $\underset{\sim}{h}(a,F)$ be the element of the cartesian product $\prod A_\gamma$ defined by

$$h(a,F)_\gamma = a \quad (\gamma = \delta), \quad = m \quad (\gamma \in F), \quad = 1 \quad \text{otherwise}.$$

Now if I is any C-ideal in $\prod A_\gamma$, define

$$a_F = \bigvee\{a \in A_\delta \mid \underset{\sim}{h}(a,F) \in I\};$$

since I is a C-ideal, we have $\underset{\sim}{h}(a_F,F) \in I$. Note also that $a_{F'} \leqslant a_F$ whenever $F' \subseteq F$, since I is a lower set. Now define $\underset{\sim}{k}(F) \in \prod A_\gamma$ by

$$k(F)_\gamma = a_F \quad (\gamma = \delta), \quad = 0 \quad (\gamma \in F), \quad = m \quad \text{otherwise};$$

then it is easily verified that $\underset{\sim}{k}(F) \in I^+$. Now let $\underset{\sim}{a} \in I^{+-}$; if some entry of $\underset{\sim}{a}$ is 0 then $\underset{\sim}{a} \in I$, so assume that this is not the case. Let F be the finite set $\{\gamma \in \Gamma - \{\delta\} \mid a_\gamma \neq 1\}$; then by comparing entries of $\underset{\sim}{a}$ with those of $\underset{\sim}{k}(F)$, we see that we must have $a_\delta \leqslant a_F$, and hence $\underset{\sim}{a} \in I$. So $I = I^{+-}$, and since I was arbitrary we have $\prod_1 A_\gamma = \bigotimes A_\gamma$.

(f) For any locale B, we can find an epimorphism $B' \to B$ in $\underset{\sim}{\text{Loc}}$ with B' Boolean (cf. [5], II 2.6). But if A is locally compact (and hence exponentiable, by [5], VII 4.11), then $A \times_1 (-)$ preserves epimorphisms; so we have an epimorphism $A \otimes B' = A \times_1 B' \to A \times_1 B$, which must factor through $A \otimes B$. □

However, not every locally compact locale is distributed:

Lemma 1.18. If X is a nonempty Hausdorff space without isolated points, then $\Omega(X)$ is not distributed.

Proof. Write A for $\Omega(X)$, and B for the (atomless) complete Boolean algebra of regular open subsets of X. Consider the set

$$I = \{(U,V) \in A \quad B \mid U \cap V = \emptyset\};$$

it is easily verified that I is a C-ideal, i.e. an element of $A \times_1 B$. However, if $(P,Q) \in I^+$, then either P or Q must be the whole of X: for if not we can choose points $x \in X - P$, $y \in X - Q$ with $x \neq y$ (since $X - Q$ is not a singleton), and then choose disjoint open neighbourhoods U, V of x, y with V regular open. Then $(U,V) \in I$, but $U \nsubseteq P$ and $V \nsubseteq Q$. So $I^{+-} = A \times B \neq I$; thus $A \otimes B \neq A \times_1 B$. □

It follows from 1.18 that part (b) of 1.17 cannot be extended to infinite products: for the two-point discrete space is distributed (as are all discrete spaces, by 1.6 and 1.17(c)), but the Cantor space is not. For Hausdorff spatial locales, we can characterize distributedness completely:

Proposition 1.19. If X is a Hausdorff space, then $\Omega(X)$ is distributed iff X is scattered.

Proof. If X is not scattered, then its perfect part is a closed subspace whose open-set locale is not distributed, by 1.18; so by 1.17(a)

$\Omega(X)$ is not distributed. Conversely, suppose X is scattered; we shall prove that $\Omega(X)$ is distributed by induction on the Cantor-Bendixson rank of X. For the empty space (the unique space of rank 0), the result is trivial. If the rank of X is a successor ordinal $\alpha+1$, then the isolated points of X form a discrete (and hence distributed) open subspace whose closed complement (the first derived set of X) is scattered of rank $\alpha$, and hence distributed by the inductive hypothesis. So $\Omega(X)$ is distributed by 1.17(d). If the rank of X is a limit ordinal $\lambda$, then for each $x \in X$ there exists $\alpha < \lambda$ such that x is an isolated point of the $\alpha$th derived set $X^{(\alpha)}$, i.e. we can find an open neighbourhood U of x such that $U \cap X^{(\alpha)} = \{x\}$. But now it follows that U has Cantor-Bendixson rank $\alpha+1 < \lambda$, and hence by the inductive hypothesis that $\Omega(U)$ is distributed; and since we can find such a neighbourhood for every $x \in X$, it follows from 1.17(c) that $\Omega(X)$ is distributed. □

Given the result (which we shall prove in the next section) that a weakly spatial locale satisfying our Hausdorff axiom is spatial, we may restate Proposition 1.19 by saying that the distributed Hausdorff locales are exactly the topologies of scattered Hausdorff spaces. (This explains our choice of the name "distributed", though there is also a deliberate echo of "distributive": generalizing 1.17(e), it can be shown that every completely distributive lattice, considered as a locale, is distributed.)

Lastly in this section, we consider <u>weakly exponentiable</u> locales, i.e. locales A such that $A \otimes (-)$ has a right adjoint. ("Weakly exponentiable" is a bad name, since we shall soon see that this property is stronger than ordinary exponentiability; but the terminology seems inescapable.)

<u>Proposition 1.20</u>. A locale is weakly exponentiable iff it is locally compact and distributed.

<u>Proof</u>. The right-to-left implication is obvious. Conversely, suppose A is weakly exponentiable; we first show that A is locally compact by adapting the argument of [5], VII 4.10. The functor $A \otimes (-)$, being a subfunctor of $A \times_1 (-)$, preserves inclusions, so its right adjoint ($A \pitchfork (-)$, say) preserves injectives, and in particular $A \pitchfork S$ is injective. So $pt(A \pitchfork S)$ is a continuous lattice; but its elements correspond bijectively to maps $A \cong A \otimes \Omega \to S$ and hence to elements of A. The proof that this correspondence is order-preserving is as in [5].

We now know that A is exponentiable as well as weakly exponentiable; and the natural transformation $A \otimes (-) \to A \times_1 (-)$ transposes to give a natural transformation $\delta: [A,-] \to A \pitchfork (-)$ between the two right adjoints. Moreover, $\delta_B$ is an isomorphism for injective B, since we may construct an inverse for it by transposing the composite

$$(A \pitchfork B) \times_1 A = (A \pitchfork B) \otimes A \xrightarrow{\text{wev}} B$$

where wev ("weak evaluation") is the counit of the adjunction, and we have used 1.17(e) and the fact that $A \pitchfork (-)$ preserves injectives. Now any locale embeds in an injective, and $[A,-]$, being a right adjoint, preserves inclusions, so $\delta_B$ is an inclusion for any B. But the fact that $\delta_B$ is monic means that, for any C, any two morphisms $A \times_1 C \rightrightarrows B$ which are equalized by $A \otimes C \to A \times_1 C$ must be equal; since this is true for all B, it follows that $A \otimes C \to A \times_1 C$ is epic (and hence an isomorphism) for all C. □

## 2. Hausdorff Locales

We recall that a locale A is said to be <u>strongly Hausdorff</u> if the diagonal map $\Delta: A \to A \times_1 A$ is a closed embedding, i.e. if every C-ideal $I \subseteq A \times A$ satisfying $I \supseteq \Delta_*(0)$ is of the form $\Delta_*(a)$ for some $a \in A$. (As in [5], III 1.3, we shall write N for the C-ideal $\Delta_*(0)$, i.e. the set $\{(x,y) \in A \times A \mid x \wedge y = 0\}$.) As is well known, this is not a conservative extension of the Hausdorff axiom for spaces (there are Hausdorff spaces whose topologies are not strongly Hausdorff), and this arises from the discrepancy between space products and locale products. Thus the weak product structure gives us an opportunity to modify the strong Hausdorff axiom so as to make it conservative; although, since $\Delta_A$ does not always factor through $A \otimes A$, we are obliged (initially, at least) to give our new definition in rather more <u>ad hoc</u> form.

<u>Definition 2.1</u>. We shall say that a locale A is <u>Hausdorff</u> if every C-ideal $I \subseteq A \times A$ which satisfies $I = I^{+-}$ and $I \supseteq N$ is of the form $\Delta_*(a)$ for some $a \in A$.

It is immediate from the definition that a strongly Hausdorff locale (in particular, a regular locale) is Hausdorff.

<u>Lemma 2.2</u>. A locale A is Hausdorff iff the following two conditions hold:

(i) The diagonal embedding $\check{A} \to A \otimes A$ is closed, where $\check{A}$ is defined as in the proof of 1.14.

(ii) $N^{+-} = \Delta_*(n)$ for some $n \in A$.

Proof. Regarded as a sublocale of $A \times_1 A$, $\check{A}$ is simply the intersection of $A \otimes A$ and the image of $\Delta_*$, i.e. it consists of those C-ideals I which satisfy both $I = I^{+-}$ and $I = \Delta_*\Delta^*(I)$. So condition (ii) above says that $N^{+-}$ is an element (necessarily the least element) of $\check{A}$, and (i) says that any element of $A \otimes A$ containing this least element is in the image of $\Delta_*$. But an element of $A \otimes A$ contains $N^{+-}$ iff it contains N, so these two conditions together are equivalent to 2.1. □

Condition (ii) of 2.2 is not always satisfied (we shall see a counterexample later), but it does hold whenever A is (weakly) spatial since we then have $N^{+-} = N$. Thus we have

Corollary 2.3. For a $T_0$-space X, $\Omega(X)$ is a Hausdorff locale iff X is a Hausdorff space.

Proof. By 1.12, 2.2 and the remark above, $\Omega(X)$ is a Hausdorff locale iff the diagonal map $\Omega(X) \to \Omega(X \times X)$ is a closed embedding. For sober spaces X, the equivalence of the two conditions is now immediate. But a $T_0$-space embeds as a subspace of its soberification, and a Hausdorff space (is sober and) cannot be expressed as the soberification of any proper subspace; so the equivalence remains valid under the weaker assumption that X is a $T_0$-space. □

Perhaps surprisingly, our Hausdorff axiom can also be expressed by a first-order condition.

Proposition 2.4. A locale A is Hausdorff iff it satisfies the following condition:

$(S_2)$ Given $a, b \in A$ with $a \neq 1$ and $a \not\leq b$, there exist $x, y \in A$ with $x \wedge y = 0$, $x \not\leq a$ and $y \not\leq b$.

Proof. Suppose A is Hausdorff, and let (a,b) be a pair of elements of A satisfying the hypotheses of $(S_2)$. Consider the C-ideal L(a,b), defined as in 1.4. By 1.4, we have $L(a,b) = L(a,b)^{+-}$; but L(a,b) cannot be in the image of $\Delta_*$, since it is not symmetric (the element (a,1) of $A \times A$ belongs to it, but (1,a) does not). So we must have $L(a,b) \not\supseteq N$; but an element (x,y) of $N - L(a,b)$ satisfies $x \wedge y = 0$, $x \not\leq a$ and $y \not\leq b$. So condition $(S_2)$ is satisfied.

Conversely, suppose A satisfies $(S_2)$ and let I be a C-ideal in $A \times A$ containing N. Condition $(S_2)$ tells us that no pair (a,b) satisfying the hypotheses of $(S_2)$ can belong to $I^+$; i.e. if $(a,b) \in I^+$ then either $a = 1$ or $a \leq b$. But by symmetry we also have $b = 1$ or $b \leq a$; so any $(a,b) \in I^+$ must satisfy one of the three conditions $a = 1$, $b = 1$ or $a = b$. And $I^+$ is an upper set, so if $a \neq 1$ and $(a,a) \in I^+$ then a must be a coatom, i.e. a maximal element of $A - \{1\}$. But any coatom is prime, so if a is a coatom the condition "$x \leq a$ or $y \leq a$" is equivalent to "$x \wedge y \leq a$". We therefore conclude that

$$I^+ = A \times \{1\} \cup \{1\} \times A \cup \{(a,a) \mid a \text{ is a coatom and } a \geq \Delta^*(I)\}.$$

It now follows easily that

$$I^{+-} = \{(x,y) \mid x \wedge y \leq a \text{ whenever } (a,a) \in I^+\}$$
$$= \Delta_*(t)$$

where t is the meet of the coatoms a such that $(a,a) \in I^+$. In particular, if $I = I^{+-}$, then I is in the image of $\Delta_*$; so A is a Hausdorff locale. □

Scholium 2.5. If A is a Hausdorff locale, then $\check{A}$ is precisely the spatial part of A. In particular, a Hausdorff locale is spatial iff it is weakly spatial; and for Hausdorff locales the transfinite construction of 1.14 converges in at most one step.

Proof. If t belongs to $\check{A}$, qua sublocale of A (i.e. if $\Delta_*(t)$ belongs to $A \otimes A$), then it follows easily from the second half of the proof of 2.4 that t must be a meet of coatoms. But the coatoms are exactly the prime elements of A, for if p is any prime element of a locale it is easy to see that we have $(p,p) \in N^+$. So A consists exactly of those elements of A which are meets of primes; i.e. it is the spatial part of A. The other two assertions follow immediately. □

Using 2.4, we may show that our class of Hausdorff locales is closed under sublocales and products (equivalently, that the full subcategory of Hausdorff locales is epireflective in Loc):

Proposition 2.6. (a) A sublocale of a Hausdorff locale is Hausdorff.

(b) A product of Hausdorff locales is Hausdorff.

Proof. (a) Let A be a Hausdorff locale, j a nucleus on A. If two elements a, b of $A_j$ satisfy the hypotheses of $(S_2)$ in $A_j$, then they continue to do so in A, so we may find $x, y \in A$ with $x \wedge y = 0$, $x \nleq a$ and $y \nleq b$. Now the elements $j(x)$ and $j(y)$ of $A_j$ satisfy $j(x) \wedge j(y) = j(0)$, $j(x) \nleq a$ and $j(y) \nleq b$, as required.

(b) Let $(A_\gamma \mid \gamma \in \Gamma)$ be a family of Hausdorff locales, and let I and J be C-ideals in $\prod A_\gamma$ such that I is not the top element of $\prod_1 A_\gamma$ and $I \nleq J$. Let $\underset{\sim}{a}$ be any element of $I - J$, and let $\{\gamma_1, \gamma_2, \ldots, \gamma_n\}$ be the finite set of indices such that $a_{\gamma_i} \neq 1$. For each $i \leq n$, let $\underset{\sim}{a}^{(i)}$ be the $\Gamma$-tuple obtained from $\underset{\sim}{a}$ on replacing each of $a_{\gamma_1}, \ldots, a_{\gamma_i}$ by 1; then we have $\underset{\sim}{a}^{(0)} = \underset{\sim}{a} \in I$ but $\underset{\sim}{a}^{(n)} = \underset{\sim}{1} \notin I$, so there must exist i with $\underset{\sim}{a}^{(i-1)} \in I$ but $\underset{\sim}{a}^{(i)} \notin I$. For any $x \in A_{\gamma_i}$, let $\underset{\sim}{h}(x)$ denote the $\Gamma$-tuple whose $\gamma_i$th entry is x and whose other entries are those of $\underset{\sim}{a}^{(i)}$. Now define

$$b = \bigvee\{x \in A_{\gamma_i} \mid \underset{\sim}{h}(x) \in I\}, \quad c = \bigvee\{x \in A_{\gamma_i} \mid \underset{\sim}{h}(x) \in J\}.$$

Since I and J are C-ideals, we have $\underset{\sim}{h}(b) \in I$ and $\underset{\sim}{h}(c) \in J$; and by construction we have $a_{\gamma_i} \leq b \neq 1$ and $a_{\gamma_i} \nleq c$ (so that $b \nleq c$). Applying $(S_2)$ in $A_{\gamma_i}$, we obtain elements x, y with $x \wedge y = 0$, $x \nleq b$ and $y \nleq c$; then $\underset{\sim}{h}(x) \notin I$ and $\underset{\sim}{h}(y) \notin J$, so that the principal C-ideals generated by these two elements satisfy the conclusions of $(S_2)$ for the pair (I,J). □

Alternative proofs of both parts of Proposition 2.6 may be given using 2.2 instead of 2.4. In each case, verifying condition (i) of 2.2 is a straightforward application of the fact that pullbacks and intersections of closed sublocales are closed; but a little more work is needed to verify (ii).

Next, we observe that 2.6(b) does not have a converse: the Hausdorffness of a nontrivial locale product $A \times_1 B$ does not imply that both factors are Hausdorff, in contrast to the situation for Hausdorff spaces (or for strongly Hausdorff locales). In fact we have

Proposition 2.7. The following conditions on a locale A are equivalent:

(i) A is Hausdorff and has no points.

(ii) $N^{+-}$ is the whole of $A \times A$.

(iii) Given $a \neq 1$ and $b \neq 1$ in A, there exist $x, y \in A$ with $x \wedge y = 0$, $x \nleq a$ and $y \nleq b$.

(iv) $A \times_1 B$ is Hausdorff for any locale B.

(v) $A \times_1 S$ is Hausdorff, where S is the Sierpiński locale.

Moreover, if these conditions hold for A, then they hold for any locale B for which there exists a locale map $B \to A$.

Proof. (i) => (ii): If A is Hausdorff and pointless, then by 2.5 $\check{A}$ is degenerate; i.e., as a sublocale of $A \times_1 A$, it consists only of the top element $A \times A$. But $N^{+-} \in \check{A}$ by 2.2(ii), so condition (ii) holds.

(ii) => (i): Conversely, if (ii) holds then $\check{A}$ is degenerate (so trivially closed in $A \otimes A$) and the conditions of 2.2 are easily verified; and by 2.5 A has no points.

(ii) <=> (iii): Condition (iii) is equivalent to the assertion that $N^{+} = A \times \{1\} \cup \{1\} \times A$ $(= (A \times A)^{+})$, so this is immediate.

We have thus established the equivalence of (i), (ii) and (iii);

before considering (iv) and (v), we prove the last assertion of the Proposition. Suppose A satisfies (iii), and let f: B → A be a locale map. If a, b ∈ B satisfy $a \neq 1_B \neq b$, then since $f^*$ preserves 1 we have $f_*(a) \neq 1_A \neq f_*(b)$, and so we can find x, y ∈ A with $x \wedge y = 0$, $x \nleq f_*(a)$ and $y \nleq f_*(b)$. Now the elements $f^*(x)$ and $f^*(y)$ of B satisfy the conclusions of (iii) for the pair (a,b).

(i) => (iv) is now immediate, since $A \times_1 B$ always has a locale map to A; and (iv) => (v) is also immediate.

(v) => (i): Since S has a point (in fact it has two), we can embed $A \cong A \times_1 \Omega$ as a sublocale of $A \times_1 S$, and so by 2.6(a) condition (v) implies that A is Hausdorff. But if A had a point then S would also be embeddable in $A \times_1 S$, which is impossible since S is not Hausdorff. □

Note that, in 2.7(v), S may be replaced by any particular locale which has a point and is not Hausdorff; also, in both (iv) and (v), the locale product $\times_1$ may be replaced by the weak product ⊗ without affecting the result. Examples of locales satisfying the conditions of 2.7 include atomless complete Boolean algebras (since complete Boolean algebras are regular as locales); in view of the results of [12], it seems worthwhile to give an example to show that not all pointless locales satisfy these conditions.

<u>Example 2.8</u>. Let $B_1$ and $B_2$ be any two pointless locales, and form their ordinal sum $A = B_1 \oplus B_2$, i.e. the ordered set obtained by identifying the top element of $B_1$ with the bottom element of $B_2$. Then A has no prime elements, since any such would be a prime element in either $B_1$ or $B_2$. But if a and b are any elements of $B_2$ other than 1, then condition (iii) of 2.7 fails for the pair (a,b). (It may be noted that A satisfies condition 2.2(i), at least if $B_1$ and $B_2$ are Hausdorff, since $\check{A}$ is degenerate; but it fails condition 2.2(ii).)

Using 2.7, we can in fact characterize Hausdorff product locales completely.

<u>Lemma 2.9</u>. Let $(A_\gamma \mid \gamma \in \Gamma)$ be a nonempty family of locales, none of which is Hausdorff. Then $\bigotimes A_\gamma$ is not Hausdorff (and hence neither is $\prod_1 A_\gamma$).

<u>Proof</u>. For each $\gamma \in \Gamma$, we can find elements $a_\gamma$, $b_\gamma$ of $A_\gamma$ with $1 \neq a_\gamma \nleq b_\gamma$, which violate condition $(S_2)$. Let $I = L(\underset{\sim}{a})$ and $J = L(\underset{\sim}{b})$ in the notation of 1.4; we shall show that they violate condition $(S_2)$ for $\bigotimes A_\gamma$. The inequalities $1 \neq I \nleq J$ are clear. Now if X and Y are two (f-closed) C-ideals satisfying $X \nleq I$ and $Y \nleq J$, then for any

$\underset{\sim}{x} \in X - I$ and $\underset{\sim}{y} \in Y - J$ we have $x_\gamma \nleq a_\gamma$ and $y_\gamma \nleq b_\gamma$ for all $\gamma$, whence $x_\gamma \wedge y_\gamma \neq 0$ for all $\gamma$, i.e. $X \cap Y$ is not the zero element $L(\underset{\sim}{0})$ of $\bigotimes A_\gamma$. □

Proposition 2.10. Let $(A_\gamma \mid \gamma \in \Gamma)$ be a family of locales. The following are equivalent:

(i) $\prod_1 A_\gamma$ is Hausdorff.

(ii) $\bigotimes A_\gamma$ is Hausdorff.

(iii) Either every $A_\gamma$ is Hausdorff, or some $A_\gamma$ is Hausdorff and pointless.

Proof. (iii) => (i) is a restatement of 2.6(b) and 2.7(iv), and (i) => (ii) follows immediately from 2.6(a).

(ii) => (iii): Let $\Gamma_1 \subseteq \Gamma$ be the set of indices such that $A_\gamma$ is Hausdorff, and $\Gamma_2 = \Gamma - \Gamma_1$, so that

$$\bigotimes_{\gamma\in\Gamma} A_\gamma \cong \Big(\bigotimes_{\gamma\in\Gamma_1} A_\gamma\Big) \otimes \Big(\bigotimes_{\gamma\in\Gamma_2} A_\gamma\Big).$$

If $\Gamma_2$ is nonempty, then the second factor above is not Hausdorff by 2.9; but if $A_\gamma$ has a point for each $\gamma \in \Gamma_1$, then the first factor has a point, and so we can embed the second factor as a sublocale of the product. So if both the conditions in (iii) fail, then $\bigotimes A_\gamma$ is not Hausdorff. □

Now we turn to considering the relationship between our Hausdorff definition and the various conservative Hausdorff axioms which have been proposed by other authors. The first one we consider was studied by Dowker and Strauss [2] (together with a number of minor variants, which we shall not list here) and also by Simmons [13]: it is actually the conjunction of two first-order conditions, of which the first is the condition which Simmons calls "conjunctivity" and Isbell [4] calls "subfitness":

(C) Given $a, b \in A$ with $a \nleq b$, there exists $c \in A$ with $a \vee c = 1$ but $b \vee c \neq 1$.

The topology of any $T_1$-space satisfies (C); for this reason (C) has been taken as a "localic $T_1$ axiom" by some authors ([1], [2]), but it is not altogether satisfactory as such, not least because it fails to be inherited by sublocales. (See also [7] for an example of a space with no closed points whose topology satisfies (C), and [12] for a rather better-behaved localic $T_1$ axiom.) The second of the two conditions, denoted $(N_2)$ by Simmons (and $(S_2')$ by Dowker and Strauss) asserts

($N_2$) Given a, b $\in$ A with a $\neq$ 1, b $\neq$ 1 but a $\vee$ b = 1, there exist x, y $\in$ A with x $\wedge$ y = 0, x $\nleq$ a and y $\nleq$ b.

Lemma 2.11. (a) A Hausdorff locale (in our sense) satisfies ($N_2$).

(b) A locale which satisfies (C) and ($N_2$) is Hausdorff.

Proof. (a) is immediate since the hypotheses of ($N_2$) imply those of ($S_2$), and the conclusions are the same.

(b) Suppose A satisfies (C) and ($N_2$), and let a, b be two elements of A satisfying the hypotheses of ($S_2$). By (C), we can find c $\in$ A with a $\vee$ c = 1 but b $\vee$ c $\neq$ 1. Then the pair (a, b $\vee$ c) satisfies the hypotheses of ($N_2$), so we can find x and y with x $\wedge$ y = 0, x $\nleq$ a and y $\nleq$ b $\vee$ c (whence y $\nleq$ b). □

Since Hausdorff spaces are $T_1$, Lemma 2.11 enables one to recover the result ([13], theorem 9) that (C) + ($N_2$) is a conservative Hausdorff axiom from our Corollary 2.3. However, not every Hausdorff locale in our sense satisfies (C):

Example 2.12. Let A be the non-conjunctive locale given by Murchiston and Stanley as example 2 in [7]: explicitly, elements of A are pairs (U,V) with U an arbitrary subset of $\mathbb{R}$, V a regular open subset of $\mathbb{R}$ and U $\subseteq$ V, the ordering on such pairs being given by inclusion in each factor. We shall verify that A satisfies condition ($S_2$). Let ($U_1$,$V_1$) and ($U_2$,$V_2$) be elements of A satisfying the hypotheses of ($S_2$). Since ($U_1$,$V_1$) is not the top element of A, we can find $x_1 \in \mathbb{R} - U_1$; and since ($U_1$,$V_1$) $\nleq$ ($U_2$,$V_2$), we can either find a point $x_2$ (necessarily different from $x_1$) in $U_1 - U_2$, or else find a nonempty open interval W contained in $\mathbb{R} - V_2$. In the former case, we may satisfy the conclusions of ($S_2$) with elements ({$x_1$},$W_1$) and ({$x_2$},$W_2$) of A, where $W_1$ and $W_2$ are disjoint open intervals around $x_1$ and $x_2$; in the latter, by choosing W small enough, we may assume that its closure does not contain $x_1$, and so satisfy the conclusions of ($S_2$) with elements of the form ({$x_1$},W') and (∅,W).

It is worth remarking that a slight modification of Example 2.12 produces a non-conjunctive locale which is embeddable in a Hausdorff spatial locale; so the conjunction (C) + ($N_2$) is not inherited by sublocales.

In [12], Rosický and Šmarda studied the condition

(S) Every semiprime element of A is a coatom,

where an element a $\in$ A is said to be semiprime if a $\neq$ 1 and (in our notation) (a,a) $\in$ $N^+$. It follows immediately from the proof of

Proposition 2.4 that our $(S_2)$ implies $(S)$, but the converse is false:

**Example 2.13.** Let $A = B \oplus \Omega$ be the ordinal sum of an atomless complete Boolean algebra and the two-element frame. In [12], example 2.3, Rosický and Šmarda showed that $A$ satisfies $(S)$ but not $(C)$; it does satisfy $(N_2)$ (vacuously, since there are no pairs $(a,b)$ satisfying the hypotheses of $(N_2)$), but it is easy to see that $(S_2)$ fails for the pair $(t,0)$, where $t$ is the unique coatom of $A$.

In [12], Rosický and Šmarda showed that $(S)$ is a conservative extension of the Hausdorff axiom for spaces, and that it is inherited by sublocales and products. More recently (and independently of the work described in this paper), Paseka and Šmarda [9] have introduced a stronger condition $(T_2)$, which they have shown to have the same properties. They formulate this condition as follows:

$(T_2)$ If $a \in A$ and $a \neq 1$, then $a = \bigvee \{y \in A \mid y \square a\}$, where $y \square a$ means that $y \leq a$ but $\neg y \nleq a$ (here $\neg y$ denotes the pseudo-complement of $y$).

**Proposition 2.14.** $(T_2)$ is equivalent to $(S_2)$.

**Proof.** Clearly $(T_2)$ holds iff, whenever we are given $a$, $b$ with $a \neq 1$ and $a \nleq b$, we can find $y \square a$ with $y \nleq b$ — equivalently, we can find $x$ and $y$ with $x \wedge y = 0$, $y \leq a$, $x \nleq a$ and $y \nleq b$. But this is just the statement of $(S_2)$ with the extra clause "$y \leq a$" among the conclusions, so it clearly implies $(S_2)$. Conversely, suppose $(S_2)$ holds and we are given $a$, $b$ with $1 \neq a \nleq b$. Since $a \wedge a = a$, we also have $a \nleq (a \to b)$, and so by $(S_2)$ we can find $x$ and $y$ with $x \wedge y = 0$, $x \nleq a$ and $y \nleq (a \to b)$ (equivalently, $y \wedge a \nleq b$). So the pair $(x, y \wedge a)$ satisfies the conclusions of $(S_2)$ plus the extra clause mentioned above, and hence $(T_2)$ holds. $\square$

Thus many of the results on $T_2$-locales in [9] may be deduced from the corresponding results on Hausdorff locales in this paper (or, if you prefer, _vice_ _versa_). In conclusion let us remark that, while it would clearly be rash to claim that our Hausdorff condition is _the_ right answer to the problem of formulating a conservative Hausdorff axiom for locales, the fact that it has been independently arrived at (from very different directions) by two separate groups of researchers does tend to support the belief that it is a condition which deserves further study.

## References

[1] B. Banaschewski and R. Harting: Lattice aspects of radical ideals and choice principles. Proc. London Math. Soc. (3) 50 (1985), 385-404.

[2] C.H. Dowker and D. Strauss: Separation axioms for frames. Colloq. Math. Soc. János Bolyai 8 (1972), 223-240.

[3] G. Gierz, K.H. Hofmann, K. Keimel, J.D. Lawson, M. Mislove and D.S. Scott: A Compendium of Continuous Lattices (Springer-Verlag, 1980).

[4] J.R. Isbell: Atomless parts of spaces. Math. Scand. 31 (1972), 5-32.

[5] P.T. Johnstone: Stone Spaces (Cambridge University Press, 1982).

[6] P.T. Johnstone: The point of pointless topology. Bull. Amer. Math. Soc. (N.S.) 8 (1983), 41-53.

[7] G.S. Murchiston and M.G. Stanley: A '$T_1$' space with no closed points, and a "$T_1$" locale which is not '$T_1$'. Math. Proc. Cambridge Philos. Soc. 95 (1984), 421-422.

[8] S. Papert: Which distributive lattices are lattices of closed sets? Proc. Cambridge Philos. Soc. 55 (1959), 172-176.

[9] J. Paseka and B. Šmarda: $T_2$-frames and almost compact frames. Preprint, J.E. Purkyně University, Brno, 1987.

[10] A. Pultr: personal communication.

[11] G.N. Raney: Tight Galois connections and complete distributivity. Trans. Amer. Math. Soc. 97 (1960), 418-426.

[12] J. Rosický and B. Šmarda: $T_1$-locales. Math. Proc. Cambridge Philos. Soc. 98 (1985), 81-86.

[13] H. Simmons: The lattice theoretic part of topological separation properties. Proc. Edinburgh Math. Soc. (2) 21 (1978), 41-48.

This paper is in final form and will not be published elsewhere.

# GENERALIZED FIBRE BUNDLES

Anders Kock

Math. Inst.
Aarhus University
Ny Munkegade
8000 Aarhus C, Denmark

As the title suggests, this is a rather formal paper: it generalizes some notions just in order to have a perfect symmetry of concepts. Thus we symmetrize the set up of our paper "Fibre bundles in general categories" |8|, and arrive at what may be called "Haefliger structures" in general categories; such structures occur in foliation theory, cf. |4| and |10|. We intend the present paper to be readable without knowledge of |6|,|7|, or |8|.

We shall work in a category $\underline{E}$ with finite inverse limits, and we shall talk about the objects of $\underline{E}$ as if they were sets. If $q: C \to B$ is a stable regular epimorphism, we may represent 'elements' of $B$ by 'elements' of $C$, provided the outcome is independent of choice of representatives.

We shall in particular consider such stable regular epimorphisms which are descent maps, in a sense we shall recapitulate, and which is a crucial concept in any fibre bundle theory, allowing one to glue objects together out of compatible local data.

I acknowledge fruitful discussions with J. Pradines and J. Duskin; some of the formalisms of the present paper are closely related to ideas occuring, in some form, in their work, |9|, and |1|, §5. Also the debt to Ehresmann's groupoid theoretic foundations of differential geometry is obvious. Finally, I would like to thank S. Eilenberg for pushing me into adopting a notation that makes calculations obvious (rather than the one of |6|,|7|,|8|). The paper is in final form.

## §1. Double equivalence relations, and pregroupoids.

A double equivalence relation is a special case of a double category, in the same way as an equivalence relation is a special case of a category. This in fact defines the notion, but let us make it more explicit:

A double equivalence relation on an object $X \in \underline{E}$ is a subobject $\Lambda \subset X^4$ (a 4-ary relation), such that

1) the two binary relations $H$ and $V$ on $X$ given by

$$H(x,y) \text{ iff } \Lambda(x,y,x,y)$$

and

$$V(x,z) \text{ iff } \Lambda(x,x,z,z),$$

respectively, are equivalence relations on $X$,

2) the binary relation $\sim_h$ on $X^2$ given by

$$(x,y) \sim_h (z,u) \text{ iff } \Lambda(x,y,z,u)$$

is an equivalence relation on $H$, and the binary relation $\sim_v$ on $X^2$ given by

$$(x,z) \sim_v (y,u) \text{ iff } \Lambda(x,y,z,u)$$

is an equivalence relation on $V$.

(In particular, $\Lambda(x,y,z,u)$ implies $(x,y)\varepsilon H$, $(z,u)\varepsilon H$, $(x,z)\varepsilon V$, and $(y,u)\varepsilon V$.)

We depict the statement $\Lambda(x,y,z,u)$ by a diagram

(1.1)

We say that a double equivalence relation is a <u>pregroupoid</u>, or a <u>pregroupoid</u> <u>structure</u> <u>on</u> $X$, if for any $x,y,z$ with $(x,y)\varepsilon H$ and $(x,z)\varepsilon V$, there exists a unique $u$ such that $\Lambda(x,y,z,u)$. Then this $u$ may be denoted $yx^{-1}z$.

This pregroupoid notion is more general than the one of |7| or |8|, since there we assume that $H = X\times X$.

<u>Example 1.1.</u> Let $H$ be the set of arrows of a groupoid, with $S$ as set of objects. Let $A\subset S$, $B\subset S$ be two subsets. We construct a pregroupoid structure on $H$ by putting $\Lambda(x,y,z,u)$ whenever

$$d_0(x) = d_0(y) \quad \varepsilon \quad A$$

$$d_1(x) = d_1(z) \quad \varepsilon \quad B$$

and

$$u = y\circ x^{-1}\circ z$$

This example is a case of a slightly more structured pregroupoid notion: let $X$ be given, together with maps $\alpha: X \to A$, $\beta: X \to B$, for some objects $A$ and $B$. A <u>pregroupoid structure on $X$ with respect to</u> $A \xleftarrow{\alpha} X \xrightarrow{\beta} B$ is just a pregroupoid structure on $X$ such that $H$ is the kernel pair of $\alpha$ and $V$ is the kernel pair of $\beta$.

In this case, the pregroupoid structure $\Lambda$ on $X$ may be given in terms of the ternary operation $yx^{-1}z$, defined whenever $\beta(z) = \beta(x)$ and $\alpha(y) = \alpha(x)$. (This is the approach of |6| and |7|, where $A = 1$, and $yx^{-1}z$ is denoted $\lambda(x,y,z)$.)

## §2. Descent; and the groupoids associated to a regular pregroupoid.

We assume some familiarity with internal category theory in $\underline{E}$, as in |5| 2.1. When we say that $\mathbf{H} = (H \rightrightarrows A)$ is a category object in $\underline{E}$, we mean that $A$ is the 'set' of objects, $H$ is the 'set' of arrows, and the two displayed maps are $d_0$ and $d_1$ (domain and codomain formation), respectively. We call $\mathbf{H}$ a <u>groupoid</u> if there is an inversion map $(\;)^{-1}: H \to H$ with the expected properties.

An <u>internal diagram</u>, or a <u>left action for</u> $\mathbf{H} = (H \rightrightarrows A)$ consists of data $\varphi: F \to A$, together with a left action by $\mathbf{H}$, so $h \cdot f$ makes sense whenever $\varphi(f) = d_0(h)$, and then $\varphi(h \cdot f) = d_1(h)$; an associative and unitary law is assumed. The information of such internal diagram may be encoded by a certain category object $\mathbf{F} = (F_1 \rightrightarrows F_0)$ (with $F_0 = F$), together with a functor $\mathbf{F} \to \mathbf{H}$ (with certain properties: a discrete opfibration). We call $\mathbf{F}$ the <u>total category</u> and $\mathbf{F} \to \mathbf{H}$ the <u>discrete opfibration</u> of the internal diagram, or of the left action. Note that the total category for an action by a groupoid is itself a groupoid.

The category of internal diagrams for $\mathbf{H}$ is denoted $\underline{E}^{\mathbf{H}}$. There is a dual notion of <u>right</u> actions for $\mathbf{H}$, they form a category $\underline{E}^{\mathbf{H}^{op}}$.

We call a left action by a groupoid <u>free</u> if the associated total category $\mathbf{F}$ is an equivalence relation, meaning that $F_1 \to F_0 \times F_0$ is mono (and symmetric - this is automatic here). And we call the action <u>principal homogeneous</u> if the associated total category $\mathbf{F}$ is codiscrete, meaning that $F_1 \to F_0 \times F_0$ is iso.

An internal functor $q: \mathbf{G} \to \mathbf{H}$ gives rise to a functor $q^*: \underline{E}^{\mathbf{H}} \to E^{\mathbf{G}}$, essentially by pulling back along $q$. If $\mathbf{H} = (H \rightrightarrows B)$ is a category, and $q: C \to B$ is a map, we may form a category $\mathbf{G} = (G \rightrightarrows C)$ by pulling $H \to B \times B$ back along $q \times q: C \times C \to B \times B$. This is the "<u>full image</u>" of $\mathbf{H}$ along $q$; it comes with a functor back to $\mathbf{G}$. In particular, if $\mathbf{H} = B$ is the discrete category on $B$ (meaning $\mathbf{H} = (B \rightrightarrows B)$), the full image of

$q: C \to B$ is just the kernel pair $\mathbb{R}_q = (R \rightrightarrows C)$ of $q$; it comes with a functor to $B$.

It is easy to see that if $q$ is a stable regular epi, then the induced functor

(2.1) $$\underline{E}/B \simeq \underline{E}^B \xrightarrow{q^*} \underline{E}^{\mathbb{R}q}$$

is full and faithful. We say that $q$ is a <u>descent map</u> if $q$ is stable regular epi, and if the functor (2.1) is an equivalence of categories. An equivalent, less sophisticated version of the notion is given in |8|. It is anyway quite standard (perhaps rather under the name <u>effective</u> descent map). Yet another simple, but non-elementary, way of describing the notion is in terms of the full and faithful left exact embedding $i: \underline{E} \to \hat{\underline{E}}$, where $\hat{\underline{E}}$ is the topos of sheaves for the canonical topology on $\underline{E}$; it takes stable regular epis to epis, and a stable regular epi $q$ is a descent map if $i(q)$ has the property that pulling back along it reflects the property of being representable (= in the image of $i$).

Using this description, one may see that descent maps are stable under composition and pull-back, and that $q \circ p$ descent $\Rightarrow q$ descent.

By a <u>descent situation</u> in $\underline{E}$ we mean a diagram

$$\begin{array}{ccccc} \cdot & \rightrightarrows & F & \longrightarrow & \cdot \\ \downarrow & & \downarrow & & \downarrow \\ R & \rightrightarrows & X & \longrightarrow & B \end{array}$$

in which the right hand square is a pull back, each of the two rows exact (kernel pair/coequalizer), and the left hand square is the discrete opfibration associated to an action of the category $R \rightrightarrows X$ on $F \to X$.

We say that an equivalence relation $\mathbb{R} = (R \rightrightarrows C)$ is a <u>descent equivalence</u> (-<u>relation</u>) if it has a coequalizer $q$, of which it is a kernel pair, and which is a descent map. In a topos, any equivalence is a descent equivalence, and any epi is a descent map.

Let $X, \Lambda$ be a pregroupoid. We call it <u>regular</u> if the equivalence relations $H$ and $V$ on $X$, derived out of $\Lambda$ as in §1, are descent equivalences. Thus there are descent maps

$$\alpha : X \to A, \quad \beta : X \to B,$$

such that $H = X \times_A X$, $V = X \times_B X$. The equivalence relation $\sim_h$ on $H = X \times_A X$ now comes about as total category of a left action of $X \times_B X \rightrightarrows X$ on $p_0 : X \times_A X \to X$, namely the one given by

$$(x,z) \cdot (x,y) := (z, yx^{-1}z),$$

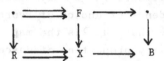

and since $X \to B$ was assumed a descent map, it follows that we have an object $X^* \to B$ in $\underline{E}/B$ participating in a descent situation

(2.2)
$$\begin{array}{ccccc} \sim_h & \rightrightarrows & X\times_A X & \xrightarrow{\beta'} & X^* \\ \downarrow & & \downarrow p_0 & & \downarrow d_0 \\ X\times_B X & \rightrightarrows & X & \xrightarrow{\beta} & B \end{array}$$

in particular

(2.3) $$X\times_B X^* \;=\; X\times_A X.$$

We shall denote $\beta'(x,y)$ by $yx^{-1}$. Note that

$$yx^{-1} \;=\; uz^{-1} \quad \text{iff} \quad u \;=\; yx^{-1}z\ .$$

We have a groupoid structure $X^* \rightrightarrows B$ on $X^*$ given by

$$d_0(yx^{-1}) \;:=\; \beta(x) \qquad d_1(yx^{-1}) \;:=\; \beta(y)$$

$$(tu^{-1})\circ(yx^{-1}) \;:=\; (tu^{-1}y)x^{-1}$$

(here we use that $\beta': X\times_A X \to X^*$ is stable regular epi, which means that we may <u>represent</u> elements in $X^*$ by elements $(x,y) \in X\times_A X$). The checking of independence of choice of representatives is easy from the definitions. We note that $d_0$ is the map $X^* \to B$ occurring in (2.2), and by the stability properties of descent maps, $d_0$ is a descent map.

The groupoid $X^*$ acts on the left on $\beta: X \to B$, the action being given by

$$(yx^{-1}).z \;:=\; yx^{-1}z.$$

For each $a \in A$, $X_a := \alpha^{-1}(a)$ is stable under the $X^*$-action, and the action of $X^*$ on $X_a$ is principal homogeneous: to any $x,y \in X_a$, there exists a unique $g \in X^*$ with $g\cdot x = y$, namely $yx^{-1}$.

By completely symmetric arguments, the equivalence $\sim_v$ on $V = X\times_B X$ comes about by a left action of $X\times_A X \to X$ on $p_0$: $X\times_B X \to X$, namely the one given by

$$(x,y)\cdot(x,z) \;:=\; (y, yx^{-1}z),$$

and since $X \to A$ was assumed a descent map, it follows that we have an object $X_* \to A$ in $\underline{E}/A$ participating in a descent situation

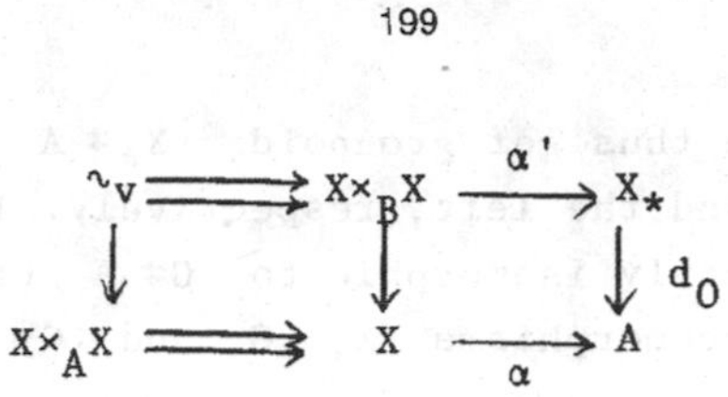

$$\begin{array}{ccccc} \sim_v & \rightrightarrows & X\times_B X & \xrightarrow{\alpha'} & X_* \\ \downarrow & & \downarrow & & \downarrow d_0 \\ X\times_A X & \rightrightarrows & X & \xrightarrow[\alpha]{} & A \end{array}$$

We denote $\alpha'(x,z)$ by $x^{-1}z$, and

$$x^{-1}z = y^{-1}u \quad \text{iff} \quad u = yx^{-1}z.$$

$X_*$ carries the structure of groupoid $X_* \rightrightarrows A$, namely

$$d_0(x^{-1}z) := \alpha(z) \qquad d_1(x^{-1}z) := \alpha(x).$$

$$(x^{-1}z)\circ(u^{-1}t) := x^{-1}(zu^{-1}t).$$

As for the $X^*$-case, $d_0 : X_* \to A$ is a descent map.

The groupoid $X_*$ acts on the right on $\alpha : X \to A$, the action being given by

$$y\cdot(x^{-1}z) := yx^{-1}z.$$

For each $b \in B$, $X_b := \beta^{-1}(b)$ is stable under the $X_*$-action, and the action of $X_*$ on $X_b$ is principal homogeneous.

Finally, the $X^*$- and $X_*$-actions commute with each other.

Note. Part of the information contained in a regular pregroupoid and its associated groupoids may be recorded in the following "butterfly" diagram, considered by Pradines |9| (for reasons related to ours):

$$\begin{array}{ccccc} X\times_A X & \searrow^{i'} & & {}^{i}\swarrow & X\times_B X \\ \beta' \downarrow & & \Lambda & & \downarrow \alpha' \\ X^* & \swarrow_{p} & & {}_{p'}\searrow & X_* \end{array}$$

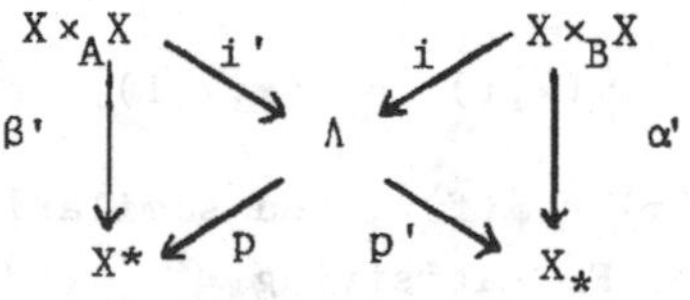

where $i(x,z) = (x,x,z,z)$, $p(x,y,z,u) = yx^{-1}$ etc., and where the diagonal sequences in a certain sense are exact sequences of groupoids (for a "diagonal" groupoid structure on $\Lambda$).

Example. Assume that the groupoid $G \rightrightarrows A$ acts in a free way on the right of some $\alpha : X \to A$. Then we can equip $X$ with structure of pregroupoid, namely by letting $\Lambda \subset X^4$ consist of the 'set' of all $(x,y,x\cdot g,y\cdot g)$ where $\alpha(x) = \alpha(y) = d_1(g)$. If $\alpha$ is a descent map, and the equivalence relation on $X$ given by the action of $G$ is a descent equivalence, with quotient $\beta : X \to B$, say, the pregroupoid is a pregroupoid with respect to $A \leftarrow X \to B$, and is regular. By the constructions above, we thus get groupoids $X_* \rightrightarrows A$ and $X^* \rightrightarrows B$ acting on $X$ from the right and the left, respectively. It is easy to see that $X_* \rightrightarrows A$ is canonically isomorphic to $G \rightrightarrows A$ (in a way compatible with the action), the isomorphisms $X_* \to G$ and $G \to X_*$ being given by

$$x^{-1}(x \cdot g) \mapsto g$$

and

$$g \mapsto x^{-1}(x \cdot g) \quad \text{(for some/any } x \in X_{d_1(g)}\text{)},$$

respectively.(Note that $X_{d_1(g)}$ is inhabited since $\alpha$ is stable regular epi.)

On the other hand, the groupoid $X^* \rightrightarrows B$ provides a new way of encoding some of the information of the $G$-action on $X$. For the case where $A = 1$, this is essentially Ehresmann's construction |2| of a groupoid $XX^{-1}$ out of a free action of a group on a space, i.e. out of a principal fibre bundle.

## §3. Generalized fibre bundles.

Assume that $A \overset{\alpha}{\leftarrow} X \overset{\beta}{\rightarrow} B$ is a regular pregroupoid. To make $\varepsilon: E \to B$ into a generalized fibre bundle for $X$ with fibre $\varphi: F \to A$ means, by definition, to give an invertible map

$$X \times_A F \xrightarrow{\sigma} X \times_B E, \tag{3.1}$$

commuting with the projections to $X$, and satisfying (3.2) below; to state (3.2), let us write $x(f)$ for the unique element in $E$ satisfying

$$\sigma(x,f) = (x, x(f))$$

(where $x \in X$, $f \in F$, and $\alpha(x) = \varphi(f)$), and similarly, we write $x^{-1}(e)$ for the unique element in $F$ satisfying

$$\sigma^{-1}(x,e) = (x, x^{-1}(e)),$$

(where $x \in X$, $e \in E$, and $\beta(x) = \varepsilon(e)$). We then have $\varepsilon(x(f)) = \beta(x)$ and $\varphi(x^{-1}(e)) = \alpha(x)$.

The condition we require on $\sigma$ is, in this notation, expressed

$$(yx^{-1}z)(f) = y(x^{-1}(z(f))) \tag{3.2}$$

whenever it makes sense, i.e. whenever

$$\varphi(f) = \alpha(z), \quad \beta(x) = \beta(z), \quad \alpha(x) = \alpha(y);$$

or, equivalently

$$(3.2') \qquad (yx^{-1}z)^{-1}(e) = z^{-1}(x(y^{-1}(e)))$$

whenever it makes sense, i.e. whenever

$$\varepsilon(e) = \beta(y),\ \beta(x) = \beta(z),\ \alpha(x) = \alpha(y).$$

Note that the data $\sigma$ of (3.1) may (in the category of sets, say) be thought of as providing for each $x \in X$ a bijection from the $\alpha(x)$-fibre of $F$ to the $\beta(x)$-fibre of $E$, in a way which is compatible with the pregroupoid structure, by (3.2). The pregroupoid of bijective maps from fibres of $F$ to fibres of $E$ is a special case of Example 1.1, by taking $H \rightrightarrows S$ there to be the groupoid of all (bijections between) sets.

A <u>generalized fibre bundle for</u> $X$ is a triple $\langle \varphi:F \to A, \varepsilon:E \to B, \sigma \rangle$, with $\sigma$ as in (3.1) and satisfying (3.2). We denote it short $\langle F,E,\sigma \rangle$. They form a category $Fib(X)$, a morphism $\langle F,E,\sigma\rangle \to \langle F',E',\sigma'\rangle$ being a pair of maps $F \to F'$, $E \to E'$, compatible with the $\varphi$'s, $\varepsilon$'s, and $\sigma$'s in the evident way. There are evident forgetful functors $Fib(X) \to \underline{E}/A$ and $Fib(X) \to \underline{E}/B$.

If $\langle F,E,\sigma \rangle$ is a fibre bundle fo $X$, there is a natural left action of the groupoid $X_* \rightrightarrows A$ on $F \to A$, and a natural left action of the groupoid $X^* \rightrightarrows B$ on $E \to B$, decribed by

$$(x^{-1}z)\cdot f := x^{-1}(z(f))$$

where $\varphi(f) = \alpha(z)$, $\beta(x) = \beta(z)$, and

$$(yx^{-1})\cdot e := y(x^{-1}(e))$$

where $\varepsilon(e) = \beta(x)$, $\alpha(x) = \alpha(y)$. These descriptions are well defined, due to (3.2) and (3.2'), respectively, and provide in fact functors

$$(3.3) \qquad Fib(X) \to \underline{E}^{X_*}, \quad Fib(X) \to \underline{E}^{X^*},$$

respectively.

When $A = 1$, one talks about <u>fibre bundles</u> instead of <u>generalized</u> fibre bundles. This is the situation considered in |8|, where we prove (Theorem 4.1) that the functors of (3.3) in this case are equivalences of categories. The proof does not depend on the assumption $A = 1$, so that we have

<u>Theorem 3.1.</u> The functors (3.3) are equivalences of categories.

In Remark 4.4 below, we shall indicate the inverse functors; their construction of course depends on constructing objects by descent.

## §4. A general descent construction.

Let $G = (G \rightrightarrows A)$ be a groupoid in $\underline{E}$, and $B$ an object of $\underline{E}$. A $\mathbf{G}$-valued cocycle on $B$ consists of a descent map $q: C \to B$ and a functor $\gamma = (\gamma_1, \gamma_0)$ from the kernel pair $\mathbf{R} = (R \rightrightarrows C)$ of $q$ to $G \rightrightarrows A$. (Sometimes we also say that $\gamma$ is a $\mathbf{G}$-valued cocycle on the covering $C$ of $B$). Such $\gamma: \mathbf{R} \to \mathbf{G}$ gives rise to a functor $\underline{E}^{\mathbf{G}} \to \underline{E}^{\mathbf{R}}$, and thus, since $q$ is a descent map, i.e. $\underline{E}^{\mathbf{R}} \simeq \underline{E}/B$, we get by composition a functor

$$\gamma^! : \underline{E}^{\mathbf{G}} \to \underline{E}/B.$$

Now $d_1: G \to A$ carries a canonical left action by $\mathbf{G}$, given by composition, so it is an object of $\underline{E}^{\mathbf{G}}$. With $\gamma$ a cocycle as above, we thus get an object $\gamma^!(d_1)$; we denote it also $\gamma^!(\mathbf{G})$. We have

Theorem 4.1. Assume that $d_1: G \to A$ (or equivalently $d_0: G \to A$), and $\gamma_0: C \to A$ are descent maps. Then $\gamma^!(\mathbf{G})$ carries canonically a structure of regular pregroupoid, with respect to maps to $A$ and $B$; and $(\gamma^!(\mathbf{G}))_* \cong \mathbf{G}$ canonically.

Proof/construction. By construction of $\gamma^!$, $\gamma^!(\mathbf{G})$ sits in a pull back diagram

$$\begin{array}{ccc} G \times_A C & \xrightarrow{q'} & \gamma^!(\mathbf{G}) \\ \downarrow & & \downarrow \beta \\ C & \xrightarrow[q]{} & B \end{array} \tag{4.1}$$

Since $q$ is descent, $q'$ is descent, and we define a map $\alpha: \gamma^! \mathbf{G} \to A$ by defining it on representatives from $G \times_A C$: $\alpha(g,c) := d_0(g)$.

Since the left hand vertical map in (4.1) comes about by pulling $d_1: G \to A$ back, and $d_1$ was assumed a descent map, it is itself a descent map, and from the stability properties of such, it follows that $\beta$ is descent. Also, $\alpha$ sits by construction in a commutative square

$$\begin{array}{ccc} G \times_A C & \xrightarrow{q'} & \gamma^!(\mathbf{G}) \\ \downarrow & & \downarrow \alpha \\ G & \xrightarrow[d_0]{} & A \end{array}$$

whose left hand vertical map comes about by pulling back $\gamma_0 : C \to A$, and since $d_0$ is descent, it follows from the stability properties of descent maps that $\alpha$ is descent. Now a (regular) pregroupoid structure can be defined on $A \leftarrow \gamma^! G \to B$, namely by the ternary operation defined (on representatives from $G \times_A C$) by

$$(4.2) \qquad (g_2,c_2)(g_1,c_1)^{-1}(g_3,c_3) := (g_2 \circ g_1^{-1} \circ \gamma_1(c_1,c_3)^{-1} \circ g_3, c_2),$$

(note that $(c_1,c_3) \in R \subset C \times C$, since $(g_1,c_1)$ and $(g_3,c_3)$ are supposed to represent elements in the same B-fibre; also $d_0(g_1) = d_0(g_2)$, since $(g_1,c_1)$ and $(g_2,c_2)$ are supposed to represent elements in the same A-fibre). We leave the further details to the reader. The isomorphism $(\gamma^!(G))_* \cong G$ is given by

$$(g_1,c_1)^{-1}(g_3,c_3) \mapsto g_1^{-1} \circ \gamma(c_1,c_3)^{-1} \circ g_3$$

in one direction, and by

$$g \mapsto (1_a,c)^{-1}(g,c)$$

in the other (for some/any $c$ with $\gamma(c) = d_1(g) = a$).

This theorem has as a special case the construction of a principal fibre bundle out of a system of coordinate transformations. The next theorem has similarly as a special case the construction of fibre bundles associated with it.

Let $G = (G \rightrightarrows A)$ and $\gamma$ be as above; then

Theorem 4.2. Assume that $\gamma_0 : C \to A$ and $d_1 : G \to A$ are descent maps. Then the functor $\gamma^! : \underline{E}^G \to \underline{E}/B$ lifts to a functor $\underline{E}^G \to \mathrm{Fib}(\gamma^!(G))$. This functor is an equivalence.

Proof/construction. Let $\varphi : F \to A$ be equipped with a left $G$-action. We construct a map

$$\gamma^! G \times_A F \xrightarrow{\sigma} \gamma^! G \times_B \gamma^! F$$

as follows:

$$\sigma((g,c),f) := ((g,c),(g \cdot f,c))$$

where $d_1(g) = c$, $\varphi(f) = d_0(g)$. It has an inverse, given by

$$\sigma^{-1}((g,c),(f',c')) := ((g,c), g^{-1} \circ \gamma(c',c) \cdot f').$$

We omit the details in checking the well-definedness of $\sigma$ and $\sigma^{-1}$, and that the $\sigma$ does indeed provide $\gamma^! F$ with structure of fibre bundle for $\gamma^! G$ with fibre $F$.

The fact that the functor is an equivalence follows from $\mathrm{Fib}(\gamma^! G) \cong \underline{E}^{(\gamma^!(G))_*}$ (Theorem 3.1), together with $(\gamma^!(G))_* \cong G$ (Theorem 4.1).

In the standard applications, the 'covering' $q: C \to B$ considered in the present § will typically be derived (with $\underline{E}$ the category of topological spaces, say) from an open covering $\{U_i \subset B \mid i \in I\}$, with $C = \coprod U_i$, the disjoint union of the $U_i$'s; then the $\gamma_1$ will be a family $\{\gamma_{ij} \mid (i,j) \in I \times I\}$ of transition maps.

However, coverings may be taken quite more general than that, (and is perhaps a novelty in our presentation), and this generality comes to work now:

Let $A \leftarrow X \to B$ be a regular pregroupoid; the we have a <u>canonical</u> $X_*$-valued cocycle on $B$, defined on the covering $\beta: X \to B$, (which is usually not of the form $\coprod U_i \to B$). This canonical cocycle is described as follows; $\gamma_0$ is just $\alpha: X \to A$; and $\gamma_1: X \times_B X \to X_*$ is given by

$$\gamma_1(x,z) = z^{-1}x.$$

Since $d_1: X_* \to A$ and $\gamma_0$ $(= \alpha)$ are descent maps, the construction in Theorem 4.1 applies, so that $\gamma^!(X_*)$ is a regular pregroupoid w.r.to $A$ and $B$. We have

<u>Proposition 4.3.</u> Let $A \leftarrow X \to B$ be a regular pregroupoid, and consider the canonical $X_*$-valued cocycle $\gamma$ on the covering $X \to B$. Then $\gamma^!(X_*) \cong X$, canonically.

<u>Proof/construction.</u> Let $(g,x)$ represent an element of $\gamma^!(X_*)$, so $g \in X_*$ and $d_1(g) = \alpha(x)$. Associate to it the element $x \cdot g \in X$. Conversely, to $x \in X$, associate the element in $\gamma^!(X_*)$ represented by $(1_{\alpha(x)}, x)$.

<u>Remark 4.4.</u> We may remark that the $\gamma^!$-construction, when applied to the canonical $X_*$-valued cocycle $\gamma$ for a regular pregroupoid $A \leftarrow X \to B$, yields the inverse for the equivalence described in Theorem 3.1. We have in fact functors

$$\underline{E}^{X_*} \xrightarrow{\ \gamma^!\ } \mathrm{Fib}(\gamma^!(X_*)) \to \mathrm{Fib}(X),$$

the first one by Theorem 4.2, and the second (isomorphism) by Proposition 4.3.

## §5. Foliations.

A groupoid $H \rightrightarrows B$ is usually called <u>transitive</u> if the map $(d_0,d_1)$: $H \to B\times B$ is epic in some strong sense, say a descent map. If $A \leftarrow X \to B$ is a regular pregroupoid in $\underline{E}$, the groupoids $X^* \rightrightarrows B$ and $X_* \rightrightarrows A$ need not be transitive: in the category of sets, for example, take $R$ to be an equivalence relation on an inhabited set $X$, let $A = X/R$, $B = X$, with obvious $\alpha$ and $\beta$, and define $\Lambda$ by

$$\Lambda(x,y,z,u) \quad \text{iff} \quad x = z, \quad y = u, \quad xRy.$$

Then $X_*$ is the discrete groupoid on $A$, whereas $X^* \rightrightarrows B$ is $R \rightrightarrows B$. So $X^*$ is transitive iff $R$ is the codiscrete equivalence relation on $X$, iff $A = 1$, iff $X_*$ is transitive.

In general, if $A \leftarrow X \to B$ is a regular pregroupoid, we have a commutative

$$\begin{array}{ccc} X\times_A X & \hookrightarrow & X\times X \\ \downarrow & & \downarrow{\scriptstyle \beta\times\beta} \\ X^* & \xrightarrow[(d_0,d_1)]{} & B\times B \end{array}$$

If $A = 1$, the top map is an isomorphism, and then the fact that $\beta\times\beta$ is descent implies that $(d_0,d_1)$ is descent, so that

<u>Proposition 5.1.</u> If $A \leftarrow X \to B$ is a regular pregroupoid, and $A = 1$, then $X^* \rightrightarrows B$ is transitive.

This is the situation which occurs for fibre bundles in contrast to the present <u>generalized</u> fibre bundles, which rather come up in foliation theory, as we shall now sketch (essentially following |4|).

A smooth foliation $\underline{F}$ of codimension $q$ on a manifold $B$ may be presented by giving an open cover $\{U_i \mid i\in I\}$ of $B$, and for each $i$, a smooth surjective submersion $f_i : U_i \to A_i \subset \mathbb{R}^q$, with connected fibres. The $A_i$'s may be assumed disjoint. The leaves of the foliation are the equivalence classes for the equivalence relation $\equiv_{\underline{F}}$ on $B$ generated by the relation $\sim$, where

$$b \sim b' \quad \text{iff} \quad (\exists i \quad \text{with} \quad b,b' \in U_i \quad \text{and} \quad f_i(b) = f_i(b')).$$

If $b \in U_i \cap U_j$, there is a unique germ $t = t_{b,i,j}$ of a diffeomorphism from $f_i(b)$ to $f_j(b)$, with $t \circ f_i$ having same germ at $b$ as $f_j$. Let $A = \bigcup A_i$, let $G \rightrightarrows A$ be the groupoid of germs generated algebraically by germs of form $t_{b,i,j}$.

The data of the $f_i$ and $t_{b,i,j}$ then provide a cocycle $\gamma$ on the covering $\coprod U_i \to B$ of $B$ with values in the groupoid $G \rightrightarrows A$. (Such a cocycle, or rather, an equivalence class of such, is a Haef-liger structure on $B$ with values in the groupoid $G \rightrightarrows A$.)

The corresponding groupoid $(\gamma^! G)^* \rightrightarrows B$ deserves the name holo-nomy groupoid of the foliation. An example of a generalized fibre bundle for the pregroupoid $A \leftarrow \gamma^! G \to B$ is the normal bundle of the foliation, which one gets by applying the $\gamma^!$-construction to the tan-gent bundle $\mathbb{R}^q \times A \to A$ of $A$.

The holonomy groupoid is a glorified version of the equivalence relation $\equiv_{\underline{F}}$, in the sense that two points in $B$ are $\equiv_{\underline{F}}$-related iff they can be connected by an arrow of the holonomy groupoid $(\gamma^! G)^*$. To prove this in a general context, we need that we out of a groupoid $H \rightrightarrows B$ can induce an equivalence relation $R$ on $B$, namely by taking the 'image' of $(d_0, d_1) : H \to B \times B$,

$$H \twoheadrightarrow R \rightarrowtail B \times B;$$

so let us assume for simplicity that $\underline{E}$ is the category of sets (or any topos).

We consider, as in §4, a groupoid $\mathbf{G} = (G \rightrightarrows A)$ with $d_0$ and $d_1$ surjective, and a $\mathbf{G}$-valued cocycle $\gamma$ defined on a covering $q: C \to B$; we assume as in §4 that $\gamma_0 : C \to A$ is surjective, so that the construc-tion in §4 provides us with a regular pregroupoid $\gamma^! \mathbf{G}$ with respect to $A$ and $B$, and with $(\gamma^! \mathbf{G})_* \cong \mathbf{G}$.

<u>Proposition 5.2.</u> The equivalence relation $R$ induced on $B$ by the groupoid $(\gamma^! \mathbf{G})^* \rightrightarrows B$ equals the relation given by: $b \sim b'$ iff

$\exists c, c' \in C$ with $q(c) = b$, $q(c') = b'$, and $\exists$ an arrow $\gamma(c) \to \gamma(c')$ in $\mathbf{G}$.

In particular, the relation thus described is an equivalence relation.

<u>Proof.</u> Let $b \sim b'$ in virtue of $g : \gamma(c) \to \gamma(c')$ in $\mathbf{G}$. Con-sider the elements $x$ and $y$ in $\gamma^! \mathbf{G}$ given by

$$x = (1_{\gamma(c)}, c) \qquad y = (g, c').$$

Then $\beta(x) = b$, $\beta(y) = b'$, and $\alpha(x) = \alpha(y) = \gamma(c)$. So $yx^{-1}$ makes sense and is an arrow in $(\gamma^! \mathbf{G})^*$ from $b$ to $b'$, so $bRb'$.

To prove the converse, we first prove

<u>Lemma 5.3.</u> Let $q(c) = b$, $q(c') = b'$. Any arrow $b \to b'$ in $(\gamma^! G)^*$ may be written in form $uz^{-1}$, for elements $u,z \in \gamma^! G$, with

(5.1) $$z = (1_{\gamma(c)}, c), \quad u = (k, c'),$$

where $k: \gamma(c) \to \gamma(c')$ is an arrow of **G**.

<u>Proof.</u> Let the given arrow be $yx^{-1}$; $x$ and $y$ may be represented (in fact uniquely) in form $x = (g,c)$, $y = (h,c')$, with $g$ and $h$ arrows of **G**

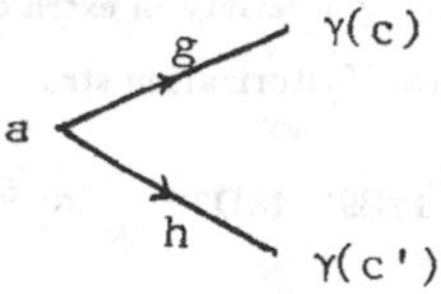


Let $k := h \circ g^{-1}$, and let $z$ and $u$ then be given by (5.1). To prove $yx^{-1} = uz^{-1}$ in $(\gamma^! G)^*$ means to prove $u = yx^{-1}z$ in $\gamma^! G$, but the recipe (4.2) for this ternary operation of $\gamma^! G$ yields $(h \circ g^{-1} \bullet \gamma(c,c)^{-1} \circ 1, c')$, which is $u$.

To finish the proof of the proposition, we may by the Lemma assume that $bRb'$ holds in virtue of an arrow $uz^{-1}$, with $u$ and $z$ as in (5.1). Then $k: \gamma(c) \to \gamma(c')$ witnesses $b \equiv b'$.

<u>Bibliography.</u>


1. J.Duskin, Free groupoids, trees, and free groups, Journ. Pure Appl. Alg. (to appear).
2. C.Ehresmann, Les connexions infinitésimales dans un espace fibré différentiable, Coll. de Topo., Bruxelles C.B.R.M. (1950), 29-55. (Oeuvres Vol I, Amiens 1984, 237-250).
3. C.Ehresmann, Catégories topologiques et catégories différentiables, Coll. Géom. Diff. Globale, Bruxelles C.B.R.M. (1959), 137-150. (Oeuvres Vol I, Amiens 1984, 237-250).
4. A.Haefliger, Gropoïdes d'holonomie et classifiant, Astérisque 116 (1984), 70-97.
5. P.T.Johnstone, Topos theory, Academic press 1977.
6. A.Kock, The algebraic theory of moving frames, Cahiers de Top. et Géom. Diff. 23 (1982), 347-362.
7. A.Kock, Combinatorial notions relating to principal fibre bundles, Journ. Pure Appl. Alg. 39 (1986), 141-151.
8. A.Kock, Fibre bundles in general categories, Aarhus Univ. Preprint Series 1986/87 No.27. To appear in Journ. Pure Appl. Alg.
9. J.Pradines, Au coeur de l'oeuvre de Charles Ehresmann et de la géometrie differentielle: Les groupoïdes différentiables. In C.Ehresmann, Oeuvres Vol I, Amiens 1984, 526-539.
10. B.L.Reinhart, Differential geometry of foliations, Springer Verlag 1983.

# CLOSURE OPERATORS WITH PRESCRIBED PROPERTIES

Jürgen Koslowski

**ABSTRACT:** The notion of closure operator on a category is explored, utilizing the approach of Dikranjan and Giuli. Conditions on the underlying factorization structure are given, which allow the construction of closure operators satisfying a variety of extra conditions.

**KEY WORDS:** closure operator, factorization structure, separated object, sheaf, closure commuting with pullbacks

**CLASSIFICATION:** 18A32, 18B99, 18D30

## 0 INTRODUCTION

The basic idea for a closure operator on a category $\mathcal{X}$ is to have for each object $X$ an extensive, isotone and idempotent operation on the partially ordered class of its subobjects. For these operations to be compatible with the structure of $\mathcal{X}$, one would like the $\mathcal{X}$-morphisms to be "continuous" in some sense with respect to them. If $\mathcal{X}$ has pullbacks (to be thought of as inverse images) of monos, this quite literally means that inverse images of closed subobjects are closed. It turns out that this notion of closure operator may be generalized in two ways. Often particular types of subobjects deserve special attention, e.g., the subspaces in topology. Hence one does not insist on just studying monos. Secondly, much of the theory can be developed without requiring the operations to be idempotent. If all partially ordered classes of subobjects under consideration are complete, any closure operator in this weak sense has an idempotent hull.

The first section refines the very elegant approach to closure operators by Dikranjan and Giuli, [DG], which no longer requires the base category to have pullbacks in order to formulate the continuity conditions mentioned earlier. We are able to clarify and strengthen some of their results. In the second part the problem of designing closure operators with certain prescribed separated objects or sheaves is addressed. The third part deals with the naturality of closure operators, i.e., with the question how to guarantee that pullbacks along certain morphisms commute with closure.

Of primary importance in the following are the notions of orthogonality and separatedness (cf. [K0] and [T0]). An $\mathcal{X}$-morphism $A \xrightarrow{a} A'$ is said to be *left orthogonal* to an $\mathcal{X}$-morphism $B \xrightarrow{b} B'$, written as $a \perp b$, if for every commutative square

$$\begin{array}{ccc} A & \xrightarrow{f} & B \\ {\scriptstyle a}\downarrow & & \downarrow{\scriptstyle b} \\ A' & \xrightarrow[f']{} & B' \end{array} \qquad (0\text{-}00)$$

there exists a unique diagonal $A' \xrightarrow{d} B$ making both induced triangles commute. This notion admits a straightforward generalization to sinks $a$ and sources $b$, and of particular importance are the cases where either the sink $a$ or the source $b$ is empty. For collections $\mathcal{A}$ of sinks and $\mathcal{B}$ of sources $\mathcal{A} \perp \mathcal{B}$ means that every pair $(a, b) \in \mathcal{A} \times \mathcal{B}$ satisfies $a \perp b$. The collection of all sources $b$ with $\mathcal{A} \perp \{b\}$ is denoted by

$\mathcal{A}^{\perp}$, and $\mathcal{B}_{\perp}$ is defined dually. The weaker notion of $a$ being *left-separated* from $b$, written as $a \perp\!\!\perp b$, is obtained by requiring the existence of at most one diagonal instead of a unique one. A relativized version of these notions is introduced in Definition 1.05.

# 1 BASIC PROPERTIES

Consider a subclass $\mathcal{M}$ of morphisms of a fixed category $\mathcal{X}$ as a full subcategory of $\mathcal{X}/\mathcal{X}$. We write $\partial_0$ and $\partial_1$ for the restrictions of the domain functor and the codomain functor to $\mathcal{M}$, respectively.

### 1.00 DEFINITION

A *closure operator* on $\mathcal{M}$ is a pair $(\delta, C)$ consisting of a functor $\mathcal{M} \xrightarrow{C} \mathcal{M}$ and a natural transformation $\mathcal{M} \xrightarrow{\delta} C$, both of which preserve codomains, i.e., $C\partial_1 = \partial_1$ and $\delta\partial_1 = \partial_1$.

Given such a closure operator $(\delta, C)$, each $\mathcal{X}$-morphism $m \in \mathcal{M}$ factors as $m\delta\partial_0 \cdot mC$.

### 1.01 DEFINITION

(0) $m \in \mathcal{M}$ is called $(\delta, C)$*-closed*, if $m\delta\partial_0$ is iso, and $(\delta, C)$*-dense*, if $mC$ is iso.

(1) $C$-**Fix** and $\delta\partial_0$-**Fix** denote the classes of $(\delta, C)$-closed and $(\delta, C)$-dense $\mathcal{M}$-objects, respectively. □

For simplicity we will assume that $m\delta\partial_0$ or $mC$ in fact are identities, if $m$ is $(\delta, C)$-closed or $(\delta, C)$-dense, respectively.

We will successively impose conditions on $\mathcal{M}$, to show precisely which properties are necessary for our results. The rather natural condition that $\mathcal{M}$ contains all isomorphisms will follow from **(C2)** below and is not needed earlier. $\mathcal{M}$ does not even have to be closed under composition. We start with

**(C0)** $\mathcal{M}$ consists of $\mathcal{X}$-monos.

Clearly now $\delta$ is pointwise epi (as well as mono), hence $(\delta, C)$ is a prereflection in the sense of Börger and Tholen (cf. [Bö], [T1]). In particular, $(\delta, C)$ is a well-pointed endofunctor, i.e., $\delta C = C\delta$, which implies that $C$-**Fix** is closed under those limits and retractions that exist in $\mathcal{M}$.

The option of choosing for $\mathcal{M}$ a proper subclass of $\mathcal{X}$-**Mono** is useful, as the next example shows:

### 1.02 EXAMPLE

If $\mathcal{X} =$ **Top**, the category of topological spaces and continuous functions, every $\mathcal{X}$-object automatically comes equipped with an idempotent closure operation on its subspaces, i.e., its regular or extremal or initial (with respect to the forgetful functor into **Set**) subobjects. Since pullbacks of closed subspaces are just the inverse images of the corresponding closed sets with the appropriate initial topology, this indeed defines a closure operator in the sense of Definition 1.00. □

To be able to discuss the denseness of $m\delta\partial_0$ in the domain of $mC$ one has to ensure that it belongs to $\mathcal{M}$. Thus one require

**(C1)** Whenever $p \cdot q$ and $q$ belong to $\mathcal{M}$, so does $p$.

Now one can interpret $\delta\partial_0$ as an endomorphisn $D$ of $\partial_0$ over $\mathcal{X}$ with $D\partial_1 = C\partial_0$. Moreover, $C$ induces a natural transformation $D \xrightarrow{\gamma} \mathcal{M}$ over $\mathcal{X}$, and the pair $(D, \gamma)$ is easily seen to be co-wellpointed. Hence $D$-**Fix** is closed under colimits and sections.

Since $D$ and $\delta$ determine each other uniquely, in the following we will just use $D$.

**1.03 DEFINITION**

$(D,C)$ is called ***proper***, if $C$ is idempotent, and ***strict***, if $D$ is idempotent.

In [DG] the term ***weakly hereditary*** is used instead of ***strict***; we suggest a slightly different but more natural use for ***weakly heraditary*** in Definition 3.01(2).

**1.04 LEMMA**

(0) $D$-**Fix** $\perp$ $C$-**Fix**.

(1) *If* $(D,C)$ *is proper, then* $D$-**Fix** $= C$-$\mathbf{Fix}_\perp \cap \mathcal{M}$.

(2) *If* $(D,C)$ *is strict, then* $D$-$\mathbf{Fix}^\perp \cap \mathcal{M} = C$-**Fix**.

(3) *If* $(D,C)$ *is strict and proper,* $D$-**Fix** *and* $C$-**Fix** *determine each other via* $\perp$.

**Proof:**

(0) For an $\mathcal{M}$-morphism $(f,g)$ from $e \in D$-**Fix** to $m \in C$-**Fix** the unique diagonal making the appropriate triangles commute is given by $d = (f,g)C\partial_0$.

(1) $e \in C$-$\mathbf{Fix}_\perp \cap \mathcal{M}$ is left orthogonal to $eC$, which implies $e \in D$-**Fix**.

(2) $m \in D$-$\mathbf{Fix}^\perp \cap \mathcal{M}$ is right orthogonal to $mD$, which implies $m \in C$-**Fix**.

(3) Just combine (2) and (3). □

In order to characterize $C$-**Fix** for a proper closure operator $(D,C)$, and $D$-**Fix** for a strict one, we relativize the notion of orthogonality (cf. [TO]). For simplicity we only formulate it for morphisms, the generalization to sinks and sources is straightforward.

**1.05 DEFINITION**

Two composable pairs $A \xrightarrow{a} A' \xrightarrow{a'} A''$ and $B \xrightarrow{b} B' \xrightarrow{b'} B''$ of $\mathcal{X}$-morphisms are called ***orthogonal***, if for every commutative square

$$\begin{array}{ccc} A & \xrightarrow{f} & B \\ {\scriptstyle a\cdot a'}\downarrow & & \downarrow{\scriptstyle b\cdot b'} \\ A'' & \xrightarrow[f'']{} & B'' \end{array} \qquad (1\text{-}00)$$

there exists a unique $A' \xrightarrow{f'} B'$ which makes the appropriate squares commute, and ***separated***, if there is at most one such $f'$. We use the symbols $\perp$ and $\perp\!\!\!\perp$ as before, and write $a \perp (b,b')$ for $(a, a\partial_1) \perp (b,b')$, and $(a,a') \perp b'$ for $(a,a') \perp (b'\partial_0, b')$.

**1.06 DEFINITION**

Let $(D,C)$ be a closure operator.

(0) A composable pair of $\mathcal{M}$-objects $X \xrightarrow{m} Y \xrightarrow{m'} Z$ with $m \cdot m' \in \mathcal{M}$ is called ***relatively*** $(D,C)$***-dense***, if $(m \cdot m')C = m'C$, and ***relatively*** $(D,C)$***-closed***, if $mD = (m \cdot m')D$.

(1) We write $C$-**Rel** and $D$-**Rel** for the collections of composable pairs which are relatively $(D,C)$-dense and relatively $(D,C)$-closed, respectively.

**1.07 PROPOSITION**

(0) $C$-**Rel** $\perp$ $D$-**Rel**.

(1) *If* $(D,C)$ *is proper, then* $C\text{-}\mathbf{Rel}^{\perp} \cap \mathcal{M} = C\text{-}\mathbf{Fix}$.

(2) *If* $(D,C)$ *is strict, then* $D\text{-}\mathbf{Fix} = D\text{-}\mathbf{Rel}_{\perp} \cap \mathcal{M}$.

**Proof:**

(0) For $(e,e') \in C\text{-}\mathbf{Rel}$, $(m,m') \in D\text{-}\mathbf{Rel}$ and an $\mathcal{M}$-morphism $(f,f'')$ from $e \cdot e'$ to $m \cdot m'$ set $f' := e'D \cdot (f,f'')C\partial_0 \cdot mC$.

(1) For $m \in C\text{-}\mathbf{Fix}$ the pair $(mD, mC) = (m\partial_0, m)$ is relatively $(D,C)$-closed, and (0) shows $m \in C\text{-}\mathbf{Rel}^{\perp}$. Conversely, if $(D,C)$ is proper, $(mD, mC)$ is relatively $(D,C)$-dense for each $m \in \mathcal{M}$. Hence $m \in C\text{-}\mathbf{Rel}^{\perp}$ implies $(mD, mC) \perp m$, i.e., $mC \cong m$.

(2) Similar. □

Notice that part (1) does not require condition (C1) to be satisfied. Moreover, if $(D,C)$ is strict or proper, the collection $\{(mD, mC) : m \in \mathcal{M}\}$ is contained in $C$-**Rel** or $D$-**Rel**, respectively, and already suffices to characterize the appropriate fixed points.

Our results so far indicate that the notions of $(D,C)$-denseness and $(D,C)$-closedness are in some sense dual to each other, which will become even more apparent in the remainder of this section. However, while $C$-**Fix** is closed under intersections since they are just special limits, the dual construction for $D$-**Fix** cannot directly be expressed in terms of colimits.

$\sqsubseteq$ denotes the standard pre-order in the $\partial_1$-fibres $\partial_1/X$, $X \in \mathcal{X}\text{-}\mathbf{Ob}$, while $\bigsqcup$ and $\bigsqcap$ stand for suprema (=unions) and infima (=intersections) in these fibres, respectively. The collection of all closure operators on $\mathcal{M}$ is pre-ordered pointwise by $\sqsubseteq$.

**1.08 LEMMA**

*Given* $m \in \mathcal{M}$ *and* $\mathcal{N} \subseteq \{p \in \mathcal{M} : m = n \cdot p \text{ and } n = nD\}$, *if* $\bigsqcup \mathcal{N}$ *exists, the unique* $q$ *with* $m = q \cdot (\bigsqcup \mathcal{N})$ *is* $C$-*dense.*

**Proof:**

Just notice that the $\bigsqcup$-sink for $\bigsqcup \mathcal{N}$ factors through $qC$. □

We will paraphrase this fact somewhat imprecisely but suggestively (we hope) as $D$-**Fix** *is closed under* $\mathcal{M}$-*unions.*

The following result provides a nice characterization of proper and strict closure operators. Notice we do not require the completeness of the $\partial_1$-fibers, where the infima and suprema are formed.

**1.09 PROPOSITION**

(0) $(D,C)$ *is proper iff every* $m \in \mathcal{M}$ *satisfies* $mC = \bigsqcap\{p \in \mathcal{M} : m = n \cdot p \text{ and } p = pC\}$.

(1) $(D,C)$ *is strict iff every* $m \in \mathcal{M}$ *satisfies* $mC = \bigsqcup\{p \in \mathcal{M} : m = n \cdot p \text{ and } n = nD\}$.

**Proof:**

(0) The functoriality of $C$ implies that $mC$ is bounded above by the infimum. Since $C$-**Fix** is closed under intersections, $mC = mCC$ iff the infimum is bounded above by $mC$.

(1) Similar; use Lemma 1.08. □

Properness and strictness are preserved under iteration of $D$ and $C$, respectively.

**1.10 LEMMA**

(0) *If $(D,C)$ is proper, so is $(D^2,C^*)$, with $mC^* = mDC \cdot mC$.*

(1) *If $(D,C)$ is strict, so is $(D^*,C^2)$, with $mD^* = mD \cdot mCD$.*

**Proof:**

(0) Let $C$ be idempotent. In order to show that $C^*$ is idempotent, it suffices to show that $mC^*D^2$ for an $\mathcal{M}$-object $X \xrightarrow{m} Y$ is iso, since

$$mC^* = mC^*D \cdot mC^*C = mC^*D^2 \cdot mC^*DC \cdot mC^*C = mC^*D^2 \cdot mC^*C^* .$$

First set $Z := mC\partial_0$ and consider the $D$-image $mC^*D \xrightarrow{(mDC,y)} Z$ of the $\mathcal{M}$-morphism $mC^* \xrightarrow{(mDC,Y)} mC$. Now set $W := mDC\partial_0$ and consider the $D$-image $mC^*D^2 \xrightarrow{(Z,k)} W$ of $mC^*D \xrightarrow{(Z,y)} mDC$. Since the $\mathcal{X}$-mono $k$ is a retraction and hence iso, $mC^*D^2$ is iso as well.

(1) Similar. □

**1.11 LEMMA**

(0) *Infima of proper closure operators are proper.*

(1) *Suprema of strict closure operators are strict.*

**Proof:**

(0) Consider a family $\mathcal{P}$ of proper closure operators with infimum $(D,C)$. For every $X \xrightarrow{m} Y$ in $\mathcal{M}$ and each $(D',C') \in \mathcal{P}$, clearly $mC'$ is $C$-closed. Thus $(D^*,C^2)$ with $mD^* = mD \cdot mCD$ is a lower bound for $\mathcal{P}$. But now $(D,C) \sqsubseteq (D^*,C^2)$ implies that $C$ and $C^2$ are isomorphic, i.e., $(D,C)$ is proper.

(1) Dually. □

In order to be able to extend a closure operator to a smallest proper one, its ***proper hull***, or to shrink it to a largest strict one, its ***strict core***, one wants the pre-ordered collection of closure operators to be complete, i.e., suprema and infima of all subcollections to exist. This is guaranteed by requiring

**(C2)** All (even class-indexed) multiple pullbacks of $\mathcal{M}$-objects exist in $\mathcal{X}$ and belong to $\mathcal{M}$.

Notice that under this condition infima and suprema of closure operators are formed pointwise, i.e., in the $\partial_1$-fibers. Moreover, since empty $\mathcal{X}$-sinks can be identified with $\mathcal{X}$-objects, $\mathcal{M}$ contains all $\mathcal{X}$-isos.

While the existence of proper hulls and strict cores is guaranteed by Lemma 1.11, for an explicit description one furthermore needs

**(C3)** Pullbacks of $\mathcal{M}$-objects exist in $\mathcal{X}$ and belong to $\mathcal{M}$.

In other words: $\partial_1$ is a fibration. Conditions **(C2)** and **(C3)** together say that $\mathcal{X}$ is $\mathcal{M}$-*complete*. They imply conditions **(C0)** and **(C1)**. Moreover, $\mathcal{M}$ is part of a factorization system for sinks; we write $\mathcal{E}$ for the collection of sinks left-orthogonal to $\mathcal{M}$, and $\mathcal{E}_0$ for $\mathcal{E} \cap \mathcal{X}$-**Mor**. In fact, $\partial_1$ is a topological functor and hence a bifibration. The left adjoint $f\exists$ to the pullback functor $\partial_1/Y \xrightarrow{f^*} \partial_1/X$ for $X \xrightarrow{f} Y$ is given by composing with $f$ and then taking the $\mathcal{M}$-component of the $(\mathcal{E},\mathcal{M})$-factorization.

An $\mathcal{M}$-morphism $m \xrightarrow{(f,g)} m'$ is cartesian or $\partial_1$-initial iff $m$ is a pullback of $m'$ along $g$, and it is co-cartesian or $\partial_1$-final iff $f \in \mathcal{E}_0$. In particular, $\partial_1$ satisfies the Beck-Chevalley condition (cf. [BR]) iff $\mathcal{E}_0$ is pullback-stable.

The special value of strict and proper closure operators on $\mathcal{M}$ lies in the fact that they bijectively correspond to factorization structures $(\mathcal{E}', \mathcal{M}')$ with $\mathcal{M}' \subseteq \mathcal{M}$.

The following result improves upon [DG] insofar as $\mathcal{X}$ need not be $\mathcal{M}$-well-powered.

### 1.12 THEOREM

(0) *For $m \in \mathcal{M}$ define $mC^\circ = \bigsqcap\{p \in \mathcal{M} : m = n \cdot p \text{ and } p = pC\}$, and let $mD^\circ$ be the unique $\mathcal{X}$-morphism with $mD^\circ \cdot mC^\circ = m$. Then $(D^\circ, C^\circ)$ is the proper hull of $(D,C)$.*

(1) *For $m \in \mathcal{M}$ define $mC_\circ = \bigsqcup\{p \in \mathcal{M} : m = n \cdot p \text{ and } n = nD\}$, and let $mD_\circ$ be the unique $\mathcal{X}$-morphism with $mD_\circ \cdot mC_\circ = m$. Then $(D_\circ, C_\circ)$ is the strict core of $(D,C)$.*

**Proof:**

(0) (C3) and the fact that $C$-**Fix** is closed under pullbacks imply that $(D^\circ, C^\circ)$ is a closure operator. Since by construction $C^\circ$-**Fix** $= C$-**Fix**, Proposition 1.04(0) shows that $C^\circ$ is idempotent. Now consider a proper closure operator $(D', C')$ with $C \sqsubseteq C'$. Clearly, $C'$-**Fix** $\subseteq C$-**Fix**, and thus by Proposition 1.09(0) one has $C^\circ \sqsubseteq C'$.

(1) In order to show that $(D_\circ, C_\circ)$ is a closure operator, consider an $\mathcal{M}$-morphism $m \xrightarrow{(f,g)} m'$. Let $((q,r),s)$ be the $(\mathcal{E},\mathcal{M})$-factorization of the sink $(mC_\circ \cdot g, m')$. (C1) implies $r \in \mathcal{M}$, hence the $\mathcal{E}$-sink $(q,r)$ factors through $rC \in \mathcal{M}$, i.e., $rC$ is iso and $r = rD$. This implies $s \sqsubseteq m'C_\circ$. The construction of $D_\circ$ together with Lemma 1.08 shows that $D_\circ$-**Fix** $= D$-**Fix**, hence $D_\circ$ is idempotent by Proposition 1.09(1). The rest is proved like in (0). □

In view of Lemma 1.04, $C^\circ$ is obtained from $C$ by taking the largest possible collection of fixedpoints for $C^\circ$, namely $C$-**Fix**, and defining $D^\circ$-**Fix** as $C$-**Fix**$_\perp \cap \mathcal{M}$. Dually, one gets $D_\circ$ by setting $D_\circ$-**Fix** $=$ $D$-**Fix** and $C_\circ$-**Fix** $= D$-**Fix**$^\perp \cap \mathcal{M}$.

### 1.13 PROPOSITION

*Let $\mathcal{M}$ be closed under composition.*

(0) *If $(D,C)$ is strict, so is $(D^\circ, C^\circ)$.*

(1) *If $(D,C)$ is proper, so is $(D_\circ, C_\circ)$.*

**Proof:**

(0) If $(D,C)$ is stirct, by Lemma 1.04(2) one has $C$-**Fix** $= D$-**Fix**$^\perp \cap \mathcal{M}$, which in particular implies that $C^\circ$-**Fix** $= C$-**Fix** is closed under composition. If $mD^\circ$ for $m \in \mathcal{M}$ factors as $mD^\circ = n \cdot p$ with $p \in C^\circ$-**Fix**, the composition $p \cdot mC^\circ$ belongs to $C^\circ$-**Fix**. Hence $p$ must be iso.

(1) Dually. □

### 1.14 LEMMA

(0) *If $(D,C)$ is strict and $D$-**Fix** is closed under composition, then $(D,C)$ is proper.*

(1) *If $(D,C)$ is proper and $C$-**Fix** is closed under composition, then $(D,C)$ is strict.*

**Proof:**

(0) Each $m \in \mathcal{M}$ factors as $m = mC^2 \cdot mCD \cdot mD$. Since both $mD$ and $mCD$ are $C$-dense, so is their composition, which implies that $mD \cdot mCD$ factors through $mD$. Hence $mCD$ must be iso.

(1) Dually. □

# 2 CONDITIONS ON CLOSURE OPERATORS

We continue with the hypotheses **(C2)** and **(C3)** of Section 1.

Certain types of closure operators have been used widely to study and characterize the epimorphisms in interesting subcategories of many familiar categories, of a topological as well as of an algebraic nature. (Throughout we assume subcategories to be full and isomorphism closed.) For a subcategory $\mathcal{Z}$ of $\mathcal{X}$ all of whose objects are right separated from $\mathcal{E}_0$, one wants to construct a closure operator $(D,C)$ such that all fixed points of $D$ are left separated from all $\mathcal{Z}$-objects. A $\mathcal{Z}$-morphism then is $\mathcal{Z}$-epi iff it is dense in the sense that its $\mathcal{M}$-component (in $\mathcal{X}$) belongs to $D$-**Fix**.

## 2.00 DEFINITION

Given a closure operator $(D,C)$, an $\mathcal{X}$-source $b$ is called $(D,C)$-*separated,* if $D$-**Fix** $\perp\!\!\perp b$, and a $(D,C)$-*sheaf,* if $D$-**Fix** $\perp b$.

The following result is well-known, cf. [DG, Section 5] and [C0, Theorem 1.12].

## 2.01 PROPOSITION

*Suppose $\mathcal{X}$ has equalizers, and $\mathcal{M}$ contains all regular monos. For a subcategory $\mathcal{Z}$ of $\mathcal{X}$ one gets a proper closure operator $(D^{\mathcal{Z}},C^{\mathcal{Z}})$ by mapping $X \xrightarrow{m} Y$ in $\mathcal{M}$ to the intersection of all equalizers of pairs $Y \xrightarrow{f,g} Z$ with $Z \in \mathcal{Z}$-**Ob** and $m \cdot f = m \cdot g$. Moreover, all $\mathcal{Z}$-objects are $(D^{\mathcal{Z}},C^{\mathcal{Z}})$-separated, and a $\mathcal{Z}$-morphism is epi iff its $\mathcal{M}$-component belongs to $D^{\mathcal{Z}}$-**Fix**.* □

The hypotheses of this proposition in fact are equivalent to $\mathcal{E}$ consisting of epi-sinks only. Thus no restrictions on $\mathcal{Z}$ are necessary. Dikranjan and Giuli call closure operators of this type *regular.*

It is also shown in [C0, Chapter 4] that $(D^{\mathcal{Z}},C^{\mathcal{Z}})$ is the largest proper closure operator which characterizes $\mathcal{Z}$-**Epi** in a strong sense (cf. Definition 2.04), hence among these it has the smallest collection of closed $\mathcal{M}$-objects. However, in general $(D^{\mathcal{Z}},C^{\mathcal{Z}})$ is not strict. Its strict core $(D^{\mathcal{Z}}_{\diamond},C^{\mathcal{Z}}_{\diamond})$ has the same dense $\mathcal{M}$-objects, and therefore is the largest proper and strict closure operator which characterizes $\mathcal{Z}$-**Epi**. In fact, it is the only such.

The question arises, whether a more direct description of $(D^{\mathcal{Z}}_{\diamond},C^{\mathcal{Z}}_{\diamond})$ is possible. The following result generalizes well-known characterizations of separated objects via closed diagonals in topology and sheaf theory to strict closure operators.

## 2.02 PROPOSITION

*Let $(D,C)$ be a closure operator. If $\mathcal{X}$ has squares, each of the following conditions on an $\mathcal{X}$-object $X$ implies the next one:*

(a) *The diagonal $X \xrightarrow{X\Delta} X \times X$ is $(D,C)$-closed.*

(b) *$X$ is $(D,C)$-separated.*

(c) *$X \xrightarrow{X\Delta} X \times X$ is a $(D,C)$-sheaf.*

*If $(D,C)$ is strict, and $X\Delta$ belongs to $\mathcal{M}$, all three conditions are equivalent.*

**Proof:**

(a) $\Rightarrow$ (b). Let $X\Delta$ be $C$-closed. Consider a parallel pair $f,g$ of $\mathcal{X}$-morphisms into $X$, and a $C$-dense $\mathcal{X}$-morphism $e$ with $e\cdot f = e\cdot g$. Since the pullback $m$ of the diagonal $X\Delta$ along $\langle f,g\rangle$ is an equalizer of $f$ and $g$, one has $e \sqsubseteq m$. But $m$ as a pullback of $X\Delta$ is $C$-closed as well, and hence must be iso, i.e., $f=g$. Thus $X$ is $(D,C)$-separated.

(b) $\Rightarrow$ (c). Consider a diagram

$$\begin{array}{ccc} Y & \xrightarrow{f} & X \\ {\scriptstyle e}\big\downarrow & & \big\downarrow{\scriptstyle X\Delta} \\ Z & \xrightarrow[g]{} & X\times X \end{array} \tag{2-00}$$

with $e \in D\text{-}\mathbf{Fix}$. If $p$ and $q$ are the projections from $X\times X$ into $X$, the compositions $g\cdot p$ and $g\cdot q$ both can serve as diagonals in (2-00), hence the separatedness of $X$ implies $g\cdot p = g\cdot q$.

(c) $\Rightarrow$ (a). If $(D,C)$ is strict and $X\Delta \in \mathcal{M}$, Lemma 1.04(2) shows that $X\Delta$ is $(D,C)$-closed. □

Now we use the fact that the polarity induced by $\perp$ on $\mathcal{M}$ allows us to construct strict and proper closure operators with prescribed closed $\mathcal{M}$-objects. Notice that the hypotheses of the next result together with (C3) imply that $\mathcal{X}$ has equalizers and that $\mathcal{M}$ contains all regular monos.

## 2.03 PROPOSITION

*If $\mathcal{X}$ has squares, and all diagonals $X \xrightarrow{X\Delta} X\times X$ belong to $\mathcal{M}$, then the closure operator $(D,C)$ generated by setting*

$$D\text{-}\mathbf{Fix} = \{\, Z\Delta : Z \in Z\text{-}\mathbf{Ob}\,\}_{\perp} \cap \mathcal{M} \qquad \text{and} \qquad C\text{-}\mathbf{Fix} = D\text{-}\mathbf{Fix}^{\perp} \cap \mathcal{M}$$

*is equal to $(D^{Z}_{\diamond}, C^{Z}_{\diamond})$.* □

While Proposition 2.01 easily generalizes to arbitrary $\mathcal{X}$-sources instead of $\mathcal{X}$-objects, only the implication (b) $\Rightarrow$ (c) of Proposition 2.02 admits a straightforward modification to this setting. We do not know whether Proposition 2.03 has an analogue for non-empty sources.

One may ask to what extent the constructions of these proper closure operators depend on the conditions on equalizers and regular monos. Fix a collection **Cond** of $\mathcal{X}$-sources, which we think of as conditions our closure operators should satisfy, either by making them separated, or by making them sheaves. Non-strict closure operators require a stronger satisfaction relation. To handle the sheaf-case, we also need to introduce a notion for sinks which is stronger than being an epi-sink, but weaker than being a coproduct.

## 2.04 DEFINITION

Let $@ \in \{\perp\!\!\!\perp, \perp\}$. We say that $(D,C)$ *$@$-satisfies* **Cond**, if $D\text{-}\mathbf{Fix} \mathbin{@} \mathbf{Cond}$, and that it *strongly $@$-satisfies* **Cond**, if $C\text{-}\mathbf{Rel} \mathbin{@} \mathbf{Cond}$.

Notice that for strict and proper closure operators the notions of satisfaction and of strong satisfaction coincide, since $C\text{-}\mathbf{Rel}$ can be replaced with the class $\{\,(mD, mC) : m \in \mathcal{M}\,\}$.

**2.05 DEFINITION**

Let $I \xrightarrow{F} \mathcal{X}$ be discrete and $(a, A)$ be an $F$-sink.

(0) The *kernel* of $a$ is the diagram consisting of all spans $IF \xleftarrow{u} X \xrightarrow{v} JF$, $I, J \in \mathcal{I}$-**Ob** (cf. [S]).

(1) $(a, A)$ is called *effective*, if $a$ is the colimit of its kernel.

**2.06 THEOREM**

(0) *If $\mathcal{E}$-sinks with members in $\mathcal{M}$ are epi-sinks,* $\mathbf{Cond}_{⫫} \cap \mathcal{M}$ *is the class of dense $\mathcal{M}$-objects for a strict and proper closure operator $(D, C)$, which is the largest such to* $⫫$*-satisfy* **Cond**.

(1) *If $\mathcal{E}$-sinks with members in $\mathcal{M}$ are effective,* $\mathbf{Cond}_{\perp} \cap \mathcal{M}$ *is the class of dense $\mathcal{M}$-objects for a strict and proper closure operator $(D, C)$, which is the largest such to $\perp$-satisfy* **Cond**.

**Proof:**

(0) For $m \in \mathcal{M}$ set $mC := \bigsqcup\{p \in \mathcal{M} : m = n \cdot p \text{ and } n ⫫ \mathbf{Cond}\}$. Since $\mathcal{M}$-unions (cf. Lemma 1.08) as $\mathcal{E}$-sinks are epi-sinks, $\mathbf{Cond}_{⫫} \cap \mathcal{M}$ turns out to be closed under $\mathcal{M}$-unions. Hence $mD$, defined as the unique $\mathcal{M}$-object with $m = mD \cdot mC$, belongs to $\mathbf{Cond}_{⫫}$. Clearly thus $D$ is an idempotent function on $\mathcal{M}$. On the other hand, for each $m \in \mathbf{Cond}_{⫫} \cap \mathcal{M}$ one has $mD = m$.

To show that $(D, C)$ is a closure operator, consider an $\mathcal{M}$-morphism $m \xrightarrow{f,g} m'$, and the $(\mathcal{E}, \mathcal{M})$-factorization $((q, r), s)$ of the sink $(mC \cdot g, m')$. By hypothesis $(q, r)$ is an epi-sink, so $mD ⫫ \mathbf{Cond}$ implies $r ⫫ \mathbf{Cond}$. This shows $r \sqsubseteq m'C$, hence $C$ is functorial.

Clearly, $D$-**Fix** is closed under composition, so $(D, C)$ is proper by Lemma 1.15.

For any closure operator $(D', C')$ which $⫫$-satisfies **Cond** one has $D'$-**Fix** $\subseteq D$-**Fix**. If $(D', C')$ is strict, by Lemma 1.04(2) this implies $C$-**Fix** $\subseteq C'$-**Fix**, and provided that $(D', C')$ is also proper, Proposition 1.09(0) shows $(D', C') \sqsubseteq (D, C)$.

(1) Similar. □

**2.07 EXAMPLE**

Consider a 5-element poset $\mathcal{X} = (\{a, b, c, d, e\}, \leq)$ with bottom $e$ and maximal elements $a$ and $b$, which are the only upper bounds of $\{c, d\}$. Set $\mathcal{M} = \mathcal{X}$-**Mor**. Clearly $\mathcal{X}$ has arbitrary intersections, but not all unions are effective. If **Cond** consists only of the empty source with domain $a$, the morphisms left orthogonal to $a$ do not constitute the set of dense morphisms for a closure operator. In fact, any closure operator for which $e$ is dense in $d$ must have $c$ dense in $b$, since $c \leq b$ is not right orthogonal to $e \leq d$ and hence cannot be closed. □

**2.08 EXAMPLE** (cf. [KO])

For a category $\mathcal{A}$ consider the Yoneda embedding $\mathcal{A} \xrightarrow{\mathcal{A}_Y} \mathcal{A}_Y = [\mathcal{A}, \mathbf{Set}^{\mathrm{op}}]^{\mathrm{op}} \cong [\mathcal{A}^{\mathrm{op}}, \mathbf{Set}]$. Set $\mathcal{X} = \mathcal{A}_Y$, and $\mathcal{M} = \mathcal{A}_Y$-**Mono**. Identifying $\mathcal{A}$ with its Yoneda image in $\mathcal{X}$, define a proper closure operator $(D^{\mathcal{A}}, C^{\mathcal{A}})$ (cf. Proposition 2.01).

The epi-reflective hull $\mathcal{A}_E$ of $\mathcal{A}$ in $\mathcal{X}$ can be constructed by mapping an $\mathcal{X}$-object $K$ to the codomain $K^e$ of the first factor in the (epi,monosource)-factorization of the source $K/\mathcal{A}_Y$-**Ob**. Now suppose that $\mathcal{A}$ is *normal*, i.e., for every $\mathcal{X}$-object $K$ the canonical functor from $K/\mathcal{A}_Y$ into $\mathcal{X}$ has a limit; denote the object-part by $K^\times$. Clearly the induced $\mathcal{X}$-morphism $K^e \xrightarrow{Km} K^\times$ is mono, so one can apply the closure operator. Denote the domain of $KmC^{\mathcal{A}}$ by $K^d$. It turns out that $K^d$ is the reflection of $K$ in the reflective hull of $\mathcal{A}$ in $\mathcal{X}$.

Let $(D,C)$ be the strict and proper closure operator constructed according to Theorem 2.06(1), where **Cond** consists just of the $\mathcal{A}$-objects. In particular, for every $\mathcal{X}$-object $K$ the $C^{\mathcal{A}}$-dense $\mathcal{X}$-mono $KmC^{\mathcal{A}}$ must be $C$-dense as well. Hence $(D,C)$ can be used to construct the same reflection as above. In fact, $(D,C)$ is the strict core of $(D^{\mathcal{A}},C^{\mathcal{A}})$.

Clearly this construction also works for other categories $\mathcal{X}$, as long as the subcategory $\mathcal{A}$ satisfies an appropriate smallness condition with respect to $\mathcal{X}$. □

We now turn to the construction of largest proper closure operators which strongly satisfy **Cond**.

## 2.09 THEOREM

(0) *If $\mathcal{E}$-sinks with members in $\mathcal{M}$ are epi-sinks,* $\mathbf{Cond}_{\perp\!\!\!\perp} \cap \mathcal{M}$ *is the class of dense $\mathcal{M}$-objects for a proper closure operator $(D,C)$, which is the largest such to strongly $\perp\!\!\!\perp$-satisfy* **Cond**.

(1) *If $\mathcal{E}$-sink with members in $\mathcal{M}$ are effective,* $\mathbf{Cond}_{\perp} \cap \mathcal{M}$ *is the class of dense $\mathcal{M}$-objects for a proper closure operator $(D,C)$, which is the largest such to strongly $\perp$-satisfy* **Cond**.

**Proof:**

(0) For $m \in \mathcal{M}$ set $mC = \bigsqcup\{p \in \mathcal{M} : m = n\cdot p \text{ and } (n,p) \perp\!\!\!\perp \mathbf{Cond}\}$. Define $mD$ to be the unique $\mathcal{M}$-object with $m = mD \cdot mC$. Since $\mathcal{M}$-unions as $\mathcal{E}$-sinks are epi-sinks, one has $(mD, mC) \perp\!\!\!\perp \mathbf{Cond}$. In particular, every fixed point of $D$ belongs to $\mathbf{Cond}_{\perp\!\!\!\perp}$. Conversely, each $\mathcal{M}$-object in $\mathbf{Cond}_{\perp\!\!\!\perp}$ clearly is a fixed point of $D$. Moreover, $C$ by construction is easily seen to be an idempotent function on $\mathcal{M}$.

An argument similar to the one used in the proof of Theorem 2.06 shows that $C$ is indeed a functor.

It is clear by construction that the relatively $(D,C)$-dense pairs of composable $\mathcal{M}$-objects are precisely the ones left-separated from **Cond**. So $(D,C)$ strongly $\perp\!\!\!\perp$-satisfies **Cond**.

Let $(D',C')$ be a closure operator which strongly $\perp\!\!\!\perp$-satisfies **Cond**, i.e., which satisfies $C'$-**Rel** $\subseteq$ $C$-**Rel**. If $(D',C')$ is proper, by Proposition 1.07(1) this implies $C$-**Fix** $\subseteq$ $C'$-**Fix**. With Proposition 1.09(0) one then has $(D',C') \sqsubseteq (D,C)$.

(1) Similar. □

The construction in Proposition 2.01 can now be seen as a special case Theorem 2.09(0). Suppose **Cond** consists just of the objects of $\mathcal{Z}$, considered as empty sources, and form $(D,C)$ according to 2.09(0). By construction $(D^{\mathcal{Z}},C^{\mathcal{Z}})$ strongly $\perp\!\!\!\perp$-satisfies **Cond**, therefore $(D^{\mathcal{Z}},C^{\mathcal{Z}}) \sqsubseteq (D,C)$. But for $X \xrightarrow{m} Y$ in $\mathcal{M}$, every equalizer used to obtain $mC^{\mathcal{Z}}$ is an upper bound for $mC$. Hence $(D,C) \sqsubseteq (D^{\mathcal{Z}},C^{\mathcal{Z}})$.

Clearly the strict cores of the proper closure operators of Theorem 2.09 are the strict and proper closure operators of Theorem 2.06.

## 2.10 EXAMPLE

It is known that none of the regular closure operators given by Proposition 2.01 on $\mathcal{X} = \mathbf{Top}$ is the ordinary topological closure, as described in Example 1.02. In particular, if $\mathcal{Z} = \mathbf{Haus}$, the operator $(D^{\mathcal{Z}},C^{\mathcal{Z}})$ which characterizes epis in **Haus** is different from the ordinary closure operator, although both agree on Hausdorff spaces. Now any closure operator $(D,C)$ which characterizes the epis in **Haus** must satisfy $D$-**Fix** $=$ $D^{\mathcal{Z}}$-**Fix**. Hence only one strict and proper closure operator characterizes the epis in **Haus**, namely $(D^{\mathcal{Z}}_{\circ},C^{\mathcal{Z}}_{\circ})$, which therefore must be the ordinary topological closure. □

# 3 NATURALITY OF CLOSURE OPERATORS

**So far we have only required pullbacks of closed $\mathcal{M}$-objects to be closed again. However, closure operators induced by Grothendieck topologies on an elementary topos have the property that closure and pullbacks commute. They can be thought of as natural transformations from the subobject functor into itself. More generally, the class of $\mathcal{X}$-morphisms with the property that closure commutes with pullbacks along them can be used to characterize closure operators.**

## 3.00 DEFINITION

Given a collection $\mathcal{L}$ of $\mathcal{X}$-morphisms, $(D,C)$ is called *$\mathcal{L}$-(co-)natural*, if $C$ preserves (co-)cartesian squares over $\mathcal{L}$-objects.

Notice that for a co-cartesian $\mathcal{M}$-morphism $(f,g)$ with $g \in \mathcal{M}$ condition (C1) forces $f$ to belong to $\mathcal{M}$, and hence to be iso, since $f$ as the first component of a co-cartesian morphism also belongs to $\mathcal{E}_0$.

## 3.01 DEFINITION

$(D,C)$ is called

(0) *cartesian*, if it is $\mathcal{X}$-**Mor**-natural;

(1) *hereditary*, if it is $\mathcal{M}$-co-natural;

(2) *weakly hereditary*, if it is $C$-**Fix**-co-natural.

This use of the term *weakly hereditary* differs slightly from [DG], but seems to be more appropriate. It does capture the idea that for every $(D,C)$-closed $X \xrightarrow{m} Y$ the closure on $\partial_1/X$ is just the relativization of the closure on $\partial_1/Y$ along $m$. For proper closure operators this in fact is equivalent to strictness.

## 3.02 PROPOSITION

(0) *A proper closure operator $(D,C)$ is strict iff it is weakly heraditary.*

(1) *Every $\mathcal{M}$-natural closure operator is strict.*

**Proof:**

(0) If $(D,C)$ is strict and proper, consider $m \in \mathcal{M}$ and a factorization $m = n \cdot p$ with $p \in C$-**Fix**. Apply $C$ to get $nC \cdot p \sqsubseteq mC$, which implies equality, i.e., $(X,p)C = (mC\partial_0, p)$ is co-cartesian. If $(D,C)$ is proper and weakly hereditary, apply $C$ to $mD \xrightarrow{(m\partial_0, mC)} m$, which is co-cartesian.

(1) Notice that the $\mathcal{M}$-morphism $mD \xrightarrow{(m\partial_0, mC)} m$ is always cartesian. □

## 3.03 EXAMPLES

(0) A Grothendieck topology can be viewed as the dense part of a proper cartesian closure operator $(D,C)$ on $\mathcal{M} = \mathcal{X}$-**Mor**. By Proposition 3.02 $(D,C)$ must be strict. In this case, $C$ is an idempotent endofunctor on the $\mathcal{X}$-category $\partial_1$, in the sense of Benabou [Be].

(1) The ordinary closure operator on **Top** (cf. Example 1.02) is hereditary, but not natural even with respect to closed embeddings. Just observe that the intersection of a dense subset with a closed one need not be dense in the relative topology of the closed set. □

Naturality and co-naturality are preserved under several constructions. All the proofs are easy.

**3.04 PROPOSITION**

*Let $(D,C)$ be $\mathcal{L}$-natural.*

(0) $(D^2,C^*)$, *with* $mC^* = mDC \cdot mC$ *is* $\mathcal{L}$-*natural.*

(1) *If* $\mathcal{L}$ *is closed under* $\mathcal{M}$-*pullbacks,* $(D^*,C^2)$ *with* $mD^* = mD \cdot mCD$ *is* $\mathcal{L}$-*natural.* □

**3.05 PROPOSITION**

*If* $(D,C)$ *is* $\mathcal{L}$-*co-natural, so is* $(D^*,C^2)$, *with* $mD^* = mD \cdot mCD$. □

**3.06 PROPOSITION**

(0) *Infima of* $\mathcal{L}$-*natural closure operators are* $\mathcal{L}$-*natural.*

(1) *Suprema of* $\mathcal{L}$-*co-natural closure operators are* $\mathcal{L}$-*co-natural.* □

We now want to construct $\mathcal{L}$-natural proper closure operators which (strongly) satisfy a collection **Cond** of $\mathcal{X}$-sources. The $\mathcal{L}$-*pullback-closure* of a collection of morphisms (composable pairs) consists of all pullbacks along $\mathcal{L}$-objects of elements of the original collection.

**3.07 DEFINITION**

Let $@ \in \{\perp\!\!\!\perp, \perp\}$. We say that $(D,C)$ $(\mathcal{L},@)$-*satisfies* **Cond**, if the $\mathcal{L}$-pullback-closure of $D$-**Fix** is contained in $\mathbf{Cond}_@$. If even the $\mathcal{L}$-pullback-closure of $C$-**Rel** is contained in $\mathbf{Cond}_@$, we say that $(D,C)$ *strongly* $(\mathcal{L},@)$-*satisfies* **Cond**.

Theorems 2.06 and 2.09 admit direct generalizations to this situation. Recall that $\partial_1/Y \xrightarrow{f^*} \partial_1/X$ is the pullback functor for $X \xrightarrow{f} Y$.

**3.08 THEOREM**

*Let* $\mathcal{E}$ *be stable under* $\mathcal{L}$-*pullbacks, and let* $\mathcal{L}$ *satisfy* **(C3)**.

(0) *If* $\mathcal{E}$-*sinks with members in* $\mathcal{M}$ *are epi-sinks,* $\{m \in \mathcal{M} : mf^* \perp\!\!\!\perp \mathbf{Cond}$ *for all* $f \in \mathcal{L}$ *with* $f\partial_1 = m\partial_1\}$ *is the class of dense* $\mathcal{M}$-*objects for an* $\mathcal{L}$-*natural strict and proper closure operator* $(D,C)$, *which is the largest such to* $\perp\!\!\!\perp$-*satisfy* **Cond**, *as well as for an* $\mathcal{L}$-*natural proper closure operator* $(D',C')$, *which is the largest such to strongly* $(\mathcal{L},\perp\!\!\!\perp)$-*satisfy* **Cond**.

(1) *If* $\mathcal{E}$-*sinks with members in* $\mathcal{M}$ *are effective,* $\{m \in \mathcal{M} : mf^* \perp \mathbf{Cond}$ *for all* $f \in \mathcal{L}$ *with* $f\partial_1 = m\partial_1\}$ *is the class of dense* $\mathcal{M}$-*objects for an* $\mathcal{L}$-*natural strict and proper closure operator* $(D,C)$, *which is the largest such to* $\perp$-*satisfy* **Cond**, *as well as for an* $\mathcal{L}$-*natural proper closure operator* $(D',C')$, *which is the largest such to strongly* $(\mathcal{L},\perp)$-*satisfy* **Cond**.

**Proof:**

(0) For $m \in \mathcal{M}$ let $mC$ be the supremum of $\{p \in \mathcal{M} : m = n \cdot p$ and $nf^* \perp\!\!\!\perp \mathbf{Cond}$ for all $f \in \mathcal{L}$ with $f\partial_1 = m\partial_1\}$, and let $mC'$ be the supremum of $\{p \in \mathcal{M} : m = n \cdot p$ and $(n,p)f^* \perp\!\!\!\perp \mathbf{Cond}$ for all $f \in \mathcal{L}$ with $f\partial_1 = m\partial_1\}$, where $(n,p)f^*$ denotes the pair $(pf^*, nk^*)$ with $k$ being the pullback of $f$ along $p$. The proof now proceeds essentially as the proofs of 2.06(0) and 2.09(0) did.

(1) Similar. □

Probably the best known application of this result is the construction of a Grothendieck topology with prescribed objects as sheaves, in case that $\mathcal{X}$ is a Grothendieck topos. However, the same principle works for constructing a Grothendieck topology with prescribed closed monos (cf. [K1]). More general categories, where all these constructions work with respect to all monos, are the ***familially regular categories*** of [S].

**ACKNOWLEDGEMENT**

I should like to thank the referee for various helpful directions which led to improvements of the paper.

## REFERENCES

[BR] J. BENABOU and J. ROUBAUD, "Monades et descente," *C. R. Acad. Sc. Paris, t.* **270**, *Série A* (1970), 96–98.

[Be] J. BENABOU, "Fibred categories and the foundations of naive category theory," *J. of Symbolic Logic* **50** (1985), 10–37.

[Bö] R. BÖRGER, *Kategorielle Beschreibung von Zusammenhangsbegriffen,* Thesis, Fernuniversität Hagen, Hagen, 1981.

[C0] G. CASTELLINI, *Closure Operators, Epimorphisms and Hausdorff Objects,* Thesis, Kansas State University 1986.

[C1] G. CASTELLINI, "Closure operators, monomorphisms and epimorphisms in categories of groups," *Cahiers Top. Geom. Diff. Cat.* **27**, *no. 2* (1986), 151–167.

[DG] D. DIKRANJAN and E. GIULI, "Closure operators I," *Topology and its Applications* **27**, *no. 2* (1987), 129–143.

[K0] J. KOSLOWSKI, *Dedekind cuts and Frink ideals for categories,* Thesis, Kansas State University 1986.

[K1] J. KOSLOWSKI, "CCCT-Hulls revisited," to appear in *Comm. Math. Univ. Carolinae.*

[S] R. STREET, "The family approach to total cocompleteness and toposes," *Trans. Am. Math. Soc.* **284** (1984), 355–369.

[T0] W. THOLEN, "Factorizations, localizations, and the orthogonal subcategory problem," *Math. Nachr.* **114** (1983), 63–85.

[T1] W. THOLEN, "Prereflections and Reflections," *Comm. in Algebra* **14** (1986), 717–740.

Jürgen Koslowski
Department of Mathematics
Vanderbilt University
Nashville, TN 37235
U.S.A.

This paper is in its final form.

# On the Unity of Algebra and Logic.

J. Lambek[1]

This article consists of two parts. In Part I it is argued that, from a categorical point of view, algebra and logic are the same, provided in algebra one admits many–sorted operations and in logic one pays attention to equality of deductions. The analogy then consists of identifying sorts with formulas and operators with deductions. In Part II we look at algebraic theories in which one of the sorts is a natural numbers object. We show that this object can be presented equationally provided every sort is equipped with a ternary Mal'cev operation. The single–sorted algebraic theory of primitive recursive functions has Mal'cev operation $(u+w)-v$.

## Part I. Comparing Algebra and Logic.

Both logic and algebra deal with arrows $A_1A_2 \dots A_n \to A_{n+1}$. In logic the $A_i$ are called *formulas* and the arrow is considered as a *deduction* of $A_{n+1}$ from the assumptions $A_1, \dots, A_n$. In algebra the $A_i$ are called *sorts* and the arrow is conceived as an *operation*. The fundamental similarity between these two situations has been obscured by two historical facts. Logicians were not interested in asking when two arrows are the same until quite recently[2] and algebraists traditionally were concerned with one sort only, so that in universal algebra the $A_i$ are usually all equal. Formally, many–sorted algebras were studied by Bénabou (1968)[3] and by Birkhoff and Lipson (1970) although, informally, they have always been with us. For example, modules are two–sorted algebras, inasmuch as the external product $AR \to R$ involves two sorts, the sort $A$ of abelian groups and the sort $R$ of rings. (I am not referring to a particular group or ring here.)

The kind of setup I have in mind here has been called a "Gentzen multicategory" by Szabo (1978), but might as well be called a "many–sorted algebraic theory", although it may differ in some technical respects from those discussed by Bénabou. Let me begin with some formal definitions.

A *multigraph* consists of two kinds of entities: arrows (also called "deductions" or "sequents" by logicians and "operations" by algebraists) and objects (also called "formulas" by logicians and "sorts" by algebraists) and two mappings:

$$\text{source} : \{\text{arrows}\} \longrightarrow \{\text{objects}\}^*,$$
$$\text{target} : \{\text{arrows}\} \longrightarrow \text{objects}.$$

Here $X^*$ denotes the free monoid generated by the class $X$. One writes

$$f : A_1 \dots A_n \to A_{n+1}$$

for

$$\text{source}\ (f) = A_1 \dots A_n\ ,\ \text{target}\ (f) = A_{n+1}.$$

A *deductive system* is a multigraph where, for each object $A$, there is given an *identity* arrow $1_A : A \longrightarrow A$ and, for given arrows $f : \Theta \to A$, $g : \Gamma A \Delta \to B$, there is given an arrow somewhat

---

1 This research was supported by the Engineering and Natural Sciences Research Council of Canada.

2 Prawitz(1971) and Lambek(1969) studied equivalence of deductions quite independently and for different reasons. A comparison between the two approaches was made by Mann (1975).

3 At the point of writing this, I have not been able to consult Bénabou's article.

inadequately denoted by $g < f >: \Gamma\Theta\Delta \to B$. (We have denoted strings of objects by capital Greek letters.) The rule

$$\frac{f : \Theta \to A \qquad g : \Gamma A \Delta \to B}{g < f >: \quad \Gamma\Theta\Delta \to B}$$

is Gentzen's ***cut***.

A ***Gentzen deductive system*** has, in addition, three ***structural*** rules:

$$\frac{f : \Gamma AB\Delta \to C}{f^i : \Gamma BA\Delta \to C} \quad \text{(interchange)},$$

$$\frac{f : \Gamma AA \to B}{f^c : \Gamma A \to B} \quad \text{(contraction)},$$

$$\frac{f : \Gamma \to B}{f^t : \Gamma A \to B} \quad \text{(thinning)}.$$

Again, outside its precise context, the notation $f^i, f^c, f^t$ is of course inadequate.

I have discussed at the Colorado meeting (1987) what happens if one drops some or all of these structural rules. Thus, my original ***syntactic calculus*** (1958) permits none, Girard's ***linear logic*** (1987) permits only interchange and Anderson and Belnap's ***relevance logic*** (1975) permits both interchange and contraction but not thinning.

A ***Gentzen multicategory*** is a Gentzen deductive system with an appropriate notion of equality between arrows. The easiest way to state this is in terms of the so-called internal language, as I have pointed out at the Colorado meeting.

A ***type language*** is given by

(a) a class of types,
(b) operation symbols $f : A_1 \ldots A_n \to A_{n+1}$ where the $A_i$ are types,
(c) countably many variables of each type,
(d) for each finite set $X$ of variables a binary relation $=_X$ to be described presently.

The class of ***terms*** is defined inductively as follows, together with an assignment of a type to each term:

(i) all variables are terms,
(ii) if $a_i$ is a term of type $A_i$, for $i = 1, \ldots, n$, and if $f : A_1 \ldots A_n \to A_{n+1}$ is an operation symbol, then $fa_1 \ldots a_n$ is a term of type $A_{n+1}$.

Suppose $a$ and $a'$ are terms of the same type $A$ containing at most the variables in the set $X$, then we are allowed to write $a =_X a'$. This relation is supposed to satisfy not only the usual properties of a congruence relation but also the following rule permitting substitution of terms for variables:

$$\text{if } a =_{X\cup\{y\}} a' \text{ then } [b/y]a =_X [b/y]a'.$$

Here it is assumed that $y$ is a variable of type $B$ not in $X$, $b$ is any term of type $B$ containing only variables from $X$ and $[b/y]a$ denotes the result of substituting $b$ for all occurrences of $y$ in $a$.

In particular, with each Gentzen deductive system we associate its ***internal language***, a type language whose operation symbols are Gentzen's sequents and where $=_X$ is the smallest congruence relation satisfying the following conditions:

$$\begin{aligned}
1_A x &=_x x,\\
g < f > x_1\dots x_m\, z_1\dots z_k\, y_1\dots y_n &=_X g\, x_1\dots x_m\, f z_1\dots z_k g y_1\dots y_n,\\
f^i x_1\dots x_m y x\, y_1\dots y_n &=_X f x_1\dots x_m x y\, y_1\dots y_n,\\
f^c x_1\dots x_m x &=_X f x_1\dots x_m x x,\\
f^t x_1\dots x_m x &=_X f x_1\dots x_m.
\end{aligned}$$

Here $X$ is supposed to denote the set of variables occurring on either side.

The Gentzen deductive system becomes a *Gentzen multicategory* if we introduce a suitable equivalence relation between operation symbols: we say operation symbols $f, g : A_1 \dots A_n \to A_{n+1}$ define the same *operation* and we write $f = g$, provided

$$f x_1 \dots x_n =_X g x_1 \dots x_n,$$

where $X = \{x_1, \dots, x_n\}$. In view of what we mean by operations being the same, the symbols $1_A, g < f >, f^i, f^c, f^t$ satisfying the above conditions are unique qua operations.

We summarize the above conditions by means of the following property of *functional completeness*: suppose $\varphi(x_1, \dots, x_n)$ is any term of type $B$ containing no variables other than $x_1, \dots, x_n$, in any order, with possible repetition and not necessarily all of them, then there exists a unique operation $f : A_1 \dots A_n \to B$ such that

$$g x_1 \dots x_n =_X \varphi(x_1, \dots, x_n),$$

where $X = \{x_1, \dots, x_n\}$.

Originally Gentzen used his deductive system (sequent calculus) to introduce logical operations such as $\top (= \mathit{true}), \wedge (= \mathit{and})$ and $\Leftarrow (= \mathit{if})$, among others.[4] Gentzen multicategories have been used to introduce corresponding operations between objects of a category, thus the terminal object, the Cartesian product and exponentiation respectively (see e.g. Lambek 1987). Gentzen had thought of $A_1 \dots A_n \to B$ as meaning $A_1 \wedge \dots \wedge A_n \to B$. In the same way, without loss in generality, one may think of a multicategory as a Cartesian category, that is, a category with canonical finite products, including the empty product.

In our view, Gentzen multicategories are the same thing as *algebraic theories*, which had previously been conceived as Cartesian categories (Lawvere 1963, Bénabou 1968). Our view seems closer to that of universal algebraists, being unhampered by irrelevant distinctions, such as between $(A \times B) \times C$ and $A \times (B \times C)$. What then are algebras? According to Lawvere and Bénabou, an algebra is a product preserving functor into the category of sets. In the present setup, an *algebra* is a multifunctor from the algebraic theory to the Gentzen multicategory of sets, in which the string $A_1 \dots A_n$ is interpreted as a Cartesian product. A *multifunctor*[5] $F$ between two multicategories sends objects to objects and arrows to arrows such that

$$\frac{f : A_1 \dots A_n \to B}{F(f) : F(A_1) \dots F(A_n) \to F(B)},$$

$$F(1_A) = 1_{F(A)},$$

$$F(g < f >) = F(g) < F(f) > .$$

---

[4] A standard reference for Gentzen sequent calculi is the book by Kleene (1952).

[5] In my 1969 article I wrote "functor" in place of "multifunctor", but on several occasions people told me they preferred "multifunctor".

We shall just look at one example here, the algebraic theory of (right) modules. There are two sorts $A$ and $R$ and operations

$$\begin{array}{ll} 0: & \to A,\ -: A \to A,\ +: AA \to A, \\ 0': & \to R,\ -': R \to R,\ +': RR \to R, \\ 1: & \to R,\ \cdot: RR \to R,\ \cdot': AR \to R. \end{array}$$

These operations satisfy a number of equations, which are most easily expressed in the internal language. We shall use $x, y, z, \ldots$ as variables of type $A$ and $u, v, w, \ldots$ as variables of type $R$. To facilitate the reading we drop the prime and write $x + y$ in place of $+xy$, etc, which necessitates occasional use of parenthesis to avoid ambiguity. We shall also omit the subscript $X$ on the equality sign. The equations are:

$$x + 0 = x,\ x + -x = 0,\ x + y = y + x,\ (x + y) + z = x + (y + z),$$

$$u + 0 = u,\ u + -u = 0,\ (u + v) + w = u + (v + w),$$

$$u \cdot 1 = u,\ 1 \cdot u = u,\ (u \cdot v) \cdot w = u \cdot (v \cdot w),$$

$$(u + v) \cdot w = (u \cdot w) + (v \cdot w),\ w \cdot (u + v) = (w \cdot u) + (w \cdot v),$$

$$(x + y) \cdot u = (x \cdot u) + (y \cdot u),\ x \cdot (u + v) = (x \cdot u) + (x \cdot v),$$

$$x \cdot (u \cdot v) = (x \cdot u) \cdot v,\ x \cdot 1 = x.$$

The *algebraic theory of modules* is assumed to be the two-sorted theory freely generated from the above data, i.e., all operation symbols are inductively defined from the given ones and no equations hold unless deducible from those listed.

An algebra of this theory is called a *(right) module*. It is given by sets $F(A)$ and $F(R)$ with operations realized as functions, e.g. $F(A) \times F(R) \to F(A)$. An equation such as $(x + y) \cdot u = (x \cdot u) + (y \cdot u)$ may now be interpreted as saying that this is so for all elements $x, y$ of $F(A)$ and $u$ of $F(R)$. Note that $F(A)$ becomes an abelian group and $F(R)$ a ring.

## Part II. Natural numbers in algebraic theories.

Recursive functions are traditionally the concern of logicians, although Eilenberg and Elgot (1970) have looked at them from an algebraic point of view. What may be surprising to some people is that the theory of primitive recursive functions may be regarded as an algebraic theory in the present sense, even a single-sorted one. This position is in fact very close to that of Goodstein (1957). Before expounding it, we shall make a little detour.

A *natural numbers object* in a Gentzen multicategory is an object $N$ equipped with arrows $0: \ \to N$ and $S : N \to N$ such that, for any object $B$ and each pair of arrows $a : \Gamma \to B$, $h : \Gamma N B \to B$, there exists a unique arrow $f : \Gamma N \to B$ satisfying two conditions, which are expressed in the internal language as follows:

$$f x_1 \ldots x_n\, 0 = a x_1 \ldots x_n,$$

$$f x_1 \ldots x_n\, Sy = h x_1 \ldots x_n y\, f x_1 \ldots x_n y,$$

where the $x_i$ are variables of type $A_i$, $\Gamma = A_1 \ldots A_n$ and subscripts on the equal sign have been omitted. (For the motivation of this definition see Kleene 1952, Lawvere 1964, Lambek and Scott 1986 and Román 1988.)

To indicate the dependence of $f$ on $a$ and $h$, we write $f$ as $R_{ah}$, so that its existence, but not its uniqueness, is expressed by the equations

$$R_{ah}x\,0 = ax,$$
$$R_{ah}x\;Sy = hxyR_{ah}xy,$$

where we have written $x$ for $x_1 \ldots x_n$. Without uniqueness, one still speaks of a *weak natural numbers object*. The uniqueness of $f$ may be expressed by the following condition for each object $B$:

$(U_B)$ if $fxSy = hxyfxy$ then $f = R_{f<0>h}$, where $f<0>x = fx0$.

Under certain conditions, the implicational condition $(U_B)$ can be replaced by equations, as was done for Cartesian closed categories in (Lambek 1986) and for Cartesian categories in (Román 1988). We shall look at their argument once more for multicategories, at the same time making it simpler and presenting an improved form of the result.

A closer examination of the above mentioned argument shows a surprising connection with an operation introduced by Mal'cev (1954) for a completely different purpose. A ternary operation $muvw$, with $u, v, w$ of type $B$, will be called a *Mal'cev operation* if

$$mvvw = w, \quad muvv = u.$$

(See Mal'cev 1954, Lambek 1957, Findlay 1960 for various applications.) Mal'cev operations are ubiquitous; for example, both sorts in the algebraic theory of modules discussed above possess one.

**Theorem.** In a Gentzen multicategory with weak natural numbers object, if there exists a Mal'cev operation $m : B^3 \to B$, $(U_B)$ can be expressed equationally thus:

$$R_{f<o>H(m,f,h)} = f,$$

where

$$H(m, f, h)xyz = m(hxyz)(hxyfxy)(fxSy).$$

(The parentheses on the right-hand side are not necessary but facilitate reading.)

**Proof.** To show the necessity of the condition, we check that

$$\begin{aligned} fxSy &= m(hxyfxy)(hxyfxy)(fxSy) \\ &= H(m, f, h)xyfxy, \end{aligned}$$

using the fact that $mvvw = w$.

To prove the sufficiency of the condition, assume that $fxSy = hxyfxy$, then

$$H(m, f, h)xyz = hxyz,$$

using $muvv = u$, hence $h = H(m, f, h)$, so that $R_{f<o>h} = f$.

Before turning our attention to the special case $(U_N)$, let us recall some operations $N^k \to N$ which can be defined in any Gentzen multicategory with weak natural numbers object. Variables will be of type $N$.

The *sum* $x + y$ is defined by $x + y = R_{ah}xy$, with $ax = 0$, $hxyz = Sz$, hence satisfies the Peano equations:

$$x + 0 = x, \; x + Sy = S(x + y).$$

The *predecessor* $Py$ is defined by $Py = R_{ah}y$, with $a = 0$, $hyz = y$, hence satisfies:

$$P0 = 0, \; PSy = y.$$

(Here $n = 0$, so $x$ does not appear.)

The *naive difference* $x \dot{-} y$ is defined by $x \dot{-} y = R_{ah}xy$, with $ax = x$, $hxyz = Pz$, hence satisfies

$$x \dot{-} 0 = x, \; x \dot{-} Sy = P(x \dot{-} y).$$

With the help of $(U_N)$ one can prove a number of well-known identities, only two of which will play a rôle here, namely (b) and (e) below. For the sake of completeness, we include the proofs, essentially the same which have been used before in different contexts (see Goodstein 1957, Pfender et al. 1982, Lambek 1985, Román 1988).[6]

**Lemma.** The following identities hold in any Gentzen multicategory with natural numbers object:

(a) $Sx \dot{-} Sy = x \dot{-} y$,
(b) $(x + y) \dot{-} y = x$,
(c) $0 + x = x$,
(d) $Sx + y = S(x + y)$,
(e) $x + y = y + x$.

**Proof.** (a) $Sx \dot{-} S0 = P(Sx \dot{-} 0) = PSx = x$, $Sx \dot{-} SSy = P(Sx \dot{-} Sy)$, hence $Sx \dot{-} Sy = R_{ah}xy$, where $ax = x$ and $hxyz = Pz$. Therefore $Sx \dot{-} Sy = x \dot{-} y$.

(b) $(x + 0) \dot{-} 0 = x \dot{-} 0 = x$, $(x + Sy) \dot{-} Sy = S(x + y) \dot{-} Sy = (x + y) \dot{-} y$ by (a), hence $(x + y) \dot{-} y = R_{ah}xy$, where $ax = x$ and $hxyz = z$. Therefore $(x + y) \dot{-} y = x$.

(c) $0 + 0 = 0$, $0 + Sy = S(0 + y)$, hence $0 + y = R_{ah}y$, where $a = 0$ and $hyz = Sz$. Therefore $0 + y = y$.

(d) $Sx + 0 = Sx$, $Sx + Sy = S(Sx + y)$, hence $Sx + y = R_{ah}xy$, where $ax = x$ and $hxyz = Sz$. But similarly $S(x + y) = R_{ah}xy$ for the same $a$ and $h$, hence $Sx + y = S(x + y)$.

(e) $0 + x = x$, $Sy + x = S(y + x)$ by (d), hence $y + x = R_{ah}z$, where $ax = x$ and $hxyz = Sz$. Therefore $y + x = x + y$.

Let us now consider the special case $B = N$ of the Theorem. We may take $muvw = (u + w) - v$, because then

$$muvv = (u + v) \dot{-} v = u$$

by (b) and

$$mvvw = (v + w) \dot{-} v = (w + v) \dot{-} v = w$$

by (e) and (b).

Note that both (b) and (e) are deducible from $(U_N)$, but only (b) was used in proving the sufficiency of the condition which replaces $(U_N)$. We therefore have:

**Corollary 1.** A weak natural numbers object satisfies $(U_N)$ if and only if the following equations hold:

$$(u + v) \dot{-} v = v,$$

$$R_{f<o>H(m,f,h)}xy = fxy,$$

[6] I take this opportunity to point out that in my paper of 1985 I should have referred to Pfender et al. (1982), about which I only learned from Román later.

where $muvw = (u+w) \dot{-} v$.[7]

By a *Kleene multimonoid* we shall mean a Gentzen multicategory with a single object which is a natural numbers object. In the definition of natural numbers object we may then confine attention to $B = N$ and $\Gamma = N^n$.

**Corollary 2.** A Gentzen multicategory with one object is a Kleene multimonoid if and only if it has operations

$$0: \quad \to N, \; S: N \to N, \; R_{ah}: N^{n+1} \to N,$$

for all $a: N^n \to N$ and $h: N^{n+2} \to N$, satisfying the following equations (where we write $x$ for $x_1 \ldots x_n$):

(1) $R_{ah}x0 = ax$,
(2) $R_{ah}xSy = hxyR_{ah}xy$,
(3) $(u+v) \dot{-} v = u$,
(4) $R_{f<o>H(m,f,h)}xy = fxy$,

for all $f: N^{n+1} \to N$, where $f<o>x = fx0$ and $muvw = (u+w) \dot{-} v$.

A Gentzen multimonoid is of course a single–sorted algebraic theory. By ***the algebraic theory of primitive recursive functions*** we mean the Kleene multimonoid which is freely generated. That is to say, its operations are defined inductively from 0 and $S$ by functional completeness and the so–called recursion scheme $R$ and only those equations hold which follow from (1) to (4). An algebra of this theory is a multifunctor to the Gentzen multicategory of sets. The canonical example of such an algebra is given by

$$F(N) = \underline{N}, \quad F(0) = \underline{0}, \quad F(S) = \underline{S},$$

where $\underline{N}$ is the set of natural numbers, $\underline{0}$ the usual zero and $\underline{S}$ the usual successor function. Every primitive recursive function will then have the form $F(f) = \underline{f}: \underline{N}^{k+1} \to \underline{N}$.

One may think of the operation $f$ as a program for computing the function $\underline{f}$. It may be of interest to study other algebras of the theory of primitive recursive functions.

Some people have asked: why the insistence on making condition $(U_N)$ equational? Aside from aesthetic considerations, there is this: multifunctors preserve equations, they don't usually preserve implications. All equations which hold in the algebraic theory hold in all models of the theory, that is, in all algebras. The possibility of replacing implications by equations had been suggested by the work of Burroni (1981).

---

[7] In my earlier argument (1965) I had implicitly used a different Mal'cev operation, namely $muvw = (u \dot{-} (v \dot{-} w)) + (w \dot{-} v)$, though without realizing it. If this ternary operation is used in Corollary 1 in place of $(u+w) \dot{-} v$, the equation $(u+v) \dot{-} v = u$ may be replaced by the simpler equation $u \dot{-} u = 0$. My original argument had been based on the observation that $fxy$ is completely determined by $fxSy \dot{-} hxyfxy$ and $hxyfxy \dot{-} fxSy$.

## REFERENCES

Anderson, A.R. and N.D. Belnap: Entailment: the logic of relevance and necessity, Princeton University Press, Princeton, N.J., 1975.

Bénabou, J.: Structures algébriques dans les catégories, Cahiers Topologie Géométrie Différentielle **10**(1968), 1-126.

Birkhoff, G. and J.D. Lipson: Heterogeneous algebras, J. Combinatorial Theory **8**(1970), 115-133.

Burroni, A.: Algèbres graphiques, 3ième colloque sur les catégories, Cahiers Topologie Géométrie Différentielle **23** (1981), 249-265.

Eilenberg, S. and C.C. Elgot: Recursiveness, Academic Press, New York 1970.

Findlay, G.D.: Reflexive homomorphic relations, Can. Math. Bull. **3**(1960), 131-132.

Girard, J.-Y.: Linear logic, J. Theoretical Computer Science **50**(1987), 1-102.

Goodstein, R.L.: Recursive number theory, North-Holland Publ. Co., Amsterdam 1957.

Kleene, S.C.: Introduction to metamathematics, Van Nostrand, New York 1952.

Lambek, J.: Goursat's theorem and the Zassenhaus lemma, Can. J. Math. **10**(1957), 45-56.

Lambek, J.: The mathematics of sentence structure, Amer. Math. Monthly **65**(1958), 154-169.

Lambek, J.: Deductive systems and categories II, Springer Lecture Notes in Mathematics **86**(1969), 76-122.

Lambek, J.: Cartesian closed categories and typed $\lambda$- calculi, in: Cousineau, Curien and Robinet (eds.), Combinators and functional programming languages, Springer Lecture Notes in Computer Science **242**(1986), 136-175.

Lambek, J.: Multicategories revisited, Proc. A.M.S. Conf. on Categories in Computer Science and Logic, 1987.

Lambek, J. and P.J. Scott: Introduction to higher order categorical logic, Cambridge Studies in advanced mathematics **7**, Cambridge University Press, Cambridge 1986.

Lawvere, F.W.: Functorial semantics of algebraic theories, Proc. Nat. Acad. Sci. U.S.A. **50**(1963), 869-872.

Lawvere, F.W.: An elementary theory of the category of sets, Proc. Nat. Acad. Sci. U.S.A. **52**(1964), 1506-1511.

Mal'cev, A.I.: On the general theory of algebraic systems, Math. Sb. N.S. **35**(1954), 3-20.

Mann, G.R.: The connection between equivalence of proofs and cartesian closed categories, Proc. London Math. Soc. **31**(1975), 289-310.

Pfender, M., R. Reiter and M. Sartorius,: Constructive arithmetics, in: Category Theory, Springer Lecture Notes in Mathematics **962**(1982), 282-236.

Prawitz, D.: Natural deduction, Almquist and Wiksell, Stockholm 1965.

Prawitz, D.: Ideas and results in proof theory, Studies in Logic **63**(1971), 235-307.

Román, L.: Cartesian categories with natural numbers object, J. Pure and Applied Algebra (1988) (to appear).

Szabo, M.: Algebra of proofs, Studies in Logic and the Foundations of Mathematics **86**, North-Holland Publ. Co., Amsterdam 1978.

Mathematics Department
McGill University
Montreal, Canada

This paper is in its final form and will not be published elsewhere.

# The Cohomology Groups of an Epimorphism

**A. Lirola, E.R. Aznar and M. Bullejos**

(*)Departamento de Algebra. Universidad de Granada (Spain)

**Abstract:** *If* $p: S \rightarrow T$ *is a* U*-split epimorphism in a monadic category* $\mathbb{C}$ *(such as groups or algebras) and* $\mathbb{G}$ *denotes the induced cotriple on* $\mathbb{C}$*, then the cohomology groups* $H^n(p,A)$ *of the epimorphism with coefficients in a abelian group object* A *in* $\mathbb{C}$ *have been defined by van Osdol ([18]) and interpreted in dimension* 1 *by Aznar and Cegarra ([1]), as isomorphism classes of* 2*-torsors which have* p *as their augmentation. In this paper it is shown that these groups are themselves "cotriple cohomology groups" for the cotriple on the category of simplicial objects of* $\mathbb{C}$ *induced by* $\mathbb{G}$ *and applied to the complex* $COSK^0(p)$ *with coefficients in the abelian group object* $K(A,1)$*. Van Osdol's long exact sequence in the first variable associated with* p *is shown to be isomorphic to the standard second variable cotriple cohomology sequence, provided by the short exact sequence* $0 \rightarrow K(A,0) \rightarrow L(A,0) \rightarrow K(A,1) \rightarrow 0$ *in this category. The interpretation of* $H^1(p,A)$ *in terms of* 2*-torsors is shown to be a consequence of the standard interpretation of* $H^1$ *as isomorphism classes of* $\underline{1}$*-torsors, combined with the properties of the functor* $\overline{W}$.

Let $U: \mathbb{C} \rightarrow \mathbb{B}$ and $F: \mathbb{B} \rightarrow \mathbb{C}$ be functors, such that F is a left adjoint for U and U is monadic, with $\mathbb{B}$ having finite inverse limits. Let $\mathbb{G} = (G=FU,\delta,\mu)$ be the cotriple induced on $\mathbb{C}$. Recall that if X is an arbitrary object in $\mathbb{C}$, one has the standard resolution of X, $\mathbb{G}.X \twoheadrightarrow X$, and that the cotriple cohomology groups of X with coefficients in an abelian group object A in $\mathbb{C}$, $H^n_G(X,A)$, $\forall n \geq 0$, are defined as the homology groups of the cochain complex

$$Hom_C(G^*X,A) = Hom_C(GX,A) \rightarrow Hom_C(G^2X,A) \rightarrow \cdots$$

which is obtained by applying the functor $Hom_C(-,A)$ to the standard resolution of X usually using the subgroups of normalized chains and then taking alternating sums in the usual way. Recall also that the underlying sets of these groups can be seen, for every $n \geq 0$, as the set of homotopy classes of simplicial n-cocycles from the standard resolution of X to the Eilenberg-Mac Lane complex, $K(A,n)$. These cotriple

(*) This paper has been partially supported by a grant from CAICYT

Proyecto de Investigación 3556-83C2-00

cohomology groups were interpreted by J. Duskin ([8]) in terms of n-torsors, obtaining natural bijections ,

$$H^n_G(X,A) \cong \mathbf{TORS}^n_U[X,A]$$

where $\mathbf{TORS}^n_U[X,A]$ denotes the set of Yoneda classes of n-torsors over X under the group object A, relative to U.

Let $p: S \twoheadrightarrow T$ be an U-split epimorphism in $\mathbb{C}$, i.e., an epimorphism in $\mathbb{C}$ such that $Up: US \to UT$ has a section in $\mathbb{B}$. The cohomology groups of this epimorphism, denoted by $H^n_G(p,A)$, were defined by van Osdol ([18]) as the homology groups of the cochain complex $C^*(p,A)$, defined in each dimension n by the short exact sequence:

$$\mathrm{Hom}_{\mathbb{C}}(G^nT,A) \rightarrowtail \mathrm{Hom}_{\mathbb{C}}(G^nS,A) \twoheadrightarrow C^{n-1}(p,A)$$

The corresponding short exact sequence of cochain complexes induces, through standard homological algebra arguments, an exact sequence of cohomology groups:

$$0 \to H^0_G(T,A) \to H^0_G(S,A) \to H^0_G(p,A) \to H^1_G(T,A) \to \cdots$$

Our first objective is to obtain a direct simplicial definition for the above cohomology groups of the epimorphism, $H^n_G(p,A)$, $\forall n\geq 0$; and by reformulating $H^n_G(S,A)$ and $H^n_G(T,A)$ in these terms as well, then obtain the above first variable sequence, as an instance of the familiar second variable long exact sequence associated with a short exact sequence of coefficients. Our second objetive will be to use the "torsor-theoretic" consequences of this point of view to obtain directly Aznar and Cegarra's interpretation of $H^1(p,A)$, as well as a result of Loday ([14]), which interpretes $H^1(p,A)$ in terms of certains 2-fold extensions, in the case of Groups.

The simplicial reformulation of these results is based on a result ([18] Th. 3.9) of van Oldol, who showed that the un-normalizaded version of the cokernel cochain complex $C^*(p,A)$ is isomorphic to the cochain complex which in dimension n is given by the group

$$\mathrm{Hom}_n(\pi_0,\pi_1) = \{f: G^{n+1}S\times_T S \to A \mid fG^{n+1}\pi_0 - fG^{n+1}\pi_1 + fG^{n+1}\pi_2 = 0\}$$

whose differential is defined by $d(f) = \sum_{i=0}^{n+1}(-1)^i fe^{n+1-i}$. Here the $\pi_i: S\times_T S\times_T S \rightrightarrows S\times_T S \rightrightarrows S$ are the canonical projections, while the $\varepsilon^n$ are defined in the usual way using the cotriple $(\mathbb{G},\varepsilon,\delta)$. The normalized version of this result uses the subgroups for which $fG^n\delta^k = 0 = fG^{n+1}\Delta$.

We now preceed to show that the complex $\text{Hom.}(\pi_0, \pi_1, A)$ is <u>itself</u> obtained from a cotriple operating on the category $\text{Simpl}(\mathbb{C})$ and that the groups $H_n(\pi_0, \pi_1, A)$ are themselves cohomology groups of this cotriple.

First note that the canonical projections $\pi_i : S \times_T S \times_T S \longrightarrow S \times_T S$ are the face maps of a simplicial object $\text{COSK}^0(p)$ in $\mathbb{C}/T$ which is obtained by the iterating the fiber products of $S$ with itself over $T$. Moreover, since $p : S \to T$ is U-split, the complex $\text{COSK}^0(p)$ admits a contracting homotopy in the category $\mathbb{B}$ above the object $U(T)$.

The cotriple $\mathscr{G}^*$ we have in mind is the obvious one obtained from $U$ and $F$ applied dimension-wise to the category $\text{Simpl}(\mathbb{C})$. This makes $\text{Simpl}(\mathbb{C}) \xrightarrow{U} \text{Simpl}(\mathbb{B})$ monadic if $U : \mathbb{C} \to \mathbb{B}$ is, and carries augmented simplicial objects to augmented ones in the obvious fashion. Thus, if we iteratively apply this cotriple to the simplicial object $\text{COSK}^0(p)$, we obtain as its standard resolution, the doubly augmented complex in $\mathbb{C}$:

$$\cdots \mathscr{G}^2\text{COSK}^0(p)/\mathscr{G}^2T \rightrightarrows \mathscr{G}\text{COSK}^0(p)/\mathscr{G}T \longrightarrow \text{COSK}^0(p)/T$$

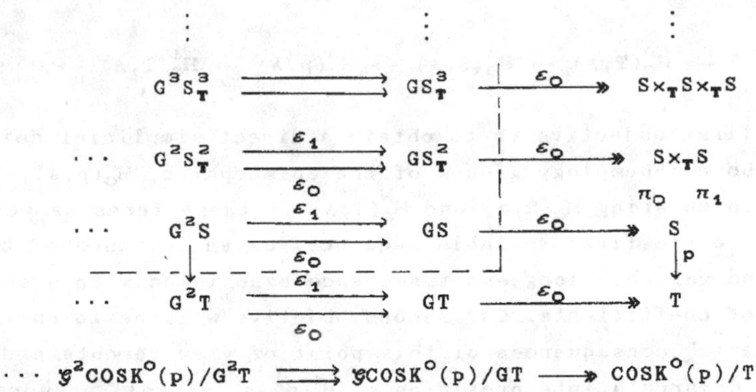

in which the rows are just the standard resolutions of their augmenting objects and the columns are obtained by applying the functor $G$ to $\text{COSK}^0(p)$. The U-splitting furnishes the columns with a contracting homotopy which, in fact, commutes with all of the $\delta_i$ and all of the $\varepsilon_i$, except for $\varepsilon_n$.

All that remains to define cohomology groups is an abelian group object in $\text{Simpl}(\mathbb{C})$, i.e. a simplicial abelian group in $\mathbb{C}$. For this we take the simplicial object $K(A,1)$, the nerve of the original abelian group object $A$ in $\mathbb{C}$. Since $A$ is abelian, $K(A,1)$ is naturally an (abelian) group in $\text{Simpl}(\mathbb{C})$,

$$K(A,1)\times K(A,1) \rightrightarrows K(A,1) \longrightarrow 1$$

and the cochain complex $\mathrm{Hom}_{\mathrm{Simpl}(\mathbb{C})}(\mathcal{G}^*(\mathrm{COSK}^{\mathrm{o}}(p),K(A,1))$ is defined. We now have the following

**Theorem 1.** Under the above assumptions, there is a natural equivalence of cochain complexes,

$$C^{*-1}{}_{(p,A)} \cong \mathrm{Hom}_{\mathrm{Simpl}(\mathbb{C})}(\mathcal{H}^*\mathrm{COSK}^{\mathrm{o}}(p),K(A,1))$$

The proof is immediate on the basis of van Osdol's theorem, for by definition, an n-cochain on this standard resolution is a simplicial map from the complex $G^{n+1}(\mathrm{COSK}^{\mathrm{o}}(p))$ into $K(A,1)$. But since $K(A,1)$ represents normalized 1-cocycles on any simplicial object in $\mathbb{C}$, this is just the group $\mathrm{Hom}_n(\pi_0,\pi_1,A)$ (in its normalized form). ∎

$K(A,1)$ is, of course, not the only simplicial abelian group available here. For our purposes we wish to consider the constant simplicial group $K(A,0)$ which is the group $A$ in each dimension, as well as the group $L(A,0)$ which is the contractible complex $\mathrm{DEC}(K(A,1))$ and is isomorphic to $\mathrm{COSK}^{\mathrm{o}}(A \to 1)$.

**Theorem 2.** There are natural equivalences of cochain complexes:

$$\mathrm{Hom}_{\mathbb{C}}(G^*S,A) \cong \mathrm{Hom}_{\mathrm{Simpl}(\mathbb{C})}(\mathcal{G}^*\mathrm{COSK}^{\mathrm{o}}(p),L(A,0))$$

$$\mathrm{Hom}_{\mathbb{C}}(G^*T,A) \cong \mathrm{Hom}_{\mathrm{Simpl}(\mathbb{C})}(\mathcal{G}^*\mathrm{COSK}^{\mathrm{o}}(p),K(A,0))$$

Again the proof is immediate since simplicial maps of any complex $X_\bullet$ into $L(A,0)$ are in bijective correspondence with the morphisms $\mathrm{Hom}_{\mathbb{C}}(X_0, A)$ (i.e. $L(A,0)$ represents 0-cochains) and we recover the cochain complex of the 0-dimensional row, $\mathrm{Hom}_{\mathbb{C}}(G^*S,A)$. Similarly, since simplicial maps of any complex $X_\bullet$ into a constant complex $K(A,0)$ are in bijective correspondence with the morphisms of $\pi_0(X_\bullet)$ (the coequalizer of the pair $X_1 \rightrightarrows X_0$) into $A$, we recover the cohomology of (-1)-dimensional row, which is, in each column, the coequalizer of $(G^{n+1}(\pi_0),G^{n+1}(\pi_1))$ (because of the contracting homotopy in columnar dimensions $\geq 0$ and via the fact that $U$ is monadic and $p$ is $U$-split in dimension -1).

The above three simplicial groups form a short exact sequence of abelian group objects in $\mathrm{Simpl}(\mathbb{C})$,

$$0 \longrightarrow K(A,0) \xrightarrow{\Delta.} L(A,0) \xrightarrow{\text{dif.}} K(A,1) \longrightarrow 0 \qquad (1)$$

with $\Delta.$ induced by the diagonal morphism $\Delta: A \to A^2$, and dif. induced by the 'difference' morphism $A^2 \to A : (a,b) \mapsto b-a$. This short exact sequence determines a short exact sequence of cochain complexes,

$$\mathrm{Hom}(\mathcal{G}^*\mathrm{COSK}^o(p),K(A,0)) \rightarrowtail \mathrm{Hom}(\mathcal{G}^*\mathrm{COSK}^o(p),L(A,0)) \twoheadrightarrow \mathrm{Hom}(\mathcal{G}^*\mathrm{COSK}^o(p),K(A,1))$$

and so, a long exact sequence in cohomology,

$$H^o_{\mathcal{G}}(\mathrm{COSK}^o(p),K(A,0)) \to H^o_{\mathcal{G}}(\mathrm{COSK}^o(p),L(A,0)) \to H^o_{\mathcal{G}}(\mathrm{COSK}^o(p).K(A,1))\cdots$$

which is, by the naturality of the isomorphisms in theorem 1 and 2, naturally equivalent to

$$\begin{array}{l} 0 \longrightarrow H^o_G(T,A) \longrightarrow H^o_G(S,A) \longrightarrow H^o_G(p,A) \\ H^1_G(T,A) \longrightarrow H^1_G(S,A) \longrightarrow H^1_G(p,A) \qquad (2) \\ H^2_G(T,A) \longrightarrow H^2_G(S,A) \longrightarrow \cdots \end{array}$$

We now turn our attention to the consequences of applying Duskin's interpretation theorem in the above context. Using it we immediately obtain

**Theorem 3.** **(Interpretation theorem)**

There are natural isomorphisms:

$$H^n_G(T,A) \cong \mathrm{TORS}^n_{U.}[\mathrm{COSK}^o(p),K(A,0)]$$

$$H^n_G(S,A) \cong \mathrm{TORS}^n_{U.}[\mathrm{COSK}^o(p),L(A,0)] \text{ , and}$$

$$H^n_G(p,A) \cong H^n_{\mathcal{G}}(\mathrm{COSK}^o(p),K(A,1)) \cong \mathrm{TORS}^n_{U.}[\mathrm{COSK}^o(p),K(A,1)]$$

(which can be also stated in terms of homotopy classes of double n-cocycles from the cotriple resolution, $\mathcal{G}.(\mathrm{COSK}^o(p))$, to the Eilenberg-MacLane complexes $K(K(A,0),n)$ ,$K(L(A,0),n)$ and $K(K(A,1),n)$ respectively).

In this paper we will investigate the meaning of the above theorem for n=0 and n=1 and thus confine ourselves to the first 8 terms of the sequence (2). We start with the identifications

$$H^0_G(T,A) \cong Hom_C(T,A)\ ,\ H^0_G(S,A) \cong Hom_C(S,A) \quad \text{and}$$

$$H^0_G(p,A) \cong H^0_{\mathcal{G}}(COSK^0(p),K(A,1)) \cong Hom_{simp(C)}(COSK^0(p),K(A,1))$$

The content of the exactness of the five exact terms of (ω) is contained in

**Theorem 4.** The set $H^0_G(p,A)$ pointed by the constant map 0 is the set of objects of a pointed groupoid $\mathbb{H}^0_G(p,A)$, whose set of arrows is the set of simplicial homotopies of the elements of $H^0_G(p,A)$. The groupoid $\mathbb{H}^0_G(p,A)$ is equivalent to the pointed subgroupoid of $TORS^1_G(T,A)$ consisting of those A-torsors above T which become trivial when pulled back over p to S. Moreover, in $\mathbb{H}^0_G(p,A)$, $H^0_G(S,A)$ may be identified with the set of homotopies whose target is 0 and $H^0_G(T,A)$ is isomorphic to the group of automorphisms of 0.

Proof.

The set $Hom_{simpl}(COSK^0(p),K(A,1))$ is the set of 1-cocycles on the "covering" defined by the epimorphism p. Since this covering is U-split one has a simplicial map $\mathbb{G}^*T \longrightarrow COSK^0(p)$ above T and by composition with this map, a 1-cocycle con $\mathbb{G}^*T$ with coefficients in A for any element of $H^0_G(p,A)$. Thus, since U in monadic, we have an A-torsor above T which becomes split (i.e. isomorphic to the product torsor defined by the 0-cocycle) when pulled back over p. Under this equivalence, the map $H^0_G(p,A) \longrightarrow H^1_G(T,A)$ is nothing more that the assignment of the 1-cocycle to the isomorphism class of the A-torsor it defines. Since the map $H^1_G(T,A) \longrightarrow H^1_G(S,A)$ corresponds to pullback over p, exactness here is a consequence of the essential surjectivity of the asserted equivalence. Homotopies of 1-cocycles are bijectively equivalent isomorphisms of the corresponding A-torsors above T; a torsor is trivial if and only if its cocycle is homotopic to the 0-cocycle, but this is exactly the result of the difference map $H^0_G(T,A) \longrightarrow H^0_G(p,A)$, which thus identifies $H^0_G(T,A)$ with the set of homotopies whose target is the 0-cocycle via the source of the homotopy. Since p is an effective epimorphism, $H^0_G(S,A)$ is isomorphic to the set whose source and target are the 0-cocycle and exactness of the remainder of the first five terms of the sequence is interpreted. ■

We now turn our attention to the consequences of the isomorphism of $H^1_G(p,A)$ with the set $TORS^1_{simpl}[COSK^0(p),(K(A,1),1)]$ of isomorphism classes of K(A,1)-torsors above $COSK^0(p)$. We will first show that any representative torsor of this set has its "total space" E. simplicially isomorphic to the nerve of a groupoid which is in fact in the monadic case under consideration here, locally equivalent as a category object to the trivial goupoid $T \times A \longrightarrow T$.

**Theorem 5.** Let NER: $GPD(\mathbb{C}) \longrightarrow Simpl(\mathbb{C})$ be the fully faithful functor whose image consists of those simplicial objects which are the nerves of groupoids of $\mathbb{C}$. Then the fully faithful functor of groupoids

$$NER\colon TORS^1_{GPD(\mathbb{C})}(COSK^0(p), K(A,1)) \longrightarrow TORS^1_{Simp(\mathbb{C})}(COSK^0(p), K(A,1))$$

induced by NER on the corresponding categories of U-split torsors is an equivalence.

**Proof.**

Let $\alpha_{..}\colon E_{..} \to K(K(A,1),1)$ be an object in $TORS^1_{Simp(\mathbb{C})}(COSK^0(p), K(A,1))$ which will be pictorically represented as a diagram,

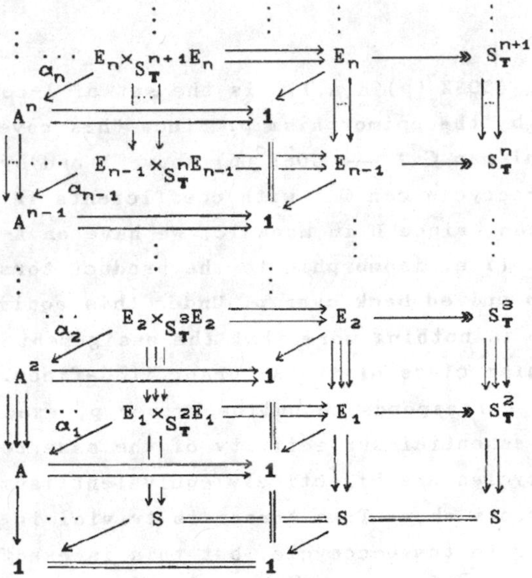


(3)

where all the horizontal planes represent torsors. The lemma will be proved if we see that the vertical complexes E. (at dimension 0), $E. \times_{COSK^0(p)} E.$ (at dimension 1), etc., are isomorphic to nerves of groupoids in $\mathbb{C}$. But, since the vertical complexes at horizontal dimensions $n \geq 1$ are built as simplicial kernels, the above condition can be reduced to seeing that E. is isomorphic to the nerve of a groupoid; but the image under U of this simplicial torsor has a splitting in $Simpl(\mathbb{B})$, so that U(E.) is isomorphic to the product complex

$$U(K(A,1)) \times U(COSK^0(p) \xrightarrow{\ \sim\ } K(U(A),1) \times COSK^0(U(p))$$

which is the nerve of a groupoid equivalent to the constant groupoid ("constant family of groups") $U(T\times A) \xrightarrow{\sim} U(T) \times U(A) \longrightarrow U(T)$ via the functor whose object map is defined by the split epimorphism $U(p)$: $U(S) \longrightarrow U(T)$. Since $U$ reflects finite inverse limits, $E.$ is the nerve of a groupoid, "locally" (i.e. after passage to $\mathbb{B}$ by $U$) equivalent to $T\times A \longrightarrow T$. ∎

**Remark.** The above theorem ([5]) is valid in any Barr exact category $\mathbb{C}$ for any $K(K(A,1),1)$ torsor above $COSK^o(p)$.

In effect, proving that $E.$ is the nerve of a groupoid is equivalent to proving that the canonical morphisms $d$ in the diagram:

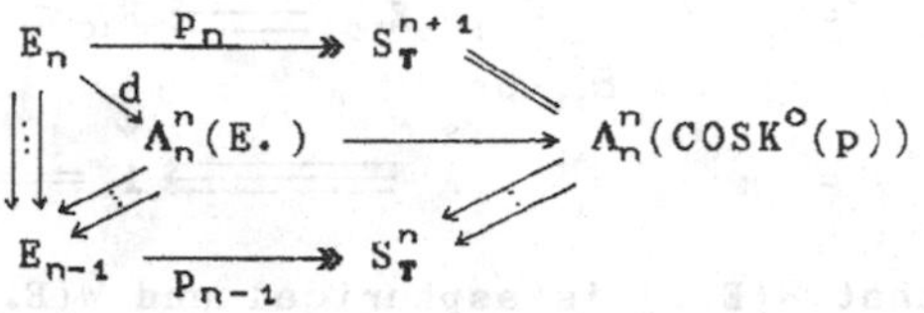


are isomorphisms in all dimensions $n\geq 2$. Let us prove first that the morphisms $d$ are epic (or equivalently, that $E.$ satisfies the Kan condition):

Given $(x_0,\ldots,x_{n-1},-) \in \Lambda_n^n(E.)$, $r=(p_{n-1}x_0,\ldots,p_{n-1}x_{n-1},-)$ is an element in $\Lambda_n^n(COSK^o(p)) = S_T^{n+1}$ and therefore there exists an $x \in E_n$ such that $p_n x = R$; but $p_n(x) = (p_{n-1}d_0x,\ldots,p_{n-1}d_{n-1}x,p_{n-1}d_nx)$ and so the pairs $(d_ix,x_i)$ are all in $E_{n-1}\times_{S_T^n}E_{n-1}$. Let us denote $\xi = (\alpha_{n-1}(d_0x,x_0), \ldots,\alpha_{n-1}(d_{n-1}x,x_{n-1}),-) \in \Lambda_n^n(K(A,1)) = A^n$; then, if we denote $\xi\cdot x$ the action of the element $\xi$ on $x$ in $E_n$, we have $\xi\cdot x \in E_n$ and $d_i(\xi\cdot x) = x_i$ for $0\leq i\leq n-1$. The proof is the same for the rest of the open horns.

Let us see now that $d$ is monic: given $(x_0,\ldots,x_{n-1},-) \in \Lambda_n^n(E.)$ let $f$ and $g$ elements in $E_n$ such that $df = dg$; this means that $d_if = x_i = d_ig$ for $0\leq i\leq n-1$ and then $(f,g) \in E_n\times_{S_T^{n+1}}E_n$. If $p_nf = (s_0,\ldots,s_n)$ and $p_ng = (s'_0,\ldots,s'_n)$, the identities $d_ip_nf = p_{n-1}d_if = p_{n-1}d_ig$, for $0\leq i\leq n-1$, oblige that $s_i = s'_i$ for $0\leq i\leq n$, since $\Lambda_n^n(COSK^o(p)) = S_T^{n+1}$. Finally, if $0\leq i\leq n-1$, $d_i\alpha_n(f,g) = \alpha_{n-1}d_i(f,g) = \alpha_{n-1}(d_if,\ d_ig) = 0$; so, $\alpha_n(f,g) = (0,\ldots,0)$ and, therefore, we conclude that $f = g$. ∎

Now, let $TORS_U^2(p,A)$ denote the category of U-split $K(A,2)$-torsors all of whose augmentations are equal to $p:S \longrightarrow T$ and whose morphisms are the identity on $S$ (as well as $T$). Then $TORS_U^2(p,A)$ is a groupoid whose group of isomorphism classes was shown in [7] to be isomorphic to $H_G^1(p,A)$. The relation of this interpretation to that advanced here is given by the application of the classifying complex functor $\bar{W}$ applied to the planes in (3) defined by $K(K(A,1),1)$ and its action on $E_{\cdot\cdot}$:

**Theorem 8.** $\bar{W}$ defines an equivalence of groupoids,

$$\mathrm{TORS}^1_{\mathrm{GPD(C)}}(\mathrm{COSK}^0(p),A) \xrightarrow{\sim} \mathrm{TORS}^2_{\mathrm{U}}(p,A)$$

Proof.

Let $\alpha_{\cdot\cdot}: E_{\cdot\cdot} \to K(K(A,1),1)$ be an object in $\mathrm{TORS}^1_{\mathrm{GPD(C)}}(\mathrm{COSK}^0(p),K(A,1))$. By applying $\bar{W}$ to the double simplicial morphism $\alpha_{\cdot\cdot}$, we obtain a simplicial morphism $\bar{W}(\alpha_{\cdot\cdot})$: $\bar{W}(E_{\cdot\cdot}) \to K(A,2)$, with $\bar{W}(E_{\cdot\cdot})$ augmentated by $p: S \twoheadrightarrow T$,

$$\begin{array}{ccccccccc}
\bar{W}(E_{\cdot\cdot}) & = & \cdots & E_{11}\times_S E_{10} & \rightrightarrows & E_{10} & \rightrightarrows & S & \twoheadrightarrow T \\
\downarrow{\scriptstyle \bar{W}(\alpha_{\cdot\cdot})} & & & \downarrow{\scriptstyle \alpha_{11}\cdot pr} & & \downarrow & & \downarrow & \\
\bar{W}(K(A,1),1) & = & \cdots & A & \rightrightarrows & 1 & = & 1 &
\end{array}$$

It is easy to see that $\bar{W}(E_{\cdot\cdot})$ is aspherical and $\bar{W}(E_{\cdot\cdot})=\mathrm{COSK}^1(\bar{W}(E_{\cdot\cdot}))$. Finally, since the square
$\begin{array}{ccc} E_{11} & \rightrightarrows & E_{10} \\ \downarrow{\scriptstyle \alpha_{11}} & & \downarrow \\ A & \rightrightarrows & 1 \end{array}$
is a pullback, so is
$\begin{array}{ccc} E_{11}\times_S E_{10} & \to & E_{10}\times_S E_{10} \\ \downarrow & & \downarrow \\ A & \longrightarrow & 1 \end{array}$.
Consequently, $\bar{W}(\alpha_{\cdot\cdot}) : \bar{W}(E_{\cdot\cdot}) \to K(A,2)$ is an exact fibration and, therefore, an object in $\mathrm{TORS}^2(p,A)$.

An inverse for this assignement is given as follows: if $\beta_{\cdot}: E_{\cdot} \to K(A,2)$ is in $\mathrm{TORS}^2(p,A)$, its attached 1-torsor and its fiber groupoid determine, respectively, the bottom horizontal plane and the vertical complex at dimension 1 of a diagram like (3), which represents an object in the category $\mathrm{TORS}^1_{\mathrm{GPD(C)}}(\mathrm{COSK}^0(p),K(A,1))$. ■

The corresponding assignment of a representative of $H^1_G(p,A)$ to its class as a 2-torsor above T now interprets the second coboundary homomorphism of the sequence (2) whose exactness can now be fitted into a pattern similar to that given in theorem 3.

In this context, it is interesting to re-establish directly the conection of theorems 3 y 4 with J. L. Loday's work ([14]) which interpreted his group Eilenberg-Mac Lane "relative 3-cocycles" using the group of congruence classes of crossed modules above T which, given the coincidence of the monadic cohomology of groups with those of Eilen-

berg-Mac Lane (after a one unit dimension shift), corresponds to our $H^1_G(p,A)$. (See also van Osdol ([18])).
Loday proves that the elements of $H^1_G(p,A)$ can be seen as 'congruence' classes of certain 2-fold extensions of T by the abelian group A, relative to p. Such an extension is an exact sequence

$$\mathbb{E} = \ 1 \longrightarrow A \longrightarrow E \xrightarrow{\rho} S \xrightarrow{p} T \longrightarrow 1$$

in which $\rho$: S $\longrightarrow$ T is a crossed module for which the restriction to A of the action of S on E is trivial (A is then a central subgroup of E). Two such extensions, $\mathbb{E}$ and $\mathbb{E}'$, are said to be "congruent" iff there exists a morphism of S-groups, f: E $\longrightarrow$ E', making commutative the following diagram:

$$\begin{array}{ccccccc} A & \rightarrowtail & E & \xrightarrow{\rho} & S & \twoheadrightarrow & T \\ \| & & \downarrow{\scriptstyle f} & & \| & & \| \\ A & \rightarrowtail & E' & \xrightarrow{\rho'} & S & \twoheadrightarrow & T \end{array}$$

f is then necessarily an isomorphism, and if $\mathcal{E}xt(p,A)$ denotes the corresponding set consisting of congruence classes of these extensions, Loday obtained a natural bijection:

$$\mathcal{E}xt(p,A) \cong H^1_G(p,A)$$

We now produce this directly by using the equivalence between crossed modules and internal groupoids in groups; Given an extension in the sense of Loday $\mathbb{E}$, the crossed module $\rho$: E $\longrightarrow$ S determines an internal groupoid in **Gr**, $E \rtimes S \rightrightarrows S$, and the whole extension, a 2-torsor over T under A,

$$\begin{array}{ccccccc} \cdots & A & \rightrightarrows & E \rtimes S & \rightrightarrows & S & \twoheadrightarrow T \\ & \beta_2 \downarrow & & & & & \\ & A & & & & & \end{array}$$

whose fiber groupoid is the above one (note that $\beta_2$ is given by $\beta_2((e_0,s_0),(e_1,s_0),(e_2,\rho(e_0)s_0)) = e_0-e_1+e_2$; see [7]). Thus we have a bijection between the set of congruence classes of extensions of p by A, and the set of Yoneda classes of 2-torsors over T under A, which have fixed augmentation p (here we too require torsor morphisms to have the morphism $id_S$ in dimension 0),

$$H^1_G(p,A) \cong \mathcal{E}xt(p,A) \cong TORS^2_U[p,A]$$

Finally, let us remark that in the more general context where $\mathbb{C}$ a Barr-exact category, p: S ↠ T a regular epimorphism and A an abelian group object in $\mathbb{C}$, we still have a short exact sequence of abelian group objects in Simpl($\mathbb{C}$) like (1) and therefore, applying Glenn's techniques, a long exact sequence,

$$0 \to Hom(COSK^{0}(p), K(A,0)) \to Hom(COSK^{0}(p), L(A,0)) \to Hom(COSK^{0}(p), K(A,1))$$

$$TORS^{1}[COSK^{0}(p), K(A,0)] \to TORS^{1}[COSK^{0}(p), L(A,0)] \to TORS^{1}[COSK^{0}(p), K(A,1)] \cdots$$

where the groups of torsors are those defined in [10]. So, associated to an epimorphism p and an abelian group object A, we have a long exact sequence as above, whose initial segment can be shown to coincide with the 6-term sequence obtained in this same context by Aznar and Cegarra in [7]:

$$0 \to Hom_{\mathbb{C}}(T,A) \to Hom_{\mathbb{C}}(S,A) \to TORS^{0}[p,A]$$

$$TORS^{1}[T,A] \to TORS^{1}[S,A] \to TORS^{1}[p,A] \to \cdots$$

Although it is tempting to conjeture that the continuation has these coincidence isomorphisms also continuing for all n, at present we can prove this completely only through dimension 1, with the higher dimensions awaiting a "descent theory" for K(A,n)-torsors in this general Barr-exact category context.

**References**

[1] Aznar,E.R. Cohomologia no abeliana en categorias de interes. Alxebra 33 (1981)

[2] Barr,M.-Beck,J. Homology and standard constructions. Lec. Not. in Math. 80. Springer (1969)

[3] Beck,J. Triples, Algebras and Cohomology. Dissert. Columbia (1967)

[4] Bullejos,M. Cohomologia no abeliana (la sucesion exacta larga). Cuadernos de Algebra. Granada (1987)

[5] Carrasco, P. Cohomologia de haces. Cuadernos de Algebra. Granada (1987)

[6] Cegarra,A.M. Cohomologia Varietal. Alxebra (1980)

[7] Cegarra,A.M.-Aznar,E.R. An exact sequence in the first variable for Torsor Cohomology: the 2-dimensional theory of obstructions. J.P. and Appl. Algebra 39, 197-250 (1986)

[8] Duskin,J. Simplicial methods and the interpretation of triple cohomology. Mem. A.M.S. (2), 163 (1975)

[9] Eilenberg,S.-Moore,J.C. Adjoint functors and triples. Ill. J. Math. 9 (1965)

[10] Glenn,P. Realization of cohomology classes in arbitrary exact categories. J. P. Appl. Algebra 25 (1), 33-107 (1982)

[11] Herrlich,H.-Strecker,G. Category Theory. Allyn&Bacon (1973)

[12] Higgins,P.J. Categories and Groupoids. Van Nostrand Reinhred. Math. Studies 32 (1971)

[13] Lirola,A. Cohomologia de torsores relativos. Mem. Lic. U. Granada (1982)

[14] Loday,J.L. Cohomologie et groupe de Steinberg relatifs. J. Algebra 54, 178-202 (1978)

[15] May,J.P. Simplicial objects in Algebraic Topology. Van Nostrand (1967)

[16] R-Grandjean,A. Homologia en categorias exactas. Alxebra 4 (1970)

[17] Rinehart,G.S. Satellites and Cohomology. J. Algebra 12, 295-329 (1969)

[18] Van Osdol,D.H. Long exact sequences in the first variable for algebraic cohomology theories. J. P. Appl. Algebra 23 (3), 271-309 (1982)

This paper is in its final form and will not be published elsewhere.

# DECKER'S SHARPER KÜNNETH FORMULA

by Saunders Mac Lane*

For chain complexes $K_1$ and $K_2$ of abelian groups, where $K_1$ has no torsion, the familiar Künneth formula ([11] and [9]) gives the integral homology of the tensor product complex $K_1 \otimes K_2$ by a short exact sequence

$$0 \to \sum_{p+q=n} H_p(K_1) \otimes H_q(K_2) \to H_n(K_1 \otimes K_2) \to \sum_{p+q=n-1} (\mathrm{Tor}(H_p(K_1), H_p(K_2)) \to 0. \quad (1)$$

The sequence splits, by a homomorphism which is not natural. This note describes additional data on the complexes $K_1$ and $K_2$ which provides a splitting for (1) natural for this data. It thus presents the homology of $K_1 \otimes K_2$ as a direct sum of graded abelian groups

$$\sum H_p(K_1) \otimes H_q(K_2) \oplus \sum \mathrm{Tor}(H_p(K_1), H_q(K_2)), \quad (2)$$

with the grading indicated in (1). This additional data was first introduced in the unpublished Ph.D. thesis of Gerald J. Decker [6]. This thesis is closely related to an earlier equally unpublished thesis of Ross Hamsher [8]. Decker's thesis is concerned with the homology of the tensor product of two or more differential graded algebras; he describes a torsion product algebra which he (and earlier, Hamsher) employ to give a more explicit description of the integral homology of Eilenberg-Mac Lane spaces. (This description is based on Cartan's analysis of constructions [5]). In the present paper Decker's technique is reorganized for the case of complexes of free abelian groups, with a view to these and other possible specific calculations.

In the sequel, all chain complexes $K$ will be complexes of abelian groups, with each $K_n$ a free abelian group. The arguments used here apply equally well if each $K_n$ is a free module over a principal ideal domain. Indeed, the crucial point of it all is a systematic use of greater common divisor calculations for the homology of such modules; the relevance of such calculations has long been recognized.

The validity of the Künneth sequence rests on the observation that there are two kinds of cycles in a tensor product $K_1 \otimes K_2$ of complexes. First, if $z_1$ and $z_2$ are cycles in $K_1$ and $K_2$ then $z_1 \otimes z_2$ is a cycle in $K_1 \otimes K_2$. This

---

*Research supported by the National Science Foundation, under grant DMS #8420698.

(geometrically evident) operation, "product of two cycles" is the source of the terms $H(K_1) \otimes H(K_2)$ in the formula (1). Second, if the homology classes of $z_1$ and $z_2$ both have the same finite order $h$ (for $h \neq 0$ in $\mathbb{Z}$), then there are chains $u_1$ and $u_2$ with $\partial u_i = hz_i$, so in this case there is an additional homology class in $K_1 \otimes K_2$, the class of the cycle

$$\rho_h(z_1,z_2) = (1/h)\partial(u_1 \otimes u_2) = z_1 \otimes u_2 + (-1)^{d_1} u_1 \otimes z_2, \quad d_1 = \dim u_1. \tag{3}$$

This cycle, of more obscure geometric provenance, has dimension $d_1 + d_2 - 1$, and is the origin of the torsion terms in the Künneth sequence. Indeed, if we write the homology class of the cycle (3) as

$$\tau_h(z_1',z_2'), \qquad z_i' = \text{homology class of } z_i, \tag{4}$$

then these symbols $\tau_h$ are linear in $z_1'$ and $z_2'$ and satisfy the "slide" relations,

$$\tau_{hk}(z_1',z_2') = \tau_h(z_1',kz_2'), \qquad \tau_{hk}(z_1',z_2') = \tau_h(kz_1',z_2') \tag{5}$$

whenever both sides are defined. Indeed, Eilenberg and Mac Lane [7] observed that the torsion product $\mathrm{Tor}(M_1,M_2)$ of two abelian groups $M_1$ and $M_2$ may be described as an abelian group with generators $\tau_h(m_1,m_2)$ for $hm_1 = 0 = hm_2$, with $h \neq 0$ and $m_i \in M_i$, with the stated relations (5).

Now the homology class (4) as defined in (3) depends on the choice of the "h-caps" $u_1$ and $u_2$ for the cycles $z_1$ and $z_2$ of homology order $h$. Altering a cap $u$ changes the class $\tau_h$ by a tensor product of cycles, and this is why the Künneth splitting is not natural. But consider the Bokštein exact sequence

$$\cdots \longrightarrow H_{n+1}(K,Z_h) \xrightarrow{\beta_h} H_n(K,Z) \xrightarrow{h} H_n(K,Z) \longrightarrow \cdots$$

for a complex $K$. A homomorphism $\psi_h$ right inverse to the Bokštein boundary

$$\psi_h\colon {}_hH_n(K,Z) \longrightarrow H_{n+1}(K,Z_h), \quad \beta_h\psi_h = 1 \tag{6}$$

(where ${}_hM = \{m \mid m \in M \text{ and } hm = 0\}$) will be called an h-cap. Consistent caps for $K$ is a system of such $\psi_h$ for all $h \neq 0$ such that the following diagrams (Decker) commute for all $h,k \neq 0$:

$$\begin{array}{ccccc} H_{n+1}(K,Z_{hk}) & \xrightarrow{\lambda} & H_{n+1}(K,Z_h) & \xrightarrow{k_*} & H_{n+1}(K,Z_{hk}) \\ \uparrow \psi_{hk} & & \uparrow \psi_h & & \uparrow \psi_{hk} \\ {}_{hk}H_n(K) & \xrightarrow[k_*]{} & {}_hH_n(K) & \xrightarrow[\nu]{} & {}_{hk}H_n(K) \end{array} \qquad (7)$$

where $k_*$ is multiplication by $k$, $\nu$ is inclusion, and $\lambda$ is induced by the projection $Z_{hk} \to Z_h$. Given consistent caps in $K_1$ and $K_2$ one may describe the torsion cycle (3) more definitely by taking each cap $u_i$ in the class of $\psi_h(z_i)$. The resulting homology class is then uniquely defined. Thus the sharp Künneth theorem: For free chain complexes of abelian groups with consistent caps, the Künneth sequence is naturally a direct sum (2).

This result also holds for an m-fold tensor product of complexes, using Bokštein's generalizations of Mac Lane's [12] triple torsion product "trip". This generality is needed for application to Cartan's constructions for $K(\pi,n)$.

For the case of complex K-theory Bödigheimer [1] has shown that the Künneth exact sequence for K-theory does split. His argument uses the same description of the torsion product by generators and relations, but is based upon an (apparently different) splitting of the universal coefficient sequences for K-theory.

2. Consistent Caps. The "consistent caps" required in (7) to split Bokštein occur in nature. Most important, all of Cartan's little constructions (tensor products of exterior and divided power algebras) have naturally given consistent caps. More generally, any chain complex $K$ of free abelian groups, finitely generated in each dimension, has such caps. For one can write $K$ (in many ways) as a direct sum of elementary complexes. An elementary complex $Z$ in some dimension, with boundary zero, has Bokštein $\beta_h = 0$, so can go uncapped. The other alternative is an elementary complex $Z(u) \to Z(v)$ with boundary $\partial u = mv$ for some integer $m$. Write $v'$ for the homology class of the cycle $v$. If $d = (m,h)$ is the g.c.d., with $m = m_0 d$ and $h = h_0 d$, $\psi_h(m_0 v') = (h_0 u)'$ gives a system of consistent caps (the only system possible for this elementary complex). For a sum of elementary complexes, one can then take the sum of these consistent caps.

I do not know whether a complex $K$ which is infinitely generated necessarily has a system of consistent caps.

For graded abelian groups $A_1,\ldots,A_n$, the n-fold tensor product $A_1 \otimes \cdots \otimes A_n$ has (counting itself) $n$ derived functors not necessarily zero. We will denote the $(i-1)^{st}$ derived functor by

$$\mathrm{Mult}_i(A_1,\ldots,A_n);$$

thus $\text{Mult}_1$ is the tensor product, while $\text{Mult}_2(A_1,A_2,A_3)$ was called "Trip" in Mac Lane [12]. In our development, we will start by descibing these "multiple" torsion products by generators and relations. Our main theorem then asserts, for chain complexes $K_i$ of free abelian groups, that there is a (non-natural) isomorphism

$$H(K_1 \otimes \cdots \otimes K_n) \cong \sum_{i=1}^{n} \text{Mult}_i(H(K_1),\ldots,H(K_n)). \tag{8}$$

Moreover, if each $K_i$ is equipped with consistent caps, this isomorphism is natural (for chain transformations respecting the consistent caps).

This result is sharper than that given by simple iteration of the ordinary Künneth exact sequence. As noted in [12] for $n = 3$, that iteration would yield a direct sum of 4 terms, compared to the three in (8).

3. Higher Torsion products. For $A_1,\ldots,A_n$ graded abelian groups, the n-fold torsion product $\text{Tor}(A_1,\ldots,A_n)$ is defined as the group with generators $\tau_h(a_1,\ldots,a_n)$ for $h \neq 0$ and $ha_i = 0$ in $A_i$, and with the following relations: Additivity in each argument $a_i$, and the "slide" relations

$$\tau_{hk}(a_1,\ldots,a_i,\ldots,a_n) = \tau_h(ka_1,\ldots,ka_{i-1},a_i,ka_{i+1},\ldots,ka_n), \tag{9}$$

the latter for all $i$ and whenever both sides are defined; i.e., when $hka_j = 0$ for all $j$ and $ha_i = 0$. This torsion product is graded by setting

$$\deg \tau_h(a_1,\ldots,a_n) = -1 + \sum_{i=1}^{n} (\deg a_i + 1), \tag{10}$$

as befits the intended application, in which $\tau_h$ is to be the class of a cycle

$$\rho_h(z_1,\ldots,z_n) = 1/h\ \partial(u_1 \otimes \cdots \otimes u_n) = \sum_{i=1}^{n} \pm u_1 \otimes \cdots \otimes \partial u_i \otimes \cdots \otimes u_n. \tag{11}$$

It will turn out, as in Mac Lane [12], that the torsion product is the (n-1)st derived functor of the additive functor $A_1 \otimes \cdots \otimes A_n$.

This torsion product is clearly additive. It may be calculated for finitely generated abelian groups from the case of cyclic groups $Z_m$ of orders $m$ by the formula

$$\theta\colon \text{Tor}(A, Z_{m_2},\ldots,Z_{m_n}) \cong {}_dA \tag{12}$$

where $d$ is the greatest common divisor $(m_2,\ldots,m_n)$ and ${}_dA = \{a \mid da = 0\}$. This formula is obtained by defining $\theta$ as

$$\theta\tau_h(a,x_2,\ldots,x_n) = (hx_2/m_2)\ldots(hx_n/m_n)a, \tag{13}$$

where the $x_i$ are integers with $hx_i \equiv 0 \pmod{m_i}$, by assumption. One checks readily that this map $\theta$ respects the defining relations for Tor. To show it an isomorphism, introduce a putative inverse $\theta'$ defined for $a$ in $A$ by setting

$$\theta' a = \tau_d(a, m_2/d, \ldots, m_n/d), \; da = 0. \tag{14}$$

Trivially $\theta\theta' = 1$. Also $\theta'\theta = 1$ on generators of the form $\tau_h(a,x_2,x_3,\ldots,x_n)$ where $h$ is a divisor of $d$. But it suffices to consider only such generators. For example, for $n = 3$ consider any generator $\tau_h(a,x_2,x_3)$, and set $(h,m_3) = e$ so that $h = h_0e$ for some $h_0$. Then $ex_3 \equiv 0$, so the defining relations yield

$$\tau_h(a,x_2,x_3) = \tau_{h_0e}(a,x_2,x_3) = \tau_e(h_0a,h_0x_2,x_3),$$

and the new subscript $e$ is now a divisor of $m_3$. Iterating this process, every generator reduces to a $\tau_h$ with $h$ a divisor of $d$. Since $\theta$ preserves the defining relations, it follows that $\theta'\theta = 1$.

This formula (12) determines the functor Tor when all the arguments are cyclic groups. However, I was unable to find a proof which is symmetric in all the arguments, and my earlier paper [12] states the result (for $n = 3$) without any detail of the calculation.

For chain complexes $K_i$ with consistent caps and for cycles $z_i$ with homology classes $a_i$ of order $h$, choice of $u_i \in \psi_h(z_i')$ and the formula (11) defines a homomorphism

$$\xi\colon \operatorname{Tor}(H(K_1),\ldots,H(K_n)) \longrightarrow H(K_1 \otimes \cdots \otimes K_n) \tag{15}$$

sending $\tau_h(a_1,\ldots,a_n)$ into the homology class of the cycle $\rho_h(z_1,\ldots,z_n)$ of (11).

4. <u>The product Trip</u>. For the tensor product $K_1 \otimes K_2 \otimes K_3$ of three chain complexes, it turns out that every homology class is a sum of classes of the following types

$$z_1 \otimes z_2 \times z_3, \qquad \rho_h(z_1,z_2,z_3),$$

$$\rho_h(z_1,z_2) \otimes z_3, \qquad \rho_h(z_1,z_3) \otimes z_2, \qquad \rho_h(z_2,z_3) \otimes z_1,$$

where each $z_i$ is a cycle of $K_i$, and, in the arguments of $\rho_h$, $hz_i$ is a boundary. Moreover, the last three types satisfy a sort of Jacobi relation

$$(-1)^{e_1}\rho_h(z_1,z_2) \otimes z_3 + (-1)^{e_2}\rho_h(z_1,z_3) \otimes z_2 + (-1)^{e_3}\rho_h(z_2,z_3) \otimes z_1 \sim 0 \quad (16)$$

with sign exponents $e_1 = |z_3| + 1$, $e_2 = |z_3|(|z_2| + 1)$, and $e_3 = (|z_1| + 1)(|z_2| + |z_3| + 1)$, where $|z_i|$ denotes the degree of the cycle $z_i$. This leads to the definition of the functor $\mathrm{Mult}_2(A_1,A_2,A_3)$ of graded abelian groups $A_i$, by generators and relations, as the quotient

$$[\mathrm{Tor}(A_1,A_2) \otimes A_3 + \mathrm{Tor}(A_1,A_3) \otimes A_2 + \mathrm{Tor}(A_2,A_3) \otimes A_1]/E \quad (17)$$

where $E$ is the set of all relations of the form (16); that is, all

$$(-1)^{e_1}\tau_h(a_1,a_2) \otimes a_3 + (-1)^{e_2}\tau_h(a_1,a_3) \otimes a_2 + (-1)^{e_3}\tau_h(a_2,a_3) \otimes a_1, \quad (18)$$

where $h \neq 0$, $a_i \in A_i$ has $ha_i = 0$ and degree $|a_i|$, while $e_1 = |a_3| + 1$, $e_2 = |a_3|(|a_2| + 1)$ and $e_3 = (|a_1| + 1)(|a_2| + |a_3| + 1)$. This functor $\mathrm{Mult}_2$ under the name "Trip" was introduced by Mac Lane in [12]; there he formulated the observation above about all the homology classes in $K_1 \otimes K_2 \otimes K_3$ as the statement that $H(K_1 \otimes K_2 \otimes K_3)$ is the (non-natural) direct sum

$$H(K_1) \otimes H(K_2) \otimes H(K_3) + \mathrm{Mult}_2(H(K_1),H(K_2),H(K_3)) + \mathrm{Tor}(H(K_1),H(K_2),H(K_3)).$$

Subsequently Bokštein in [4] obtained the analogous result for a tensor product of $n$ chain complexes. This result uses functors $\mathrm{Mult}_i$ of $n$ graded abelian groups $A_i$ defined in terms of shuffles of the integers $\{1,2,\ldots,n\}$. Thus an $(i,n-i)$ shuffle is an ordered partition $\{s_1 < \ldots < s_i; r_1 < \ldots < r_{n-i}\}$ of these integers into two disjoint parts, labelled $s$ and $r$. Then $\mathrm{Mult}_i$ for $2 \leq i < n$ is defined as

$$\mathrm{Mult}_i(A_1,\ldots,A_n) = [\textstyle\sum \mathrm{Tor}(A_{s_1},\ldots,A_{s_i}) \otimes A_{r_1} \otimes \cdots \otimes A_{r_{n-i}}]/E, \quad (19)$$

where the direct sum $\sum$ is taken over all $(i,n-i)$ shuffles $(s,r)$, while the set $E$ includes a number of relations of the form

$$[\sum_{j=1}^{i+1}(-1)^{(|a_j|+1)|\tau_h(a_{j+1},\dots,a_{i+1})|}\tau_h(a_1,\dots,\hat{a}_j,\dots,a_{i+1})\otimes a_j]\otimes a_{i+2}\otimes\cdots\otimes a_n . \quad (20)$$

Here $a_j \in A_j$ has degree $|a_j|$, while $ha_j = 0$ if $1 \leq j < i+1$ and $|\tau_h(a)| = |a|$, $|\tau_h(-)| = 1$, in agreement with the signs already appearing in (18) above; in this and similar formulas, the term $\hat{a}_j$ is to be omitted. Moreover, $E$ is to include exactly all the relations obtained from this one by replacing the two rows of indices $(1,2,\dots,i+1)$, $(i+2,\dots,n)$ by any $(i+1,n-i-1)$ shuffle of the set $\{1,\dots,n\}$.

This defines $\mathrm{Mult}_i$ for $2 \leq i < n$. To complete the definition, set

$$\mathrm{Mult}_1(A_1,\dots,A_n) = A_1 \otimes \cdots \otimes A_n, \dots \quad (21)$$

$$\mathrm{Mult}_n(A_1,\dots,A_n) = \mathrm{Tor}(A_1,\dots,A_n) \quad (22)$$

In these terms, our result will be

Theorem. For chain complexes $K_i$ of free abelian groups with consistent caps there is an isomorphism, natural for chain complexes with consistent caps,

$$H_m(K_1 \otimes \cdots \otimes K_n) \cong \sum_{i=1}^{n} \sum_{p_1+\dots+p_n=m-i+1} \mathrm{Mult}_i(H_{p_1}(K_1),\dots,H_{p_n}(K_n)). \quad (23)$$

For complexes without consistent caps, there is still such an isomorphism, which is not natural (i.e., which arises as does the usual non-natural splitting of the Künneth sequence).

From this theorem it will follow that $\mathrm{Mult}_i$ as just now defined is the $(i-1)^{st}$ derived function of the n-fold tensor product $A_1 \otimes \cdots \otimes A_n$. For take for each abelian group $A_i$ (concentrated in dimension 0) a short free resolution $0 \to R_i \to F_i \to A_i \to 0$. By definition, the $j^{th}$ derived functor is then the homology $H_j(K_1 \otimes \cdots \otimes K_n)$, where $K_i$ is the complex with boundary $\partial: R_i \to F_i$, dimension 1 to dimension 0. Since $H_0(K_i) = A_i$, this derived functor, by the theorem, is exactly $\mathrm{Mult}_{j+1}$.

In establishing this theorem we will use a homomorphism

$$\xi_i: \mathrm{Mult}_i(H(K_1),\dots,H(K_n)) \longrightarrow H(K_1 \otimes \cdots \otimes K_n) \quad (24)$$

defined for $1 < i < n$ in an evident way on the generators (19) of the group $\mathrm{Mult}_i$. Specifically, each factor $\mathrm{Tor}(H(K_{s_1}),\dots,H(K_{s_i}))$ is sent into

$H(K_{s_1} \otimes \cdots \otimes K_{s_i})$ by the map $\xi$ of (15); the resulting cycle (sum of chains) is multiplied by a cycle of $H(K_{r_1}) \otimes \cdots \otimes H(K_{r_s})$, and the resulting cycle is then mapped into a cycle of $H(K_1 \otimes \cdots \otimes K_n)$ by the usual rule for permuting the product of two chains $u$ and $v$. (Note that the cycles in question, like $\rho_h$, are sums of chains, so the rule must be applied to each summand of this sum. It is routine to check that this map $\xi_i$ does respect the relation $E$ used to define $\mathrm{Mult}_i$.

5. The Filtration. Even without a choice of caps the map $\xi$ of (15) is well defined "modulo products". This is just the generalization of the usual splitting of the ordinary Künneth exact sequence. For example, with three factors, consider three cycles $z_i$ of order $h$ and with caps $u_i$, $hu_i = hz_i$. In the corresponding torsion cycle

$$\rho_h(z_1,z_2,z_3) = z_1 \otimes u_2 \otimes u_3 \pm u_1 \otimes z_2 \otimes u_3 \pm u_1 \otimes u_2 \otimes z_3$$

change the cap $u_1$ to $v_1 = u_1 + w_1$, where $\partial w_1 = 0$. Then $\rho_h$ changes by

$$w_1 \otimes (z_2 \otimes u_3 \pm u_2 \otimes z_3) = w_1 \otimes \rho_h(z_2,z_3);$$

in other words, changes by a product of cycles. This suggests the introduction of a filtration

$$0 \subset F_1 \subset F_2 \subset \cdots \subset F_n = H(K_1 \otimes \cdots \otimes K_n) \tag{25}$$

defined as follows. First, $F_1$ is the image of $H(K_1) \otimes \cdots \otimes H(K_n)$ under the product. Now consider an $(i,n-i)$ shuffle $s_1 < \cdots < s_i$; $r_1 < \cdots < r_{n-i}$ of the $n$ integers $\{1,\ldots,n\}$. The product of homology classes followed by the appropriate signed permutation then yields a homomorphism

$$H(K_{s_1} \otimes \cdots \otimes K_{s_i}) \otimes H(K_{r_1}) \otimes \cdots \otimes H(K_{r_{n-i}}) \longrightarrow H(K_1 \otimes \cdots \otimes K_n);$$

we define $F_i$ in the filtration (25) to be the union of the images of these homomorphisms, for all $(i,n-i)$ shuffles. In particular, for $i = 1$ this does give $F_1$ exactly as first described: The union of all products of $n$ homology classes, one from each factor complex $K$.

For the direct sum $\sum \mathrm{Mult}_i(H(K_1,\ldots,H(K_n))$ we now take a corresponding filtration $0 \subset G_1 \subset \cdots \subset G_n$, defined by

$$G_i = \textstyle\sum_{j\leq i} \mathrm{Mult}_i . \tag{26}$$

It is then clear that the maps $\xi_i$ (that is, the map $\sum\xi_i$) carries the filtration $G_i$ into the filtration $F_i$. Thus to prove the sharp Künneth theorem stated above, we need only show that this map for each $i$ yields an isomorphism

$$\eta_i : \underset{R_i}{\underset{\|}{G_i/G_{i-1}}} \cong \underset{Q_i}{\underset{\|}{F_i/F_{i-1}}}$$

on the filtration quotients, which we label as $R_i$ and $Q_i$. For that matter, it will be clear that this isomorphism (like $\xi$) does not depend on the choice of caps, consistent or otherwise. It is a natural transformation between two functors of $K_1,\ldots,K_n$.

6. <u>The Demonstration</u>. In this map $\eta_i : R_i \to Q_i$ each functor $R_i$ or $Q_i$, like the functors $\otimes$ and Tor, is defined by means of generators and relations; hence each functor commutes with direct limits. Since each complex $K_i$ is a direct limit (in fact, a union) of its finitely generated subcomplexes, it suffices to prove that each $\eta_i$ is an isomorphism when each of its arguments $K_j$ is finitely generated in each dimension, and bounded below. Now homology is an additive functor of complexes, in the usual sense that the injections of the direct sum $K \oplus K'$ yield an "additivity" isomorphism $H_n(K) \oplus H_n(K') \cong H_n(K \oplus K')$. Also the definitions of Tor and of tensor require additivity in each argument, so it follows that each of the functors $R_i$ and $Q_i$ is additive in each of its arguments. But classically each complex $K$ of finitely generated free abelian groups is a direct sum of elementary complexes, so it suffices by additivity to prove that $\eta_i$ is an isomorphism when each argument $K$ is an elementary complex.

There are two types of elementary complexes: The complex $Z$ whose only non-zero group is the group $Z$ in one dimension, say in dimension 0, and the complex $E^m$ with one-generator free abelian groups $Z(u)$ and $Z(x)$ in successive dimensions (say dimensions 1 and 0) and boundary $\partial u = mx$ for some integer $m$

$$E^m:\ \partial\colon Z(u) \longrightarrow Z(x), \qquad \partial u = mx. \tag{27}$$

Now suppose first that one of the complexes, say the complex $K_1$, is the trivial complex $Z$. Then the classical Künneth theorem for two factor complexes $Z$ and $K_2 \times \cdots \times K_n$ yields a natural isomorphism

$$H_k(K_2 \otimes \cdots \otimes K_n) \cong Z \otimes H_k(K_2 \otimes \cdots \otimes K_n) \cong H_k(Z \otimes K_2 \otimes \cdots \otimes K_n).$$

For the filtration quotients $Q_i = F_i/F_{i-1}$ this gives a corresponding isomorphism

$$Q_i(K_2,\ldots,K_n) \cong Q_i(Z,K_2,\ldots,K_n).$$

On the other hand, consider the functor $\mathrm{Mult}_j(A_1,\ldots,A_n)$ when the first abelian group $A_1$ is $Z$. In the definition $\sum \mathrm{Tor}(A_{s_1},\ldots,A_{s_j}) \otimes A_{r_1} \otimes \cdots \otimes A_{r_{n-j}}/E$ all terms where $A_1 = Z$ appears inside Tor must vanish, by the definition of Tor. Hence $r_1 = 1$, $A_{r_1} = Z$ and may be dropped, so the summation is now taken over all $(j,n-j-1)$ shuffles $(s,r)$ of the integers $(2,\ldots,n)$. The identity $E$ reduce similarly, so that one gets an isomorphism

$$\mathrm{Mult}_j(Z,A_2,\ldots,A_n) \cong \mathrm{Mult}_j(A_2,\ldots,A_n)$$

which in its turn implies a corresponding isomorphism for complexes:

$$R_i(Z,K_2,\ldots,K_n) \cong R_i(K_2,\ldots,K_n).$$

After perhaps several applications of these isomorphisms all elementary complexes $Z$ are removed. Hence it will suffice to demonstrate that each $\eta_i$ is an isomorphism when each of the complexes $K_i$ is of type $E^m$ for some $m$.

Next we reduce to the "primary" case. For consider an elementary complex $E^m$ when $m = kk'$ is the product of two relatively prime integers $k$ and $k'$; we claim that $E^m$ then reduces to $E^k \oplus E^{k'}$. For take $\partial u = kx$ and $\partial u' = k'x'$ in $E^k$ and $E^{k'}$, and choose integers $t$ and $t'$ with $tk + t'k' = 1$. Then replace the basis $u$, $u'$ in dimension 1 of $E^k \oplus E^{k'}$ by

$$b_1 = k'u + ku', \quad c_1 = -tu + t'u',$$

and similarly $x$, $x'$ in dimension 0 by

$$b_0 = x + x' \quad c_0 = -tkx + t'k'x',$$

with a change by a transformation of determinant 1 in each case. Then $\partial b_1 = c_0$ and $\partial c_1 = c_0$. Hence $E^k \oplus E^{k'}$ is isomorphic to $E^m \oplus E^1$. Since $E^1$ has homology zero, the projection $E^m \oplus E^1 \to E^m$ is a homology isomorphism. This means that each elementary complex $E^m$ may be replaced by $E^k + E^{k'}$, where $kk' = m$ and $k$ and $k'$ are relatively prime. Iterating this replacement reduces us to elementary complexes $E^k$ where each $k$ is a prime power. (This is of course essentially the classical reduction of finite abelian groups to the "primary" case).

Now consider $\eta_i: R_i(K_1,\ldots,K_n) \to Q_i(K_1,\ldots,K_n)$ when all the arguments $K_j$ are primary elementary complexes, with at least two different primes involved. In this case, we claim that both terms $R_i$ and $Q_i$ are zero. First $Q_i$ is a subquotient of $H(K_1 \otimes \cdots \otimes K_n)$. But the homology is zero, for if $K_1 = E^m$ and $K_2 = E^{m'}$ with $(m,m') = 1$ one readily calculates that $H(K_1 \otimes K_2) = 0$, and hence the homology of the whole tensor product is zero. On the other side $R_i$ is built up from $\otimes$ and torsion products. But

$$Z_m \otimes Z_{m'} = Z_{(m,m')} \quad \text{and} \quad \mathrm{Tor}(Z_m, Z_{m'}) = Z_{(m,m')}$$

so all these terms vanish when the g.c.d $(m,m') = 1$.

We are left with the case in which all the complexes $K_i$ are elementary complexes $E^{m_i}$ with every $m_i$ a power of one fixed prime $p$. We can then reorder the complexes $K_i$ so that each $m_i$ divides the next one; write $m_i = d_i m_1$ for $i > 1$. In this final case we explicitly calculate each side of the morphism $\eta_i: R_i \to Q_i(K_1,\ldots,K_n)$. One easily finds that $H(K_1,\ldots,K_n)$ for $K_i = E^{m_i}$ has in dimension 0 homology $Z_m$ with generator $x_1 \otimes \cdots \otimes x_n$ and in dimension i homology the direct sum of $(i,n-i-1)$ copies of $Z_m$, $m = m_1$, one each with generators

$$\rho_m(x_1, d_{s_1}x_{s_1},\ldots,d_{s_i}x_{s_i}) \otimes x_{r_1} \otimes \cdots \otimes x_{r_{n-i-1}} \tag{28}$$

for each $(i,n-i-1)$ shuffle $s, r$ of the integers $\{2,\ldots,n\}$. For the group $R_i = \mathrm{Mult}_i(H(K_1),\ldots,H(K_n))$ one finds exactly the corresponding generators, with $\rho_k$ replaced by $\tau_k$. Since the map $\eta_i$ takes $\tau_k$ to $\rho_k$, this establishes the fact that $\eta_i$ is an isomorphism.

It remains to show why $R_i = \mathrm{Mult}_i(H(K_1),\ldots,H(K_n))$ has exactly the corresponding generators, but of course with $\rho_{m_1}$ replaced by $\tau_{m_1}$. Now $H(K_i) = A_i$ is a cyclic abelian group of order $m_i = d_i m_1$ with generator, say, $g_i$. We will then show that $\mathrm{Mult}_i(A_1,\ldots,A_n)$ is a direct sum of cyclic groups of order $m = m_1$ and with generators the symbols

$$\tau_m(g_1, d_{t_1}g_{t_1},\ldots,d_{t_{i-1}}g_{t_{i-1}}) \otimes g_{u_1} \otimes \cdots \otimes g_{u_{n-i}} \tag{29}$$

for all $(i-1,n-i)$ shuffles $(t,u)$ of the set $\{2,\ldots,n\}$ of subscripts. (This matches the term with index $i-1$ in (28)). Indeed, an element $\tau_k(a_1,\ldots,a_i)$ with $a_i = x_i g_i$, $x_i$ an integer, can be expressed, according to the calculation (13) and (14) of $\mathrm{Tor}(A_1,\ldots,A_i)$, as

$$\tau_h(x_1g_1, x_2g_2,\ldots,x_ig_i) = x_1(hx_2/m_2)\cdots(hx_i/m_i)\,\tau_m(g_1, d_2g_2,\ldots,d_ig_i) \tag{30}$$

Now for $\mathrm{Mult}_i$ we have for each $(i+1,n-i-1)$ shuffle $(s,r)$ the relations

$$0 = \sum_{j=1}^{i+1}(-1)^j \tau_h(a_{s_1},\ldots,\hat{a}_{s_j},\ldots,a_{s_{i+1}}) \otimes a_{s_j} \otimes a_{r_1} \otimes \cdots \otimes a_{r_{n-i+1}} \tag{31}$$

Writing each $a_i$ as $a_i = x_i g_i$, applying the formula (30) to each term and then getting rid of fractions $(hx_i/m_i)$ and writing $d_{i,j}$ for $d_{s_i}/d_{s_j}$ reduces each of these relations to the form

$$\begin{aligned} &m_{s_2}\tau_{m_{s_2}}(g_{s_2},d_{3,2}g_{s_3},\ldots,d_{i+1,2}g_{s_{i+1}}) \otimes g_{s_1} \otimes g' \\ &= \sum_{j=2}^{i+1}(-1)^i d_{j,2}\tau_{m_{s_1}}(g_{s_1},d_{2,1}g_{s_2},\ldots,\widehat{d_{2,j}g_{s_j}},\ldots,d_{i+1,1}g_{s_{i+1}}) \otimes g_{s_j} \otimes g', \end{aligned} \tag{32}$$

where $g'$ is short for $g_{r_1} \otimes \cdots \otimes g_{n-i-1}$. This reduction shows first that the relation (31) for the same shuffle but with different arguments $a_i$ reduces to just one relation, the one above for that shuffle. Hence $\mathrm{Mult}_i$ is generated by the $\tau_m$ and $\otimes$ symbols appearing in this relation; because of the presence of a factor $g_1$, they are all of order $m = m_1$. In this relation, consider first those shuffles where $s_1 = 1$. The equation then expresses

$$\tau_{m_{s_2}}(g_{s_2},d_{3,2}g_{s_3},\ldots,d_{i+1,2}g_{s_{i+1}}) \otimes g_{s_1} \otimes g' \tag{33}$$

in terms of the right hand side of (32), where all the generators have the form $\tau_{m_1}(g_1,\ldots) \otimes \cdots$; that is, are among the intended generators (29). Now substitute the values so obtained for (33) in all the given relations (i.e., for the remaining shuffles), a tedious calculation shows that all the relations then hold. Thus $\mathrm{Mult}_i(A_1,\ldots,A_n)$ has the indicated independent generators (29), each of order $m$. The proof of the theorem is complete.

We note that the proof depends essentially upon the lack of symmetry (in the arguments $x_1,\ldots x_i$) exhibited in the formula (30) for the torsion product of cyclic groups.

7. <u>Some Comments</u> It may seem strange that the descriptions of Mult involve just one $\tau_h$, multiplied by a tensor product of cycles - especially since there might also be generators which are the tensor products of two or more such $\tau_h$. The explanation is easy: For the product of two cycles $\rho_h$, written as in (3), one readily checks the identity

$$\rho_h(z_1,z_2) \otimes \rho_h(z_3,z_4) = (-1)^{1+d_3}\rho_h(z_1,z_2,z_3) \otimes z_4 + (-1)^{d_3(d_1+1)}\rho_h(z_1,z_2,z_4) \otimes z_3$$

where $d_i$ = degree $z_i$. There are similar formulas for products of two longer $\rho_h$'s.

Bokštein's paper [3] claimed to give an example in which the Künneth sequence for $H(K \otimes L)$ does not split when $K$ but not $L$ is free. However, it has already been noted in the review of Kelly [10] (Math. Reviews vol. 27 #2538) that the example appears to be erroneous. The trouble resides in a slipped sign for $\partial(u \otimes v)$.

One purpose of this study is to get more illuminating formulas for the integral homology $H(\pi,n;Z)$ of an Eilenberg-Mac Lane space. Thus Cartan [5] introduced certain homology operations $\alpha$ and used corresponding groups $D_\alpha(\pi) = \pi/p\pi$ for $\alpha$ of first type or $D_\alpha(\pi) = {}_p\pi$ for $\alpha$ of second type, $p$ a prime. In the stable range he proved

$$H_{n+k}(\pi,n;Z) = A_k(\pi) = \sum_\alpha D_\alpha(\pi) \quad 0 \leq k < n$$

with sum taken over suitable homology operations $\alpha$. For $A_k = A_k(\pi)$ Decker has a corresponding result for the quadratic range, $0 \leq k < n$,

$$H_{2n+k}(\pi,n;Z) = Q_{n,k}(\pi) \oplus L_{nk}(\pi)$$

$$+ \sum_{\substack{r+s=k\\ r<s}} A_r \otimes A_s + \sum_{\substack{r+s=k-1\\ r<s}} \mathrm{Tor}(A_r,A_s).$$

Here $L_{n,k}(\pi)$ is a sum of certain $D_\alpha(\pi)$, so is linear in $\pi$, while $Q_{n,k}(\pi)$ is a basic quadratic functor; for example

$$Q_{n,2q}(\pi) = \Gamma_4(A_q(\pi)) \quad \text{when } n+q \equiv 0 \pmod 2,$$

$$Q_{n,2q}(\pi) = \Lambda_2(A_q(\pi)) \quad \text{when } n+q \equiv 1 \pmod 2.$$

Here $\Gamma_4$ is Whiteheads universal quadratic functor, while $\Lambda_2$ is the exterior square - $\Lambda_2(A) = A \otimes A/\langle a \otimes a\rangle$ for all $a \in A$.

There must be similar formulas in the cubic range, presumably such involving Trip - and so on. The case above indicates that the basic quadratic functor of a group $\pi$ for this purpose are

$$\Gamma_4(\pi),\ \Lambda_2(\pi),\ \pi \otimes \pi,\ \mathrm{Tor}(\pi,\pi)$$

and their compositions with linear functors of $\pi$.

The University of Chicago

## References


[1] Bödigheimer, C.-F., Splitting the Künneth Sequence in K-Theory I and II. Math. Annalen 242 (1979), 159-171 and 251 (1980), 249-252.

[2] Bokštein, M., Complete Modular Spectrum of Cohomology rings of a Tychonof product (Russian) Mat. Sb. (M.S.) 51 (93) (1960), 73-98; erratum 53 (95) (1961), 261-263.

[3] Bokštein, M., On splittings in Künneth's formula (Russian) Dokl. Akad. Nauk. SSSR 146 (1962), 270-273.

[4] Bokštein, M., A multiple Künneth formula (Russian) Izv. Akad. Nauk. SSSR Ser. Mat. 27 (1963), 467-482.

[5] Cartan, H., Seminaire Ecole Normal Sup. 1954/55. Algebre d'Eilenberg-Mac Lane et Homotopie. Reprinted, Benjamin, 1967.

[6] Decker, Gerald John, The Integral Homology Algebra of an Eilenberg-Mac Lane Space. Dissertation. The University of Chicago, June 1974.

[7] Eilenberg, S. and S. Mac Lane, On the Groups $H(\pi,n)$, I. Ann. of Math. 58 (1953), 55-106.

[8] Hamsher, Ross, Eilenberg-Mac Lane Algebras and their Computation; An Invariant Description of $H(\pi,1)$. Dissertation. The University of Chicago, August 1973.

[9] Hilton, P. and U. Stammbach, A course in Homological Algebra. Springer Verlag 1971, 338 pp.

[10] Kelly, G. M., Observations on the Künneth Theorem. Proc. Cambridge Philos. Soc. 59 (1963), 575-587.

[11] Mac Lane, S., Homology. Springer Verlag 1963, 422 pp.

[12] Mac Lane, S., Triple Torsion Products and Multiple Künneth Formulas, Math. Annalen 140 (1960), 51-64.

[13] Whitehead, J.H.C., A Certain Exact Sequence, Ann. of Math. 52 (1950), 51-110.


This paper is in final form and will not be published elsewhere.

Note added in proof.

The referee has reminded me of a somewhat neglected paper by F. P. Palermo (reference below) which determines the homology ring of the product K $\otimes$ L of twqo differential graded Z-algebras K and L. His determination is made in terms of the whole Bokštein spectrum and does not use the torsion product or a set of consistent caps, so does not appear connected to the results above. Palermo's work makes use of an earlier paper of Whitehead, which did use splitting of the Bokštein (reference below).

F. P. Palermo, The cohomology ring of product complexes, Trans. Amer. Math. Soc., 86(1957), 174--196.

J. H. C. Whitehead, On simply connected 4-dimensional polyhedra, Comment. Math. Helv., 22(1949), 48--92.

# ON A CATEGORICAL ANALYSIS OF ZADEH GENERALIZED SUBSETS OF SETS I

M.M. MAWANDA*

Institut de Mathématique, Université Catholique de Louvain
2, chemin du Cyclotron, B-1348 Louvain-la-Neuve, Belgium.

**Abstract**

On a categorical analysis of Zadeh generalized subsets of sets I.

How can fuzzy sets be viewed as generalized subsets of actual sets ? An answer to this question is given via a categorical analysis of the synchronic approach of fuzzy set theory. Starting from an abstract definition of a category with fuzzy subsets, we establish necessary and sufficient conditions on a poset to be that of truth-values for such a category. These conditions justify Zadeh's original choice of the unit segment as a set of truth-values. Some defects concerning universal constructions in a non trivial category with fuzzy subsets are mentioned and a natural best toposophical approximation of such a category is proposed.

## Introduction

Introduced by L.A. Zadeh [10] and J.A. Goguen [2] as a basic ingredient in a generalization of set theory, fuzzy sets are commonly used as generalized subsets of sets. This presentation is known as the synchronic approach (Negoita [7]) of fuzzy set theory accepts ipso facto the classical sets and is thus in opposition with the underlying principle. These papers aim to show how fuzzy sets can be viewed as generalized subsets of usual sets. The main theorem gives minimal conditions on the mathematical structure of the set of truth-values for allowing the synchronic approach of fuzziness.

The basic object of this paper is a categorical analysis of Zadeh generalized characteristic maps for which the real unit segment $I = [0,1]$ is the set of truth-values. If a subset of X is a map $\varphi$ from X to I, such a subset is viewed in the usual category of sets as $\{(x,\varphi(x)) \mid \varphi(x) \neq 0\}$, which is a subset of $X = I^*$ where $I^* = ]0,1]$. In that way, any usual set X is viewed as a new set $X = X \times I^*$ in the ideal category $\mathbb{C}$ with fuzzy subsets and $\mathbb{C}$ itself is obtained as a solution to the problem of finding a suitable pair of adjoint functors related to a monad whose endofunctor is precisely $- \times I^*$. From this intuition, we introduce in §2 the concept of a category with fuzzy subsets, without any assumption on the mathematical structure of the set L of truth-values.

* Supported by Projet FDS "Théorie des Topos" C.A.C. 216/1459.

In §1, we introduce the concept of subset related to a category whose objects are sets. Such a category is called a category of sets. Some examples of categories of sets are given and we show how the concept of subset of a set is determined by what is adopted as a map between sets. In §3, the mathematical structure of the set L of truth-values is analysed. It is showed that L must be a monoïd with a strict unit 1, an absorbant element 0 such that elements of $L^* = L - \{0\}$ are left regular and such that L is an $\wedge$-semilattice for the relation $\alpha \leqslant \beta$ iff $\exists\tau(\alpha = \beta\tau)$. These properties follow from the observation that any category $\mathbb{C}$ with fuzzy subsets is equivalent to a category of sets $\mathbb{S}(M)$ constructed from a monoid M which can be identified with $L^*$.

In §4, the main theorem exhibits those conditions on L which are necessary and sufficient to make $\mathbb{S}(L^*)$ a category with fuzzy subsets. The last paragraph show how set like operations follow naturally from further properties of L. Call a non trivial category with fuzzy subsets, a category $\mathbb{C}$ with fuzzy subsets such that the set L of truth-values is not the trivial boolean algebra $2 = \{0,1\}$. It is shown that such a category has not finite limits and finite colimits. Finally, we show that any category with fuzzy subsets can be embedded in a "good topos" in which defects concerning universal constructions are corrected. In the case of a non trivial category with fuzzy subsets, the internal logic of the related toposophical best approximation is not classical. It satisfies De Morgan's law and the various real number objects are isomorphic and essentially classical.

The present paper has been presented at the University of Sussex during the thirty-fifth meeting of the Peripathetic Seminar on Sheaves and Logic and at the Louvain-la-Neuve Séminaire de Théorie des Catégories. I want to thank F. Borceux and G. Van Den Bossche for pertinent remarks concerning a preliminary version of this paper and I. Moerdijk who suggested to find how the topos of action on $L^*$ is a best approximation of $\mathbb{S}(L^*)$. I also want to thank T. Lucas and P. Lecouturier for their help when I wrote the final version. It is not easy to find much information on fuzzy set theory in Mathematical libraries. Therefore, specially thanks to B. Banaschewski for his invitation at McMaster University ; it is at the McMaster University Library that I collected my first informations on the foundations of fuzzy set theory. I finally thank M.K. Luhandjuzula, responsable of the Séminaire sur la programmation floue at Tizi-Ouzou. The main question "How fuzzy set can be viewed as subsets of usual sets ?" raised where I participated in his seminar on diachronic and synchronic approches of fuzziness.

## 1. What are subsets of sets ?

1.1. In the view point of the theory of categories, the concept of subset of a set X can designate the subobject of X in a category for which the class of objects is the collection $\mathbb{S}$et of sets. If the usual concept of map between sets is adopted as morphism and $\mathbb{S}$ denotes the resulting category then a subset of X in $\mathbb{S}$ is exactly the usual concept of subset of X. It is easy to see that usual maps can be substituted by functional relations, injections or $\alpha$-maps for a non zero cardinal $\alpha$, to obtain the same concept of subset. By an $\alpha$-map, we mean a $f : X \to Y$ such that $|f^{-1}(y)| \leq \alpha$ for any $y \in Y$ ; where $|A|$ denotes the cardinal of A. If one adopts identities, bijections or surjections as morphisms between sets, it is easy to see that the resulting categories have the same concept of subset. In each of these cases, any set has a single subset. Here is an accurate definition of subset.

1.2. A *category of sets* is any category $\mathbb{C}$ for which $\mathbb{S}$et is the class of objects. Two categories of sets differ in the nature of morphisms and then so are different universes. In the sequel, a category of sets will be defined by giving the nature of morphisms which we will call maps and composition of such maps. As we want to use the usual nomenclature of $\mathbb{S}$ for the same concepts in a category of sets, we use the adjective ordinary for the concepts in $\mathbb{S}$. Then so, $\mathbb{S}$ is the ordinary category of sets for which morphisms are ordinary maps. For example, monomorphisms in $\mathbb{S}$ are ordinary injections. The discrete category of sets is the category $\mathbb{S}$et for which morphisms are identities. If X is a set and $\mathbb{C}$ a category of sets, a subobject of X in $\mathbb{C}$ is called a *subset* of X in $\mathbb{C}$. In the next points we give some examples of categories of sets and related concept of subset.

1.3. Let $\mathbb{C}$ be a category of sets. The category $M(\mathbb{C})$ of sets for which maps are $\mathbb{C}$-monomorphisms has trivially the same subsets as $\mathbb{C}$. The category $J(\mathbb{C})$ for which maps are $\mathbb{C}$-isomorphisms is such that any set has a single subset. Let $E(\mathbb{C})$ be the category of sets for which maps are $\mathbb{C}$-epimorphisms. It is not always trivial to describe in general subsets in $E(\mathbb{C})$. With some assumptions on $\mathbb{C}$, it is possible to show that subsets in $E(\mathbb{C})$ are subsets in $J(\mathbb{C})$. This is possible when $\mathbb{C}$ is balanced and has an atom object. Let $\mathbb{D}(A,B)$ denotes the sets of $\mathbb{D}$-morphisms from A to B where $\mathbb{D}$ is a category. An object A of $\mathbb{D}$ is said to be an atom object of $\mathbb{D}$ if for any pair (X,Y) of $\mathbb{D}$-objects, the map

$$\mathbb{D}(X,Y) \to S(\mathbb{D}(A,X),\ \mathbb{D}(A,Y))$$

$$f \rightsquigarrow \mathbb{D}(A,f)$$

$$\text{where} \qquad \mathbb{D}(A,f)(u) = f \circ u$$

is a bijection and this correspondence is natural both in X and Y.

1.4. Any poset on the collection $\mathbb{S}$et determines a category of sets $\mathbb{S}et_{\leq}$ for which

$$Set_{\leq}(X,Y) = \begin{cases} \{(X,Y)\} & \text{if } X \leq Y \\ \phi & \text{otherwise.} \end{cases}$$

For any set X, let $[[X]] = \{Y \mid Y \leq X\}$. A subset of X is an equivalence class of $[[X]]$ modulo the equivalence relation on $[[X]]$ induced by the poset on $[[X]]$. Let $X \leq Y$ iff there is an ordinary injection of X in Y. If Card($\mathbb{S}$) denotes the category of sets related to this poset, it follows from Cantor - Bernstein theorem and the axiom of choice that a subset of X in Card($\mathbb{S}$) is any cardinal least than $|[X]|$. The category of sets related to the converse poset is not locally small.

## 2. Categories with fuzzy subsets.

2.1. Let $\mathbb{C}$ be a fixed category. An *extension* (on subobjects) is a category $\mathbb{D}$ equipped with a functor $i : \mathbb{C} \to \mathbb{D}$ such that the following properties hold :

(1) i preserves monos

(2) i is faithful

(3) if $f : i(X) \to i(Y)$ and $g : i(Y) \to i(Z)$ are such that there is $u : X \to Z$ such that $i(u) = g \circ f$ then there is $r : X \to Y$ such that $f = i(r)$.

Let $\mathbb{P}_{\mathbb{E}}(A)$ denote the class of subobjects of an object A of the category $\mathbb{E}$ then the following is an outcome of definition :

**PROPOSITION 1 :** *For an extension* $\mathbb{D}$ *of* $\mathbb{C}$ *the correspondence from* $P_{\mathbb{C}}(X)$ *to* $\mathbb{P}_{\mathbb{D}}(i(X))$ *which sends* $\overline{m}$ *to* $\overline{i(m)}$ *is an injection* ($\overline{m}$ *denotes the subobject for which* m *is a representant*). *For any retract* $r : i(X) \to i(Y)$ *there is* $k : Y \to X$ *such that* $r \circ i(k) = 1$.

The functor i will be called the *inclusion* of $\mathbb{C}$ in $\mathbb{D}$.

2.2. An extension $\mathbb{D}$ of $\mathbb{C}$ will be called a *simple extension* if the inclusion of $\mathbb{C}$ in $\mathbb{D}$ is representative (i.e. any object D of $\mathbb{D}$ is isormorph to i(C) for some C in $\mathbb{C}$). It is easy to see that the full subcategory of an extension of $\mathbb{C}$ for which objects are images of objects of $\mathbb{C}$ by the inclusion is a simple extension of $\mathbb{C}$. But there is another interesting simple extension associated to an extension. Let $i(\mathbb{C})$ denote the category for which objects are the same as $\mathbb{C}$, $i(\mathbb{C})(X,Y) = \{(X,f,Y) \; f : i(X) \to i(Y)\}$ and the obvious composition. It is easy to see that $i(\mathbb{C})$ is a simple extension of $\mathbb{C}$ and it is equivalent to the previous full subcategory of the extension.

**PROPOSITION 2 :** *For any extension* $\mathbb{D}$ *of* $\mathbb{C}$ *the following properties hold* :

*(1)* $i(\mathbb{C})$ *is a simple extension equivalent to the full subcategory of* $\mathbb{D}$ *consisting of images by* i *of objects of* $\mathbb{C}$.

*(2) If* $\mathbb{D}$ *is a simple extension then* $\mathbb{D}$ *is equivalent to* $i(\mathbb{C})$.

2.3. Let $\mathbb{D}$ be an extension of $\mathbb{C}$ with inclusion is denoted by i. We say that $\mathbb{D}$ is an *universal extension* of $\mathbb{C}$ if i has a right adjoint. That means that for any $\mathbb{D}$-object D there is a universal morphism $iC(D) \xrightarrow{f} D$ from i to D. The right adjoint

$C : \mathbb{D} \to \mathbb{C}$ will be called the *concretion* of $\mathbb{D}$ in $\mathbb{C}$ and $C \circ i$ also denoted by $T : \mathbb{C} \to \mathbb{C}$ will be called the *endofunctor associated to the universal extension.* This endofunctor is the trace in $\mathbb{C}$ of the universal extension. T will play the role of the functor $- xL^*$ in the definition of category with fuzzy subsets. Of course, we want that T satisfies a property satisfied by product functors. Let $F : A \to A$ be an endofunctor in the category **A** with direct products. We say that F commutes with product functors if for any object X in A, there is a natural isomorphism $\alpha_X : F(X\times -) \to X \times F(-)$ which is natural in X. If **A** has a terminal object 1, then it follows that F is isomorphic to $- \times F(1)$. Consequently such a function can be replaced by $- \times M$ where $M = F(1)$.

2.4. Recall that if $\mathbb{E}$ is a locally small category which has pullbacks, then subobjects in $\mathbb{E}$ define a contravariant functor $\mathbb{P} : \mathbb{E} \to \mathbb{S}$. $\mathbb{P}(X)$ is the set of subobjects of X and $f^{-1}(\overline{m})$ is defined by pulling back along f the subobject $\overline{m}$. A contravariant functor $F : \mathbb{E} \to \mathbb{S}$ is said to be representable if there is an object H in $\mathbb{E}$ such that $F \xrightarrow{\sim} \mathbb{E}(-,H)$. As in 1.3, for any morphism $f : A \to B$ and $u : B \to H$ we define $\mathbb{E}(f,H)(u) = u \circ f$.

2.5. Let $\mathbb{C}$ be a subcategory of $\mathbb{D}$. If $\mathbb{I}$ is a small category, $\mathbb{C}$ is said to be an $\mathbb{I}$-complete subcategory of $\mathbb{D}$ if for any functor $F : \mathbb{I} \to \mathbb{D}$ which factors through $\mathbb{C}$ and such that F has a limit in $\mathbb{D}$, this limit can be taken in $\mathbb{C}$. That means $\underrightarrow{\lim}\, F$ and morphisms of structure can be taken in $\mathbb{C}$. Let A be an object of the category $\mathbb{D}$. If $\mathbb{C}$ is the monoid $\mathbb{D}(A,A)$ viewed as a full subcategory of $\mathbb{D}$, A is said to be an $\mathbb{I}$-*complete object* of $\mathbb{D}$. A complete object in $\mathbb{D}$ will be an $\mathbb{I}$-complete object of $\mathbb{D}$ for any small category $\mathbb{I}$. It is easy to see that the terminal object of a category $\mathbb{D}$ is always a complete object of $\mathbb{D}$. A functor $F : \mathbb{C} \to \mathbb{D}$ preserves $\mathbb{I}$-complete (resp. complete) objects if F(A) is $\mathbb{I}$-complete (resp. complete) for any $\mathbb{I}$-complete (resp. complete) object A. We are now able to define categories with fuzzy subsets.

2.6. **DEFINITION** : *A* *__category with fuzzy subsets__* *is a universal simple extension* $\mathbb{C}$ *of the ordinary category* $\mathbb{S}$ *of sets such that the following properties hold :*

*(1) the endofunctor* T *associated to the universal extension commutes with product functors*

*(2) the inclusion* i *preserves* $\mathbb{3}$*-complete objects where* $\mathbb{3}$ *is the category with three objects and five arrows*

*(3)* $\mathbb{C}$ *has pullbacks and* $\mathbb{P} \circ i$ *is representable where* $\mathbb{P}$ *is the contravariant functor induced by subobjects in* $\mathbb{C}$.

Recall that (3) supposes that $\mathbb{C}$ must be locally small. The representant L of

$\mathbb{P} \circ i$ will be called the *set of truth-values*. In the following paragraph, we set out the mathematical structure of L.

## 3. Mathematical structure of the set of truth-values.

3.1. In the following, $\mathbb{1}$ will denote a singleton set. Let $\mathbb{C}$ be a category with fuzzy subsets. From 2.6 (1) and 2.3, the endofunctor T can be taken as $- \times M$ where $M = T(\mathbb{1})$ It follows from well known fact in category theory (see Mac-Lane [6] Ch. VI) that the universal extension determine a monad for which $- \times M$ is the endofunctor. A description of this monad shows that M is equipped with an operation $M \times M \to M$, $(\alpha,\beta) \mapsto \alpha\beta$

and an element 1 such that $(M,.,1)$ is a monoid. For any set X the unit map is $X \to X \times M$, $x \mapsto (x,1)$ and the multiplication map is $X \times M \times M \to X \times M$, $(x,\alpha,\beta) \mapsto (x,\alpha\beta)$.

3.2. From 2.2 the category with fuzzy subsets $\mathbb{C}$ is equivalent to the category of sets $i(\mathbb{C})$. As the universal extension determines a monad $(T,\eta,u)$, one can consider the Kleisli construction to obtain a category of sets $\mathbb{S}_T$ (see also Mac-Lane [6]) for which morphisms from X to Y are ordinary maps from X to TY. If $X \rightsquigarrow Y$ and $g : Y \rightsquigarrow Z$ are maps in $S_T$ then $g \circ f : \mu_Z \circ Tg \circ f$. This category is equipped with a pair of adjoint functors $\mathbb{S} \underset{C_T}{\overset{i_T}{\rightleftarrows}} \mathbb{S}_T$ such that $i_T(f) = \eta_Y f$, $C_T(f) = \mu_Y \circ Tf$ and $T = C_T \circ i_T$. Let now i and C denote also the inclusion of $\mathbb{S}$ in $i(\mathbb{C})$ and the concretion of $i(\mathbb{C})$ in $\mathbb{S}$. As $(\mathbb{S}_T, i_T, C_T)$ is an initial object between solutions of pair of adjoint functors which induce the previous monad, there is a unique functor $K : \mathbb{S}_T \to i(\mathbb{C})$ such that $i = K \circ i_T$ and $K \circ C_T = C$. It follows directly that $K(X \overset{f}{\rightsquigarrow} Y) = (X, \varepsilon_{i(Y)} \circ i(f), Y)$ where $\varepsilon_{i(Y)}$ is the counit. Let $G : i(\mathbb{C}) \to \mathbb{S}_T$ be the trivial functor such that :

$$G(X,f,Y) = C(f)\eta_X.$$

It is an easy exercise to show that $K \circ G = 1$ and $G \circ K = 1$. Consequently $i(\mathbb{C})$ and $\mathbb{S}_T$ are equivalent. Viewed as a category with fuzzy subsets, $\mathbb{S}_T$ has $C_T$ as the concretion functor and $i_T$ as the inclusion functor. If T is replaced by $- \times M$ as in 3.1 then $\mathbb{S}_T$ is the category of sets $\mathbb{S}(M)$ for which a map from X to Y is a pair $(f,\varphi)$ such that $f : X \to Y$ and $\varphi : X \to M$ are ordinary maps. The composition $(g \circ f, \psi f * \varphi)$ of $(f,\varphi) : X \rightsquigarrow Y$ and $(g,\psi) : Y \rightsquigarrow Z$ is defined by :

$$(\psi f * \varphi)(x) = \psi(f(x))\varphi(x)$$

We have then proved the following :

**THEOREM 1** : *Any category with fuzzy subsets* $\mathbb{C}$ *is equivalent to the category of sets* $\mathbb{S}(M)$ *for a monoid* $M = T(\mathbb{1})$ *where* T *is the endofunctor associated to the universal extension and is isomorphic to* $- \times M$.

3.3. The previous theorem allows one to use the category $\mathbb{S}(M)$ instead of $\mathbb{C}$. Let us now consider the representant L of $P \circ i$. It is clear that $L \cong \mathbb{P}(i(\mathbb{1}))$ where i will designate the inclusion of $\mathbb{S}$ in $\mathbb{S}(M)$ and C the concretion. In order to obtain further information about the mathematical structure of L, we need some technical lemmas. M is a precommutative monoid if the following holds :

$$\forall \alpha,\beta \in M \quad \exists \gamma,\tau \in M \quad \alpha\gamma = \beta\tau$$

**LEMMA 1** : *If* M *is a precommutative monoid and* $X \overset{(f,\varphi)}{\rightsquigarrow} \mathbb{1}$ *a monomorphism in* $\mathbb{S}(M)$ *then* X *has at most one element.*

**Proof** : Suppose that $x_1, x_2 \in X$. As M is precommutative, there are $\alpha,\beta \in M$ such that $\varphi(x_1)\alpha = \varphi(x_2)\beta$. It follows from the fact that $(f,\varphi)$ is a mono that $x_1 = x_2$. □

**LEMMA 2** : *The trivial equivalent following conditions hold :*

*(1) Elements of* M *are left regular*

*(2) Any map from* $\mathbb{1}$ *to* $\mathbb{1}$ *in* $\mathbb{S}(M)$ *is a mono.*

**Proof** : It follows from 2.1 (3) that the unit 1 of M is strict (i.e. $\alpha\beta = 1$ implies $\alpha = 1 = \beta$). As $\mathbb{1}$ is $\exists$-complete in $\mathbb{S}$, so is for the same object in $\mathbb{S}(M)$. Then for any $\alpha \in M$, there is a maximal pair $(\gamma,\theta) \in M \times M$ such that $\alpha\tau = \alpha\gamma$ because $\mathbb{S}(M)$ has pullbacks. Pre-order on $M \times M$ is defined by $(\sigma,\varepsilon) \leq (\sigma',\varepsilon')$ iff there is $\rho$ such that $\sigma = \sigma'\rho$ and $\varepsilon = \varepsilon'\rho$. From $\alpha 1 = \alpha 1$, it follows that $\tau = \gamma = 1$. □

3.4. It follows from lemma 1 and lemma 2 that if 0 denotes the empty subset of $\mathbb{1}$ in $\mathbb{S}(M)$ then $L \cong \mathbb{P}(i(\mathbb{1})) \cong \mathbb{S}(M)(\mathbb{1},\mathbb{1}) \cup \{0\}$. But it is clear that $\mathbb{S}(M)(\mathbb{1},\mathbb{1}) = M$. If $L^* = L \setminus \{0\}$ then M is exactly $L^*$. It follows that L can be seen as obtained from the monoid $L^*$ by adjonction of an absorbant 0. We can then state the following :

**THEOREM 2** : *For any category with fuzzy subsets* $\mathbb{C}$, *the set* L *of truth-values is a non trivial monoïd with an absorbant* 0, *a strict unit* 1, *every non zero element of* L *is left regular and the relation* $\exists\tau(\alpha = \beta\tau)$ *is an ordering such that* L *is an* $\wedge$*-semilattice. Furthermore, if* $L^* = L \setminus \{0\}$ *then* $\mathbb{C}$ *is equivalent to the category of sets* $\mathbb{S}(L^*)$.

**Proof** : From theorem 1 and previous lemmas, it suffices to prove that $L^*$ is an $\wedge$-semilattice. It is clear that the relation is reflexive and transitive. Antisymetry and the fact that 1 is strict follows from lemma 2. As $\mathbb{1}$ is $\exists$-complete and $\mathbb{S}(L^*)$ has pullbacks, it follows that $L^*$ is an $\wedge$-semilattice. □

From theorem 2, we obtain the following corollary which is a way to a representation result about categories with fuzzy subsets :

**COROLLARY 1.** *Any category with fuzzy subsets is equivalent to a category of sets* $\mathbb{S}(L^*)$ *for a monoid* $L^*$ *satisfying :*

*(1)* 1 *is a strict unit*

*(2) every element is left regular*

*(3) the relation* $\exists\tau \in L^*\ (\alpha = \beta\tau)$ *is an* $\wedge$*-ordering i.e.* $L^*$ *is an* $\wedge$*-semilattice.*

*Furthermore the set* L *of truth-values is obtained from* $L^*$ *by adjunction of an absorbant.*

A consequence of this is the following assertion, where a no trivial category with fuzzy subsets means a category with fuzzy subsets which is not equivalent to the ordinary category of sets and $[a] = \{ab \mid b \in L^*\}$.

**COROLLARY 2 :** *The following properties are equivalent for a category with fuzzy subsets* $\mathbb{C}$ *where* L *is the set of truth-values and* $L^* = L \setminus \{0\}$ :

*(i)* $\mathbb{C}$ *is no trivial*

*(ii)* $L^*$ *is a no trivial monoid*

*(iii) there is* $a \in L^*$, $[a]$ *is infinite*

*(iv) for any* $a \in L^*$, $[a]$ *is infinite*

*(v)* $L^*$ *is infinite.*

## 4. Representation theorem

4.1. The objectif of this paragraph is to prove that conditions of corollary 1 of theorem 2 are sufficient for $\mathbb{S}(L^*)$ to be a category with fuzzy subsets. That completes the proof of the following representation theorem for categories with fuzzy subsets :

**THEOREM 3** (Main theorem) *A category* $\mathbb{C}$ *is a category with fuzzy subsets if and only if* $\mathbb{C}$ *is equivalent to a category of the form* $\mathbb{S}(L^*)$, *where* $L^*$ *is a monoid satisfying the following properties :*

*(1)* 1 *is a strict unit*

*(2) every element is left regular*

*(3)* $\exists\tau \in L^*(\alpha = \beta\tau)$ *is an* $\wedge$*-ordering on* $L^*$.

*The set* L *of truth-values is obtained from* $L^*$ *by adjunction of an absorbant* element.

Throughout the paragraph, $L^*$ is a monoid such that (1), (2) and (3) in the Main theorem hold.

4.2. If $\alpha \in L^*$, we will write $c_\alpha$ for the constant map to $L^*$ with value $\alpha$ for any x of the domain. Without any assumption on the monoid, it is easy to see that $i : \mathbb{S} \to \mathbb{S}(L^*)$

such that $i(X \xrightarrow{f} Y) = (f, c_1) : X \rightsquigarrow Y$ is faithful and preserves monos. The condition 2.1 (3) holds because $L^*$ satisfies (1). Without any hypothesis on the monoid, the functor $C : \mathbb{S}(L^*) \to \mathbb{S}$ such that $C(X \rightsquigarrow Y) = X \times L^* \to Y \times L^*$

$$(f,\varphi) \qquad (x,\lambda) \mapsto (f(x), \varphi(x)\lambda)$$

is trivially a right adjoint to $i$. The endofunctor $- \times L^*$ associated with this universal simple extension commutes with product functors. It suffices to show that :

- $i$ preserves $\mathbb{S}$-complete objects
- $\mathbb{S}(L^*)$ has pullbacks and $\mathbb{P} \circ i$ is representable to complete the proof of the Main Theorem.

4.3. Let $F : \mathbb{I} \to \mathbb{S}(L^*)$ be a functor, where $\mathbb{I}$ is a small category. For an $\mathbb{I}$ -morphism $u : i \to j$, we write $Fu = (F_1u, F_2u)$. It follows that $F_1 : \mathbb{I} \to \mathbb{S}$ such that $F_1(i) = F(i)$, is a functor. Recall that if $i \xrightarrow{u} j \xrightarrow{v} k$ are $\mathbb{I}$- morphisms then $F_2(vu) = (F_2v \circ F_1u) * F_2u$. Let $\mathbb{L}^*$ denote the category associated with the monoid $L^*$. Suppose that $\mathbb{L}^*$ has all $\mathbb{I}$-limits. For any $x \in K$ choose $\lim\limits_{\mathbb{I}} F_x = (\beta_i(x))$:
Now consider the following diagram in $\mathbb{S}(L^*)$

$$(K \underset{(\gamma_i, \beta_i)}{\rightsquigarrow} F(i)).$$

It is clear that the previous diagram is an inductive cone from K to F. It is easy to see that $(K \underset{(\gamma_i, \beta_i)}{\rightsquigarrow} F(i))$ is the limit of F in $\mathbb{S}(L^*)$. That implies the following assertion :

**PROPOSITION 3** : *The following properties are equivalent for a small category* $\mathbb{I}$.

*(i)* $\mathbb{L}^*$ *has all* $\mathbb{I}$*-limits*

*(ii)* $\mathbb{L}$ *is* $\mathbb{I}$*-complete and* $\mathbb{S}(L^*)$ *has all* $\mathbb{I}$*-limits.*

4.4. Let $(f,\varphi) : X \rightsquigarrow Y$ be a monomorphism in $\mathbb{S}(L^*)$ and $x_1, x_2 \in X$ be such that $f(x_1) = f(x_2)$. From the $\wedge$-ordering property, there exist $\alpha, \beta \in L^*$ such that $\varphi(x_1)\alpha = \varphi(x_2)\beta$. Since $(f,\varphi)$ is mono, it follows that $(x_1,\alpha), (x_2,\beta)$ as morphisms from $\mathbb{1}$ to X are equal. So, f is an ordinary injection. Let now f be an ordinary injection and $(f,\varphi) : X \rightsquigarrow Y$ be a morphism. Suppose that $A \underset{(u',\psi')}{\overset{(u,\psi)}{\rightrightarrows}} X$ are such that $(f,\varphi)(u,\psi) = (f,\varphi)(u',\psi')$. It follows that $u = u'$ and for any $a \in A$ $\varphi u(a)\psi(a) = \varphi u(a)\psi(a)$. That implies $\psi = \psi'$. From these considerations we can assert that ;

**PROPOSITION 4** : *In* $\mathbb{S}(L^*)$, *the following properties are equivalent for a morphism* $(f,\varphi)$ :

*(i)* $(f,\varphi)$ *is a monomorphism*

*(ii)* $f$ *is an ordinary injection.*

4.5. Proof of the Main Theorem : In 4.2 we state the fact that $\mathbb{S}(L^*)$ is a universal simple extension such that the endofunctor $- \times L^*$ commutes with product functors. Let $\alpha, \beta \in L^*$, by the $\wedge$-ordering property there are $\gamma, \tau \in L^*$ such that $\alpha\gamma = \beta\tau = \alpha \wedge \beta$. If $\gamma', \tau' \in L^*$ are such that $\alpha\gamma' = \beta\tau'$ then $\alpha\gamma' \leq \alpha \wedge \beta$. It follows that there is $\sigma \in L^*$ such that $\alpha\gamma = \alpha\gamma\sigma = \beta\tau\sigma = \beta\tau'$. Since elements of $L^*$ are left regular, it follows that $\gamma' = \gamma\sigma$ and $\tau\sigma = \tau'$. Uniqueness of such a $\sigma$ follows from left regularity Then $\mathbb{L}^*$ has $\mathbb{B}$-limits. From proposition 3, we conclude that $\mathbb{L}$ is $\mathbb{B}$-complet and $\mathbb{S}(L^*)$ has pullbacks. The fact that the empty set is $\mathbb{B}$-complete is obvious. Thus the inclusion i preserves $\mathbb{B}$-complete objects. From Proposition 4, the map $\theta_X : \mathbb{P}(i(X)) \to \mathbb{S}(X,L)$ defined by

$$\theta_X(\overline{(f,\varphi)})(x) = \begin{cases} \varphi(x) & \text{if } \exists x' f(x') = x \\ 0 & \text{otherwise} \end{cases}$$

is a bijection, where $L = L^* \dot{\cup} \{0\}$. The correspondence $\theta_X$ is trivially natural on X. □

5. The best toposophical approximation of a category with fuzzy subsets.

5.1. It is clear that the empty set is an initial object of $\mathbb{S}(L^*)$ and for any family $(X_i)_{i\in I}$ of sets indexed by a set I, the following diagram $(X \xrightarrow{(\alpha_i, c_1)} \coprod X_i)$ where $\alpha_i X \hookrightarrow \coprod X_i$ is the canonical inclusion in $\mathbb{S}$, is a direct sum. Furthermore the following assertion is trivially true :

**PROPOSITION 5 :** *Any category with fuzzy subsets has all direct sums and they are disjoint and universal.*

5.2. Let $\mathbb{T}$ be an algebraic theory and let $\mathrm{Mod}_{\mathbb{T}}$ denote the category of $\mathbb{T}$-models in $\mathbb{S}$. Suppose that $F : \mathbb{S} \to \mathbb{S}$ is a representable contravariant functor and $U : \mathrm{Mod}_{\mathbb{T}} \to \mathbb{S}$ is the forgetful functor. Let A denote the representative of F. If F factors throught U then A is a $\mathbb{T}$-model since $F(\mathbb{1}) \cong \mathbb{S}(1,A) \cong A$. Conversely, suppose that A is a $\mathbb{T}$-model. Fix a set X and consider $A^X$. For any constant $a \in A$ let $C_a : X \to A$ be defined by $C_a(x) = a$. If O is a J-operation in A then define $O^J : (A^X)^J \to A$ by $O^J(f_i)_{i\in J}(x) = O(f_i(x))_{i\in I}$. It follows that $A^X$ is also a $\mathbb{T}$-model. As $F(X) \cong A^X$, it is easy to see that the previous construction of a $\mathbb{T}$-model structure on $A^X$ is natural in X. So F factors throught U. The previous remarks mean :

**PROPOSITION 6 :** *If* F *is a representable contravariant functor with representative* A *then* F *factors through* U *iff* A *is a* $\mathbb{T}$*-model. The following assertion is a corollary of the previous one.*

**COROLLARY :** *In any category* $\mathbb{C}$ *with fuzzy subsets the set like operations on subsets* *are just determined by the algebraic structure of the set of truth-values.*

It follows from this corollary and the Main theorem that given a set X, one can form the intersection of two subsets $\varphi : X \to L$ and $\psi : X \to L$, it is $\varphi \wedge \psi : X \to L$ where $(\varphi \wedge \psi)(x) = \varphi(x) \wedge \psi(x)$ for any $x \in X$. In the case of Zadeh's set I of truth-values, which is a complete lattice, all set like operations (arbitrary intersection and union, pseudo-complementation) on subsets of a given set X exist. They are given by :

$$(\vee\varphi_i)(x) = \vee\varphi_i(x) \; ; \; (\wedge\varphi_i)(x) = \wedge\varphi_i(x).$$

$$\neg(\varphi)(x) = \begin{cases} 1 \text{ if } \varphi(x) = 0 \\ 0 \text{ otherwise.} \end{cases}$$

The existence of the pseudo-complementation follows from the classical Heyting Algebra structure on I.

It is important here to notice that the set like operations are not the only ones which can exist naturally on the subsets. For example in the case of $\mathbb{S}(I^*)$ these are two other operations. For any subsets $\varphi, \psi : X \to L$, define $\varphi.\psi$ and $\varphi \oplus \psi$ by :

$$(\varphi,\psi)(x) = \varphi(x)\psi(x) \quad \text{and} \quad (\varphi \oplus \psi)(x) = \varphi(x) + \psi(x) - \varphi(x)\psi(x).$$

The importance of . and $\oplus$ in I are well known in probability theory.

5.3. In view of previous remarks 5.1 and 5.2 one could think that a category with fuzzy subsets is similar to the ordinary category of sets when the set L of truth-values has good algebraic properties. The following disastrous proposition shows that non trivial categories with fuzzy subsets are far from looking like the ordinary category of sets :

**PROPOSITION 7 :** *The following properties are equivalent, for a category $\mathbb{C}$ with fuzzy subsets :*

*(i) $\mathbb{C}$ is trivial*

*(ii) $\mathbb{C}$ is finitely complete*

*(iii) $\mathbb{C}$ is finitely cocomplete.*

<u>Proof</u> : It is clear that (i) implies both (i) and (iii). Suppose that $\mathbb{S}(L^*)$ satisfies (ii) and that $\alpha, \beta \in L^*$. The terminal object A is trivially non empty. If $a \in A$ then $\mathbb{I} \overset{(a,\alpha)}{\underset{(a,\beta)}{\rightrightarrows}} A$ are equal. So $L^*$ is trivial.

Suppose that $\mathbb{S}(L^*)$ satisfies (iii) and $\alpha, \beta \in L^*$. Now consider $\mathbb{I} \xrightarrow{(a,\gamma)} A$ a quotient of $\mathbb{I} \overset{\alpha}{\underset{\beta}{\rightrightarrows}} \mathbb{I}$. It follows that $\alpha = \beta$ because elements of $L^*$ are left regular.

5.4. It is now clear that no non-trivial categories with fuzzy subsets are not similar to $\mathbb{S}$. There are some universal constructions which are not possible in such categories. This observation about the limitations of the synchronic approach corresponds to an analogous limitation of the diachronic approach (A. Pitts [ 8 ], M. Barr [ 1 ]) where a category of fuzzy sets fails to be a topos unless the basic Heyting algebra is boolean. We want to show that any category with fuzzy subsets can be embedding in a "good topos".

5.5. In 3.2 we observed that for a category $\mathbb{C}$ with fuzzy subsets, $\mathbb{C}$ is equivalent to the Kleisli category of the monoid induced by the universal extension. Let $\mathbb{C}$ be equivalent to $\mathbb{S}(L^*)$ where $L^*$ is the monoid of Theorem 3. We can now consider the Einlenberg-Moore category of the monad. Let us denote this category by $L^*$-$\mathbb{S}$et. It is clear that an object of $L^*(\mathbb{S})$ is a set X equipped with an action of $L^*$. A morphism $f : X \to Y$ is an action-preserving map (i.e. $f(\alpha x) = \alpha f(x)$ for all $\alpha \in L^*$ and $x \in X$). The category $L^*$-$\mathbb{S}$et is trivially the topos of presheaves on the monoid $L^*$. Let $j : \mathbb{S}(L^*) \to L^*$-$\mathbb{S}$et denote the unique functor such that if $(i^*, C^*)$ is the suitable pair of adjoint functors associated to $L^*$-$\mathbb{S}$et then $i^* = j \circ i$ and $C = C^* \circ j$. It is a routine to see that $j(X)$ is $X \times L^*$ equipped with the trivial action $\alpha(x,\beta) = (x,\alpha\beta)$ and that $j(X \underset{(f,\varphi)}{\rightsquigarrow} Y)(x,\alpha) = (f(x),\alpha\varphi(x))$ for $x \in X$ and $\alpha \in L^*$. From left regularity of elements of $L^*$, it follows that j is full and faithful. Now, $\mathbb{S}(L^*)$ is equivalent to the full subcategory $L^*(\mathbb{S})$ of $L^*$-$\mathbb{S}$et whose objects are those of the form $X \times L^*$ equipped with the trivial action $\alpha(x,\beta) = (x,\alpha\beta)$. It follows that :

**THEOREM 4** : *Any category $\mathbb{C}$ with fuzzy subsets is equivalent to the supdense full subcategory $L^*(\mathbb{S})$ of the topos $L^*$-Set for which $L^* = L \setminus \{0\}$, where L is the set of truth-values. Furthermore any object of $L^*$-$\mathbb{S}$et is a quotient of an object of $L^*(\mathbb{S})$. If the elements of $L^*$ are regular (for example $L^*$ is a commutative monoid) then $L^*$-$\mathbb{S}$et is an extensions of $L^*$-$\mathbb{S}$et.*

Proof : The fact that $L^*(\mathbb{S})$ is a sup-dense subcategory of $L^*$-$\mathbb{S}$et follows from 0.12 of [ 3 ] and the fact that the representable functor is an object in $L^*(\mathbb{S})$. It is easy to see that $L^*(\mathbb{S})$ is an I-cocomplete subcategory of $L^*$-$\mathbb{S}$et for any discrete small category I. By the usual construction of colimits via the quotient of two parallel morphisms whose target is a direct sum, it follows that objects in $L^*$-$\mathbb{S}$et are quotients of objects of $L^*(\mathbb{S})$. The fact that $L^*$-$\mathbb{S}$et is an extension on subobjects of $L^*(\mathbb{S})$ when elements of $L^*$ are regular, is trivial.

5.6. Theorem 4 can be understood as a statement about a "best toposophical approximation" of a category with fuzzy subsets. It is important to remark that for non trivial categories with fuzzy subsets the related best approximation topos is not boolea

It satisfies De Morgan's law (see Proposition 1.1 of [4]) and number objects (natural numbers, integers, rationals and reals) are trivially "classical". Cauchy reals, Dedekind reals and MacNeille reals coincide. The corresponding object of real numbers is a geometric field.

BIBLIOGRAPHY

++++++++++++

[1] M. Barr, "Fuzzy Set Theory and Topos Theory", Canad. Math. Bull. Vol. 29(4), (1986), 501-508.

[2] J.A. Goguen, "Concept representation in natural and artificial languages : Axioms and Applications for fuzzy sets", International Journal of Man-Machine Studies (1974)6, 513-561.

[3] P.T. Johnstone, "Topos Theory", Academic Press, (1977).

[4] P.T. Johnstone, "Conditions related to De Morgan's Law", Springer Lecture Notes in Math. 753, (1977), 479-491.

[5] J. LAMBEK, "Completions of Categories", Springer Lecture Notes in Math. 24, (1966).

[6] S. Mac Lane, "Categories for Working Mathematician", Graduate Texts in Mathematics n°5, Springer-Verlag, 1971.

[7] C.V. Negoita, "Fuzzy sets in topoi", Fuzzy Sets and Systems 8 (1984) n°1, 93-99.

[8] A.M. Pitts, "Fuzzy sets do not forms a topos", Fuzzy Sets and Systems 8(1984)n°1, 101-104.

[9] L.N. Stout, "Topoi and categories of fuzzy sets", Fuzzy Sets and Systems 12(1984) n°2, 169-1984.

[10] L.A. Zadeh, "Fuzzy sets" Information and Control 8, (1965), 338-353.

# GLOBAL AND INFINITESIMAL OBSERVABLES(·)

*G.C. Meloni and E. Rogora*

## Contents



## 1. Introduction.

We work in a topos $\mathcal{X}$ of "smooth spaces" in the sense that we assume that $\mathcal{X}$ satisfies the usual conditions required for the development of *Categorical Dynamics* (see [L1], [L2] and [LS]) or *Synthetic Differential Geometry* (see [Ko]).

For every space $X$ (i.e. for every object in the topos) we introduce (§3) the object

$$\mathcal{F}_n(X) = \sum_{a,b \in \Re^n} X^{[a,b]}$$

of ***n-dimensional figures of rectangular type*** and we define an ***n-dimensional global observable*** as an "additive" function on $\mathcal{F}_n(X)$. By a fundamental result on $n$-dimensional integrals, the value of an $n$-dimensional global observable on a figure is obtained by integrating a function

$$o : X^{[0,0]} \times \Re^n \longrightarrow \Re$$

over the domain of that figure. We call ***n-dimensional infinitesimal element of observable*** (or briefly ***element of observable***) a function of this kind, to be seen as ***generalization*** of the notion of ***differential form.*** This interpretation depends on the fact that we can see $[0,0]$ as the object of the elements infinitely close to $0$. The function $o$ is determined by taking the "principal part" of the restriction of $\mathcal{O}$ to first order infinitesimal figures. This correspondence, between $o$ and $\mathcal{O}$, is bijective (§4) and this fact displays the relationship between the global and the infinitesimal aspects of observables. In classical differential geometry the same bijection works

(·) Lavoro eseguito con il contributo del CNR, Contratto no. 87.00987.01.

between differential forms and "additive" functions defined on the object of "oriented submanifolds" (see also §5). The notion of element of observable, anyway, is more general because it embodies, for example, quadratic forms, time dependent forms, germs determined forms and so on. In this more general context it is still true that the relationship between infinitesimal and global aspects allows us to choose the best framework for approaching a specific problem (e.g. cohomology) and to pass freely from one representation to the other in order to understand constructions and perform operations in the simplest way.

For example, by composing an arbitrary global observable $\mathcal{O}$ with the boundary operator $\partial$ we obtain a simple operation $(\mathcal{O} \longmapsto \mathcal{O} \circ \partial)$ between global observables. The corresponding operation in the infinitesimal context is the *generalization*, to the elements of observable, of the *exterior derivative* of differential forms. To have clear in mind both pictures makes Stokes "theorem" a simple definition even at this level of generality (§4).

The bijection between global observables and elements of observable also works between a certain number of subclasses of some interest. In particular we *characterize* the global observables which correspond to differential forms. We will do this in three steps that can be summarized in the following way (see §5).

If $\mathcal{O}$ and $o$ are a global observable and the corresponding element of observable, then:

i) $o : X^{[0,0]} \times \Re^n \longrightarrow \Re$ is independent of $\Re^n$ (in the sense that it factors through the projection $X^{[0,0]} \times \Re^n \longrightarrow X^{[0,0]}$) if and only if $\mathcal{O}$ is invariant under reparametrizations (of figures) which are translations;

ii) if $i : W \hookrightarrow [0,0]$ is a subobject of $[0,0]$ then $o$ factors through the image $(X^W)_*$ of $X^i : X^{[0,0]} \to X^W$ if and only if $\mathcal{O}$ satisfies the property mentioned in i) and for every $\phi$, $\psi$ figures and every $\mathbf{x}$ belonging to the domain of both figures, if

$$\forall \mathbf{w} \in W \quad \phi(\mathbf{x}+\mathbf{w}) = \psi(\mathbf{x}+\mathbf{w}),$$

then

$$\forall \mathbf{h} \in D^n \quad \mathcal{O}(\mathbf{x}, \mathbf{x}+\mathbf{h}, \phi) = \mathcal{O}(\mathbf{x}, \mathbf{x}+\mathbf{h}, \psi).$$

iii) $o : (X^{D^n})_* \to \Re$ is multihomogeneous and alternating if and only if $\mathcal{O}$ satisfies the condition stated in ii), with $W = D^n$, and is invariant under any reparametrization of figures.

The last assertion must be made meaningful in a proper way by introducing a suitable theorem for the changement of variables under the integral sign. This is exactly what we discuss in the last section.

Since the *composition with* $\partial$ preserves the properties mentioned in i), ii) and iii), Stokes "theorem" is also valid for elements of observable satisfying such properties.

The results on differential forms and those of the last section depend on a lemma (Lemma 27, §5) which states that the value of a differential form on an infinitesimal figure varies with the reparametrization of the figure according to the determinant of the gradient at zero of the reparametrizing function. The proof of this lemma constitutes the most thecnical part in the paper.

## 2. Order and Integration in $\Re^n$

To get the notion of *bounded* figure we introduce a preorder relation in $\Re$. About this relation we assume (see [Ko]) compatibility with the ring structure and the following infinitesimal condition

$$h \text{ nilpotent} \quad \Longrightarrow \quad h \le 0 \quad \text{and} \quad 0 \le h.$$

We do not require antisymmetry otherwise the last assumption would imply, for every $h$ nilpotent, $h = 0$.

Although the next definition looks a little strange, actually it reveals itself very useful in the continuation.

**Definition 1.** For every $a, b \in \Re$ we define the *"unoriented" interval* $[a,b]$ by putting

$$[a,b] = \{x \in \Re \mid a \leq x \leq b \text{ or } b \leq x \leq a\}$$

**Observation 2.** The assumption about infinitesimal elements made on the order structure implies:

i) for all $h_1, h_2$ nilpotents

$$[a,b] = [a + h_1, b + h_2],$$

ii) the interval $[a,a]$ is not a singleton because it contains, at least, all the elements of the form $a + h$, with $h$ nilpotent.

**Definition 3.** If $n \geq 0$, $\mathbf{a}, \mathbf{b} \in \Re^n$, $\mathbf{a} = (a_1, \ldots, a_n)$ and $\mathbf{b} = (b_1, \ldots, b_n)$ we define the *"unorienteded" n-dimensional interval* $[\mathbf{a}, \mathbf{b}]$ by putting

$$[\mathbf{a}, \mathbf{b}] = [a_1, b_1] \times \cdots \times [a_n, b_n].$$

The theory of integration, introduced in [MC] following a suggestion of Lawvere (see [L1], pag. 19) and ulteriorly worked out in [Ca], [Co] and [Ro], is based on the following axiom which can be stated (and partially developed) independently of the axiom for differentiation (for another approach to integration see [Ko]).

**Axiom 4.** (for one dimensional integration) For every $f \in \Re^{[a,b]}$ there exists exactly one $I_f \in \Re^{[a,b]^2}$ satisfying the following conditions:

i) (additivity on domain) for every $x, y, z \in [a,b]$

$$I_f(x,y) + I_f(y,z) = I_f(x,z)$$

ii) (first order behaviour) for every $h \in D$

$$I_f(x, x+h) = h \cdot f(x).$$

Obviously $I_f(x,y)$ is, by definition, the integral of $f$ between $x$ and $y$, that is we put

$$\int_x^y f := I_f(x,y).$$

A certain number of properties of integrals (e.g. linearity) follows using solely the previous axiom. Furthermore, if we assume both axiom 4 and the usual axiom for differentiation we can easily prove the fundamental theorem of calculus.

In order to develop the theory of $n$-dimensional integration we need to fix some notations.

**Notations 5.**

i) if $\mathbf{u} = (u_1, \ldots, u_{n-1}) \in \Re^{n-1}$, $1 \leq i \leq n$ and $x \in \Re$, we will write $\mathbf{u} *_i x$ to denote the element $(u_1, \ldots, u_{i-1}, x, u_i, \ldots, u_{n-1})$ of $\Re^n$.

ii) if $\mathbf{a} = (a_1, \ldots, a_n) \in \Re^n$ and $1 \leq i \leq n$ we will write $i\mathbf{a}$ to denote the element $(a_1, \ldots, a_{i-1}, a_{i+1}, \ldots, a_n)$ of $\Re^{n-1}$.

**Definition 6.** Two pairs of points $(\mathbf{a}, \mathbf{b}_1)$, $(\mathbf{b}_2, \mathbf{c})$ of $\Re^n$ are said to be *adjacent along the i-th coordinate* if there exist $\mathbf{u}, \mathbf{v} \in \Re^{n-1}$ and $x, y, z \in \Re$ such that

$$\mathbf{a} = \mathbf{u} *_i x, \ \mathbf{b}_1 = \mathbf{v} *_i y, \ \mathbf{b}_2 = \mathbf{u} *_i y, \ \mathbf{c} = \mathbf{v} *_i z.$$

By assuming axiom 4 and the usual axiom for differentiation we can prove (see [MC] and [Ro]) the higher dimensional analogue of Axiom 4.

**Theorem 7.** (for $n$-dimensional integration) For every $f \in \Re^{[\mathbf{a},\mathbf{b}]}$ $(\mathbf{a}, \mathbf{b} \in \Re^n)$ there exists exactly one $I_f \in \Re^{[\mathbf{a},\mathbf{b}]^2}$ satisfying the following conditions:

i) (additivity along any coordinate) if $1 \leq i \leq n$ and $(\mathbf{x}, \mathbf{y}_1)$, $(\mathbf{y}_2, \mathbf{z})$ $(\in [\mathbf{a},\mathbf{b}]^2)$ are adjacent along the $i$-th coordinate then

$$I_f(\mathbf{x}, \mathbf{y}_1) + I_f(\mathbf{y}_2, \mathbf{z}) = I_f(\mathbf{x}, \mathbf{z})$$

ii) (first order behaviour) if $\mathbf{h} \in D^n$ then

$$I_f(\mathbf{x}, \mathbf{x}+\mathbf{h}) = h_1 \cdot \ldots \cdot h_n \cdot f(\mathbf{x}).$$

Furthermore, such a function $I_f$ can be expressed, as usual, by means of iterated 1-dimensional integrals.

Obviously we put

$$\int_{\mathbf{x}}^{\mathbf{y}} f := I_f(\mathbf{x}, \mathbf{y}).$$

**Observation 8.** Every function $A \in \Re^{[\mathbf{a},\mathbf{b}]^2}$ which is additive along any coordinate direction determines, by differentiation axiom, a unique function $f \in \Re^{[\mathbf{a},\mathbf{b}]^2}$ such that, for any $\mathbf{h} \in D^n$, $h_1 \cdot \ldots \cdot h_n \cdot f(\mathbf{x}) = A(\mathbf{x}, \mathbf{x}+\mathbf{h})$. Therefore, theorem 7 means that the map $A \mapsto f$ is a bijection.

For 1-dimensional integrals there is a very simple proof of the following theorem for changing variables under the integral sign.

**Theorem 9.** If $a, b, a', b' \in \Re$, $\phi \in [a', b']^{[a,b]}$ and $f \in \Re^{[a',b']}$ then

$$\int_{\phi(a)}^{\phi(b)} f = \int_a^b (f \circ \phi) \cdot \phi'.$$

A generalization of this theorem to $n$ variables is not so easy, even to formulate. In fact the image of an $n$-dimensional interval under a reparametrizing function is not, in general, an $n$-dimensional interval and we do not know, till now, what it means to integrate over sets different from intervals. We will deal with this problem in full generality in §6, but for special kinds of change of variables we can get, also for $n$-dimensional integrals, an easy theorem.

**Definition 10.** An *I-function* is a function $\phi \in [\mathbf{a}', \mathbf{b}']^{[\mathbf{a},\mathbf{b}]}$ of the form

$$\phi(\mathbf{x}) = (\phi_1(x_{\sigma(1)}), \ldots, \phi_n(x_{\sigma(n)}))$$

where $\sigma : \{1, \ldots, n\} \to \{1, \ldots, n\}$ is a bijection and, for every $i = 1, \ldots, n$,

$$\phi_i : [a_{\sigma(i)}, b_{\sigma(i)}] \longrightarrow [a'_i, b'_i]$$

is an arbitrary function.

**Theorem 11.** If $\phi$ $(\in [\mathbf{a}', \mathbf{b}']^{[\mathbf{a},\mathbf{b}]})$ is an $I$-function and $f \in \Re^{[\mathbf{a}',\mathbf{b}']}$ then

$$\int_{\phi(\mathbf{a})}^{\phi(\mathbf{b})} f = \int_{\mathbf{a}}^{\mathbf{b}} (f \circ \phi) \cdot \mathit{det}(\mathit{Grad}\, \phi).$$

The proof is a trivial application of theorem 9.

## 3. Figures and Observables

Let $X$ be any object of the given category $\mathcal{X}$ of "smooth spaces". This means that, as suggested in [L1] and partially developed for example in [CM], we work out the theory without restrictions on the kind of space.

**Definition 12.** For any $n \geq 0$, the space $\mathcal{F}_n(X)$ of *n-dimensional figures of rectangular type* is defined by

$$\mathcal{F}_n(X) = \sum_{\mathbf{a},\mathbf{b} \in \Re^n} X^{[\mathbf{a},\mathbf{b}]}.$$

An $n$-dimensional figure is then a triple $\Phi = \langle \mathbf{a}, \mathbf{b}, \phi \rangle$ with $\mathbf{a}, \mathbf{b} \in \Re^n$ and $\phi \in X^{[\mathbf{a},\mathbf{b}]}$. The pair $\langle \mathbf{a}, \mathbf{b} \rangle$ specifies the boundary of $\phi$ and the order in the pair specifies *some features* of the orientation of the figure. We need to take explicitly into account the boundary of $\phi$ because this information is not completely contained in the interval itself.

**Notation 13.** Starting from any figure $\Phi = \langle \mathbf{a}, \mathbf{b}, \phi \rangle$ we can get new figures by *restriction*. In fact, for every $\mathbf{x}, \mathbf{y} \in [\mathbf{a}, \mathbf{b}]$ we can consider the new figure

$$\Phi' = \langle \mathbf{x}, \mathbf{y}, \phi|_{[\mathbf{x},\mathbf{y}]} \rangle.$$

In order to make notations simpler we still indicate with $\langle \mathbf{x}, \mathbf{y}, \phi \rangle$

**Definition 14.** Let $\langle \mathbf{a}, \mathbf{b}_1 \rangle, \langle \mathbf{b}_2, \mathbf{c} \rangle$ be pairs adjacent along the $i$-th coordinate. Given any $\phi \in X^{[\mathbf{a},\mathbf{b}_1] \cup [\mathbf{b}_2, \mathbf{c}]}$ we say that $\Phi_1 = \langle \mathbf{a}, \mathbf{b}_1, \phi \rangle$ and $\Phi_2 = \langle \mathbf{b}_2, \mathbf{c}, \phi \rangle$ are *summable along the i-th coordinate direction* and we define their $i$-th sum as

$$\Phi_1 +_i \Phi_2 := \langle \mathbf{a}, \mathbf{b}, \phi \rangle.$$

In this way we have defined $n$ partial sums on $\mathcal{F}_n(X)$.

**Definition 15.** A function $\mathcal{O} : \mathcal{F}_n(X) \to \Re$ is an *n-dimensional global Observable* on $X$, or simply an *Observable*, if it is a morphism for all the partial sums defined on $\mathcal{F}_n(X)$, i.e. if $1 \le i \le n$ and $\Phi_1, \Phi_2$ summable along the $i$-th coordinate imply

$$\mathcal{O}(\Phi_1 +_i \Phi_2) = \mathcal{O}(\Phi_1) + \mathcal{O}(\Phi_2).$$

To introduce the fundamental notion of boundary of a figure we need to pass from the object of figures to the abelian group generated by them. As we want to maintain the above defined partial sums we are led to the following

**Definition 16.** The abelian group of *aggregates of n-dimensional figures of rectangular type* in $X$ is the quotient of the abelian group freely generated by $\mathcal{F}_n(X)$ with respect to the relations

$$\Phi_1 + \Phi_2 \sim \Phi_1 +_i \Phi_2$$

where $\Phi_1$ and $\Phi_2$ are any two figures which are summable along the $i$-th coordinate direction. We denote this group with $F_n(X)$.

**Observation 17.** There is a bijective correspondence between $n$-dimensional global Observables on $X$ and additive morphisms from $F_n(X)$ to $\Re$. We will identify, also notationally, the elements of these two sets.

**Definition 18.** Let $\phi \in X^{[\mathbf{a},\mathbf{b}]}$, $\mathbf{a} = \langle a_1, \ldots, a_n \rangle$ and $\mathbf{b} = \langle b_1, \ldots, b_n \rangle$. We define $2n$ functions $\phi a_i$, $\phi b_i$ (for $i = 1, \ldots, n$), from $[\check{i}\mathbf{a}, \check{i}\mathbf{b}]$ to $X$, by putting

$$\phi a_i = \{\mathbf{x} \mapsto \phi(\mathbf{x} *_i a_i)\}$$
$$\phi b_i = \{\mathbf{x} \mapsto \phi(\mathbf{x} *_i b_i)\}.$$

**Definition 19.** The *boundary map* is the function $\partial : \mathcal{F}_{n+1}(X) \to F_n(X)$ given by

$$\partial(\langle \mathbf{a}, \mathbf{b}, \phi \rangle) = \sum_{1 \le i \le n} (-)^i [\langle \check{i}\mathbf{a}, \check{i}\mathbf{b}, \phi a_i \rangle - \langle \check{i}\mathbf{a}, \check{i}\mathbf{b}, \phi b_i \rangle].$$

This function preserves the relations which define $F_n(X)$ and hence it extends uniquely to a morphism of abelian groups $\partial : F_{n+1}(X) \to F_n(X)$. This morphism verifies the identity $\partial^2 = 0$.

**Definition 20.** The *coboundary map* is the function $\delta : \mathcal{O}^n(X) \to \mathcal{O}^{n+1}(X)$ defined by $\delta\mathcal{O} := \mathcal{O} \circ \partial$.

## 4. Infinitesimal Elements of Observable and Bijection Theorem.

We will show in this section that any global observable $\mathcal{O}$ is obtained by "glueing" a function of infinitesimal figures.

**Definition 21.** Let $W$ be any subobject of the infinitesimal interval $[0, 0]$. Given any function $\phi \in X^{[\mathbf{a},\mathbf{b}]}$ we define $\phi^{\cdot}_W(\mathbf{x})$, the *W-localization* of $\phi$ at $\mathbf{x}$, as the element of $X^W$ given by $(\phi^{\cdot}_W(\mathbf{x}))(\mathbf{w}) = \phi(\mathbf{x} + \mathbf{w})$, for all $\mathbf{w} \in W$.

Given an n-dimensional observable $\mathcal{O}$ and a figure $\Phi = \langle a, b, \phi \rangle$, consider the function $\mathcal{O}_\phi : [a,b]^2 \to \Re$ defined by $\mathcal{O}_\phi(x,y) = \mathcal{O}(\langle x,y,\phi \rangle)$. This function is additive along any coordinate direction, hence by observation 8 there exists a unique function $o_\phi : [a,b] \to \Re$ such that:

$$\mathcal{O}(\langle x,y,\phi \rangle) = \mathcal{O}_\phi(x,y) = \int_x^y o_\phi(z).$$

If the functions $\phi$ and $\phi'$ of two figures $\Phi$ and $\Phi'$ agree on a common subinterval $[a', b']$ of the two domains, the corresponding functions $o_\phi$ and $o_{\phi'}$ have equal restrictions to $[a', b']$ by bijectivity of the map in observation 8. This allows us to define a new function

$$o : \sum_{a \in \Re^n} X^{[a,a]} \to \Re$$

by $o(\tau) = o_\psi(z)$ (here $\psi$ is any map whose restriction to $[0,0]$ is $\tau$).

The function $o$ may be seen equivalently as a function $o : X^{[0,0]} \times \Re^n \to \Re$ and this explicit splitting of the domain will show to be useful later.

The link between $o$ and $\mathcal{O}$ is given by the equations

$L_1$)

$$\mathcal{O}(x,y,\phi) = \int_x^y o(\phi^{\cdot}_{|0,0|}(z), z)$$

$L_2$)

$$\forall h \in D^n \quad h_1 \cdot \ldots \cdot h_n \cdot o(\tau, x) = \mathcal{O}(x, x+h, \tau_x)$$

(here $(\tau_x)(y) = \tau(y - x)$).

**Definition 22.** We call a function $o : X^{[0,0]} \times \Re^n \to \Re$ an *n- dimensional infinitesimal element of observable* on $X$ or simply an *element of observable* and we call the set $\sum_{a \in \Re^n} X^{[a,a]}$ *space of infinitesimal n-dimensional figures* in $X$. We have therefore the following "bijection theorem"

**Theorem 23.** There is a bijection between n-dimensional global observables and n-dimensional infinitesimal elements of observable on any space $X$, and this bijection is explicitly given by the equations $L_1$ and $L_2$.

This bijection theorem allows us to pass freely from the global to the infinitesimal representation of an observable and then to perform constructions and understand definitions on observables in the most convenient framework. For example the construction of the coboundary of an observable corresponds, at infinitesimal level, to a notion of *exterior derivative* which reduces to the usual definition when specialized to infinitesimal observables which are differential forms. We will use the symbol $d\,o$ to denote the exterior derivative of an infinitesimal observable $o$. Hence $d\,o$ and $\delta\mathcal{O}$ are linked by the same equations $L_1$ and $L_2$ which link $o$ and $\mathcal{O}$.

By defining $\int_\Phi o := \mathcal{O}(\Phi)$ we get therefore

$$\int_\Phi d\,o = (\delta\mathcal{O})(\Phi) = \mathcal{O}(\partial\,\Phi) = \int_{\partial\Phi} o$$

that is Stokes "theorem".

On the other hand many constructions are much easier from the "infinitesimal" point of view. A simple example is given by fixing a function $f : \Re \to \Re$ and considering the map between elements of observable induced by $f$ according to the formula $o \mapsto f \circ o$.

To finish this paragraph we stress again that the way to construct $\mathcal{O}$ from $o$ is by glueing together the values of $o$ around any point, and the device employed for glueing them together is the integral. Here is tacitly understood another element, namely the differential form $dx_1 \wedge \ldots \wedge dx_n$ with respect to which the integral is defined. The explicit consideration of the role played by this further element will allow us to establish in the last paragraph a theorem of change of variables for n-dimensional integrals.

## 5. Infinitesimal Elements of Observable with further properties.

In this paragraph we will characterize the properties of those observables which bijectively correspond to elements of observable with further properties, the most important of which is that of being a differential form.

**Proposition 24.** The element of observable $o : X^{[0,0]} \times \Re^n \to \Re$ is independent on $\Re^n$ if and only if the corresponding observable $O$ is invariant under reparametrizations of figures which are translations.

The proof of this proposition is an immediate consequence of the theorem for changing variables with respect to I-functions (translations are I-functions).

**Proposition 25.** Let $W$ be any subobject of $[0,0]$ (with inclusion $i : W \hookrightarrow [0,0]$). Then $o : X^{[0,0]} \to \Re$ factors through the image $(X^W)_*$ of $X^i : X^{[0,0]} \to X^W$ if and only if $O$ is invariant under translations and has the following property:

For any pair of figures $\Phi,\Psi$ and for any $\mathbf{x}$, if $\phi^{\cdot}_W(\mathbf{x}) = \psi^{\cdot}_w(\mathbf{x})$, then for all $\mathbf{h} \in D^n$ $O(\langle \mathbf{x},\mathbf{x}+\mathbf{h},\phi\rangle) = O(\langle \mathbf{x},\mathbf{x}+\mathbf{h},\psi\rangle)$.

The proof is a consequence of the equalities $L_1$ and $L_2$ and of the properties of the integral. Special cases of these results, in dimension one and with $W = D$, was proved in [Ca] and [Co].

Let $W$ be any subobject of $[0,0]$ which contains $D^n$, $\Gamma$ any subobject of $[\mathbf{0},\mathbf{0}]^W$ such that for every $\gamma$ in $\Gamma$ $\gamma(\mathbf{0}) = \mathbf{0}$ and $det_\mathbf{0} : [\mathbf{0},\mathbf{0}]^W \to \Re$ the function which assigns the determinant of the Jacobian matrix at zero to any $\gamma \in [\mathbf{0},\mathbf{0}]^W$.

We define:

$\omega\Gamma$, as the set of all those $o$ which factor through $(X^W)_*$ and such that

$$o(\tau \circ \gamma) = o(\tau) \cdot det_\mathbf{0}(\gamma).$$

$G$, as the set of all the functions $g \in \Re^{n[\mathbf{a},\mathbf{b}]}$ defined over intervals and such that

$$\forall \mathbf{x} \text{ in the domain of } g,\ \dot{g}_w(\mathbf{x}) \in \Gamma.$$

$\Omega\Gamma$, as the set of all the observables which satisfy proposition 25 with respect to W and such that

$$\forall g \in G \qquad O(\langle \mathbf{a},\mathbf{b},\phi \circ g\rangle) = \int_\mathbf{a}^\mathbf{b} o(\phi^{\cdot}_W(g(\mathbf{x})) \cdot det_\mathbf{0}(C^W_\mathbf{x}(g))$$

(here $C^W_\mathbf{x}(g)$ is the element of $[\mathbf{0},\mathbf{0}]^W$ defined by $(C^W_\mathbf{x}(g))(\mathbf{w}) = g(\mathbf{x}+\mathbf{w}) - g(\mathbf{x})$ ).

Then we have the following easy

**Proposition 26.**

$$o \in \omega\Gamma \qquad \text{if and only if} \qquad O \in \Omega\Gamma.$$

The meaning of the condition for $O$ to belong to $\Omega\Gamma$ is that $O$ is invariant under reparametrizations of figures whose W-localization at any point of their domain is in $\Gamma$. We will postpone the discussion of this point until the next paragraph and prove now the lemma on which the discussion will be based.

**Lemma 27.** For every $\tau' \in (X^{D^n})_*$ let $\tau$ be an extension of $\tau'$ defined on the whole interval $[0,0]$ and $\gamma$ any function $\gamma \in [0,0]^{D^n}$ such that $\gamma(\mathbf{0}) = \mathbf{0}$. If $o : (X^{D^n})_* \to \Re$ is multihomogeneous and alternating (a differential form), then

$$o(\tau \circ \gamma) = o(\tau') \cdot det_\mathbf{0}(\gamma).$$

To prove this lemma we need to decompose a generic element of $[\mathbf{0},\mathbf{0}]^{D^n}$ and in order to do that we will deal with the structure of the iterated tangent bundles of $\Re^n$.

On the iterated tangent bundle $\Re^{n D^n}$ of $\Re^n$ are defined n natural isomorphisms

$$i : (\Re^n)^{D^n} \to (\Re^n)^{(D^{n-1})^D}$$

given by

$$\gamma \mapsto (i\gamma) = [h \mapsto [\mathbf{h} \mapsto \gamma(\mathbf{h} *_i h)]].$$

Also, on any affine space $P$ with translation space $V$ ($P^D \cong P \times V$) we can define the addition of those vectors which have the same base point by computing their sum in V.

Therefore, as $(\Re^n)^{D^{n-1}}$ is affine and $(\Re^n)^{D^n}$ is isomorphic in n ways to its tangent bundle, it is possible to add two elements $\gamma$ and $\bar{\gamma}$ of $(\Re^n)^{D^n}$ such that

$$(i\gamma)(0) = (i\bar{\gamma})(0)$$

for some $i$ by using the sum on the tangent bundle of $(\Re^n)^{D^{n-1}}$ and pulling it back to $(\Re^n)^{D^n}$ along the i-th isomorphism. In other words we can define the i-th addition for functions $\gamma$, $\bar{\gamma}$ that satisfy the above condition by the formula

$$\gamma +_i \bar{\gamma} := i^{-1}(i\gamma + i\bar{\gamma}).$$

By differentiation axiom any *n-vector* $\gamma \in (\Re^n)^{D^n}$ is determined by its Taylor coefficients $\partial_H \gamma$ (here $H$ is any subset of $\{1, \ldots, n\}$,) through the formula

$$\forall \mathbf{h} \in D^n \quad \gamma(\mathbf{h}) = \sum_{H \subseteq \{1,\ldots,n\}} (\prod_{i \in H} h_i)\partial_H \gamma.$$

By reffering to their Taylor expansions, we can say that $\gamma$ and $\bar{\gamma}$ are summable with respect to the index $i$ if and only if $\partial_H \gamma = \partial_H \bar{\gamma}$ for all those subsets $H$ of the set $\{1, \ldots, n\}$ such that $i \notin H$. The Taylor coefficients of $\gamma +_i \bar{\gamma}$ are then $\partial_H \gamma$ if $i \notin H$ and $\partial_H \gamma + \partial_H \bar{\gamma}$ if $i \in H$. It is an easy consequence of a well known result in S.d.G. that any function $L : (\Re^n)^{D^n} \to \Re$ that is homogeneous with respect to the i-th product (meaning $L(c_i \gamma) = c \cdot L(\gamma)$, where $c_i \gamma = [\mathbf{h} \mapsto \gamma((h_1, \ldots, c \cdot h_i, \ldots, h_n))]$ ) is additive with respect to the i-th sum, that is $L(\gamma +_i \bar{\gamma}) = L(\gamma) + L(\bar{\gamma})$.

Given $o$ and $\tau'$ as in the lemma, let us consider the function $T : [\mathbf{0}, \mathbf{0}]^{D^n} \to \Re$ defined by $T(\gamma) = o(\tau \circ \gamma)$. A simple computation shows that $T(c_i \gamma) = c \cdot T(\gamma)$ and therefore $T(\gamma +_i \bar{\gamma}) = T(\gamma) + T(\bar{\gamma})$.

Consider now $\gamma \in [\mathbf{0}, \mathbf{0}]^{D^n}$ and its Taylor coefficients $\partial_H \gamma$. Let us suppose that every index appears only one time. Then we do not need to go any further. Otherwise there exists an index which appears at least two times. Let $i$ be the least index with this property. Decompose $\gamma$ as a sum of pieces each with only one derivative containing the i-th coordinate (the sum is then the i-th sums). After a finite number of steps we will have $\gamma$ decomposed as the sum of pieces in which each index $i$ appears only one time in the Taylor expansion of any piece, and the sum is made in a suitable order and with respect to suitable indexes. Therefore we have

$$T(\gamma) = \sum T(\gamma_j) = \sum o(\tau \circ \gamma_j)$$

where $\gamma_j$ are the pieces obtained above. Antisimmetry and homogeneity of $o$ imply that the only term different from zero in the sum is $o(\tau \circ \gamma_P)$, where $\gamma_P$ is the term whose Taylor coefficients contain only the first-order derivatives of $\gamma$. By using antisimmetry and homogeneity once again we get

$$o(\tau \circ \gamma_P) = o(\tau') \cdot det_{\mathbf{O}} \gamma_P = o(\tau') \cdot det_{\mathbf{O}} \gamma.$$

that proves the lemma.

**Corollary 28.** An observable $\mathcal{O}$ corresponds to a differential form if and only if it satisfies the hypothesis in proposition 24 and in proposition 25, with $W = D^n$, and is invariant under any reparametrization of figures.

## 6.Theorem of Change of Variables

Let us now go back to the problem of justifying in what sense

$$\int_{\mathbf{a}}^{\mathbf{b}} o(\phi_W^{\cdot}(g(\mathbf{x}))) \cdot det_{\mathbf{O}}(C_{\mathbf{x}}^W(g))$$

is the value of the observable $\mathcal{O}$ on the restriction of the figure $\Phi$ to the image of $g$ ($\phi$, $g$ and $\mathcal{O}$ just as before proposition 26).

We need first to extend the notion of observable to figures that are more general than those of rectangular type. We want to be able to deal with figures which are obtained by restricting rectangular type figures to any image of intervals.

**Definition 29.** A figure of this kind, called an *n-dimensional figure*, is a quadruple $\langle \mathbf{a}, \mathbf{b}, g, \Phi \rangle$, where $\Phi$ is an n-dimensional figure of rectangular type ($\Phi = \langle \mathbf{a}', \mathbf{b}', \phi \rangle$), $\mathbf{a}, \mathbf{b}$ are elements of $\Re^n$, and $g$ is a function $g \in [\mathbf{a}', \mathbf{b}']^{[\mathbf{a},\mathbf{b}]}$.

We want to define the value of an observable $O$ on an n-dimensional figure in such a way that this value is obtained by "glueing" $o(\phi^{\cdot}_{W}(\mathbf{x}))$ on the image of $g$, generalizing what we have shown to be true for rectangular type figures (see comment at the end of §4).

For this we make explicit the following definition.

**Definition 30.** Given any $p : (X^W)_* \to \Re$, any figure $\Gamma = \langle \mathbf{a}, \mathbf{b}, \gamma \rangle$, $\Gamma \in \mathcal{F}_n(X)$ and any function $f \in \Re^{\gamma([\mathbf{a},\mathbf{b}])}$, by *glueing* or *integrating* $f$ with respect to $p$ on the image of $\gamma$, we mean to find a function $A$, the "glue" or "integral", such that the following two conditions are satisfied:

i) $A : [\mathbf{a}, \mathbf{b}]^2 \to \Re$ is additive along any coordinate direction

ii) $\forall \mathbf{h} \in D^n \quad A(\mathbf{x}, \mathbf{x} + \mathbf{h}) = h_1 \cdot \ldots \cdot h_n \cdot p(\dot{\gamma}_w(\mathbf{x})) \cdot f(\gamma(\mathbf{x}))$.

The element of observable which should glue $o(\phi^{\cdot}_{D^n}(\mathbf{x}))$ to give the value of $O$ on a figure should be $dx_1 \wedge \ldots \wedge dx_n$ because we have shown that this element of observable is the one which works for rectangular type figures (see the end of §4 considering the fact that $((dx_1 \wedge \ldots \wedge dx_n)((id^{\cdot}_w)(\mathbf{y})) = 1)$. The next proposition, whose proof depends on lemma 27, is the first step in showing that this is what happens.

**Proposition 31.** The function $A$ defined by

$$A(\mathbf{y}, \mathbf{z}) = \int_{\mathbf{y}}^{\mathbf{z}} f(\gamma(\mathbf{x})) \cdot det_O(C^{D^n}_{\mathbf{x}}(\gamma))$$

glues $f$ on the image of $\gamma$ with respect to $dx_1 \wedge \ldots \wedge dx_n$ .

Proposition 31 provides a justification for the next definition which makes explicit what we mean by integrating a function over the image of an interval and therefore can be considered as the theorem for changing variables under the integral sign for n-dimensional integrals.

**Definition 32.** *The integral of $f$ over* $\gamma([\mathbf{a}, \mathbf{b}])$ is given by $A(\mathbf{a}, \mathbf{b})$, or

$$\int_{\gamma([\mathbf{a},\mathbf{b}])} f = \int_{\mathbf{a}}^{\mathbf{b}} f(\gamma(\mathbf{x})) \cdot det_O(C^{D^n}_{\mathbf{x}}(\gamma)).$$

If we take now, for $f$ the function $o(\phi^{\cdot}_{W}(\mathbf{x}))$ and for $\gamma$ the function $g$ as in definition 29 and successive considerations, proposition 31 shows that $dx_1 \wedge \ldots \wedge dx_n$ glues $o(\phi^{\cdot}_{W}(\mathbf{x}))$ on the image of $g$, that is, we can state the following definition for the value of an observable on a figure, in agreement with the considerations made after definition 29.

**Definition 33.** The *value of an observable $O$ on a figure* $\langle \mathbf{a}, \mathbf{b}, g, \Phi \rangle$ is given by

$$\int_{\mathbf{a}}^{\mathbf{b}} o(\phi^{\cdot}_{W}(g(\mathbf{x}))) \cdot det_O(G^W_{\mathbf{x}}(g)).$$

Another way to formulate proposition 26 is then

$$o \in \Omega\Gamma \quad \text{if and only if} \quad \forall g \in G \quad O(\langle \mathbf{a}, \mathbf{b}, g, \phi \rangle) = O(\langle \mathbf{a}, \mathbf{b}, \phi \circ g \rangle)$$

that means that $O$ is invariant under any reparametrization in $G$.

By defining the value of an observable on a figure we must take into account the orientation induced by a reparametrizing function. The following note gives an idea of what this amounts to.

**Note 34.** Let $\sigma_{ij} : [\mathbf{a}, \mathbf{b}] \to [\mathbf{a}, \mathbf{b}]$ the function defined by

$$\sigma(\langle \cdots x_i \cdots x_j \cdots \rangle) = \langle \cdots x_j \cdots x_i \cdots \rangle.$$

Then

$$O(\langle \mathbf{a}, \mathbf{b}, \phi \rangle) = O(\langle \mathbf{a}, \mathbf{b}, \phi, id \rangle) = -O(\langle \mathbf{a}, \mathbf{b}, \phi, \sigma \rangle).$$

## REFERENCES


[Ca] A. Casella, "Sui fondamenti della geometria e dell'analisi" Tesi di Laurea in Matematica, Univ. degli Studi di Milano, a.a 1979–80.

[CM] S. Chiodo and G.C. Meloni, "Proprietà infinitesimali e algebre di Lie" Istituto Lombardo (Rend. Sc.), A 116, 65–72 (1982).

[Co] F. Costa, "Teorie algebriche per la termostatica" Tesi di Laurea in Matematica, Univ. degli Studi di Milano, a.a. 1979–80.

[Ko] A. Kock, "Synthetic Differential Geometry" London Math Soc. Publ. Series **51**, Cambridge Univ. Press, 1981.

[L1] F.W. Lawvere, "Categorical Dynamics" in Topos Theoretic Methods in Geometry, Math Inst. Var. Publ. Series N. **50** (1979), Aahrus.

[L2] F.W. Lawvere, "Toward the description in a smooth topos of the dynamically possible motions and deformations of a continous body" Cahiers de Top. et Géom. Différentielle, XXI-4 (1980) 337–392.

[LS] F.W. Lawvere and S.H. Schanuel (eds.), "Categories in Continuum Physics" Lect. Notes in Math. 1174, Springer, 1986.

[MC] G.C. Meloni and A. Casella, "Introduzione alle quantità pure con infinitesimi" Quaderno n. 42/S, Istituto di Matematica "F. Enriques", Milano, Gennaio 1980.

[Ro] E. Rogora, "Strutture categoriali per la fisica" Tesi di Laurea in Matematica, Univ. degli Studi di Milano, a.a. 1984–85.



Gian Carlo Meloni
Dipartimento di Matematica
Universita' degli Studi di Milano
Via C. Saldini 50
20133 Milano
ITALY

Electronic Mail Address:
mcvax!i2unix!ccaumi!meloni

Enrico Rogora
Departement of Mathematics
S.U.N.Y. at Buffalo
Diefendorf Hall
14214 Buffalo N.Y.
U.S.A.

Electronic Mail Address:
V999PCT4@UBVMSA


This paper is in final form and will not be published elsewhere.

# Toposes and Groupoids

IEKE MOERDIJK*
Department of Mathematics
University of Chicago

The aim of this paper is to explain to what extent the category of Grothendieck toposes can be described in terms of groupoids (in the category of locales). In the first section, I describe how every groupoid $G$ gives rise to a topos $BG$, and in section 2, I discuss some of the functorial properties of this constrution $G \mapsto BG$. After having introduced two completeness properties of groupoids, we will see that toposes can be obtained by localizing groupoids (§4) or by considering geometric morphisms as obtained by tensoring with something analogous to a bimodule (§5). In section 6 I briefly discuss the fundamental group of a topos.

This paper provides a summary of my earlier papers [M2], [M3], [M4]. Since the proofs given there are often long and technical, and involve extensive use of change-of-base methods, I believe it is worthwhile to present these results all together in a more directly accessible way, and save the reader from being distracted by perhaps less digestible technicalities.

## 1. Equivariant sheaves.

We will be concerned with groupoids in the category of spaces (i.e., locales), briefly called *continuous groupoids*. If $G$ is such a continuous groupoid, we write $G_0$ (resp.$G_1$) for the space of objects (morphisms) of $G$, $d_0$ for the domain $d_1$ for the codomain, $m$ for composition and $s$ for the map associating the identity-morphism to a given object. So $G$ is given as a diagram of locales

$$G_1 \underset{G_0}{\times} G_1 \;\begin{matrix}\xrightarrow{\pi_0}\\ \xrightarrow{m}\\ \xrightarrow{\pi_1}\end{matrix}\; G_1 \;\begin{matrix}\xrightarrow{d_0}\\ \xleftarrow{s}\\ \xrightarrow{d_1}\end{matrix}\; G_0$$

having the usual properties. In the particular case where $G_0 = 1$, we have a group-object in locales, or a continuous group. A *homomorphism* of groupoids $G \xrightarrow{f} H$ consists of two continuous maps $G_0 \xrightarrow{f_0} H_0$ and $G_1 \xrightarrow{f_1} H_1$ satisfying the usual identities. A groupoid $G$ is called *open* if $d_0$ and $d_1 : G_1 \to G_0$

*Research supported by a Huygensfellowship of the Z.W.O.

are both open maps (this implies that $m$ is open as well). *In this paper, all groupoids are assumed to be open.*

If $G$ is a continuous groupoid, a $G$-space is a space $E$ over $G_0$ equipped with a (contravariant) action of $G$ on the right; i.e., there are maps $E \xrightarrow{p} G_0$ and $E \underset{G_0}{\times} G_1 \overset{\cdot}{\longrightarrow} E$ satisfying the usual conditions. (Rather than expressing these by commutative diagrams, we just write the equations in set-theoretic language; so the domain of the action map $E \underset{G_0}{\times} G_1 \overset{\cdot}{\longrightarrow} E$ is the set of pairs $(e, g)$ with $d_1 g = p(e)$, and the equations are $p(e \cdot g) = d_0 g$, $e \cdot s(p(e)) = e$, $(e \cdot g) \cdot h = e \cdot (g \circ h)$ where $g \circ h = m(g, h)$.) A *morphism* of $G$-spaces $E \to E'$ is a map of spaces (locales) over $G_0$ which preserves the action. This defines a category ($G$-spaces).

A $G$-space $E$ is *open* if its projection $E \xrightarrow{p} G_0$ is an open map; this implies that the action $E \underset{G_0}{\times} G_1 \overset{\cdot}{\longrightarrow} E$ is open (recall that $G$ is assumed to be an open groupoid).

A $G$-space is called *étale* if $E \xrightarrow{p} G_0$ is a local homeomorphism. This means that $E$ is a *sheaf* on $G_0$, such that the morphisms of $G$ act on the fibers of $E$: if $x \xrightarrow{g} y$ is a point of $G_1$, the action defines a map

$$g^* : E_y \to E_x, \quad g^*(e) = e \cdot g$$

1.1 PROPOSITION. *The category of étale $G$-spaces or equivariant $G$-sheaves, is a Grothendieck topos.*

This topos is denoted by $BG$, and called the *classifying topos* of $G$.

We remark that the definition of the topos $BG$ also makes sense if $G$ is just a *continuous category* (a category object in locales = spaces), rather than a groupoid.

Let me give some examples.

(1) The easiest case is where $G$ is an abstract group ($G_0 = 1$ and $G_1$ is a discrete space). The topos $BG$ is then simply the category of $G$-sets; i.e., objects are sets $X$ equipped with an action $X \times G \overset{\cdot}{\longrightarrow} X$, and morphisms are functions which preserve the action. Simple as they are, these toposes arises naturally in many contexts. $BG$ is a $K(G,1)$-topos, and it classifies $G$-torsors, so (writing [-,-] for isomorphism classes of geometric morphisms)

$$[\mathcal{E}, BG] \cong H^1(\mathcal{E}, G)$$

(see [**JW**]). Moreover for any abstract group, $G$ is an *atomic* topos [**BD**].
(2) If $G$ is a topological group, $BG$ is the category of *continuous G-sets*, i.e., sets $X$ equipped with an action $X \times G \xrightarrow{\cdot} X$ which is continuous if $X$ is given the discrete topology. Continuity of the action is equivalent to the requirement that all *stabilizer subgroups*

$$S_x = \{g \in G \mid x \cdot g = x\}$$

are open. For a topological group $G$, $BG$ is still an atomic topos, since one can simply write a given continuous $G$-set $X$ as the sum of its orbits: for $x \in X$, the orbit $\mathcal{O}(x) = \{x \cdot g \mid g \in G\}$ is the smallest subobject of $X$ containing $x$. If $U \subset G$ is an open subgroup, the set of right cosets $G/U$ is an object of $BG$, and clearly

$$\mathcal{O}(x) \cong G/S_x$$

as objects of $BG$. Consequently, the full subcategory of objects of the form $G/U$ ($U$ an open subgroup of $G$) equipped with the atomic Grothendieck topology (all maps are covers) is a site for $BG$.

Toposes of continuous $G$-sets form the natural setting for so-called permutation-models in set-theory ([**Fo**], [**Fr**]). On the other hand, it is hard to say what a topos $BG$ classifies, for a general topological group $G$. This is related to the fact that many different topological groups $G$ determine the same topos $BG$; see [**M3**], and section 6 below.

(3) Let $G$ be a profinite topological group, with a fundamental system of open normal subgroups $\{U_i\}$. So $G_i = G/U_i$ is a finite group and $G = \varprojlim G_i$ is a filtered inverse limit of finite groups and surjections. An object $X$ of $BG$, i.e., a continuous $G$-set, can be written as a union

$$X = \bigcup X_i$$

where $X_i$ is a $G_i$-set, and the actions of $G_i$ on $X_i$ for the different $i$ satisfy an obvious compability requirement: simply let $X_i = \{x \in X \mid S_x \supseteq U_i\}$.

The cohomology of $BG$ is precisely the Galois cohomology of $G$:

$$H^*(BG, A) \cong H^*_{\mathrm{Gal}}(G, A)$$

for any (discrete) $G$-module $A$ (see [**S**]), and $BG$ is still a $K(G,1)$-topos, i.e. the profinite fundamental group of $BG$ is $G$, and the other homotopy groups vanish (see e.g. [**AM**]).

(4) More generally, if $\underline{G} = \{G_i\}_i$ is a filtered inverse system of discrete groups and *surjective* homomorphisms (Grothendieck calls this a *strict* progroup), one may consider sets $X$ which can be written as a union $X = \bigcup X_i$ of compatible $G_i$-sets $X_i$, just like in the preceding example (see [**SGA4**, p. 319]). Such sets $X$ again form a topos $B\underline{G}$, which is still a $K(\pi, 1)$-topos if one interprets the fundamental group as a progroup [**AM**]. $B\underline{G}$ is an example of a topos of equivariant sheaves: although the inverse limit $\varprojlim G_i$ may be trivial as a topological group, it is not so when one computes the inverse limit

$$G = \varprojlim G_i$$

in the category of locales. One thus obtains a *prodiscrete* continuous group $G$, and $BG$ is the same as $B\underline{G}$; in fact

$$BG = B(\varprojlim G_i) \simeq \varprojlim BG_i = B\underline{G}.$$

(This example is discussed in detail in [**M4**]; see also section 6 below.)

(5) A well-known particular case of (2) above is the Schanuel topos $B(\mathrm{Aut}(\mathbb{N}))$, where $\mathrm{Aut}(\mathbb{N})$ is the group of permutations of the natural numbers with the usual (product) topology. If you compute a site for the Schanuel topos consisting of coset-objects $\mathrm{Aut}(\mathbb{N})/U$ for $U$ a (basic) open subgroup, you will find that the Schanuel topos is precisely sheaves on (the opposite category of) the category of finite sets and monomorphisms, equipped with the atomic topology.

One may also put all finite sets and inclusions together, and look at the topological monoid $\mathrm{Mono}(\mathbb{N})$ of monomorphisms $\mathbb{N} \to \mathbb{N}$. Let $B(\mathrm{Mono}(\mathbb{N}))$ be the topos of sets equipped with a continuous action of $\mathrm{Mono}(\mathbb{N})$. It is an instructive exercise to verify that the inclusion $\mathrm{Aut}(\mathbb{N}) \hookrightarrow \mathrm{Mono}(\mathbb{N})$ induces an equivalence

$$B(\mathrm{Aut}(\mathbb{N})) \simeq B(\mathrm{Mono}(\mathbb{N})).$$

This is an instance of a general phenomenon, discussed in §3 below.

$B(\mathrm{Aut}(\mathbb{N}))$ classifies the notion of an infinite decidable set. From the point of view of homotopy and cohomology however, nothing like the situation in (3) and (4) holds, since $B(\mathrm{Aut}(\mathbb{N}))$ is contractible (see [**JW**]).

(6) Naturally, if $X$ is a space,

$$\mathrm{Sh}(X) = BX,$$

where on the right, $X$ stands for the trivial groupoid whose only morphisms are identities. More generally, if $G$ is a topological group acting as a group of transformations on a topological space $X$, by $X \times G \xrightarrow{\cdot} X$ say, one can construct a topological groupoid

$$X_G = (X \times G \underset{\pi_1}{\rightrightarrows} X)$$

i.e., X is the space of objects of $X_G$, $X \times G$ the space of morphisms, and $d_0(x,g) = x \cdot g$, $d_1(x,g) = x$, etc. Then $BX_G$ is precisely the topos of sheaves $E$ on $X$ equipped with an action by the group $G$ which lifts the original action of $G$ on $X$, i.e.,

$$\begin{array}{ccc} E \times G & \longrightarrow & E \\ \downarrow & & \downarrow \\ X \times G & \longrightarrow & X \end{array}$$

commutes.

(7) Let $M$ be a foliated manifold, and let $\mathrm{Hol}(M)$ be the holonomy groupoid of $M$: its space of objects is $M$, and a morphism from $x$ to $y$ in this groupoid is a homotopy class of paths $I \xrightarrow{\alpha} L$, where $L \subset M$ is the leaf of $x$ and $y$ (if $x$ and $y$ are on different leaves, there are no morphisms from $x$ to $y$). This set has the structure of a manifold, so $\mathrm{Hol}(M)$ is a continuous (differentiable) groupoid. $B\mathrm{Hol}(M)$ is the category of sheaves on $M$ which are locally constant on each leaf.

(8) An étendue is a topos $\mathcal{E}$ such that for some cover $B \twoheadrightarrow 1$ in $\mathcal{E}$, $\mathcal{E}/B$ is equivalent to $\mathrm{Sh}(G_0)$ for some space $G_0$. The diagram

$$B \times B \times B \;\substack{\longrightarrow\\ \longrightarrow\\ \longrightarrow}\; B \times B \;\substack{\longrightarrow\\ \longleftarrow\\ \longrightarrow}\; B \tag{1}$$

is a (trivial) groupoid in $\mathcal{E}$, and gives rise to a continuous groupoid

$$G = (G_1 \underset{d_1}{\overset{d_0}{\rightrightarrows}} G_0) \tag{2}$$

where $\mathrm{Sh}(G_1) = \mathrm{Sh}(G_0) \underset{\mathcal{E}}{\times} \mathrm{Sh}(G_0) = \mathcal{E}/(B \times B)$; i.e., the functor $X \mapsto \mathrm{Sh}(X)$ from spaces (locales) to toposes sends this groupoid (2) to the groupoid

$$\mathcal{E}/(B \times B \times B) \;\substack{\longrightarrow\\ \longrightarrow\\ \longrightarrow}\; \mathcal{E}/(B \times B) \rightrightarrows \mathcal{E}/B$$

obtained from (1) by slicing. Thus, an object of $BG$ is an object $E \xrightarrow{p} B$ of $\mathcal{E}/B$, equipped with an isomorphism $\theta : E \times B \to B \times E$ over $B \times B$, which can be described in set-theoretic notation (using the internal logic of $\mathcal{E}$!) as follows: for points $b_1, b_2$ of $B$ there is an isomorphism

$$\theta_{b_1,b_2} : p^{-1}(b_1) \xrightarrow{\sim} p^{-1}(b_2)$$

such that

$$\theta_{b,b} = id, \quad \theta_{b_2,b_3} \circ \theta_{b_1,b_2} = \theta_{b_1,b_3}. \tag{3}$$

But then clearly $E \xrightarrow{p} B$ is just a projection: let $E/\sim$ be the quotient of $E$ obtained by identifying $e$ with $\theta_{p(a),b}(e)$ for any $e \in E$ and $b \in B$. Then, writing $[e]$ for the equivalence class of $e$ in $E/\sim$, the map

$$E \xrightarrow{\varphi} (E/\sim) \times B, \qquad \varphi(e) = ([e], p(e))$$

is an isomorphism, with (well-defined, by (3)) inverse map $\psi$ defined by $\psi([e], b) = \theta_{p(e),b}(e)$. Thus, we obtain an equivalence

$$BG \simeq \mathcal{E}.$$

(Notice that $G$ is a groupoid whose domain and codomain maps are local homeomorphisms.)

The situation described in (8) is discussed in SGA4, exposé IV. The equivalence $BG \simeq \mathcal{E}$ is a very simple case of a result due to A. Joyal and M. Tierney, which asserts that any topos is of the form $BG$:

2.1 REPRESENTATION THEOREM. *([**JT**]) For any Grothendieck topos $\mathcal{E}$ there exists an open continuous groupoid $G$ such that $\mathcal{E}$ is equivalent to $BG$.*

## 2. Basic Properties.

In this section I will describe some of the elementary functorial properties. Detailed proofs of the results in this section can be found in [**M2**,I,§4-6].

First of all, the construction of the topos $BG$ can be performed over any base topos. More precisely, if $\mathcal{E}$ is a topos and $G$ is a continuous groupoid in $\mathcal{E}$, one may consider étale $G$-spaces inside $\mathcal{E}$; these form a topos $B(\mathcal{E}, G)$ over $\mathcal{E}$, and we write

$$B(\mathcal{E}, G) \xrightarrow{\gamma} \mathcal{E}.$$

for the canonical geometric morphism. A most important fact is that one can use change-of-base methods in the context of toposes of the form $B(\mathcal{E}, G)$, since this construction is stable:

2.1 STABILITY THEOREM. *Let $\mathcal{F} \xrightarrow{f} \mathcal{E}$ be a geometric morphism, and let $G$ be a continuous groupoid in $\mathcal{E}$. Then $f^{\#}(G)$ is a continuous groupoid in $\mathcal{F}$, and there is a canonical equivalence*

$$B(\mathcal{F}, f^{\#}G) \simeq \mathcal{F} \underset{\mathcal{E}}{\times} B(\mathcal{E}, G).$$

In this theorem, $f^{\#}$ denotes the pullback functor from spaces in $\mathcal{E}$ to spaces in $\mathcal{F}$.

In what follows, we fix an arbitrary base topos $\mathcal{S}$, and just write $BG$ for $B(\mathcal{S}, G)$. Some basic properties of the topos $BG$ follow from properties of $G$; e.g.

2.2 PROPOSITION. *Let $G$ be a continuous groupoid in the base topos $\mathcal{S}$.*

(1) *If $G_0$ is an open space, then $BG \xrightarrow{\gamma} \mathcal{S}$ is open.*

(2) *If $G_0$ is locally connected and $d_0, d_1 : G_1 \rightrightarrows G_0$ are both open, then $BG \xrightarrow{\gamma} \mathcal{S}$ is locally connected.*

(3) *If $G_0$ is an open space and $G_1 \xrightarrow{(d_0,d_1)} G_0 \times G_0$ is an open map, then $BG \xrightarrow{\gamma} \mathcal{S}$ is atomic.*

(For open maps, see **[J2]**, **[JT]**; for locally connected maps see **[BP]**, **[M1**, Appendix]; for atomic maps see **[BD]**.) To prove 2.2, one may use properties of the projection map

$$\mathrm{Sh}(G_0) \xrightarrow{\pi_G} BG$$

whose inverse $\pi_B^*$ is the forgetful functor. Here we have

2.3 PROPOSITION. (1) *If $d_0, d_1 : G_1 \rightrightarrows G_0$ are both open, so is $Sh(G_0) \xrightarrow{\pi_G} BG$.*

(2) *If $d_0, d_1$ are both locally connected, so is $Sh(G_0) \xrightarrow{\pi_G} BG$.*

(3) *If $d_0, d_1 : G_1 \rightrightarrows G_0$ are both local homeomorphisms, then $Sh(G_0) \xrightarrow{\pi_G} BG$ is atomic.*

One can apply 2.2 to get sharper forms of the representation Theorem 1.2 (2.4(4) appears in SGA4, *loc. cit*).

2.4 COROLLARY. *Let $\mathcal{E}$ be a topos over $\mathcal{S}$. Then*

(1) *$\mathcal{E} \to \mathcal{S}$ is open if and only if $\mathcal{E} \simeq BG$ for some continuous groupoid with $G_1 \rightrightarrows G_0 \to 1$ all open maps.*

(2) *$\mathcal{E} \to \mathcal{S}$ is (connected) locally connected if and only if $\mathcal{E} \simeq BG$ for a continuous groupoid with $G_1 \rightrightarrows G_0 \to 1$ all (connected) locally connected maps.*

(3) *$\mathcal{E} \to \mathcal{S}$ is (connected) atomic if and only if $\mathcal{E} \simeq BG$ for a continuous groupoid $G$ with $G_0 \to 1$ open (and surjective) and $G_1 \xrightarrow{(d_0,d_1)} G_0 \times G_0$ open (and surjective).*

(4) *$\mathcal{E}$ is an étendue if and only if $\mathcal{E} \simeq BG$ for a groupoid $G$ with both $G_1 \rightrightarrows G_0$ local homeomorphisms.*

The construction is functorial in $G$: if $G \xrightarrow{\varphi} H$ is a continuous homomorphism, the pullback of an étale $H$-space $E \xrightarrow{p} H_0$ along $G_0 \xrightarrow{\varphi_0} H_0$ has an obvious induced action by $G$, and this gives the inverse image of a geometric morphism

$$B(\varphi) : BG \to BH.$$

$G \xrightarrow{\varphi} H$ is called *open* if both $G_0 \xrightarrow{\varphi_0} H_0$ and $G_1 \xrightarrow{\varphi_1} H_1$ are open maps. Moreover, imitating the usual categorical notions, $\varphi$ is called *essentially surjective* if $G_0 \underset{H_0}{\times} H_1 \xrightarrow{d_1\pi_2} H_0$ is an open surjection, *full* if $G_1 \to H_1 \underset{(H_0\times H_0)}{\times} (G_0 \times G_0)$ is an open surjection, and *fully faithful* if

$$\begin{array}{ccc} G_1 & \longrightarrow & H_1 \\ \downarrow & & \downarrow \\ G_0 \times G_0 & \longrightarrow & H_0 \times H_0 \end{array}$$

is a pullback; $\varphi$ is an *essential equivalence* if $\varphi$ is open, fully faithful and essentially surjective.

The basic properties are

2.5 THEOREM. *Let $B\varphi : BG \to BH$ be the geometric morphism induced by a continuous homomorphism $G \xrightarrow{\varphi} H$.*

(1) *If $\varphi_0$ is open, $B\varphi$ is open.*

(2) *If $\varphi$ is essentially surjective then $B\varphi$ is surjective.*

(3) *If $\varphi$ is essentially surjective and full, then $B\varphi$ is connected.*

(4) *If $\varphi$ is open and full, then $B\varphi$ is atomic.*

(5) *If $\varphi$ is an essential equivalence, $B\varphi$ is an equivalence of toposes.*

*And of a somewhat different nature:*

(6) *If $\varphi_0$ and $G_1 \rightrightarrows G_0$ are locally connected and $H_1 \rightrightarrows H_0$ are open, then $B\varphi$ is locally connected.*

(7) *If $\varphi_0$ and $G_1 \rightrightarrows G_0$ are local homeomorphisms and $H_1 \rightrightarrows H_0$ are open, then $B\varphi$ is atomic.*

## 3. Completions of Continuous Groupoids.

Let $G$ be a continuous groupoid. I will describe a continuous groupoid $\hat{G}$ and a continuous category $\gamma(G)$, together with maps

$$G \xrightarrow{\theta} \hat{G} \subset \gamma(G),$$

such that these all define the same topos:

$$BG \xrightarrow{\sim} B\hat{G} \xrightarrow{\sim} B\gamma(G).$$

**3.1 Construction of $\gamma(G)$** (cf [**M2**,II,§3]). Consider the lax pullback

$$(1) \qquad \begin{array}{ccc} \mathcal{L} & \xrightarrow{d_1} & \mathrm{Sh}(G_0) \\ \downarrow d_0 & \swarrow\!\!\swarrow & \downarrow \pi_G \\ \mathrm{Sh}(G_0) & \xrightarrow{\pi_G} & BG \end{array}$$

$\mathcal{L}$ is actually a localic topos, i.e., there is a unique locale $\gamma(G)_1$ such that $\mathcal{L} \simeq \mathrm{Sh}(\gamma(G)_1)$. $\gamma(G)$ is the category with $\gamma(G)_0 = G_0$ as space of objects, $\gamma(G)_1$ as space of morphisms, and $d_0, d_1$ from diagram (1) as domain and codomain.

By the universal property of (1), points of $\gamma(G)_1$ are triples $(x, y, \alpha)$, where $x$ and $y$ are points of $G_0$, and $\alpha : \mathrm{ev}_y \to \mathrm{ev}_x$ is a natural transformation ($\mathrm{ev}_x : \mathrm{Sh}(G_0) \to$ Sets takes the fiber at $x$). $d_0(x, y, \alpha) = x$, $d_1(x, y, \alpha) = y$. It is clear how to define composition in $\gamma(G)$ on *points*. By change of base (Yoneda lemma) this actually defines the structure of a continuous category on $\gamma(G)$.

The universal property of (1) gives a homomorphism of continuous categories

$$G \xrightarrow{\theta} \gamma(G)$$

which is the identity on objects; on morphisms, $\theta$ sends a point $g$ of $G_1$ to the triple $(d_0 g, d_1 g, g^*)$.

3.2 PROPOSITION. $G \xrightarrow{\theta} \gamma(G)$ *induces an equivalence of toposes*

$$BG \xrightarrow{\sim} B\gamma(G).$$

**3.3 Construction of $\widehat{G}$.** $\widehat{G}$ is the subcategory of $\gamma(G)$ with the same objects, but only the isomorphisms of $\gamma(G)$ as arrows: so $\widehat{G}$ is in fact a continuous groupoid, and $G \xrightarrow{\theta} \gamma(G)$ factors through $\widehat{G} \subset \gamma(G)$. $\widehat{G}_1$ can be directly described by the pullback

$$\begin{array}{ccc} \mathrm{Sh}(\widehat{G}_1) & \longrightarrow & \mathrm{Sh}(G_0) \\ \downarrow & \swarrow\!\!\swarrow & \downarrow \\ \mathrm{Sh}(G_0) & \longrightarrow & BG \end{array} \tag{2}$$

We call $G$ *étale complete* if $G \xrightarrow{\theta} \widehat{G}$ is an isomorphism.

3.4 PROPOSITION. $G \xrightarrow{\theta} \widehat{G}$ *induces an equivalence of toposes*

$$BG \xrightarrow{\sim} B\widehat{G}$$

This is a consequence of the descent theorem for open geometric morphisms ([JT]).

3.5 EXAMPLE: (cf [M3]) Let $G$ be a topological group with a countable base at the identity element $e$. Then

$$\gamma(G) = \varprojlim_{U} G/U$$

where $U$ ranges over open subgroups (ordered by inclusion), and $G/U$ is the set of right cosets $Ux$ (the quotient topology on $G/U$ is the discrete one). $\gamma(G)$ is a topological monoid, with multiplication defined by the formula

$$(\bar{x} \cdot \bar{y})_U = Ux_U \cdot y_{x_U^{-1} U x_U}$$

where we write a point of $\gamma(G)$ as a sequence $\bar{x} = \{U \cdot x_U\}_U$. For instance, for the group $\mathrm{Aut}(S)$ of isomorphisms of some infinite set $S$, we have

$$\gamma(\mathrm{Aut}(S)) = \{1-1 \text{ maps } S \to S\}$$

Notice also that if $G = \varprojlim G_i$ is profinite, then $\gamma(G) \cong G$; this holds in fact for arbitrary localic prodiscrete groups (§1, example (4); and §6 below).

## 4. Toposes as a localization of groupoids.

Let [Top] denote the category of Grothendieck toposes (over some fixed base topos) and isomorphisms classes of geometric morphisms. I will describe how [Top] can be considered as a localization of a category of continuous groupoids. (Detailed proofs are given in [**M2**, I,§7].)

First of all, the category of continuous groupoids is a 2-category in a natural way: 1-cells $G \to H$ are continuous homomorphisms, and if $G \overset{\varphi}{\underset{\psi}{\rightrightarrows}} M$ are two such, a 2-cell $\alpha : \varphi \Rightarrow \psi$ is the localic analogue of a natural transformation, i.e., a continuous map $G_0 \xrightarrow{\alpha} H_1$ such that $d_0\alpha = \varphi$, $d_1\alpha = \psi$, and

$$\begin{array}{ccc} G_1 & \xrightarrow{(\alpha d_1, \varphi_1)} & H_1 \times_{H_0} H_1 \\ {\scriptstyle (\psi, \alpha d_0)}\downarrow & & \downarrow {\scriptstyle m} \\ H_1 \times_{H_0} H_1 & \xrightarrow{m} & H_1 \end{array}$$

commutes. Let [Groupoids] denote the category of continuous groupoids and isomorphism classes of continuous homomorphisms. Let $ECG \subseteq$ [Groupoids] be the full subcategory given by the étale complete groupoids.

4.2 Proposition. *The class $\underline{E}$ of isomorphism classes of essential equivalences (cf §2) admits a calculus of right fractions (in the sense of [**GZ**]) in the category [Groupoids], as well as in the subcategory $ECG$ of étale complete groupoids.*

4.2 Localization theorem. *The functor $B$ from continuous groupoids to toposes induces an equivalence*

$$ECG[\underline{E}^{-1}] \xrightarrow{\sim} [Top].$$

4.3 REMARK: In the proof of 4.2, one uses the following construction. If $BG \xrightarrow{f} BH$ is any geometric morphism, define a space $K_0$ by the pullback

$$\begin{array}{ccc} \mathrm{Sh}(K_0) & \xrightarrow{\psi_0} & & \mathrm{Sh}(H_0) \\ \downarrow{\scriptstyle \varphi_0} & & & \downarrow{\scriptstyle \pi_H} \\ \mathrm{Sh}(G_0) & \xrightarrow{\pi_G} & BG \xrightarrow{f} & BH \end{array}$$

and make $K_0$ into a groupoid by defining $K_1$ as the pullback

$$\begin{array}{ccc} K_1 & \xrightarrow{\varphi_1} & G_1 \\ \downarrow & & \downarrow \\ K_0 \times K_0 & \longrightarrow & G_0 \times G_0 \end{array}$$

so that we obtain an essential equivalence $K \xrightarrow{\varphi} G$. If $H$ is étale complete, then $f$ gives a homomorphism $K \xrightarrow{\psi} H$ such that

$$\begin{array}{ccc} BK & & \\ {\scriptstyle B\varphi}\downarrow & \searrow{\scriptstyle B\psi} & \\ BG & \xrightarrow{f} & BH \end{array}$$

commutes (up to natural isomorphism).

We remark here that it can be shown that $K_0$ can be equipped with an action of $G$ on the right and one of $H$ on the left, and that the induced functor

$$BH \to BG, \quad E \mapsto E \underset{H}{\otimes} K_0$$

is naturally isomorphic to the inverse image $f^*$ of the given geometric morphism $f$. Thus, every geometric morphism comes from tensoring by a "bispace", a space with two actions as above. However, this particular construction does not take care of 2-cells, i.e., natural transformations between geometric morphisms. In the next section, a more careful construction will be given that does take these 2-cells into account.

## 5. Geometric Morphisms as Tensor Products.

In this section I will describe the category Top($BH, BG$) of geometric morphisms and natural transformations between their inverse image functors, completely in terms of the completions $\gamma H$ and $\gamma G$, and spaces equipped with an action by each of these completions. The reader can find detailed proofs in [**M2**,II,§4–6].

**5.1 Bispaces.** Let $G$ and $H$ be continuous groupoids, with completions $\gamma G$ and $\gamma H$ (cf. 3.1). A *$\gamma G$- $\gamma H$-bispace* (or briefly *bispace*) is a space $R$ which is at the same time a left $\gamma G$-space and a right $\gamma H$-space, such that the two actions commute with each other. So there are projections $p_G : R \to G_0$ and $p_H : R \to H_0$, and action maps $\gamma(G)_1 \underset{G_0}{\times} R \xrightarrow{*} R$ (pullback along $\gamma(G)_1 \xrightarrow{d_0} G_0$), $R \underset{H_0}{\times} \gamma(H)_1 \xrightarrow{\cdot} R$ (pullback along $\gamma(H)_1 \xrightarrow{d_1} H_0$) satisfying the usual identities for an action (covariant for ., contravariant for *), as well as three compatibility conditions expressed by the following commutative diagrams:

(1) $$\gamma(G)_1 \underset{G_0}{\times} R \underset{\pi_2}{\overset{*}{\rightrightarrows}} R \xrightarrow{p_H} H_0$$

(2) $$R \underset{H_0}{\times} \gamma(H)_1 \underset{\pi_1}{\overset{\cdot}{\rightrightarrows}} R \xrightarrow{p_G} G_0$$

(3) $$\begin{array}{ccc} \gamma(G)_1 \underset{G_0}{\times} R \underset{H_0}{\times} \gamma(H)_1 & \xrightarrow{1\times .} & \gamma(G)_1 \underset{G_0}{\times} R \\ \Big\downarrow {\scriptstyle *\times 1} & & \Big\downarrow {\scriptstyle *} \\ R \underset{H_0}{\times} \gamma(H)_1 & \xrightarrow{\cdot} & R \end{array}$$

A *homomorphism* of bispaces $R \to R'$ is a continuous map of spaces which i compatible with both actions. This defines a category

$$(\gamma G\text{- } \gamma H\text{-bispaces}).$$

**5.2 Tensor products.** If $R$ is a bispace as above, and $E$ is a (right $\gamma(G)$-space, the tensor product $E \underset{\gamma(G)}{\otimes} R$ is defined by the usual coequalize liagram

$$(4)\qquad E \underset{G_0}{\times} \gamma(G)_1 \underset{G_0}{\times} R \underset{.\times R}{\overset{E\times *}{\rightrightarrows}} E \underset{G_0}{\times} R \twoheadrightarrow E \underset{\gamma G}{\otimes} R$$

This coequalizer can be pretty unmanageable, mainly because it need not be table. However, if the actions of $\gamma(G)$ on $E$ and $R$ are both given by open naps, then (4) is stable.

**5.3 Open bispaces, flat bispaces.** A bispace $R$ as in 5.1 is *open* if $\cdot_H : R \to H_0$ is open, both actions $\gamma(G)_1 \underset{G_0}{\times} R \xrightarrow{*} R$ and $R \underset{H_0}{\times} \gamma(H)_1 \xrightarrow{\cdot} R$ are pen, and the diagonal action $\gamma(G)_1 \underset{G_0}{\times} \gamma(G)_1 \underset{G_0}{\times} R \xrightarrow{\mu} R \underset{H_0}{\times} R$ ("$\mu(\xi,\xi',r) = \xi * r, \xi' * r$)") is open.

LEMMA. *If $R$ is an open bispace and $E$ is an étale $G$-space (which can be onsidered as an étale $\gamma(G)$-space, cf 3.2) then $E \underset{\gamma(G)}{\otimes} R$ is an étale $H$- (or $(H)$-) space.*

So an open bispace $R$ induces a functor

$$g(R)^* = - \underset{\gamma G}{\otimes} R : BG \to BH$$

$R$ is called *flat* (on the left) if $g(R)^* = - \underset{\gamma G}{\otimes} R$ is left-exact. We denote the full ubcategory of ($\gamma G$-$\gamma H$-bispaces) consisting of the flat ones by

$$\underline{\text{Flat}}(\gamma G, \gamma H).$$

o we obtain a functor

$$\underline{\text{Flat}}(\gamma G, \gamma H) \xrightarrow{g} \underline{\text{Top}}(BH, BG).$$

.4 THEOREM. *The functor $g$ has a fully faithful right adjoint*

$$\underline{R} : \underline{Top}(BH, BG) \to \underline{Flat}(\gamma G, \gamma H).$$

The construction of $\underline{R}$ is easy enough to describe: given a geometric morphism $BH \xrightarrow{f} BG$, $\underline{R}(f)$ is constructed as the lax pullback

$$\begin{array}{ccccc} \mathrm{Sh}(\underline{R}(f)) & & \longrightarrow & & \mathrm{Sh}(G_0) \\ \downarrow & & & & \downarrow{\scriptstyle \pi_G} \\ \mathrm{Sh}(H_0) & \xrightarrow{\pi_H} & BM & \xrightarrow{f} & BG \end{array} \tag{5}$$

(this lax pullback is a localic topos, so determines a unique space $\underline{R}(f)$). By the universal property of (5), $\underline{R}(f)$ can be equipped with the structure of a $\gamma G$-$\gamma H$-bispace. It follows from 5.4 as stated that tensoring with $\underline{R}(f)$ defines a functor which is naturally isomorphic to $f^*$ (the counit of $g \dashv \underline{R}$ is an isomorphism):

5.5 COROLLARY. *For every geometric morphism $BH \xrightarrow{f} BG$ there is a natural isomorphism of functors $BG \to BH$:*

$$f^* \simeq - \underset{\gamma G}{\otimes} \underline{R}(f).$$

Let us call a flat bispace $R$ complete if $\eta_R : R \to \underline{R}g(R)$ is an isomorphism. One can then form a large bicategory whose objects are continuous groupoids $G$, whose 1-cells $G \to H$ are complete flat $\gamma G$-$\gamma H$-bispaces, and whose 2-cells are homomorphisms of such. The tensor product = composition of 1-cells is given by first taking the tensor product of flat bispaces and then completing; i.e., for $G \xrightarrow{R} H \xrightarrow{S} K$, we define

$$R \hat{\otimes} S =_{def} \underline{R}g(R \underset{\gamma(H)}{\otimes} S) \simeq \underline{R}(g(S) \circ g(R))$$

5.6 COROLLARY. *This bicategory of continuous groupoids and complete fla[t] bispaces is equivalent to the dual of the bicategory of toposes and geometri[c] morphisms.*

## 6. Pointed atomic toposes, Galois toposes, and the fundamenta[l] group.

In this section, I come back to toposes of the form $BG$ for a continuous grou[p] $G$. Such a topos is connected, atomic, and has a canonical point $\mathcal{S} \xrightarrow{p_G} B[G]$ whose inverse image $p_G^*$ is the forgetful functor ($\mathcal{S}$ is an arbitrary base topos). It is proved in [**JT**] that the converse also holds:

6.1 **THEOREM.** *(see [JT]) For any atomic connected $\mathcal{S}$-topos $\mathcal{E} \to \mathcal{S}$ with a point $\mathcal{S} \xrightarrow{p} \mathcal{E}$, there exists a continuous group $G$ in $\mathcal{S}$ such that $\mathcal{E} \simeq BG$ as $\mathcal{S}$-toposes, and $p$ corresponds to the canonical point $p_G$ of $BG$ under the equivalence.*

If $(\mathcal{E}, p)$ and $(\mathcal{F}, q)$ are pointed $\mathcal{S}$-toposes, I write $\underline{\text{Top}}.((\mathcal{F}, q), (\mathcal{E}, p))$ for the category of pointed maps; i.e. the objects are pairs $(f, \alpha), \mathcal{F} \xrightarrow{f} \mathcal{E}$ over $\mathcal{S}$ and $\alpha : fq \xrightarrow{\sim} p$ (over $\mathcal{S}$), and the maps $\beta : (f, \alpha) \to (f', \alpha')$ are natural transformations $f^* \xrightarrow{\beta} f'^*$ compatible with the $\alpha$'s, i.e., for each $E \in \mathcal{E}$,

$$\begin{array}{ccc} b^* f^*(E) & \xrightarrow[\sim]{\alpha_E} & p^*(E) \\ {\scriptstyle q^*(\beta_E)} \downarrow & \swarrow {\scriptstyle \sim\ \alpha'_E} & \\ q^* f'^*(E) & & \end{array}$$

commutes. Note that if $q$ is faithful, there can be at most one such $\beta$, and $\beta$ is an isomorphism. So for continuous groups $G$ and $H$, the 2-cells define an equivalence relation on $\underline{\text{Top}}.((BH, p_H), (BG, p_G))$, with quotient denoted by $[(BH, p_H), (BG, p_G)]_.$.

I will also consider the category $\underline{\text{Top}}_+((\mathcal{F}, q), (\mathcal{E}, p)$ whose objects are maps of pointed toposes $(f, \alpha$ as above, and whose morphisms $(f, \alpha) \to (f', \alpha')$ are natural transformations $f^* \to f'^*$ (no compatibility with $\alpha$ and $\alpha'$ required).

From 5.4 one can deduce the following result (see [**M3**]).

6.2 **THEOREM.** *For continuous groups $G$ and $H$, the canonical functor*

(1) $$Hom(\gamma H, \gamma G) \to [(BH, p_H), (BG, p_G)].$$

*is an isomorphism of sets, and*

(2) $$\underline{Hom}(\gamma H, \gamma G) \to \underline{Top}_+((BH, p_H), (BG, p_G))$$

*is an equivalence of categories.*

In (1), $\text{Hom}(\gamma H, \gamma G)$ is the set of continuous homomorphisms; in (2), $\underline{\text{Hom}}(\gamma H, \gamma G)$ is the same set made into a category, where for homomorphisms $\varphi$ and $\psi : \gamma H \to \gamma G$, a map $\varphi \to \psi$ is a point $\overline{g}$ of $\gamma(G)$ such that $\varphi(x) \cdot \overline{g} = \overline{g} \cdot \psi(x)$.

I now briefly discuss how Grothendieck's theory of the fundamental group (see [**SGA1**]), interpreted as a progroup as in [**AM**], fits into this context. For details I refer the reader to [**M4**]; some of what follows has independently been considered by J. Kennison, see [**K**].

A prodiscrete group is a continuous group $G$ which is the inverse limit (in locales!) of a filtered inverse system of discrete groups (and surjective homomorphisms, as one may without loss of generality assume). Since clearly for a prodiscrete group $G$, $G \cong \hat{G} \cong \gamma(G)$, we have

6.3 COROLLARY. *The embedding of prodiscrete groups into pointed toposes is fully faithful; in fact*

$$\underline{Hom}(H, G) \simeq \underline{Top}_{+}((BH, p_H), (BG, p_G))$$

*whenever $G$ and $H$ are continuous groups with $G$ prodiscrete.*

The image of the embedding in 6.3 can be characterized as follows. Recall that an atom $A$ in a topos $\mathcal{E}$ is *normal* if it is an $\mathrm{Aut}_{\mathcal{E}}(A)$-torsor in $\mathcal{E}$. A *Galois topos* is a pointed connected atomic topos which is generated by its normal atoms. From [**M3**,§3], I quote:

6.4 THEOREM. *A pointed topos $(\mathcal{E}, p)$ is a Galois topos if and only if there exists a prodiscrete continuous group $G$ such that $\mathcal{E} \simeq BG$ ($p$ corresponding to $p_G$ as in 6.1).*

Let $\mathcal{E} \xrightarrow{\gamma} \mathcal{S}$ be a connected locally connected topos over $\mathcal{S}$. An object $E$ of $\mathcal{E}$ is *locally constant* if there are an $S$ in $\mathcal{S}$ and a $V \twoheadrightarrow 1$ in $\mathcal{E}$ such that $E \times V \cong \gamma^*(S) \times V$ over $V$. The connected locally constant objects form a normal atomic site, and if one closes off under sums, one obtains a topos $\pi_1(\mathcal{E})$, and the inclusion $\pi_1(\mathcal{E}) \subset \mathcal{E}$ is the inverse image of a surjective geometric morphism

$$\varphi : \mathcal{E} \twoheadrightarrow \pi_1(\mathcal{E}).$$

If $\mathcal{E}$ has a point $p$, $\pi_1(\mathcal{E})$ has a point $\varphi(p)$, and therefore by 6.4 there exists a prodiscrete group $\pi_1(\mathcal{E}, p)$ such that

$$(\pi_1(\mathcal{E}), \varphi(p)) \simeq (B\pi_1(\mathcal{E}, p), \overline{p}),$$

where $\overline{p} = p_{\pi_1(\mathcal{E},p)}$ is the canonical point of $B(\pi_1(\mathcal{E}, p))$.

6.5 PROPOSITION. *Let $(\mathcal{E},p)$ be a connected locally connected pointed topos. $(\mathcal{E},p) \xrightarrow{\varphi} (B\pi_1(\mathcal{E},p),\overline{p})$ is the universal map into a Galois topos; i.e., for every Galois topos $\mathcal{G}$, composition with $\varphi$ defines an equivalence*

$$\underline{Top}\big(B(\pi_1(\mathcal{E},p)),\mathcal{G}\big) \xrightarrow{\sim} \underline{Top}(\mathcal{E},\mathcal{G})$$

*(which restricts to equivalences*

$$\underline{Top}_+\big(B(\pi_1(\mathcal{E},p)),\mathcal{G}\big) \simeq \underline{Top}_+(\mathcal{E},\mathcal{G})$$

*and*

$$\underline{Top}_{\bullet}(B\pi_1(\mathcal{E},p),\mathcal{G}) \simeq \underline{Top}_{\bullet}(\mathcal{E},\mathcal{G})).$$

In the parenthetical remark, we have suppressed the points from the notation. Putting 6.3, 6.4, and 6.5 together, we obtain

6.6 COROLLARY. *Let $(\mathcal{E},p)$ be a connected locally connected topos. For any prodiscrete group $G$ there is an equivalence (natural in $G$)*

$$\underline{Top}_+\big((\mathcal{E},p),(BG,p_G)\big) \simeq \underline{Hom}(\pi_1(\mathcal{E},p),G)$$

Since for a discrete group $G$, any two points of $BG$ are isomorphic, we conclude

6.7 COROLLARY. *Let $(\mathcal{E},p)$ be a connected locally connected pointed topos. (a) For any discrete group $G$,*

$$\underline{Top}(\mathcal{E},BG) \simeq \underline{Hom}(\pi_1(\mathcal{E},p),G)$$

*(b) For any abelian group $A$,*

$$H^1(\mathcal{E},A) \cong Hom(\pi_1(\mathcal{E},p),A)$$

Note that (b) follows from (a) and the fact that $BA$ classifies $A$-torsors, i.e., $H^1(\mathcal{E},A) \cong \mathrm{Top}(\mathcal{E},BA)$ (cf. eg **[JW]**, **[J1]**).

# REFERENCES

[SGA4] M. Artin, A. Grothendieck, J. L. Verdier, "Théorie des Topos et Cohomologie des Schémas," SLN 269, 1972.

[AM] M. Artin, B. Mazur, "Etale Homotopy," SLN 100, 1969.

[BD] M. Barr, R. Diaconescu, *Atomic toposes*, JPAA **17** (1980), 1–24.

[BP] M. Barr,R. Paré, *Molecular toposes*, JPAA **17** (1980), 127–152.

[Fo] M. Fourman, *Sheaf models for set theory*, JPAA **19** (1980), 91–101.

[Fr] P. Freyd, *The axiom of choice*, JPAA **19** (1980), 103-125.

[GZ] P. Gabriel, M. Zisman, "Calculus of Fractions and Homotopy Theory," Springer-Verlag, 1967.

[SGA1] A. Grothendieck, "Revêtements Étales et Groupe Fondamental," SLN 224, 1971.

[J1]] P. T. Johnstone, "Topos Theory," Academic Press, 1977.

[J2] ____________, *Open maps of toposes*, Manuscripta Math **31** (1980), 217-247.

[JT] A. Joyal, M. Tierney, *An extension of the Galois theory of Grothendieck*, Memoirs AMS **309** (1984).

[JW] A. Joyal, G. Wraith, *$K(\pi, n)$-toposes*, in Contemporary Mathematics **30** (1984).

[K] J. Kennison, *The fundamental group of a molecular topos*, JPAA **46** (1987), 187–215.

[M1] I. Moerdijk, *Continuous fibrations and inverse limits of toposes*, Comp. Math. **58** (1986), 45–72.

[M2] _________, *The classifying topos of a continuous groupoid, I and II*, (part I is to appear in Transactions AMS).

[M3] _________, *Morita equivalence of continuous groups*, (to appear in Math. Proc. Cambridge Phil. Soc.).

[M4] _________, *Prodiscrete groups and Galois toposes,*, submitted.

[S] J. P. Serre, "Cohomologie Galoisienne," SLN 5, 1964.

University of Chicago, Chicago, IL 60637

This paper is in final form and will not be published elsewhere.

# COMPONENTAL NUCLEI*

S. B. Niefield and K. I. Rosenthal
Department of Mathematics
Schenectady, N.Y. 12308 USA

## Introduction

In a pair of papers [8], [9] in the 1960's, D. Kirby studied closure operations on the lattice Idl(R) of ideals of a commutative ring with identity. Among the examples were what he called componental operations. These were intended to generalize the notion of the S-component

$$A_S = \{x \in R | sx \in A \text{ for some } s \in S\} = S^{-1}A \cap R$$

of an ideal A relative to a multiplicative set S.

There are some interesting insights to be gained by considering rings and lattices (in particular, locales) simultaneously. We shall adopt this point of view and consider Kirby's ideas in the context of locales and nuclei. A nucleus $j$ on a locale $L$ will be called componental if it can be represented as a sup of open nuclei in the assembly $NL$. Such a $j$ is determined by the set $S_j = \{a \in L | j(a) = 1\}$. We shall proceed to generalize Kirby's work to coherent locales. In addition, we shall show how componental nuclei capture the notion of a Gabriel topology, in that every Gabriel topology can be represented as $S_j$, for some componental $j$. Furthermore, the componental nuclei on a coherent locale form a locale. This is related to the well-known ring theoretic result that the Gabriel topologies on a ring form a locale [3],[13]. We conclude the paper with some examples of componental nuclei on an arbitrary locale, with particular emphasis on those nuclei which are determined by families of prime elements of the locale.

For the basics of locale theory, the reader is referred to [6].

## 1. Preliminaries

Let $L$ be a locale.

**Definition 1.1** If $S \subseteq L$, then $c \in L$ is called S-prime if $s \to c = c$, for all $s \in S$.

Let $L_S$ denote the set of S-prime elements of $L$. Clearly, $L_S = \bigcap_{s \in S} L_{s \to -}$. Thus, $L_S$ is a sublocale of $L$, being the intersection of sublocales. The corresponding nucleus is given by $j_S = \sup\{s \to - | s \in S\}$, where the sup is computed in the assembly $NL$ of nuclei on $L$. We shall see (Proposition 2.4) that, with certain conditions on $L$ and $S$, this sup can be computed pointwise. None the less, one has the usual explicit description of $j_S(a)$ arising from the sublocale, namely $j_S(a) = \inf\{b | a \le b, b \text{ S-prime}\}$. If $PL$ denotes the power set of $L$, $S \mapsto j_S$ defines an order-preserving map $PL \to NL$. Now, $j \mapsto S_j = \{a \in L | j(a) = 1\}$ clearly defines an order-preserving map $NL \to PL$. To see that the operations $j_{(\ )}$ and $S_{(\ )}$ are adjoint to each other, we first present a lemma, which will also prove useful later on.

---

This paper is in its final form and will not be published elsewhere.

* The research for this paper was supported by National Science Foundation RUI Grants No. 8506039 and 8701717.

Lemma 1.2 Let $k$ be a nucleus on a locale $L$. Then,

(i) $b \to k(a) = k(b) \to k(a)$, for all $a,b \in L$

(ii) $k(b \to a) \leq b \to k(a)$, for all $a,b \in L$.

Proof (i) see [11, Prop. 2.6].

(ii) Since $b \wedge (b \to a) \leq a$, applying $k$, we get $k(b) \wedge k(b \to a) \leq k(a)$. Thus, $k(b \to a) \leq k(b) \to k(a)$. The latter equals $b \to k(a)$, by part (i). ///

In fact, with regard to (i), it is shown that a function $k : L \to L$ is a nucleus if and only if it satisfies (i) (see [11]).

Theorem 1.3 The functor $j_{(\,)} : PL \to NL$ is left adjoint to $S_{(\,)} : NL \to PL$.

Proof It suffices to show that $S \subseteq S_{j_S}$ and $j_{S_j} \leq j$, for all $S \in PL$, $j \in NL$. Let $a \in S$. Clearly, if $a \leq b$ and $b$ is $S$-prime, then $a \to b = b$ implies $b = 1$, and thus, $a \in S_{j_S}$. So, $S \subseteq S_{j_S}$. Now, if $j \in NL$, then to show that $j_{S_j} \leq j$, it suffices to show that $j(a)$ is $S_j$-prime, for all $a \in L$. Using Lemma 1.2, if $b \in S_j$, then $b \to j(a) = j(b) \to j(a) = 1 \to j(a) = j(a)$. ///

Definition 1.4 Let $j \in NL$. Then $j_{S_j}$ is called the core of $j$ and is denoted by $\tilde{j}$. If $j = \tilde{j}$, then $j$ is called componental.

Definition 1.5 Let $S \subseteq L$. Then $S_{j_S}$ is called the closure of $S$, and is denoted by $\overline{S}$. If $S = \overline{S}$, then $S$ is called closed.

From the adjunction, we have $\tilde{j} \leq j$ and $S \subseteq \overline{S}$, for all $j \in NL$, $S \in PL$. In addition, $j_S$ is componental for all $S$, since $j_S = j_{\overline{S}}$, and $S_j$ is closed for all $j$, since $S_j = S_{\tilde{j}}$. Also, $S$ is closed precisely when $a \in S \Leftrightarrow j_S(a) = 1$.

We record the following proposition.

Proposition 1.6 If $j \in NL$, then $\tilde{j} = \sup\{s \to - \mid j(s) = 1\}$, where the sup is computed in $NL$.

Proof $L_{\tilde{j}} = L_{j_{S_j}} = \bigcap_{s \in S_j} L_{s \to -}$. ///

Thus, componental nuclei are sups of open nuclei, or equivalently, they correspond to sublocales which are intersections of open sublocales. Isbell called these nuclei "fitted" in [5]. As mentioned in the introduction, the terminology "componental" is derived from ring theory.

## 2. Coherent Locales and Gabriel Topologies

If $R$ is a commutative ring with 1 and $S$ is a multiplicative set, one can form the ring of quotients $R_S$, and consider the usual embedding of $\mathrm{Spec}(R_S)$ as a subspace of $\mathrm{Spec}(R)$. To see the connection with the localic approach, we identify these spectra with the locales $\mathrm{Rad}(R_S)$ and $\mathrm{Rad}(R)$ of radical ideals of $R_S$ and $R$, respectively. Then the morphism $i : \mathrm{Rad}(R_S) \to \mathrm{Rad}(R)$ corresponding to the embedding is given by $i_*(B) = B \cap R$, for $B \in \mathrm{Rad}(R_S)$ and $i^*(A) = S^{-1}A$, for $A \in \mathrm{Rad}(R)$. Thus, the nucleus which identifies $\mathrm{Rad}(R_S)$ as a sublocale of $\mathrm{Rad}(R)$ is $A \mapsto S^{-1}A \cap R = A_S$ the $S$-component of $A$. Now, if $\hat{S} = \{A \in \mathrm{Rad}(R) \mid A \cap S \neq \emptyset\}$, then it can be shown that

$j_{\hat{S}}(A) = A_S$, and it follows that $L_{\hat{S}} \cong Rad(R_S) \cong O(SpecR_S)$.

In ring theory, the most general notion of localization comes from considering Gabriel topologies [16]. By the work of Hochster [4], every coherent locale is of the form Rad(R), for some commutative ring R with 1. We shall thus consider componental nuclei on coherent locales, and generalize Kirby's work to show that the notion of closed sets, as well as Kirby's "closed" sets, both coincide with that of Gabriel topologies. This not only provides a different (and perhaps simpler) perspective on Gabriel topologies in the commutative ring case, but it also suggests that componental nuclei and closed sets may provide a suitable notion of "localization" for noncommutative locales.

We begin by presenting a slight generalization of a result of Kirby [9], as well as providing a simpler proof.

<u>Proposition 2.1</u> Let L be an algebraic locale. If j is a nucleus on L satisfying $j(b \to a) = b \to j(a)$, for all b compact in L and for all $a \in L$, then j is componental.

<u>Proof</u> It suffices to show that $L_j = L_{\tilde{j}}$. We always have $L_j \subseteq L_{\tilde{j}}$. For the reverse containment, suppose $a \in L_{\tilde{j}}$, i.e. a is $S_j$-prime. It suffices to show that $b \to a \in S_j$, for all compact $b \leq j(a)$, since $j(a) = \sup\{b | b \leq j(a), b \text{ compact}\}$ $\leq \sup\{(b \to a) \to a | b \leq j(a), b \text{ compact}\}$, and $(b \to a) \to a = a$ whenever $b \to a \in S_j$ and a is $S_j$-prime. Thus, $j(a) = a$, and so $a \in L_j$, as desired. To show $b \to a \in S_j$:

$$\begin{aligned} 1 = j(a) \to j(a) &= (\sup\{b | b \leq j(a), b \text{ compact}\}) \to j(a) \\ &= \inf\{b \to j(a) | b \leq j(a), b \text{ compact}\} \\ &= \inf\{j(b \to a) | b \leq j(a), b \text{ compact}\}. \end{aligned}$$

Therefore, $j(b \to a) = 1$ (or equivalently, $b \to a \in S_j$), for all compact $b \leq j(a)$. ///

We would like to obtain a converse to this result, thus characterizing componental nuclei. First, we need a definition and a few preliminary lemmas.

<u>Definition 2.2</u> Let L be an algebraic locale. A subset S of L is <u>Kirby closed</u> if

(1) S is an upper set

(2) $\sup_{\alpha \in I} b_\alpha \in S$, and $c_\alpha \in S$, for all $\alpha \in I \Rightarrow \sup_{\alpha \in I} (b_\alpha \wedge c_\alpha) \in S$, whenever $\{b_\alpha\}_{\alpha \in I}$ is a family of compact elements of L.

Note that choosing all $c_\alpha$'s equal, it follows that a Kirby closed set is closed under finite meets. Thus, combining this with (1), we see that every Kirby closed set is a filter.

<u>Proposition 2.3</u> Let L be an algebraic locale. If $S \subseteq L$ is closed, then S is Kirby closed.

<u>Proof</u> Suppose S is closed. Then $S = \{a \in L | j_S(a) = 1\}$, which is clearly an upper set. For (2), suppose $\{b_\alpha\}_{\alpha \in I}$ is a family of compact elements of L, $\sup_{\alpha \in I} b_\alpha \in S$, and $\{c_\alpha\}_{\alpha \in I} \subseteq S$. We must show that $j_S(\sup(b_\alpha \wedge c_\alpha)) = 1$, i.e. if $\sup(b_\alpha \wedge c_\alpha) \leq x$ and x is S-prime, then $x = 1$. Suppose $c_\alpha \wedge b_\alpha \leq x$, for all $\alpha \in I$, where x is S-prime. Then $b_\alpha \leq c_\alpha \to x$ and $c_\alpha \to x = x$, since $c_\alpha \in S$ and x is S-prime. Thus, $b_\alpha \leq x$, for all $\alpha \in I$, and so $\sup_{\alpha \in I} b_\alpha \leq x$. But, $\sup_{\alpha \in I} b_\alpha \in S$ and S is a filter, so $x \in S$.

Since $x \in S$ and $x$ is S-prime, it follows that $x = 1$. ///

Proposition 2.4 Let $L$ be an algebraic locale such that the compact elements are closed under binary meets, and let $S \subseteq L$ be Kirby closed. Then $j_S(a) = \sup_{b \in S}(b \to a)$, for all $a \in L$.

Proof Let $b' \in S$ and consider $b' \to \sup_{b \in S}(b \to a)$. It suffices to show that this equals $\sup_{b \in S}(b \to a)$, for then the latter element will be S-prime, and hence, $j_S(a) \leq \sup_{b \in S}(b \to a)$. The opposite inequality holds since $j_S(a)$ is S-prime and $b \to a \leq b \to j_S(a) = j_S(a)$, for all $b \in S$.

Let $c \leq b' \to \sup_{b \in S}(b \to a)$, where $c$ is compact and write $b' = \sup_{\alpha \in I} d_\alpha$, where the $d_\alpha$'s are compact. Then

$$c \leq (\sup_{\alpha \in I} d_\alpha) \to (\sup_{b \in S}(b \to a)) \leq \inf_{\alpha \in I} (d_\alpha \to \sup_{b \in S}(b \to a))$$

Thus, $c \wedge d_\alpha \leq \sup_{b \in S}(b \to a)$, for all $\alpha \in I$. Since $\{b \to a | b \in S\}$ is directed and $c \wedge d_\alpha$ is compact, it follows that $c \wedge d_\alpha \leq b_\alpha \to a$, for some $b_\alpha \in S$. Thus, $c \wedge d_\alpha \wedge b_\alpha \leq a$, for all $\alpha \in I$, and so $c \wedge \sup_{\alpha \in I}(d_\alpha \wedge b_\alpha) \leq a$. Since $S$ is Kirby closed, we know $b_0 = \sup_{\alpha \in I}(b_\alpha \wedge d_\alpha) \in S$. Thus, $c \leq b_0 \to a \leq \sup_{b \in S}(b \to a)$. Hence, $b' \to \sup_{b \in S}(b \to a) \leq \sup_{b \in S}(b \to a)$. ///

We can now characterize componental nuclei by the equation of Proposition 2.1.

Theorem 2.5 Let $L$ be an algebraic locale in which the compact elements are closed under binary meets. Then a nucleus $j$ on $L$ is componental if and only if $j(b \to a) = b \to j(a)$, for all compact $b \in L$ and for all $a \in L$.

Proof ($\Leftarrow$) This is just Proposition 2.1.

($\Rightarrow$) If $j$ is componental, then $j = j_S$, where $S = \{a \in L | j(a) = 1\}$. Since $S$ is Kirby closed by Proposition 2.3, applying Proposition 2.4, we have $j(a) = \sup_{s \in S}(s \to a)$. Since $j(b \to a) \leq b \to j(a)$ always holds by Lemma 1.2 (ii), it remains to prove the opposite inequality. Suppose $c < b \to j(a)$, where $c$ and $b$ are compact in $L$ and $a \in L$. Then $c \wedge b \leq \sup_{s \in S}(s \to a)$, and since $c \wedge b$ is compact and $\{s \to a | s \in S\}$ is directed, it follows that $c \wedge b \leq s_0 \to a$, for some $s_0 \in S$. Thus, $c \wedge b \wedge s_0 \leq a$, and therefore, $c \wedge s_0 \leq b \to a$. Thus $j(c \wedge s_0) \leq j(b \to a)$. But, $j(c \wedge s_0) = j(c) \wedge j(s_0) = j(c) \wedge 1 = j(c)$. Thus, $c \leq j(c) \leq j(b \to a)$. Since $c \leq b \to j(a)$ implies $c \leq j(b \to a)$, for all $c$ compact and $L$ is algebraic, it follows that $b \to j(a) \leq j(b \to a)$. ///

We are now ready to use this characterization to show that the notion of componental nucleus is equivalent to that of Gabriel topology.

Definition 2.6 Let $L$ be an algebraic locale. A subset $S$ of $L$ is called a Gabriel topology on $L$ if

(1) $S$ is a filter on $L$

(2) $b \in S$ and $c$ is compact in $L \Rightarrow c \to b \in S$

(3) $b \in S$ and $c \to a \in S$, for all compact $c \leq b \Rightarrow a \in S$

Proposition 2.7 Let $L$ be an algebraic locale in which the compact elements are closed under binary meets. If $S$ is closed in $L$, then $S$ is a Gabriel topology.

Proof If $S$ is closed, then $S = \{a \in L \mid j(a) = 1\}$, for some componental nucleus $j$. Clearly, $S$ is a filter. Suppose $b \in S$ and $c$ is compact. Then $j(c \to b) = c \to j(b)$, by Theorem 2.5. But, $j(b) = 1$, and so $j(c \to b) = 1$. Hence, $c \to b \in S$, verifying (2) of Definition 2.6. For (3), suppose that $b \in S$ and $c \to a \in S$, for all compact $c \leq b$. Then $1 = j(c \to a) = c \to j(a)$, and so $c \leq j(a)$, for all such $c$. Since $L$ is algebraic, we know $b = \sup\{c \mid c \leq b, c \text{ compact}\} \leq j(a)$. Hence, $1 = j(b) \leq j(j(a)) = j(a)$, and so $a \in S$, as desired. ///

We can now prove the main theorem of this section demonstrating the equivalence of "closed", "Kirby closed", and "Gabriel topology".

Theorem 2.8 Let $L$ be a coherent locale. The following are equivalent for $S \subseteq L$.

(1) $S = S_j$, for some nucleus $j$ on $L$.

(2) $S = S_j$, for some componental nucleus $j$ on $L$.

(3) $S$ is closed.

(4) $S$ is Gabriel topology on $L$.

(5) $S$ is Kirby closed.

Proof (1) $\iff$ (2) $\iff$ (3) follows from the remarks after Definition 1.5.

(3) $\Rightarrow$ (4) by Proposition 2.7.

(4) $\Rightarrow$ (5) If $S$ is a Gabriel topology, it is clearly an upper set. For (2) of Definition 2.2, suppose $\{c_\alpha\}_{\alpha\in I} \subseteq S$ and $\sup_{\alpha\in I} b_\alpha \in S$, where $\{b_\alpha\}_{\alpha\in I}$ is a family of compact elements of $L$. We must show that $a = \sup_{\alpha\in I}(b_\alpha \wedge c_\alpha) \in S$. By (3) of the definition of Gabriel topology, it suffices to show that $d \to a \in S$, for all compact $d \leq \sup_{\alpha\in I} b_\alpha$. From (2) of the same definition, we know that $b_\beta \to c_\beta \in S$, for all $\beta \in I$, since $c_\beta \in S$ and $b_\beta$ is compact. Thus, each $b_\beta \to a \in S$, since $S$ is an upper set and $b_\beta \to a \geq b_\beta \to \sup_{\alpha\in I}(b_\alpha \wedge c_\alpha) \geq \sup_{\alpha\in I}(b_\beta \to (b_\alpha \wedge c_\alpha)) \geq b_\beta \to (b_\beta \wedge c_\beta) = b_\beta \to c_\beta$. Now, since $d \leq \sup_{\alpha\in I} b_\alpha$ and $d$ is compact, we know that $d \leq b_{\beta_1} \vee \ldots \vee b_{\beta_n}$, for some $\beta_1,\ldots,\beta_n \in I$. Thus, $d \to a \geq (b_{\beta_1} \vee \ldots \vee b_{\beta_n}) \to a = (b_{\beta_1} \to a) \wedge \ldots \wedge (b_{\beta_n} \to a)$. But, this is in $S$, since each $b_{\beta_i} \to a \in S$ and $S$ is closed under finite meets. Since $S$ is an upper set, it follows that $d \to a \in S$, as desired.

(5) $\Rightarrow$ (3) Suppose $S$ is Kirby closed. To see that $S$ is closed, it suffices to show that $j_S(a) = 1$ implies $a \in S$. If $j_S(a) = 1$, then $\sup_{b\in S}(b \to a) = j_S(a) = 1$, by Proposition 2.4. Since $1$ is compact, it follows that $1 = (b_1 \to a) \vee \ldots \vee (b_n \to a) \leq (b_1 \wedge \ldots \wedge b_n) \to a$, for some $b_1,\ldots,b_n \in S$. Thus, $b_1 \wedge \ldots \wedge b_n \leq a$. Since a Kirby closed set is closed under finite meets, we know $b_1 \wedge \ldots \wedge b_n \in S$. Therefore, $a \in S$, as desired. ///

Note that we needed $1$ to be compact only for (5) $\Rightarrow$ (3).

This approach using closure operators provides a new perspective on Gabriel topologies in the commutative ring case. The non-commutative case will hopefully lend itself to a similar treatment in the context of quantales and quantic nuclei, as discussed in [11].

A close connection between Gabriel topologies and torsion theories on the module category of a ring is well known [16]. In particular, there is a torsion theoretic spectrum, which has been extensively studied by Golan [3], as well as by Simmons [13]. It follows, from their work, that the Gabriel topologies on a ring form a locale. This fact was also established in a more general setting by Borceux and Kelly [12]. The corresponding result for componental nuclei can easily be established using Theorem 2.5.

Theorem 2.9 If $L$ is an algebraic locale in which the compact elements are closed under finite meets, then the set $CL$ of componental nuclei on $L$ forms a locale.

Proof For completeness, we shall verify that $CL$ is closed under infima. By Theorem 2.5, it suffices to show that if $\{j_\alpha\}_{\alpha\in I} \subseteq CL$, then $(\inf j_\alpha)(b \to a) = b \to (\inf_{\alpha\in I} j_\alpha)(a)$, for all compact $b \in L$ and for all $a \in L$. Since each $j_\alpha$ is componental, we know $j_\alpha(b \to a) = b \to j_\alpha(a)$. Thus, $(\inf_{\alpha\in I} j_\alpha)(b \to a) = \inf_{\alpha\in I} j_\alpha(b \to a) = \inf_{\alpha\in I}(b \to j_\alpha(a)) = b \to \inf_{\alpha\in I} j_\alpha(a) = b \to (\inf_{\alpha\in I} j_\alpha)(a)$.

To see that $CL$ is a locale, we need but observe that if $j,k,\ell \in CL$, then $j \wedge k \leq$ if and only if $j \leq \widehat{k \to \ell}$, where $k \to \ell$ denotes the implication in the assembly $NL$. ///

## 3. Some Examples

In this section, we shall consider componental nuclei on an arbitrary locale $L$. In particular, we shall investigate their relationship with prime elements and consider when the componental nuclei are determined by families of primes.

It is clear, from the definition, that an open nucleus is componental. In §2, we characterized componental nuclei on coherent locales in terms of implication (Theorem 2.5). For an arbitrary locale, it turns out that sub-open nuclei are componental. These nuclei were investigated by Johnstone in [7], where several different characterizations were presented. First, we recall their definition.

Definition 3.1 A nucleus $j$ on $L$ is called sub-open if $j(a \to b) = j(a) \to j(b)$, for all $a,b \in L$.

Proposition 3.2 If $j \in NL$ is sub-open, then $j$ is componental.

Proof Since $\tilde{j} \leq j$, it suffices to show that $L_{\tilde{j}} \subseteq L_j$. If $a \in L_{\tilde{j}}$, then $a$ is $S_j$-prime. Since $j$ is sub-open, $j(j(a) \to a) = j(j(a)) \to j(a) = 1$, and so $j(a) \to a \in S_j$. Hence, $(j(a) \to a) \to a = a$, since $a$ is $S_j$-prime. But, $j(a) \leq (j(a) \to a) \to a$ yields $j(a) \leq a$, and so $a \in L_j$. ///

It follows that the nucleus $\neg\neg$ is componental, since it is sub-open [7, Corollary 1.8].

Now, let us turn our attention to closed subsets of $L$. Fix $a \in L$, for the following three examples.

1) It is not hard to see that $\{c \in L \mid c \to a = a\} = \{c \in L \mid (c \to a) \to a = 1\} = S_{(-\to a)\to a}$. Since $(-\to a) \to a$ is a nucleus, it follows that this is a closed set. In fact if $S$ is any closed set and $a$ is $S$-prime, then $S \subseteq S_{(-\to a)\to a}$.

2) One can also show that $S_{a\vee -} = \{c \in L \mid c \vee a = 1\}$, for the closed nucleus $a \vee -$

is closed and $S_{a \vee -} \subseteq S_{(-\to a)\to a}$.

3) Consider $\{c \in L | c \not\leq a\}$. Then $a$ is prime if and only if $\{c \in L | c \not\leq a\} = \{c \in L | c \to a = a\} = S_{(-\to a)\to a}$. Thus, if $p$ is prime, then $S_p = \{c \in L | c \not\leq p\}$ is a closed set. This closed set corresponds to the notion of "localization at p" in the localic setting. The sublocale $L_S$ will be denoted by $L_p$. We record the following proposition to emphasize the ring theoretic motivation.

Proposition 3.3 If $L = \mathrm{Rad}(R)$, where $R$ is a commutative ring with 1, and $P$ is a prime ideal of $R$, then $L_P \cong \mathrm{Rad}(R_P)$, where $R_P$ denotes the localization of $R$ at $P$.

More generally, one can consider a family $P$ of prime elements of $L$. Let $S_P = \bigcap_{p \in P} S_p$. Since $S_{(\ )}$ is a right adjoint, it preserves infs and thus, an intersection of closed sets is closed. It is well known [16, VI §6] that if $R$ is a commutative Noetherian ring, then every Gabriel topology $S$ on $R$ is of the form $S_P$, for some family $P$ of prime ideals. We shall consider the question of when is every closed set $S$ in a locale $L$ of the form $S_P$, and in doing so we will obtain a slight generalization of the ring theoretic result in the context of radical ideals.

The following lemma is a straightforward consequence of the definition of $S_P$.

Lemma 3.4 If $S$ is a closed set in a locale $L$ and $P = \{p \in L | p \notin S,\ p \text{ prime}\}$, then $S = S_P$ if and only if for all $a \notin S$, there exists $p \in P$ such that $a \leq p$.

Lemma 3.5 If $j$ is a componental nucleus on $L$ and $p$ is a prime element of $L$, then $j(p) = p$ or $j(p) = 1$ (i.e. $p \in L_j$ or $p \in S_j$).

Proof Since $j$ is componental, $j = j_S$ where $S = \{a \in L | j(a) = 1\}$. If $p$ is prime and is also S-prime, then $j(p) = p$. If $p$ is not S-prime, then $d \to p \neq p$, for some $d \in S$. Since $d \wedge (d \to p) \leq p$, we must have $d \leq p$. But, $S$ is an upper set, and so $p \in S$, i.e. $j(p) = 1$. ///

Proposition 3.6 If $S$ is a closed subset of a locale $L$ and $L_S$ is a spatial sublocale of $L$, then $S = S_P$, where $P = \{p \in L | p \text{ prime},\ p \notin S\}$.

Proof Suppose that $L_S$ is spatial and $a \notin S$. Note that by Lemma 3.5 $P = \{p \in L | p \text{ prime},\ p \in L_S\}$. Then $j_S(a) = \inf\{p | a \leq p,\ p \in P\}$. Since $j_S(a) \neq 1$, this inf is nonempty. Thus, there exists $p \in P$, such that $a \leq p$. But, $p \notin S$, and so by Lemma 3.5, we know $S = S_P$. ///

We now introduce a definition which will allow us to state our desired result.

Definition 3.7 A locale $L$ is totally spatial if every sublocale of $L$ is spatial.

Totally spatial locales were investigated and characterized by Niefield and Rosenthal in [12]. Topologically, they correspond to weakly scattered spaces. They have also been considered by Simmons [14], [15].

Corollary 3.8 If $L$ is a totally spatial locale, then every closed $S$ is of the form $S_P$, for some family $P$ of prime elements of $L$.

Corollary 3.9 If $R$ is a commutative ring with 1 such that every proper radical ideal is an irredundant meet of prime ideals, then every closed $S \subseteq O(\mathrm{Spec}\ R)$ is of the form $S_P$, for some family $P$ of prime ideals of $R$.

Proof  Niefield and Rosenthal have shown [12] that a locale is totally spatial if and only if every proper element is an irredundant meet of prime elements. ///

Any commutative Noetherian ring satisfies the hypotheses of Corollary 3.9.

We obtain a partial converse to Corollary 3.8.

Proposition 3.10  Let $L$ be a locale in which every prime element is maximal. If every closed $S \subseteq L$ is of the form $S_P$, for some family $P$ of prime elements, then $L$ is totally spatial.

Proof  It follows from [12] that it suffices to show that if $a \neq 1$, there is a prime $p$ with $(p \to a) \to a = p$. Let $S = \{b \in L | b \to a = a\} = \{b \in L | (b \to a) \to a = 1\}$. By assumption $S = S_P$, for some family $P$ of primes. Since $a \notin S$, by Lemma 3.4, there exists $p \notin S$ such that $a \leq p$. Since $p$ is maximal and $(p \to a) \to a \neq 1$, it follows that $(p \to a) \to a = p$. ///

If $X$ is a $T_1$ sober space, then $O(X)$ is totally spatial if and only if $X$ is scattered [12]. We can now apply this last result to Stone duality for Boolean algebras. It is known that the Stone space of a Boolean algebra $B$ is scattered if and only if $B$ is superatomic, that is, every Boolean algebra quotient of $B$ has an atom [1]. Thus, we obtain a new angle on superatomicity.

Corollary 3.11  A Boolean algebra $B$ is superatomic if and only if every closed $S \subseteq \mathrm{Idl}(B)$ is of the form $S_P$ for some family $P$ of maximal ideals.

Proof  Apply Corollary 3.8 and Proposition 3.10. ///

REFERENCES


1. G. W. Day, Superatomic Boolean algebras, Pac. Jour. of Math. Vol. 23. No. 3, (1967), 479-489.
2. F. Borceux and G.M.Kelly, Locales of localizations, Jour. of Pure and Appl. Alg. 46, (1987), 1-34.
3. J.S.Golan, Localization of Non-Commutative Rings, Marcel-Dekker, (1975).
4. M.Hochster, Prime ideal structure in commutative rings, Trans. Amer. Math. Soc. 142, (1969), 43-60.
5. J.R.Isbell, Atomless parts of spaces, Math. Scand. 31 (1972), 5-32.
6. P.T.Johnstone, Stone Spaces, Cambridge Univ. Press (1982).
7. P.T.Johnstone, Open maps of toposes, Manuscripta Math. 31, (1980), 217-247.
8. D.Kirby, Components of ideals in a commutative ring, Ann. Mat. Pure Appl. (4), 71 (1966), 109-125.
9. D.Kirby, Closure operations on ideals and submodules, J. London. Math. Soc. 44 (1969) 283-291.
10. M.Larsen and P.McCarthy, Multiplicative Theory of Ideals, Pure and Appl. Math. Vol. 43, Academic Press (1971).
11. S.Niefield and K.Rosenthal, Constructing locales from quantales, (preprint).
12. S.Niefield and K.Rosenthal, Spatial sublocales and essential primes, Top. and its Appl. 26(1987), 263-269.
13. H. Simmons, Torsion theoretic points and spaces, Proc. Roy. Soc. Edinburgh 96A (1984), 345-361.
14. H.Simmons, Ranking techniques for modular lattices (preprint).
15. H.Simmons, Two sided multiplicative lattices and ring radicals, (preprint).
16. B. Stenstrom, Rings of Quotients, Springer-Verlag (1975).

# REPRESENTATION THEOREMS FOR P-CATEGORIES

G. ROSOLINI*

Categories of partial maps have been an object of study since the early stages of category theory, but, oddly enough, very little was developed in relation to partial recursive functions. It was with the work of DiPaola & Heller [1986] that a general theory of what they called "recursion categories" took shape and interesting results were obtained. The theory is based on the remarkably simple notion of dominical category as an abstraction for a category of partial maps on a category with binary products, and is well-suited for the treatment of essential constructors like range and coproduct which play the parts of image and definition by cases respectively.

Recall that a partial map $[m, \phi] : A \rightharpoonup B$ between objects of a category A is an equivalence class of pairs $(m: D \rightarrowtail A, \phi: D \to B)$ consisting of a monomorphism and a map in A with the same source under the equivalence defined by isomorphic variations of $D$. If A is locally small and well-powered, and has pullbacks of monos, then partial maps form a locally small category Ptl(A) where composition is defined by means of an obvious pullback. More generally, if $\mathcal{M}$ is a given class of monomorphisms of A containing the identities and closed under pullbacks and composition, then a category of partial maps $\mathcal{M}$-Ptl(A) can be defined in the same fashion as Ptl(A) with the further condition on maps $[m, \phi] : A \rightharpoonup B$ that $m$ be in $\mathcal{M}$.

In Rosolini [1985], the author proved a representation theorem which showed that every dominical category can be fully embedded into a category of partial maps over a category with binary products. It was immediately clear that not every category of partial maps was dominical, and that a more general notion ought to be considered. Moreover, Montagna [1986] has shown that there are syntactic categories of partial maps connected to the theory of recursive functions which are not dominical. So when completely algebraic notions like those of a p-category and of a partially cartesian closed category were proved to characterise categories of partial maps (*cf.* Rosolini [1986], and also Robinson & Rosolini [1986] and Curien & Obtułowicz [1986]), the question arose of what part of the theory of recursion could be extended.

In this paper we shall be concerned with ranges and coproducts in the context of p-categories. After recalling in section 1 the representation of a p-category as a full subcategory of one of the form $\mathcal{M}$-Ptl(A), we argue in section 2 that the existence of ranges provides a certain factorisation system on the category A as one would expect (we said before that ranges are a replacement for images in the partial map setting). But this is not enough to characterise p-categories with ranges and we prove that the further request that the factorisation system be stable under pullback is necessary and sufficient. We then apply the representation theorem to strengthen a result of DiPaola & Heller [1986]. In section 3 we turn to coproducts in the partial map setting: again a complete characterisation of p-categories with coproducts is given. By means of this some new interesting results about coproducts in a p-category are proved.

*Department of Pure Mathematics and Mathematical Statistics, 16 Mill Lane, Cambridge CB2 1SB, England, and Dipartimento di Matematica, Università, 43100 Parma, Italy.

The paper is in final form and will not be published elsewhere.

## 1 Representation of p-categories

As we mentioned above there have been various attempts to give an algebraic description for the notion of a category of partial maps (some references are listed at the end of the paper). If existing at all, such a notion will include any subcategory of a category of partial maps closed under the required algebraic operations. The problem has been solved by various authors independently: here we shall sketched briefly the solution proposed by Rosolini [1986] (reported also in Robinson & Rosolini [1986]) for categories of partial maps over a category with binary products.

1.1 DEFINITION A *p-category* is a category C endowed with a bifunctor $\times : \mathsf{C} \times \mathsf{C} \to \mathsf{C}$ which is called *product*, a natural transformation $\Delta : (-) \to (- \times -)$ and two families of natural transformations $\{p_{-,Y} : (- \times Y) \to (-) \mid Y \in \operatorname{ob} \mathsf{C}\}$ and $\{q_{X,-} : (X \times -) \to (-) \mid X \in \operatorname{ob} \mathsf{C}\}$, satisfying the identities

$$p_{X,X}\Delta_X = \mathrm{id}_X = q_{X,X}\Delta_X \qquad (p_{X,Y} \times q_{X,Y})\Delta_{X\times Y} = \mathrm{id}_{X\times Y}$$
$$p_{X,Y}(\mathrm{id}_X \times p_{Y,Z}) = p_{X,Y\times Z} \qquad p_{X,Z}(\mathrm{id}_X \times q_{Y,Z}) = p_{X,Y\times Z}$$
$$q_{X,Y}(p_{X,Y} \times \mathrm{id}_Z) = q_{X\times Y,Z} \qquad q_{X,Z}(q_{X,Y} \times \mathrm{id}_Z) = q_{X\times Y,Z}.$$

Finally we require that the associativity and commutativity isomorphisms $\alpha$ and $\tau$ defined by

$$\alpha_{X,Y,Z} = ((\mathrm{id}_X \times p_{Y,Z}) \times q_{Y,Z}q_{X,Y\times Z})\Delta_{X\times(Y\times Z)} : X \times (Y \times Z) \to (X \times Y) \times Z$$

and

$$\tau_{X,Y} = (q_{X,Y} \times p_{X,Y})\Delta_{X\times Y} : X \times Y \to Y \times X$$

are natural in all variables—though their components need not be.

Note that any category of partial maps $\mathcal{M}$-Ptl(A) on a category A with binary products is a p-category (it is easy to see that, if A has products, then $\mathcal{M}$ is closed under products of monos). The product bifunctor on $\mathcal{M}$-Ptl(A) is thus induced by the product on A and is defined by

$$[m, \phi] \times [n, \psi] = [m \times n, \phi \times \psi],$$

and determines a p-category structure on $\mathcal{M}$-Ptl(A). Hence any full subcategory of $\mathcal{M}$-Ptl(A) closed under products of objects is a p-category. In fact this is the most general situation as stated in the following theorem. With some regret we coin the word *p-subcategory* for a subcategory of a p-category which is closed under the p-structure.

1.2 THEOREM *Given a p-category* C, *there are a category* D *with binary products and a family* $\mathcal{D}$ *of monomorphisms in* D *such that* C *is equivalent to a full p-subcategory of* $\mathcal{D}$-Ptl(D).

We refer the reader to Robinson & Rosolini [1986] for a detailed proof of 1.2, contenting ourselves with recalling the construction of the category D of *domains* and the description of the embedding functor.

For any map $\phi : X \to Y$ in C, let $\operatorname{dom}\phi : X \to X$ be the composite $p_{X,Y}(\mathrm{id}_X \times \phi)\Delta_X : X \to X$. Following DiPaola & Heller [1986] we call this map the *domain* of $\phi$. (Notice that in the p-category $\mathcal{M}$-Ptl(A) the domain of $[m, \phi] : A \rightharpoonup B$ is the partial map $[m, m] : A \rightharpoonup A$.) One checks easily that dom is an idempotent operator, and domains form a collection of idempotent maps of C closed under composition and product. The category D is obtained by *formally* adding domains of maps in C: the objects of D are the domains themselves, and a map $f:\delta \to \varepsilon$ between domains $\delta: X \to X$ and $\varepsilon: Y \to Y$ is a map $f: X \to Y$ in C such that

$$\operatorname{dom} f = \delta \quad \text{and} \quad \varepsilon f = f.$$

The intuition about a map $f:\delta \to \varepsilon$ in D is that $f$ has to be defined on all of $\delta$ taking values into $\varepsilon$.

Next we need to define the family $\mathcal{D}$ of subobjects of D: notice that a pair of domains $\delta, \delta': X \to X$ such that $\delta = \delta\delta'$ induces a monic $\delta: \delta \rightarrowtail \delta'$ in D. Let $\mathcal{D}$ consist of all monos in D of this form. In particular, for $\phi: X \to Y$ in C, one has that $\operatorname{dom}\phi: \operatorname{dom}\phi \rightarrowtail \mathrm{id}_X$ is in $\mathcal{D}$. The class is closed under pullbacks as the pullback of $\delta: \delta \rightarrowtail \delta'$ along $f: \eta \to \delta'$ in D can be explicitly written as

$$\begin{array}{ccc} \operatorname{dom}(\delta f) & \xrightarrow{\ \delta f\ } & \delta \\ {\scriptstyle \operatorname{dom}(\delta f)} \big\downarrow & & \big\downarrow {\scriptstyle \delta} \\ \eta & \xrightarrow{\ f\ } & \delta' \end{array}$$

—in such a situation, we shall call the top row map the *corestriction* of $f$ to $\delta$. Finally, the category C is embedded into $\mathcal{D}$-Ptl(D) by taking $\phi: X \to Y$ to $[\operatorname{dom}\phi, \phi]: X \rightharpoonup Y$.

The category $\mathcal{D}$-Ptl(D) is the completion of C with respect to the splitting of all domain idempotents, *cf.* Freyd [1974]. Therefore, the embedding is an equivalence of categories exactly when all domain idempotents of C split.

The representation theorem allows us to infer that any Horn sentence in the language of p-categories which holds in all categories of the form $\mathcal{M}$-Ptl(A) holds also in an arbitrary p-category. We shall apply this remark in the next sections to extend properties of ranges and coproducts from the dominical case to the more general one of p-categories. The reader is referred to the main source DiPaola & Heller [1986] for the theory of dominical categories and to Robinson & Rosolini [1986] for further comments on the relationship between dominical categories and p-categories.

## 2 Ranges

From now on we consider a fixed p-category C and denote its category of domains by D. The following definition is a direct extension of that by the same name given in DiPaola & Heller [1986].

2.1 DEFINITION A *range* of a map $\phi: X \to Y$ in C is a domain $\varepsilon: Y \to Y$ such that $\varepsilon\phi = \phi$ and

$$\forall \psi, \theta: Y \to Z\ [\psi\phi = \theta\phi \Longrightarrow \psi\varepsilon = \theta\varepsilon.]$$

In case all maps have ranges, we say that C *has ranges*.

Since $\varepsilon$ is a domain and $\varepsilon\phi = \phi$, one has that in the category of domains $\phi\colon \mathrm{dom}\,\phi \to \varepsilon$, and it is easy to deduce from the definition that the map is epic. But this does not capture the notion of range to its full extent in D. The next lemma shows just another property enjoyed by ranges.

Maps of the form $\phi\colon \mathrm{dom}\,\phi \twoheadrightarrow \varepsilon$ where $\varepsilon$ is the range of $\phi$ in C will be the main subject for the rest of the section, and it will be handy to have a special symbol $\mathcal{R}$ for the class that they form.

**2.2 LEMMA** *Suppose that in the commutative diagram*

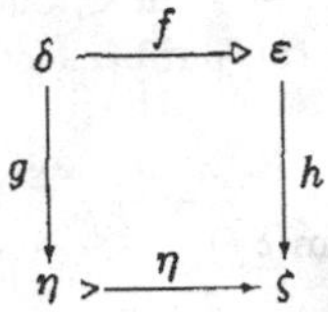


*of arrows in* D *$f$ is in $\mathcal{R}$ and $\eta$ is in $\mathcal{D}$ (a domain). Then there exists a unique map from $\varepsilon$ into $\eta$ in* D, *namely $h\colon \varepsilon \to \eta$, making the two triangles commute.*

*Proof.* Need only to show that $h$ is defined into $\eta$. But $\eta h f = \eta\eta g = \eta g = h f$, thus $\eta h = \eta h\varepsilon = h\varepsilon = h$. □

Notice that the lemma can be restated as saying that ranges are orthogonal to domains. Notice that it also follows easily that a range is unique.

We intend to characterise $\mathcal{R}$ in terms of the category D and the family of monos $\mathcal{D}$. Recall from Freyd & Kelly [1972] that an epimorphism $f\colon \delta \twoheadrightarrow \varepsilon$ in D is *extremal* (with respect to $\mathcal{D}$) if, whenever a diagram

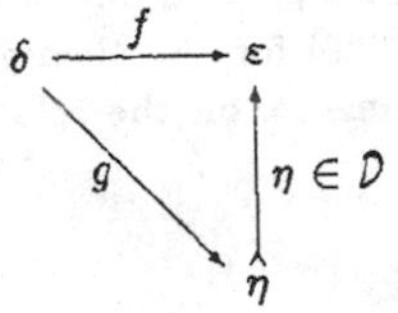


commutes, then $\eta$ is the identity on $\varepsilon$. It follows immediately from 2.2 that maps in $\mathcal{R}$ are extremal. The next proposition describes the ranges (hence the maps in $\mathcal{R}$) as those extremal epimorphisms which retain such property when pulled back along a map in $\mathcal{D}$.

**2.3 PROPOSITION** *Let $\phi\colon X \to Y$ be a map in* C. *The domain $\varepsilon\colon Y \to Y$ is the range of $\phi$ in* C *if and only if $\phi\colon \mathrm{dom}\,\phi \to \varepsilon$ is extremal in* D *as well as all its corestrictions $\delta\phi\colon \mathrm{dom}(\delta\phi) \to \delta\varepsilon$ along domains $\delta\colon Y \to Y$.*

*Proof.* Suppose $\varepsilon$ is the range of $\phi$ in C. We already know that $\phi:\mathrm{dom}\,\phi \to \varepsilon$ is extremal. So let $\delta$ be a domain on $Y$. It is easy to see from the definition 2.1 that $\delta\varepsilon$ is the range of $\delta\phi$, and hence that $\delta\phi:\mathrm{dom}(\delta\phi) \to \delta\varepsilon$ is extremal. Conversely, suppose that the conditions are verified. Firstly $\varepsilon\phi = \phi$ by definition of morphism in D. Secondly suppose $\psi,\theta:Y \to Z$ in C are such that $\psi\phi = \theta\phi$. Let $\delta = \mathrm{dom}\,\psi$ and $\gamma = \mathrm{dom}\,\theta$. By 1.2 (or by direct computation) one has that $\delta\phi = \gamma\phi$, and by hypothesis, $\delta\phi:\mathrm{dom}(\delta\phi) \to \delta\varepsilon$ is extremal. Thus $\varepsilon\delta\gamma\phi = \delta\varepsilon\gamma\phi = \gamma\phi$ in C, and $\delta\phi = \gamma\phi:\mathrm{dom}(\delta\phi) \to \delta\varepsilon$ is the composite $\mathrm{dom}(\delta\phi) \xrightarrow{\gamma\phi} \gamma\delta\varepsilon \xrightarrow{\gamma\delta\varepsilon} \delta\varepsilon$ as $\delta\gamma\phi = \gamma\phi$. By extremality of $\delta\phi:\mathrm{dom}(\delta\phi) \to \delta\varepsilon$ it follows that $\gamma\delta\varepsilon:\gamma\delta\varepsilon \rightarrowtail \delta\varepsilon$ is the identity: in other words, $\gamma\delta\varepsilon = \delta\varepsilon$ in C. Similarly $\delta\gamma\varepsilon = \gamma\varepsilon$, and hence $\delta\varepsilon = \gamma\varepsilon$. Since $\delta\phi:\mathrm{dom}(\delta\phi) \twoheadrightarrow \delta\varepsilon$ is epi, it now follows that $\psi\varepsilon = \theta\varepsilon$. □

We are now in a position to prove the first representation theorem for p-categories with ranges.

2.4 THEOREM *The p-category* C *has ranges if and only if the category* $\mathcal{D}$-Ptl(D) *has ranges. When this is the case, the embedding of* C *into* $\mathcal{D}$-Ptl(D) *preserves them.*

*Proof.* The equivalence follows from 2.3. The final statement follows from how a range is determined in $\mathcal{D}$-Ptl(D). □

2.5 REMARK From the results in Freyd & Kelly [1972] it follows that $\mathcal{D}$ is closed under retracts, hence all the diagonal maps $\langle\delta,\delta\rangle: \delta \rightarrowtail \delta\times\delta$ are in $\mathcal{D}$. But given any pair of maps $f,g:\varepsilon \to \delta$, the pullback of the diagonal $\langle\delta,\delta\rangle$ along $f\times g$ (which exists because D has pullbacks of monos in $\mathcal{D}$) is the equalising pair of $f$ and $g$. Hence D has equalisers and they are in $\mathcal{D}$. Therefore D has all finite limits.

A useful hypothesis which can be made on ranges is that they are *stable* under product: the range of $\phi\times\psi$ is the product of the ranges of $\phi$ and $\psi$. The following characterisation is straightforward.

2.6 PROPOSITION *Ranges are stable under product in the p-category* C *if and only if the family* $\mathcal{R}$ *is closed under products of maps in* D. *When this is the case, ranges are stable under product in* $\mathcal{D}$-Ptl(D).

Not so trivial is the consequence that $\mathcal{R}$ is closed under all pullbacks.

2.7 THEOREM *Suppose ranges are stable under product in* C, *and consider a pullback diagram*

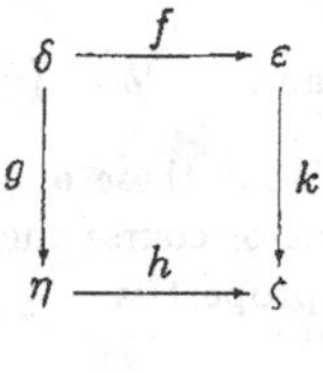

*in* D. *If h is in $\mathcal{R}$, then so is f.*

*Proof.* Note that $k$ equals the composite of its graph $\langle \varepsilon, k\rangle : \varepsilon \to \varepsilon \times \varsigma$ with the second projection. Hence $f$ is obtained by pulling $\varepsilon \times h : \varepsilon \times \eta \rightharpoonup \varepsilon \times \varsigma$ back along a domain. But $\varepsilon \times h$ is in $\mathcal{R}$ by 2.6, and so is its corestriction to $\langle \varepsilon, k\rangle$ by 2.3. □

We can now prove a strong representation theorem for p-categories with ranges which are stable under product.

2.8 THEOREM *A p-category has ranges which distribute over product if and only if it is equivalent to a full p-subcategory of $\mathcal{M}$-Ptl(A) where A is a category with finite limits and $\mathcal{M}$ is the monic part of a proper factorisation system $(\mathcal{E}, \mathcal{M})$ where both $\mathcal{E}$ and $\mathcal{M}$ are closed under pullback along maps in* A.

*Proof.* Follows from 2.3 and 2.7. □

Theorem 2.8 can be read as a completeness result and can be applied to prove that a certain property of morphisms in a p-category with ranges stated by means of a Horn sentence holds in all p-categories with stable ranges. To do that one needs only to check that the given property holds in those of the form $\mathcal{M}$- Ptl(A) using also the further conditions stated in 2.8 (and that all domains split).

We end this section using this criterion to explain how results about stable ranges in dominical categories are extended to the general case.

2.9 REMARK It is useful to translate two "axioms of choice" for dominical categories which appear in DiPaola & Heller [1986] into statements about the relative category of domains D. The weak axiom of choice of *loc.cit.* (already when stated for a p-category) is equivalent to the request that the family $\mathcal{D}$ consists of all subobjects of D. And the (strong) axiom of choice is equivalent to the further condition that all maps in $\mathcal{R}$ have a section. Therefore either holds in C if and only if it holds in $\mathcal{D}$- Ptl(D).

The family $\mathcal{D}$ induces a indexed poset $\mathcal{D}$ : D → Pos over D where $\mathcal{D}(X)$ is the set of subobjects of $X$ in D represented by a mono in $\mathcal{D}$, and $\mathcal{D}^f : \mathcal{D}(Y) \to \mathcal{D}(X)$ is the inverse image along $f : X \to Y$. This is obviously all very general. And, as usual with factorisation systems, in the case of a p-category with ranges these induce an indexed left adjoint $\mathcal{R} \dashv \mathcal{D}$. Furthermore they satisfy Beck-Chevalley condition when ranges in the p-category are stable under products by 2.7—Barbieri Viale [1986] carries out a detailed study of a very similar situation of indexed posets.

The two operators can be extended to the category $\mathcal{D}$-Ptl(D): the posets are the same $\mathcal{D}X$, but for a partial map $[\delta, \phi] : X \rightharpoonup Y$ one has to let

$$[\delta, \phi]^*(\varepsilon) = \mathcal{R}^\delta(\mathcal{D}^\phi(\varepsilon)) \qquad \text{and} \qquad [\delta, \phi]_*(\eta) = \mathcal{R}^\phi(\mathcal{D}^\delta(\eta)) = \mathcal{R}^\phi(\delta \cap \eta).$$

It takes a two second checking to see that these operators restrict on C to their homonyms defined in DiPaola & Heller [1986]. But of course they do not form an adjoint pair anymore, though they still enjoy many algebraic properties.

Because of this all the results in §5 of *loc.cit.* can be generalised to p-categories with ranges stable under products by checking that they hold in a p-category of the form $\mathcal{M}$-Ptl(A). We should point out that it is now easy to prove Proposition 6.7 without the hypothesis that the axiom of choice holds in C.

## 3 Coproducts

In this section we shall pursue a result similar to Theorem 2.8 for p-categories with coproducts.

3.1 PROPOSITION *Suppose the coproduct $X+Y$ exists in* C. *Then, given any $\phi : X \to Z$ and $\psi : Y \to Z$, one has that* $\mathrm{dom}\binom{\phi}{\psi} = \mathrm{dom}\,\phi + \mathrm{dom}\,\psi$.

*Proof.* The assertion follows from the three identities below:

$$\Delta = \binom{i\times i}{j\times j}(\Delta+\Delta), \quad \left(\mathrm{id}\times\binom{\phi}{\psi}\right)\binom{i\times i}{j\times j} = \binom{i\times \mathrm{id}}{j\times \mathrm{id}}(\mathrm{id}\times\phi + \mathrm{id}\times\psi), \quad p\binom{i\times \mathrm{id}}{j\times \mathrm{id}} = p+p. \qquad \Box$$

3.2 COROLLARY *If the p-category* C *has coproducts, then* D *has coproducts and $\mathcal{D}$ is closed under sums.*

3.3 COROLLARY *A p-category* C *has coproducts if and only if so does the category $\mathcal{D}$*-Ptl(D). *When this is the case, the embedding of* C *into $\mathcal{D}$*-Ptl(D) *preserves them.*

This provides us with the following characterisation of p-categories with coproducts.

3.4 THEOREM *A p-category has coproducts if and only if it is a full p-subcategory, closed under coproducts, of a p-category $\mathcal{M}$*-Ptl(A) *where* A *is a category with coproducts and $\mathcal{M}$ is closed under sums of monos.*

Like in the previous section we can apply 3.4 to obtain results about an arbitrary p-category with coproducts by deducing them from the standard case of $\mathcal{M}$-Ptl(A) as long as these can be stated using Horn sentences.

3.5 REMARK Many results in §7 of DiPaola & Heller [1986] can be analysed from the new perspective of 3.3. It is a remarkable fact that we never needed a distributive law of product over coproduct for any of the properties proved above. In particular, condition (i) in the definition of +-dominicality is redundant. Also Proposition 7.6 in *loc.cit.* follows from Proposition 2.1.1 in Freyd & Kelly [1972] by means of the representation theorem 3.4. Proposition 7.7 can be now strengthen to hold also under the weaker assumption that domains split in C (which is tantamount to saying that C is equivalent to $\mathcal{D}$-Ptl(D)); therefore Proposition 8.15 holds without the assumption of dominicality.

We conclude with an interesting property about canonical injections in a coproduct when the p-category is pointed. Recall from Robinson & Rosolini [1986] that a *point* in a p-category is a family of maps $0_{X,Y} : X \to Y$ such that $\phi 0 = 0 = 0\psi$ and $\phi \times 0 = 0$ for any morphisms $\phi, \psi$ in C. It is easy to see that any $0_{X,X} = \mathrm{dom}\, 0_{X,X}$ is a strict initial object in the category of domains D.

3.6 COROLLARY *Suppose* C *is a pointed p-category. Then in* D *injections into coproducts are domains and they are disjoint.*

*Proof.* In D one has $\delta \simeq \delta + 0 \rightarrowtail \delta + \varepsilon$; hence $\delta \rightarrowtail \delta + \varepsilon$ is a domain by 3.1. Disjointness is trivial now, as meet is given by composition. □

# REFERENCES


ASPERTI, A. & LONGO, G.

[1986] **Categories of partial morphisms and the relation between type-structures,** Nota Scientifica S-7-85, Dipartimento di Informatica, Università di Pisa, 1986

BARBIERI VIALE, L.

[1986] *Lattice-theoretic aspects of doctrines and hyperdoctrines,* in Rend. Accad. Naz. Sci. XL Mem. Mat. **104** (1986) 93-102

CARBONI, A.

[1986] *Bicategories of partial maps,* to appear in Cahiers Top. et Géom. Diff., 1986

CURIEN, P.-L. & OBTUŁOWICZ, A.

[1986] *Partiality and cartesian closedness,* typescript, 1986

DIPAOLA, R. & HELLER, A.

[1987] *Dominical categories: recursion theory without elements,* in Journ. Symb. Logic **52** (1987) 594-635

FREYD, P.J.

[1974] *Allegories,* mimeographed notes, 1974

FREYD, P.J. & KELLY, G.M.

[1972] *Categories of continuous functors, I,* in J. Pure Appl. Alg. **2** (1972) 169-191

HOEHNKE, H.J.

[1977] *On partial algebras,* in Col. Math. Soc. J. Bolyai **29** (1977) 373-412

LONGO, G. & MOGGI, E.

[1984] *Cartesian closed categories and partial morphisms for effective type structures,* in **International Symposium on Semantics of Data Types** (edited by G. Kahn, D.B. McQueen & G. Plotkin), Lecture Notes in Computer Science 173, Springer-Verlag, Berlin (1984) 235-255

MOGGI, E.

[1986] *Categories of partial maps and $\lambda_p$-calculus,* in **Category Theory and Computer Programming** (edited by D. Pitt, S. Abramsky, A. Poigné & D. Rydeheard), Lecture Notes in Computer Science 240, Springer-Verlag, Berlin (1986) 242-251

MONTAGNA, F.

[1986] **'Pathologies' in two syntactic categories of partial maps**, Rapporto matematico 151, Università di Siena, 1986

OBTUŁOWICZ, A.

[1986] *The logic of categories of partial functions and its applications*, in Diss. Math. **141** (1986)

ROBINSON, E.P. & ROSOLINI, G.

[1986] *Categories of partial maps*, to appear in Inform. and Comp., also Quaderno di Matematica 18, Università di Parma, 1986

ROSOLINI, G.

[1985] *Domains and dominical categories*, in Riv. Mat. Univ. Parma (4) **11** (1985) 387-397

[1986] **Continuity and effectiveness in topoi**, D.Phil. thesis, University of Oxford, 1986

VOGEL, H.J.

[1979] *On Birkhoff algebras in dht-symmetric categories*, in Col. Math. Soc. J. Bolyai **28** (1979) 759-779

## On the category of compact convex sets and related categories

by

Zbigniew Semadeni (Warsaw)

**Introduction.** It is known (see [6], #23.3.5) that $\mathcal{A}(I^m, I)$ is a free object with $m$ generators in the category **Bcaf** of spaces $\mathcal{A}(K, I)$ as objects and nonnegative linear unit-preserving maps as morphisms. Here $I = [-1,1]$, $K$ is a compact convex set, $\mathcal{A}(K,K')$ denotes the space of all continuous affine maps $K \longrightarrow K'$, and $m$ is a cardinal number. On the other hand, it is easy to see that

$$\mathcal{A}(I^m, \mathbb{R}) \equiv \ell^1(m+1), \tag{1}$$

where $\equiv$ denotes an isometrical isomorphism of Banach spaces. The space $\ell^1(m+1)$ may be regarded as a free Banach space with $m+1$ generators.

The purpose of this paper is to discuss isomorphism (1) in the context of six categories: that of compact convex sets, that of compact convex sets with distinguished (base) points, that of symmetric compact convex sets, and the three dual categories. In particular, the role of the extra dimension (i.e., $m+1$ rather than $m$) is explained in terms of some canonical functors between these categories.

The author is indebted to prof. Dieter Pumplün for inspiring discussions and valuable comments.

**Preliminaries.** In this paper, all vector spaces are over the field $\mathbb{R}$ of reals. If $K$ is a *compact convex set*, i.e., a compact convex subset of a locally convex Hausdorff topological vector space $E$, then $\mathcal{A}(K)$ denotes the space of all continuous affine functions $K \longrightarrow \mathbb{R}$ (a map is called *affine* iff it preserves the convex combinations). $\mathcal{A}(K)$ is a closed subspace of the space $\mathcal{C}(K)$ of all continuous functions $K \longrightarrow \mathbb{R}$. The symbol Cmpcnvx will denote the category of compact convex sets and continuous affine maps. If $\phi : K \longrightarrow K'$ is a morphism in this category, then $\mathcal{A}(\phi) : \mathcal{A}(K') \longrightarrow \mathcal{A}(K)$ denotes the corresponding linear operator defined as $\mathcal{A}\phi.g = g \circ \phi$.

Bcaf denotes the category in which an object means any closed subspace $H$ of some space $\mathcal{C}(X)$, $X$ compact, such that the constant function $\mathbf{1}_X$ is in $H$, while a morphism from $H$ to some $G \subseteq \mathcal{C}(Y)$ means a bounded linear operator $\Phi : H \longrightarrow G$ satisfying $\|\Phi\| \leq 1$ and $\Phi(1_X) = 1_Y$. $\mathcal{K}(H)$ will denote the *state space* of K, i.e., the set

$$\mathcal{K}(H) = \{\xi \in H^* : \xi \geq 0,\ \xi(1_X) = 1\},$$

which is convex and compact in the weak$^*$ topology. For each morphism $\Psi : H \longrightarrow G$ in **Bcaf**, $\mathcal{K}(\Psi)$ denotes the corresponding weakly$^*$ continuous affine map from $\mathcal{K}(G)$ to $\mathcal{K}(H)$. Both canonical maps

$$K \longrightarrow \mathcal{K}(\mathcal{A}(K)) \quad \text{and} \quad H \longrightarrow \mathcal{A}(\mathcal{K}(H))$$

are isomorphisms of the respective categories, which may be regarded as dual to each other ([6], #23.3.2).

If B is a Banach space, then $\mathrm{O}B$ and $\mathrm{O}^*B$ will denote the unit ball of B and of $B^*$, respectively, the latter being provided with the weak$^*$ topology.

The letter I stands for the interval $[-1,1]$ (i.e., the unit ball of $\mathbb{R}$). $I^m$ denotes the product of m copies of I with the product topology and the obvious structure of a convex set, which make it a product in **Cmpcnvx**. Consequently, $\mathcal{A}(I^m)$ is a coproduct of m copies of $\mathcal{A}(I)$ in **Bcaf**. The cardinal number m is understood in von Neumann's sense, i.e., m is the same as the set of all ordinal numbers $\tau$ of cardinality less than m.

$\mathcal{A}(I^m)$ may be regarded as a **Bcaf**-free object with m generators, that is, there is canonical map $\pi$ from the set m into

$$\mathcal{A}(I^m, I) = \{f \in \mathcal{A}(I^m) : \|f\| \leq 1\} = \mathrm{O}\mathcal{A}(I^m)$$

satisfying the following condition: for each map $\xi$ from m into any space $\mathcal{A}(K, I)$ (in the category of sets) there exists a unique **Bcaf**-morphism $\Theta : \mathcal{A}(I^m, I) \longrightarrow \mathcal{A}(K, I)$ such that $\Theta\pi = \xi$. Specifically, to each $\tau$ in m we assign the coordinate projection $\pi_\tau : I^m \longrightarrow I$ regarded as an element of $\mathcal{A}(I^m, I)$.

**Lemma**. *Let* B *be a Banach space. Each* f *in* $\mathcal{A}(\mathrm{O}^*B)$ *is of the form* $f(\xi) = \xi(b) + s$ *for* $\xi$ *in* $\mathrm{O}^*B$, *where* $b \in B$ *and* $s = f(0) \in \mathbb{R}$. *Moreover*, $\|f\| = \|b\| + |s|$.

This follows from the Krein-Šmulyan theorem applied to the linear extension of the functional $g(\xi) = f(\xi) - f(0)$ (see e.g., [2]p. 434; cf. [6], #23.1.11(C)).

**Proposition 1.** *Every* $f$ *in* $\mathcal{A}(I^m)$ *is of the form*

$$(2) \qquad f = s_\infty \cdot \mathbf{1} + \sum_{\tau<m} s_\tau \pi_\tau ,$$

*i.e.*, $f(x) = s_\infty + \sum s_\tau x_\tau$ *for* $x = (x_\tau)$ in $I^m$; *moreover*,

$$\| f \| = |s_\infty| + \sum |s_\tau|.$$

Proof. The cube $I^m$ may be identified with the unit ball of $\ell^\infty(m)$ provided with the weak$^*$ topology of the space conjugate to $\ell^1(m)$. Hence the conclusion follows immediately from the preceding lemma. □

Isomorphism (1) is now defined as $\Gamma(f) = \left( s_\infty, (s_\tau)_{\tau<m} \right)$ for $f$ of the form (2).

Both vector spaces $\mathcal{A}(I^m)$ and $\ell^1(m+1)$ have natural pointwise-defined orders. Isomorphism (1), however, is not order-preserving. Indeed, an element $f$ in $\mathcal{A}(I^m)$ is nonnegative if and only if $f(x) \geq 0$ for each $x$ in $I^m$; consequently, condition $f \geq 0$ means that $s_\infty + \sum s_\tau x_\tau \geq 0$ for all $x = (x_\tau)$ satisfying $|x_\tau| \leq 1$, i.e., that

$$\sum |s_\tau| \leq s_\infty .$$

On the other hand, an element $s = (s_\infty, (s_\tau))$ in $\ell^1(m+1)$ is nonnegative iff $s_\infty \geq 0$ and $s_\tau \geq 0$ for $\tau$ in $m$.

Let us note the commutativity of the diagram:

$$\begin{array}{ccc} m & \xrightarrow{\ \pi\ } & \mathcal{A}(I^m) \\ \delta \big\downarrow & & \big\downarrow \Gamma \\ \ell^1(m) & \xrightarrow{\ \Lambda\ } & \ell^1(m+1) \end{array}$$

where $\delta_\tau$ is the $\tau$-th unit vector and $\Lambda$ is the obvious isometrical embedding (with zero as the extra coordinate).

**Compact convex sets with base points.** By an object of the category $\mathrm{Cmpcnvx}_\bullet$ we shall mean a pair $(K; x_o)$ where $K$ is a compact convex set and $x_o \in K$; a morphism from $(K; x_o)$ to $(K'; x'_o)$ is a continuous affine map $\varphi : K \longrightarrow K'$ such that $\varphi(x_o) = x'_o$.

Let $\mathcal{A}_o(K; x_o)$ denote the space of all continuous real-valued affine functions $f$ such that $f(x_o) = 0$. It is a subspace of $\mathcal{A}(K)$ and its codimension equals 1; indeed, each function $f$ in $\mathcal{A}(K)$ is of the form

$f = f_o + c1_K$ where $f_o \in \mathcal{A}_o(K;x_o)$, $c = f(x_o)$ and $1_K$ denotes the constant function 1 on K.

If $\varphi : (K;x_o) \longrightarrow (K';x'_o)$ is a morphism in $\text{Cmpcnvx}_\bullet$, then

$$\mathcal{A}_o(\varphi) : \mathcal{A}_o(K';x'_o) \longrightarrow \mathcal{A}_o(K;x_o)$$

is the corresponding linear operator $\mathcal{A}_o(\varphi).g = g\circ\varphi$; thus, $\mathcal{A}_o(\varphi)$ is the restriction of $\mathcal{A}(\varphi)$ to the subspace $\mathcal{A}_o(K';x'_o)$. Note that the norm $\|\mathcal{A}_o(\varphi)\|$ may be smaller than 1. Indeed, let e.g., $K = [0,1]$, $K' = [0,2]$, both with the base point 0. The identical embedding $\varepsilon : K \to K'$ is a morphism; clearly $\mathcal{A}_o(\varepsilon)$ is the restriction operator $g \longmapsto g|_K$ and $\|\mathcal{A}_o(\varepsilon)\| = \frac{1}{2}$.

By an object of the category $\text{Bcaf}_o$ we mean any closed subspace $H_o$ of some space

$$\mathscr{C}_o(X;x_o) = \{ f \in \mathscr{C}(X) : f(x_o) = 0 \},$$

where X is compact and $x_o \in X$. One may also think of $H_o$ as a subspace of the space $\mathscr{C}_o(Y)$ of functions continuous on a locally compact space Y and vanishing at infinity; Y is compact if and only if the point at infinity is isolated. For each object $H_o$ of $\text{Bcaf}_o$ the space

(3) $$H = \{ f_o + c1_X : f_o \in H_o,\ c \in \mathbb{R} \}$$

is a Bcaf-object. By a $\text{Bacf}_o$-morphism from a subspace $H_o$ of $\mathscr{C}_o(X;x_o)$ to a subspace $G_o$ of $\mathscr{C}_o(Y;y_o)$ we mean a linear operator $\Phi_o : H_o \longrightarrow G_o$ such that

(4) $$\forall_{\alpha,\beta\in\mathbb{R}} \left( \forall_{x\in X}\ \alpha \le h(x) \le \beta \right) \implies \left( \forall_{y\in Y}\ \alpha \le \Phi_o h.(y) \le \beta \right).$$

Note that (4) is equivalent to the condition:

(4') $$\forall_{\beta\in\mathbb{R}} \left(\forall_{x\in X}\ h(x) \le \beta \right) \implies \left(\forall_{y\in Y}\ \Phi_o h.(y) \le \beta\right)$$

and is also equivalent to an analogous condition with $\alpha$ only. Bear in mind that $h(x_o) = 0$ implies that $\alpha \le 0$ and $\beta \ge 0$.

If $H_o$ is an object of $\text{Bcaf}_o$, define

(5) $$\mathcal{K}_o(H_o) = \{\xi \in H_o^* : \forall_{\beta\in\mathbb{R}}\ h \le \beta 1_X \implies \xi(h) \le \beta \}.$$

Thus, $\mathcal{K}_o(H_o)$ is a convex weakly$^*$ compact subset of $H_o^*$. Condition $h(x_o) = 0$ implies that the functional 0 belongs to $\mathcal{K}_o(H_o)$. However, conditions $\xi \ge 0$ and $\|\xi\| \le 1$ do not imply that $\xi \in \mathcal{K}_o(H_o)$. Let us also note that $\mathbb{R}$ may be identified with $\mathcal{A}_o([0,1];0)$; hence $\mathcal{K}_o(H_o)$ may be regarded as the set of all $\mathbf{Bcaf}_o$-morphisms from $H_o$ to $\mathbb{R}$. Let us consider the following examples, where $H = \mathcal{A}(K)$, $H_o = \mathcal{A}_o(K;x_o)$.

**(a)** Let $K = [0,1]$, $x_o = 0$. Then $\dim H = 2$, each h in H is of the form $h(x) = a(1-x) + bx$, each $\xi$ in $H^*$ is of the form $\xi(h) = \lambda a + \mu b$; such a $\xi$ is in $\mathcal{K}(H)$ if and only if $0 \le \lambda \le 1$ and $\mu = 1-\lambda$; hence $\mathcal{K}(H)$ is isomorphic to $[0,1]$. On the other hand, $\dim H_o = 1$, each $h_o$ in $H_o$ is of the form $h_o(x) = ax$, each $\xi$ in $\mathcal{K}_o(H_o)$ is of the form $\xi(h) = \lambda a$ with $0 \le \lambda \le 1$; hence $\mathcal{K}_o(H_o)$ is isomorphic to the object $([0,1];\ 0)$. A linear operator $\Phi_o$ on $H_o$ satisfies (4) if and only if $\|\Phi_o\| \le 1$ and $\Phi_o \ge 0$.

**(b)** Let K be compact convex and *symmetric*, i.e., let $x \in K$ imply that $-x \in K$. If $h \in H_o$ and $x_o = 0$, then the condition $\alpha \le h(x) \le \beta$ is equivalent to $\|h\| \le \min(|\alpha|, \beta)$. No function in $H_o$ is nonnegative; therefore each linear operator on $H_o$ is nonnegative. A linear operator $\Phi_o$ on $H_o$ satisfies (4) if and only if $\|\Phi_o\| \le 1$. Each $\xi$ in $H_o^*$ such that $\|\xi\| \le 1$ is the evaluation at some point of K and can be extended to a multiplicative linear functional on $\mathcal{C}(K)$ (Phelps [3]).

**(c)** Let $K = [-\frac{1}{2}, 1]$, $x_o = 0$. Then each h in $H_o$ is of the form $h(x) = ax$ and each linear functional $\xi$ on $H_o$ is of the form $\xi(h) = \lambda a$. If the condition $\forall_{x\in K}\ h(x) \le \beta$ is to imply that $\lambda a \le \beta$, then $\lambda$ must be in $[-\frac{1}{2}, 1]$. For each h in $H_o$ let h' denote the affine extension of h to $[-1,1]$. If $\eta$ is any linear functional on $H_o$ with $\|\eta\| \le 1$ then there exists an $x_o$ in $[-1,1]$ such that $\eta(h) = h'(x_o)$ for h in $H_o$; $\eta$ belongs to $\mathcal{K}_o(H_o)$ if and only if $x_o \in [-\frac{1}{2}, 1]$.

**Proposition 2.** *Let* $1_Y \in G \subseteq \mathcal{C}(Y)$ *and let* $\Phi_o : H_o \longrightarrow G$ *be a linear operator satisfying condition* (4). *Then there exists a unique* Bcaf-*morphism* $\Phi : H \longrightarrow G$ *such that* $\Phi|_{H_o} = \Phi_o$. □

If we apply condition (4) to the case $H \subseteq \mathcal{C}(X)$, $G \subseteq \mathcal{C}(Y)$, $\mathbf{1}_X \in H$, $\mathbf{1}_Y \in G$, then it may serve as a description of morphisms in **Bcaf**, i.e., it is equivalent to $\Phi \ge 0$ and $\Phi(\mathbf{1}_X) = \mathbf{1}_Y$. Consequently, Proposition 2 implies that each $\xi_o \in \mathcal{K}_o(H_o)$ can be extended to a unique $\xi$ in $\mathcal{K}(H)$. The operation of restriction $\xi \longmapsto \xi|_{H_o}$ gives the canonical bijection

(6) $$\Theta : \mathcal{K}(H) \longrightarrow \mathcal{K}_o(H_o).$$

Obviously, $\mathcal{A}_o$ is a contravariant functor from $\mathbf{Cmpcnvx}_{\bullet}$ to the cate gory $Bcaf_o$. Now, if $\Phi : H_o \to H'_o$ is a morphism in $Bcaf_o$, let $\mathcal{K}_o(\Phi)$ denote the restriction of the adjoint operator $\Phi^* : H'^*_o \to H^*_o$ to the subset $\mathcal{K}_o(H'_o)$ of $H'^*_o$. It is clear that $\mathcal{K}_o$ is a contravariant functor from $Bcaf_o$ to $Cmpcnvx_\bullet$.

**Proposition 3.** *Let* $(K;x_o)$ *be an object in* $Cmpcnvx_\bullet$ *and let* $H_o = \mathcal{A}_o(K;x_o)$. *If* $x \in K$, *let* $\hat{x}(h) = h(x)$ *for* $h$ *in* $H_o$. *Then the map*

(7) $$\hat{} : K \longrightarrow \mathcal{K}_o(H_o)$$

*is a weakly** *continuous affine bijection. Moreover,* $\hat{x}_o = 0$.

Proof. The argument is routine. The space $\mathcal{A}_o(K;x_o)$ clearly separates the points of K and hence map (7) is one-to-one. We also claim that map (7) is onto. Indeed, let

$$Q = \{\hat{x} : x \in K\}.$$

Then $Q$ is a compact convex subset of $H^*_o$. Suppose $\mathfrak{z} \in H^*_o \setminus Q$. Then there exists an h in $H_o$ such that $\hat{x}(h) \le 1$ for $\hat{x}$ in Q and $\mathfrak{z}(h) > 1$. Hence, $h(x) \le 1$ for x in K and $\mathfrak{z} \notin \mathcal{K}_o(H_o)$.□

**Proposition 4.** *Let* $H_o \subseteq \mathcal{C}_o(X;x_o)$, $K_o = \mathcal{K}_o(H_o)$. *If* $h_o \in H_o$, *let* $\hat{h}_o(\xi_o) = \xi_o(h_o)$ *for* $\xi_o$ *in* $K_o$. *Then the canonical map*

(8) $$\hat{} : H_o \longrightarrow \mathcal{A}_o(K_o;\ 0)$$

*is a linear isometrical order-preserving bijection.*

Proof. Map (8) is obviously linear, of norm $\le 1$, and nonnegative. We shall now prove the following: *for each* $h_o$ *in* $H_o$ *there exist* $\xi_o$, $\eta_o$ *in* $K_o$ *such that*

(9) $$\xi_o(h_o) = \alpha,\quad \eta_o(h_o) = \beta,\ \textit{where}\ \alpha = \inf_X h_o,\quad \beta = \sup_X h_o.$$

Let H be given by (3). Then there exist $\xi, \eta$ in $\mathcal{K}(H)$ such that $\xi(h_o) = \alpha$ and $\eta(h_o) = \beta$ (cf. [6], #23.2.5). It is now enough to let $\xi_o$, $\eta_o$ be restrictions of $\xi$, $\eta$ to $H_o$.

Consequently, map (8) is an isometry and $\hat{h} \ge 0$ implies that $h \ge 0$. It remains to be shown that (8) is onto. Let $\mathfrak{h}$ be in $\mathcal{A}_o(K_o;0)$, i.e., $\mathfrak{h}_o : K_o \to \mathbb{R}$ is affine, weakly* continuous and $\mathfrak{h}_o(0) = 0$. We claim that there exists an $h_o$ in $H_o$ such that $\mathfrak{h}_o = \hat{h}_o$. Define $\mathfrak{h} = \mathfrak{h}_o \circ \Theta$, where $\Theta$ is given by (6). Then $\mathfrak{h} \in \mathcal{A}(\mathcal{K}(H))$. Now, let $\hat{x}_o$ be the functional in $\mathcal{K}(H)$ defined as $\hat{x}_o(h) = h(x_o)$ for h in H. Then

$$\mathfrak{h}(\hat{x}_o) = \mathfrak{h}_o(\Theta(\hat{x}_o)) = \mathfrak{h}_o(0) = 0.$$

Moreover, there exists an h in H such that $\mathfrak{h}(\xi) = \xi(h)$ for h in H. We claim that $h \in H_o$. Indeed, $h(x_o) = \hat{x}_o(h) = \mathfrak{h}(\hat{x}_o) = 0$ and hence, if $\xi_o$ is in $K_o$, then

$$\mathfrak{h}_o(\xi_o) = \mathfrak{h}_o(\Theta\xi) = \mathfrak{h}(\xi) = \xi(h) = \xi_o(h), \text{ i.e., } \mathfrak{h}_o = \hat{h}_o. \ \square$$

**Corollary**. *The composite functors*

$$\mathcal{K}_o\mathcal{A}_o : \text{Cmpcnvx}_\bullet \longrightarrow \text{Cmpcnvx}_\bullet \quad \textit{and} \quad \mathcal{A}_o\mathcal{K}_o : \text{Bcaf}_o \longrightarrow \text{Bcaf}_o$$

*are naturally equivalent to the corresponding identity functors.*

**Relations between categories** Cmpcnvx **and** $\text{Cmpcnvx}_\bullet$. Let us consider the following diagram of categories:

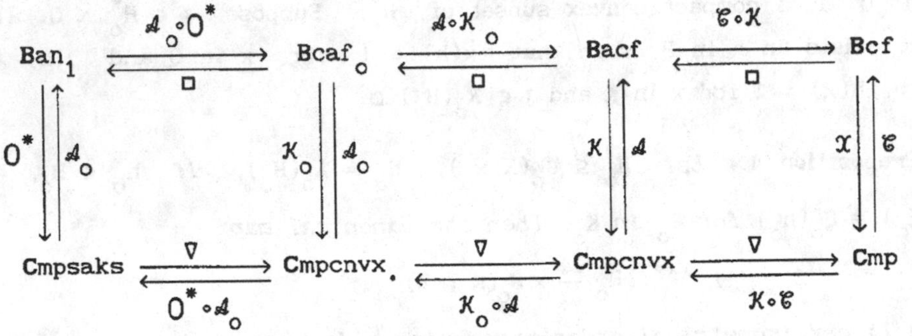


The right-hand square of this diagram is well known and will not be discussed here (see [5] and [6]). We only recall that Bcf denotes the category of spaces $\mathcal{C}(X)$ and unit-preserving Banach-algebra homomorphisms; $\mathfrak{X}$ is the Gelfand functor (i.e., the maximal-ideal-space functor) to the category Cmp of compact spaces; $\mathcal{K}\circ\mathcal{C}$ is the Choquet-Bauer simplex functor.

The diagram is similar to that in [5]; the difference is that the pair $\text{Cmpcvx}_\bullet$ and $\text{Bcaf}_o$ was absent there and we now put it in the right place.

Symbols □ and ∇ stand for the obvious forgetful functors; only the functor from Bcaf to $\text{Bcaf}_o$ should be commented on. To each object H $\leq \mathcal{C}(X)$, X compact, $1_X \in H$, we assign the space $H_o^{\sim}$ of extensions of functions of H to the one-point compactification $\gamma X = X \cup \{\infty\}$ letting

$h(\infty) = 0$ at the isolated extra point $\infty$. We now have to be careful with the description of $\mathcal{K}$ and $\mathcal{K}_o$: in (4') and (5) we have to distinguish between the condition $h(x) \le \beta$ for $x$ in $X$ which characterizes $\mathcal{K}(H)$ and the condition $h(x) \le \beta$ for $x$ in $\gamma X$ which characterizes $\mathcal{K}_o(H_o^{\sim})$. Note that if we apply the construction (3) to the space $H_o^{\sim}$ we do not get the initial space H but a space $H^{\sim}$ containing the image of H as a subspace of codimension 1. The functor

(10) $$\mathcal{A}\circ\mathcal{K}_o : \mathbf{Bcaf}_o \longrightarrow \mathbf{Bcaf}$$

assigns to each space $\mathcal{A}_o(K;x_o)$ the corresponding space $\mathcal{A}(K)$ and assigns to each space $H_o \subseteq \mathcal{C}_o(X;x_o)$ the space H in (3). From Proposition 2 it follows that functor (10) is a left adjoint of $\square$ : $\mathbf{Bcaf} \longrightarrow \mathbf{Bcaf}_o$. On the other hand, the functor

(11) $$\mathcal{K}_o\circ\mathcal{A} : \mathbf{Cmpcnvx} \longrightarrow \mathbf{Cmpcnvx}_{\bullet}$$

may be described as follows: to each compact convex set K, identified with its canonical image $\hat{K} = \mathcal{K}(H)$, where $H = \mathcal{A}(K)$, we assign the set $\mathcal{K}_o(H_o^{\sim})$ which may be identified with the set

(12) $$\mathcal{K}_o(H) = \{\xi \in H^* : \xi \ge 0, \|\xi\| \le 1\} = \mathrm{conv}(0, \hat{K}),$$

i.e., with the intersection of the unit ball of $H^*$ with the cone generated by $\hat{K}$; the distinguished point 0 of $\mathcal{K}_o(H)$ is the vertex of the cone. It is clear that each affine function on $\hat{K}$ can be extended to a unique affine function on set (12).

**Symmetrization of compact convex sets.** We begin by discussing the relations between the category $\mathbf{Bcaf}_o$ and the category $\mathbf{Ban}_1$ of Banach spaces and linear contractions. There is an obvious forgetful functor

(13) $$\square : \mathbf{Bcaf}_o \longrightarrow \mathbf{Ban}_1.$$

On the other hand, each Banach space B is $\mathbf{Ban}_1$-isomorphic to a subspace $H_o$ of $\mathcal{C}_o(X;x_o)$, where $X$ is the weakly* compact unit ball $O^*B$ with the distinguished element 0. Then, by the Krein-Šmulyan theorem, the canonical map

(14) $$B \longrightarrow \mathcal{A}_o(O^*B;\ 0)$$

is a linear isometrical bijection. Moreover, it is clear that each $\mathbf{Ban}_1$-morphism $B \to \mathcal{A}_o(K;\ x_o)$ can be uniquely factored through map (14), and each $Ban_1$-morphism from $\mathcal{A}_o(O^*B;0)$ to $\mathcal{A}_o(K;x_o)$ is automatically a $Bcaf_o$-morphism (since condition $h(\xi) \le \beta$ for $\xi$ in $O^*B$ implies that $\beta \ge 0$ and is equivalent to $\|h\| \le \beta$; note that for each $h_o$ in $\mathcal{A}_o(O^*B;0)$ the numbers $\alpha, \beta$ defined by (9) satisfy $-\alpha = \beta$). Consequently, the functor

(15) $$\mathcal{A}_o \circ O^* : Ban_1 \longrightarrow Bcaf_o$$

is a left adjoint of (13).

Cmpsaks is the category of compact Saks spaces and continuous affine zero-preserving maps. A compact Saks space (or a $\gamma$-compact space with mixed topology, see [1]) is a pair $(S,\tau)$ where S is the unit ball of a Banach space E and $\tau$ is a locally compact Hausdorff topology on E such that $(S,\tau)$ is compact. Another description of such pairs $(S,\tau)$ (not referring to a norm) can be found in [7]: S is a convex symmetric subset of a vector space E, $\tau$ is a compact topology on S such that 0 has a fundamental system of convex symmetric neighborhoods, and the map $(x,y) \longmapsto \frac{1}{2}(x+y)$ is continuous.

The compositions $\mathcal{A} \circ O^*$ and $O^* \circ \mathcal{A}$ are naturally equivalent to the corresponding identities, hence Cmpsaks may be regarded as the category dual to $Ban_1$ ([4], p. 37).

If K is a compact convex set with a base point $x_o$ then the Saks space $O^*\mathcal{A}_o(K;x_o)$ is a compact convex symmetric set with base point 0. The canonical map

$$K \longrightarrow O^*\mathcal{A}_o(K;x_o)$$

is a homeomorphic affine embedding and the base point $x_o$ is sent to 0; $O^*\mathcal{A}_o(K;x_o)$ may be called the *symmetric hull* of K with respect to the point $x_o$ chosen as the center. A routine verification shows that

$$O^* \circ \mathcal{A}_o : Cmpcnvx_\bullet \longrightarrow Cmpsaks$$

is a left adjoint of the corresponding forgetful functor $\nabla$.

We now go back to our original question. Let $\ell^1(m)$ be a free Banach space. Acting with functor (15) we get the space

$$\mathcal{A}_o(O^*\ell^1(m)) \equiv \mathcal{A}_o(I^m) \equiv \ell^1(m).$$

Acting on the resulting space with functor (10) we get

$$\mathcal{A} \circ \mathcal{K}_o(\mathcal{A}_o(I^m)) \equiv \mathcal{A}(I^m) \equiv \ell^1(m+1).$$

Finally, acting with the functor $\mathscr{C}\circ\mathscr{K}$ : Bcaf $\longrightarrow$ Bcf we get the space

$$\mathscr{C}\circ\mathscr{K}\circ\mathscr{A}(I^m) \equiv \mathscr{C}(I^m).$$

Note that if we replaced $I = [-1,1]$ by $I_+ = [0,1]$, then the spaces $\mathscr{A}(I^m)$ and $\mathscr{A}(I_+^m)$ would be Bcaf-isomorphic and (2) would be valid, but the formula for the norm of $f$ would be different and isomorphism (1) would be less evident.

This paper is in final form and will not be published elsewhere.

## References

[1] A. Alexiewicz and Z. Semadeni, *A generalization of two norm spaces. Linear functionals*, Bull. Acad. Polon. Sci., série sci. math., astr. phys. 6 (1958), 135-139.

[2] N. Dunford and J.T. Schwartz, *Linear operators I. General theory*, New York 1958.

[3] R. R. Phelps, *A problem of Hewitt on restrictions of multiplicative linear functionals*, Studia Math. 25 (1964), 1-3.

[4] Z. Semadeni, *Projectivity, injectivity and duality*, Dissertationes Math. 25 (1963), 1-47.

[5] ———, *The Banach-Mazur functor and related functors*, Comment. Math. Prace Mat. 14 (1970), 173-182.

[6] ———, *Banach spaces of continuous functions*, vol. 1, Warsaw 1971.

[7] L. Waelbroeck, *Le complètè et le dual d'un espace localement convexe*, Bull. Soc. Math. Belg. 16 (1964), 393-406.

Address of the author:

Instytut Matematyczny
Uniwersytet Warszawski
PKiN, IX p.
00-901 Warszawa
Poland

# On the category of compact convex sets and related categories

by

Zbigniew Semadeni (Warsaw)

**Introduction.** It is known (see [6], #23.3.5) that $\mathcal{A}(I^{m}, I)$ is a free object with $m$ generators in the category Bcaf of spaces $\mathcal{A}(K, I)$ as objects and nonnegative linear unit-preserving maps as morphisms. Here $I = [-1,1]$, K is a compact convex set, $\mathcal{A}(K, K')$ denotes the space of all continuous affine maps $K \longrightarrow K'$, and $m$ is a cardinal number. On the other hand, it is easy to see that

$$(1) \qquad \mathcal{A}(I^{m}, R) \equiv l^{1}(m+1),$$

where $\equiv$ denotes an isometrical isomorphism of Banach spaces. The space $l^{1}(m+1)$ may be regarded as a free Banach space with $m+1$ generators.

The purpose of this paper is to discuss isomorphism (1) in the context of six categories: that of compact convex sets, that of compact convex sets with distinguished (base) points, that of symmetric compact convex sets, and the three dual categories. In particular, the role of the extra dimension (i.e., $m+1$ rather than $m$) is explained in terms of some canonical functors between those categories.

The author is indebted to prof. Dieter Pumplün for inspiring discussions and valuable comments.

**Preliminaries.** In this paper, all vector spaces are over the field R of reals. If K is a *compact convex set*, i.e., a compact convex subset of a locally convex Hausdorff topological vector space E, then $\mathcal{A}(K)$ will denote the space of all continuous affine functions $K \longrightarrow R$ (a map is called *affine* iff it preserves the convex combinations). $\mathcal{A}(K)$ is a closed subspace of the space $\mathcal{C}(K)$ of all continuous functions $K \longrightarrow \mathbf{R}$. The symbol Cmpcnvx will denote the category of compact convex sets and continuous affine maps. If $\phi : K \longrightarrow K'$ is a morphism in this category, then $\mathcal{A}(\phi) : \mathcal{A}(K') \longrightarrow \mathcal{A}(K)$ denotes the corresponding linear operator defined as $\mathcal{A}\phi.g = g \circ \phi$.

**Bcaf** denotes the category in which an object means any closed subspace $H$ of some space $\mathscr{C}(X)$, $X$ compact, such that the constant function $\mathbf{1}_X$ is in H, while a morphism from H to some $G \subseteq \mathscr{C}(Y)$ means a bounded linear operator $\Phi : H \longrightarrow G$ satisfying $\|\Phi\| \leq 1$ and $\Phi(\mathbf{1}_X) = \mathbf{1}_Y$. $\mathscr{K}(H)$ will denote the *state space* of K, i.e., the set

$$\mathscr{K}(H) = \{\xi \in H^* : \xi \geq 0,\ \xi(\mathbf{1}_X) = 1\},$$

which is convex and compact in the weak$^*$ topology. For each morphism $\Psi : H \longrightarrow G$ in **Bcaf**, $\mathscr{K}(\Psi)$ denotes the corresponding weakly$^*$ continuous affine map from $\mathscr{K}(G)$ to $\mathscr{K}(H)$. Both canonical maps

$$K \longrightarrow \mathscr{K}(\mathscr{A}(K)) \quad \text{and} \quad H \longrightarrow \mathscr{A}(\mathscr{K}(H))$$

are isomorphisms of the respective categories, which may be regarded as dual to each other ([6], #23.3.2).

If B is a Banach space, then $\mathrm{O}B$ and $\mathrm{O}^*B$ will denote the unit ball of B and of $B^*$, respectively, the latter being provided with the weak$^*$ topology.

The letter I stands for the interval $[-1,1]$ (i.e., the unit ball of R). $I^{\mathfrak{m}}$ denotes the product of $\mathfrak{m}$ copies of I with both the product topology and the obvious structure of a convex set, which make it a product in **Cmpcnvx**. Consequently, $\mathscr{A}(I^{\mathfrak{m}})$ is a coproduct of $\mathfrak{m}$ copies of $\mathscr{A}(I)$ in **Bcaf**. The cardinal number $\mathfrak{m}$ is understood in von Neumann's sense, i.e., $\mathfrak{m}$ is the same as the set of all ordinal numbers $\tau$ of cardinality less than $\mathfrak{m}$.

$\mathscr{A}(I^{\mathfrak{m}})$ may be regarded as a **Bcaf**-free object with $\mathfrak{m}$ generators, that is, there is canonical map $\pi$ from the set $\mathfrak{m}$ into

$$\mathscr{A}(I^{\mathfrak{m}}, I) = \{f \in \mathscr{A}(I^{\mathfrak{m}}) : \|f\| \leq 1\} = \mathrm{O}\mathscr{A}(I^{\mathfrak{m}})$$

satisfying the following condition: for each map $\xi$ from $\mathfrak{m}$ into any space $\mathscr{A}(K,I)$ there exists a unique **Bcaf**-morphism $\Theta : \mathscr{A}(I^{\mathfrak{m}}, I) \longrightarrow \mathscr{A}(K,I)$ such that $\Theta\pi = \xi$. Specifically, to each $\tau$ in $\mathfrak{m}$ we assign the coordinate projection $\pi_\tau : I^{\mathfrak{m}} \longrightarrow I$ regarded as an element of $\mathscr{A}(I^{\mathfrak{m}}, I)$.

**Lemma.** *Let* B *be a Banach space. Each* f *in* $\mathscr{A}(O^*B)$ *is of the form* $f(\xi) = \xi(b) + s$ *for* $\xi$ *in* $O^*B$, *where* $b \in B$ *and* $s = f(0) \in \mathbb{R}$. *Moreover,* $\|f\| = \|b\| + |s|$.

This follows from the Krein-Šmulyan theorem applied to the linear extension of the functional $g(\xi) = f(\xi) - f(0)$ (see e.g., [2], p. 434; cf. [6], #23.1.11(C))

**Proposition 1.** *Every* f *in* $\mathscr{A}(I^{\mathfrak{m}})$ *is of the form*

$$(2) \qquad f = s_\infty \cdot 1 + \sum_{\tau < \mathfrak{m}} s_\tau \pi_\tau,$$

*i.e.*, $f(x) = s_\infty + \sum s_\tau x_\tau$ *for* $x = (x_\tau)$ in $I^{\mathfrak{m}}$; *moreover,*

$$\|f\| = |s_\infty| + \sum |s_\tau|.$$

Proof. The cube $I^{\mathfrak{m}}$ may be identified with the unit ball in $l^\infty(\mathfrak{m})$ provided with the weak$^*$ topology of the space conjugate to $l^1(\mathfrak{m})$. Therefore the conclusion follows immediately from the preceding lemma. □

Isomorphism (1) is now defined as $\Gamma(f) = \left( s_\infty, (s_\tau)_{\tau<\mathfrak{m}} \right)$ for f of the form (2).

Both $\mathscr{A}(I^{\mathfrak{m}})$ and $l^1(\mathfrak{m}+1)$ have natural pointwise-defined vector-space orders. Isomorphism (1), however, is not order-preserving. Indeed, an element f in $\mathscr{A}(I^{\mathfrak{m}})$ is nonnegative if and only if $f(x) \geq 0$ for each x in $I^{\mathfrak{m}}$; consequently, condition $f \geq 0$ means that $s_\infty + \sum s_\tau x_\tau \geq 0$ for all $x = (x_\tau)$ satisfying $|x_\tau| \leq 1$, i.e., that

$$\sum |s_\tau| \leq s_\infty.$$

On the other hand, an element $s = (s_\infty, (s_\tau))$ in $l^1(\mathfrak{m}+1)$ is nonnegative iff $s_\infty \geq 0$ and $s_\tau \geq 0$ for $\tau$ in $\mathfrak{m}$.

Let us note the commutativity of the diagram:

$$\begin{array}{ccc}
\mathfrak{m} & \xrightarrow{\ \pi\ } & \mathscr{A}(I^{\mathfrak{m}}) \\
\Big\downarrow \delta & & \Big\downarrow \Gamma \\
l^1(\mathfrak{m}) & \xrightarrow{\ \Lambda\ } & l^1(\mathfrak{m}+1)
\end{array}$$

where $\delta_\tau$ is the $\tau$-th unit vector and $\Lambda$ is the obvious isometrical embedding (with zero as the extra coordinate).

**Compact convex sets with base points.** An object of the category $\mathbf{Cmpcnvx}_\bullet$ is a pair $(K; x_o)$ where K is a compact convex set and $x_o \in K$; a morphism from $(K; x_o)$ to $(K'; x'_o)$ is a continuous affine map $\varphi : K \longrightarrow K'$ such that $\varphi(x_o) = x'_o$.

$\mathcal{A}_o(K;x_o)$ will denote the space of all continuous affine maps $f : K \longrightarrow K'$ such that $f(x_o) = 0$. It is a subspace of $\mathcal{A}(K)$ and its codimension equals 1; indeed, each f in $\mathcal{A}(K)$ is of the form $f = f_o + c1_K$ where $f_o \in \mathcal{A}_o(K;x_o)$, $c = f(x_o)$ and $\mathbf{1}_K$ denotes the constant function 1 on K.

If $\varphi : (K;x_o) \longrightarrow (K';x'_o)$ is a morphism in $\mathbf{Cmpcnvx}_\bullet$ then

$$\mathcal{A}_o(\varphi) : \mathcal{A}_o(K';x'_o) \longrightarrow \mathcal{A}_o(K;x_o)$$

is the corresponding linear operator $\mathcal{A}_o(\varphi).g = g\circ\varphi$; thus, $\mathcal{A}_o(\varphi)$ is the restriction of $\mathcal{A}(\varphi)$ to the subspace $\mathcal{A}_o(K';x'_o)$. Note that the norm $\|\mathcal{A}_o(\varphi)\|$ may be smaller than 1. E.g., let $K = [0,1]$, $K' = [0,2]$, both with base point 0. The identical embedding $\varepsilon : K \longrightarrow K'$ is a morphism, $\mathcal{A}_o(\varepsilon)$ is the restriction operator $g \longmapsto g|_K$ and $\|\mathcal{A}_o(\varepsilon)\| = 1/2$.

By an object of the category $\mathbf{Bcaf}_o$ we mean any closed subspace $H_o$ of some space

$$\mathcal{C}_o(X;x_o) = \{ f \in \mathcal{C}(X) : f(x_o) = 0 \},$$

where X is compact and $x_o \in X$. One may also think of $H_o$ as a subspace of the space $\mathcal{C}_o(Y)$ of functions continuous on a locally compact space Y and vanishing at infinity; Y is compact if and only if the point at infinity is isolated. For each object $H_o$ of $\mathbf{Bcaf}_o$ the space

(3) $$H = \{ f_o + c1_X : f_o \in H_o, c \in \mathbf{R} \}$$

is an object of $\mathbf{Bcaf}$. By a $\mathbf{Bacf}_o$-morphism from a subspace $H_o$ of $\mathcal{C}_o(X;x_o)$ to a subspace $G_o$ of $\mathcal{C}_o(Y;y_o)$ we mean a linear operator $\Phi_o : H_o \longrightarrow G_o$ such that for each $\alpha$ and $\beta$ in R

(4) $$(\forall_{x\in X}\ \alpha \le h(x) \le \beta\ ) \Longrightarrow (\forall_{y\in Y}\alpha \le \Phi_o h.(y) \le \beta).$$

Note that (4) is equivalent to the condition:

$$(4') \qquad \forall_{\beta\in R}(\forall_{x\in X}\ h(x) \le \beta\ ) \Longrightarrow (\forall_{y\in Y}\ \Phi_o h.(y) \le \beta)$$

and is also equivalent to an analogous condition with $\alpha$ only. Bear in mind that $h(x_o) = 0$ implies that $\alpha \le 0$ and $\beta \ge 0$.

If $H_o$ is an object of $\mathrm{Bcaf}_o$, define

$$(5) \qquad \mathcal{K}_o(H_o) = \{\xi \in H_o^* : \forall_{\beta\in R}\ h \le \beta 1_X \Longrightarrow \xi(h) \le \beta\}.$$

Thus, $\mathcal{K}_o(H_o)$ is a convex weakly$^*$ compact subset of $H_o^*$. Condition $h(x_o) = 0$ implies that the functional $0$ belongs to $\mathcal{K}_o(H_o)$. However, conditions $\xi \ge 0$ and $\|\xi\| \le 1$ do not imply that $\xi \in \mathcal{K}_o(H_o)$. Let us also note that $R$ may be identified with $\mathcal{A}_o([0,1];0)$; hence $\mathcal{K}_o(H_o)$ may be regarded as the set of all $\mathrm{Bcaf}_o$-morphisms from $H_o$ to $R$. Let us consider the following examples, where $H = \mathcal{A}(K)$, $H_o = \mathcal{A}_o(K;x_o)$.

(a) Let $K = [0,1]$, $x_o = 0$. Then $\dim H = 2$, each $h$ in $H$ is of the form $h(x) = a(1-x) + bx$, each $\xi$ in $H^*$ is of the form $\xi(h) = \lambda a + \mu b$; such a $\xi$ is in $\mathcal{K}(H)$ if and only if $0 \le \lambda \le 1$ and $\mu = 1-\lambda$; hence $\mathcal{K}(H)$ is isomorphic to $[0,1]$. On the other hand, $\dim H_o = 1$, each $h_o$ in $H_o$ is of the form $h_o(x) = ax$, each $\xi$ in $\mathcal{K}_o(H_o)$ is of the form $\xi(h) = \lambda a$ with $0 \le \lambda \le 1$; hence $\mathcal{K}_o(H_o)$ is isomorphic to $([0,1];\ 0)$. A linear operator $\Phi_o$ on $H_o$ satisfies (4) if and only if $\|\Phi_o\| \le 1$ and $\Phi_o \ge 0$.

(b) Let $K$ be compact convex and *symmetric*, i.e., let $x \in K$ imply that $-x \in K$. If $h \in H_o$ and $x_o = 0$, then the condition $\alpha \le h(x) \le \beta$ is equivalent to $\|h\| \le \min(|\alpha|,\beta)$. No function in $H_o\setminus\{0\}$ is nonnegative; hence each linear operator on $H_o$ is nonnegative. A linear operator $\Phi_o$ on $H_o$ satisfies (4) if and only if $\|\Phi_o\| \le 1$. Each $\xi$ in $H_o^*$ such that $\|\xi\| \le 1$ is the evaluation at some point of $K$ and can be extended to a multiplicative linear functional on $\mathcal{C}(K)$ (Phelps [3]).

(c) Let $K = [-1/2,\ 1]$, $x_o = 0$. Then each $h$ in $H_o$ is of the form $h(x) = ax$ and each linear functional $\xi$ on $H_o$ is of the form $\xi(h) = \lambda a$. If the condition $\forall_{x\in K}\ h(x) \le \beta$ is to imply that $\lambda a \le \beta$, then $\lambda$ must be in $[-1/2,\ 1]$. For each $h$ in $H_o$ let $h'$ denote the affine extension of $h$ to $[-1,1]$. If

$\eta$ is any linear functional on $H_o$ with $\|\eta\| \le 1$ then there exists an $x_o$ in $[-1,1]$ such that $\eta(h) = h'(x_o)$ for h in $H_o$; $\eta$ belongs to $\mathcal{K}_o(H_o)$ if and only if $x_o \in [-1/2, 1]$.

**Proposition 2.** *Let* $1_Y \in G \subseteq \mathcal{C}(Y)$ *and let* $\Phi_o : H_o \longrightarrow G$ *be a linear operator satisfying condition (4). Then there exists a unique* Bcaf-*morphism* $\Phi : H \longrightarrow G$ *such that* $\Phi|_{H_o} = \Phi_o$. □

If we apply condition (4) to the case $H \subseteq \mathcal{C}(X)$, $G \subseteq \mathcal{C}(Y)$, $1_X \in H$, $1_Y \in G$, then it may serve as a description of morphisms in Bcaf, i.e., it is equivalent to $\Phi \ge 0$ and $\Phi(1_X) = 1_Y$. Consequently, Proposition 2 implies that each $\xi_o \in \mathcal{K}_o(H_o)$ can be extended to a unique $\xi$ in $\mathcal{K}(H)$. The operation of restriction $\xi \longmapsto \xi|_{H_o}$ gives the canonical bijection

(6) $$\Theta : \mathcal{K}(H) \longrightarrow \mathcal{K}_o(H_o).$$

Obviously, $\mathcal{A}_o$ is a contravariant functor from $\mathrm{Cmpcnvx}_\bullet$ to the category $\mathrm{Bcaf}_o$. Now, if $\Phi : H_o \longrightarrow G_o$ is a morphism in $\mathrm{Bcaf}_o$, let $\mathcal{K}_o(\Phi)$ denote the restriction of the adjoint operator $\Phi^* : G_o^* \longrightarrow H_o^*$ to the subset $\mathcal{K}_o(G_o)$ of $G_o^*$. It is clear that $\mathcal{K}_o$ is a contravariant functor from $\mathrm{Bcaf}_o$ to $\mathrm{Cmpcnvx}_\bullet$.

**Proposition 3.** *Let* $(K;x_o)$ *be an object in* $\mathrm{Cmpcnvx}_\bullet$ *and let* $H_o = \mathcal{A}_o(K;x_o)$. *If* $x \in K$, *let* $\hat{x}(h) = h(x)$ *for* h *in* $H_o$. *Then the map*

(7) $$\hat{\ } : K \longrightarrow \mathcal{K}_o(H_o)$$

*is a weakly** *continuous affine bijection. Moreover,* $\hat{x}_o = 0$.

Proof. The argument is routine. The space $\mathcal{A}_o(K;x_o)$ clearly separates the points of K and hence map (7) is one-to-one. We also claim that map (7) is onto. Indeed, let

$$Q = \{\hat{x} : x \in K\}.$$

Then Q is a compact convex subset of $H_o^*$. Suppose $\mathfrak{z} \in H_o^* \setminus Q$. Then there exists an h in $H_o$ such that $\hat{x}(h) \le 1$ for $\hat{x}$ in Q and $\mathfrak{z}(h) > 1$. Hence, $h(x) \le 1$ for x in K and $\mathfrak{z} \notin \mathcal{K}_o(H_o)$. □

**Proposition 4.** *Let* $H_o \subseteq \mathscr{C}_o(X;x_o)$, $K_o = \mathscr{K}_o(H_o)$. *If* $h_o \in H_o$, *let* $\hat{h}_o(\xi_o) = \xi_o(h_o)$ *for* $\xi_o$ *in* $K_o$. *Then the canonical map*

(8) $$\hat{\ }: H_o \longrightarrow \mathscr{A}_o(K_o;\, 0)$$

*is a linear isometrical order-preserving bijection.*

Proof. Map (8) is obviously linear, of norm $\le 1$, and non-negative. We shall now prove the following:

*for each* $h_o$ *in* $H_o$ *there exist* $\xi_o$, $\eta_o$ *in* $K_o$ *such that*

(9) $$\xi_o(h_o) = \alpha, \quad \eta_o(h_o) = \beta, \text{ where } \alpha = \inf_X h_o,\ \beta = \sup_X h_o.$$

Let H be given by (3). Then there exist $\xi,\eta$ in $\mathscr{K}(H)$ such that $\xi(h_o) = \alpha$, $\eta(h_o) = \beta$ (cf. [6], #23.2.5). It is now enough to let $\xi_o$, $\eta_o$ be restrictions of $\xi$, $\eta$ to $H_o$.

Consequently, map (8) is an isometry and $\hat{h} \ge 0$ implies that $h \ge 0$. It remains to be shown that (8) is onto. Let $\mathfrak{h}$ be in $\mathscr{A}_o(K_o;0)$, i.e., $\mathfrak{h}_o : K_o \to \mathbf{R}$ is affine, weakly$^*$ continuous and $\mathfrak{h}_o(0) = 0$. We claim that there exists an $h_o$ in $H_o$ such that $\mathfrak{h}_o = \hat{h}_o$. Define $\mathfrak{h} = \mathfrak{h}_o \circ \Theta$, where $\Theta$ is given by (6). Then $\mathfrak{h} \in \mathscr{A}(\mathscr{K}(H))$. Now, let $\hat{x}_o$ be the functional in $\mathscr{K}(H)$ defined as $\hat{x}_o(h) = h(x_o)$ for h in H. Then

$$\mathfrak{h}(\hat{x}_o) = \mathfrak{h}_o(\Theta(\hat{x}_o)) = \mathfrak{h}_o(0) = 0.$$

Moreover, there exists an h in H such that $\mathfrak{h}(\xi) = \xi(h)$ for h in H. We claim that $h \in H_o$. Indeed, $h(x_o) = \hat{x}_o(h) = \mathfrak{h}(\hat{x}_o) = 0$ and hence, if $\xi_o \in K_o$, then

$$\mathfrak{h}_o(\xi_o) = \mathfrak{h}_o(\Theta\xi) = \mathfrak{h}(\xi) = \xi(h) = \xi_o(h), \text{ i.e., } \mathfrak{h}_o = \hat{h}_o. \quad \square$$

**Corollary.** *The composite functors*

$$\mathscr{K}_o\mathscr{A}_o : \mathrm{Cmpcnvx}_{\bullet} \longrightarrow \mathrm{Cmpcnvx}_{\bullet} \quad \text{and} \quad \mathscr{A}_o\mathscr{K}_o : \mathrm{Bcaf}_o \longrightarrow \mathrm{Bcaf}_o$$

*are naturally equivalent to the corresponding identity functors.*

**Relations between categories Cmpcnvx and Cmpcnvx$_{\bullet}$.** Let us consider the following diagram of categories:

$$\begin{array}{ccccccc}
\mathrm{Ban}_1 & \underset{\square}{\overset{\mathcal{A}_0\circ O^*}{\rightleftarrows}} & \mathrm{Bcaf}_0 & \underset{\square}{\overset{\mathcal{A}\circ\mathcal{K}_0}{\rightleftarrows}} & \mathrm{Bacf} & \underset{\square}{\overset{\mathcal{C}\circ\mathcal{K}}{\rightleftarrows}} & \mathrm{Bcf} \\
O^* \downarrow\uparrow \mathcal{A}_0 & & \mathcal{K}_0 \downarrow\uparrow \mathcal{A}_0 & & \mathcal{K}\downarrow\uparrow\mathcal{A} & & \mathcal{X}\downarrow\uparrow\mathcal{C} \\
\mathrm{Cmpsaks} & \underset{O^*\circ\mathcal{A}_0}{\overset{\nabla}{\rightleftarrows}} & \mathrm{Cmpcnvx}_* & \underset{\mathcal{K}_0\circ\mathcal{A}}{\overset{\nabla}{\rightleftarrows}} & \mathrm{Cmpcnvx} & \underset{\mathcal{K}\circ\mathcal{C}}{\overset{\nabla}{\rightleftarrows}} & \mathrm{Cmp}
\end{array}$$

The right-hand square of this diagram is well known and will not be discussed here (see [5] and [6]). We only recall that Bcf denotes the category of spaces $\mathcal{C}(X)$ and unit-preserving Banach-algebra homomorphisms; $\mathcal{X}$ is the Gelfand functor (i.e., the maximal-ideal-space functor) to the category Cmp of compact spaces; $\mathcal{K}\circ\mathcal{C}$ is the Choquet-Bauer simplex functor.

The diagram is similar to that in [5]; the difference is that the pair Cmpcvx and $\mathrm{Bcaf}_0$ was absent there and we now put it in the right place.

Symbols $\square$ and $\nabla$ stand for the obvious forgetful functors; only the functor from Bcaf to $\mathrm{Bcaf}_0$ should be commented on. To each object $H \leq \mathcal{C}(X)$, $X$ compact, $1_X \in H$, we assign the space $\tilde{H}_0$ of extensions of functions of $H$ to the one-point compactification $\gamma X = X \cup \{\infty\}$ letting $h(\infty) = 0$ at the isolated extra point $\infty$. We now have to be careful with the description of $\mathcal{K}$ and $\mathcal{K}_0$: in (4') and (5) we have to distinguish between the condition $h(x) \leq \beta$ for $x$ in $X$ which characterizes $\mathcal{K}(H)$ and the condition $h(x) \leq \beta$ for $x$ in $\gamma X$ which characterizes $\mathcal{K}_0(\tilde{H}_0)$. Note that if we apply the construction (3) to the space $\tilde{H}_0$ we do not get the initial space H but a space $\tilde{H}$ containing the image of H as a subspace of codimension 1. The functor

(10) $$\mathcal{A}\circ\mathcal{K}_0 : \mathrm{Bcaf}_0 \longrightarrow \mathrm{Bcaf}$$

assigns to each space $\mathcal{A}_0(K;x_0)$ the corresponding space $\mathcal{A}(K)$ and assigns to each space $H_0 \leq \mathcal{C}_0(X;x_0)$ the space H in (3). From Proposition 2 it follows that functor (10) is a left adjoint of $\square$ : Bcaf $\longrightarrow$ $\mathrm{Bcaf}_0$. On the other hand, the functor

$$(11) \qquad \mathcal{K}_o \circ \mathcal{A} : \text{Cmpcnvx} \longrightarrow \text{Cmpcnvx}.$$

may be described as follows: to each compact convex set K, identified with its canonical image $\hat{K} = \mathcal{K}(H)$, where $H = \mathcal{A}(K)$, we assign the set $\mathcal{K}_o(\tilde{H}_o)$ which may be identified with the set

$$(12) \qquad \mathcal{K}_o(H) = \{\xi \in H^* : \xi \geq 0, \ \|\xi\| \leq 1\} = \text{conv}(0, \hat{K}),$$

i.e., with the intersection of the unit ball of $H^*$ with the cone generated by $\hat{K}$; the distinguished point 0 of $\mathcal{K}_o(H)$ is the vertex of the cone. It is clear that each affine function on $\hat{K}$ can be extended to a unique affine function on set (12).

**Symmetrization of compact convex sets.** We begin by discussing the relations between the category $\text{Bcaf}_o$ and the category $\text{Ban}_1$ of Banach spaces and linear contractions. There is an obvious forgetful functor

$$(13) \qquad \square : \text{Bcaf}_o \longrightarrow \text{Ban}_1 .$$

On the other hand, each Banach space B is $\text{Ban}_1$-isomorphic to a subspace $H_o$ of $\mathscr{C}_o(X;x_o)$, where X is the weakly$^*$ compact unit ball $O^*B$ with the distinguished element 0. Then, by the Krein-Šmulyan theorem, the canonical map

$$(14) \qquad B \longrightarrow \mathcal{A}_o(O^*B;\ 0)$$

is a linear isometrical bijection. Moreover, it is clear that each $\text{Ban}_1$-morphism $B \rightarrow \mathcal{A}_o(K;\ x_o)$ can be uniquely factored through map (14), and each $\text{Ban}_1$-morphism from $\mathcal{A}_o(O^*B;0)$ to $\mathcal{A}_o(K;x_o)$ is automatically a $\text{Bcaf}_o$-morphism (since condition $h(\xi) \leq \beta$ for $\xi$ in $O^*B$ implies that $\beta \geq 0$ and is equivalent to $\|h\| \leq \beta$; note that for each $h_o$ in $\mathcal{A}_o(O^*B;\ 0)$ the numbers $\alpha, \beta$ defined by (9) satisfy $-\alpha = \beta$). Consequently, the functor

$$(15) \qquad \mathcal{A}_o \circ O^* : \text{Ban}_1 \longrightarrow \text{Bcaf}_o$$

is a left adjoint of (13).

Cmpsaks is the category of compact Saks spaces and continuous affine zero-preserving maps. A compact Saks space (or a $\gamma$-compact space with mixed topology, see [1]) is a pair $(S,\tau)$ where S is the unit ball of a Banach space E and $\tau$ is a locally compact Hausdorff topology on E such that $(S,\tau)$ is compact. Another description of such pairs $(S,\tau)$ (not referring to a norm) can be found in [7]: $S$ is a convex symmetric subset of a vector space $E$, $\tau$ is a compact topology on $S$ such that 0 has a fundamental system of convex symmetric neighborhoods, and the map $(x,y) \longmapsto (x+y)/2$ is continuous.

The compositions $\mathcal{A}\circ\mathbf{O}^*$ and $\mathbf{O}^*\circ\mathcal{A}$ are naturally equivalent to the corresponding identities, hence Cmpsaks may be regarded as the category dual to $\mathrm{Ban}_1$ ([4], p. 37).

If $K$ is a compact convex set with a base point $x_o$ then the Saks space $\mathbf{O}^*\mathcal{A}_o(K;x_o)$ is a compact convex symmetric set with base point 0. The canonical map

$$K \longrightarrow \mathbf{O}^*\mathcal{A}_o(K;x_o)$$

is a homeomorphic affine embedding and the base point $x_o$ is sent to 0; $\mathbf{O}^*\mathcal{A}_o(K;x_o)$ may be called the *symmetric hull* of $K$ with respect to the point $x_o$ chosen as the center. A routine verification shows that

$$\mathbf{O}^*\circ\mathcal{A}_o : \mathrm{Cmpcnvx}_\bullet \longrightarrow \mathrm{Cmpsaks}$$

is a left adjoint of the corresponding forgetful functor $\nabla$.

We now go back to our original question. Let $l^1(m)$ be a free Banach space. Acting with functor (15) we get the space

$$\mathcal{A}_o(\mathbf{O}^* l^1(m)) \equiv \mathcal{A}_o(I^m) \equiv l^1(m).$$

Acting on the resulting space with functor (10) we get

$$\mathcal{A}\circ\mathcal{K}_o(\mathcal{A}_o(I^m)) \equiv \mathcal{A}(I^m) \equiv l^1(m+1).$$

Finally, acting with the functor $\mathcal{C}\circ\mathcal{K}$ : Bcaf $\longrightarrow$ Bcf we get the space

$$\mathcal{C}\circ\mathcal{K}\circ\mathcal{A}(I^m) \equiv \mathcal{C}(I^m).$$

Note that if we replaced $I = [-1,1]$ by $I_+ = [0,1]$, then the spaces $\mathcal{A}(I^m)$ and $\mathcal{A}(I_+^m)$ would be Bcaf-isomorphic and (2) would be valid, but the formula for the norm of $f$ would be different and isomorphism (1) would be less evident.

## References


[1] A. Alexiewicz and Z. Semadeni, *A generalization of two norm spaces. Linear functionals*, Bull. Acad. Polon. Sci. série sci. math., astr. phys. 6 (1958), 135-139.

[2] N. Dunford and J. T. Schwartz, *Linear operators I. General theory*, New York 1958.

[3] R. R. Phelps, *A problem of Hewitt on restrictions of multiplicative linear functionals*, Studia Math. 25 (1964), 1-3.

[4] Z. Semadeni, *Projectivity, injectivity and duality*, Dissertationes Math. 25 (1963), 1-47.

[5] ———, *The Banach-Mazur functor and related functors*, Comment. Math. Prace Mat. 14 (1970), 173-182.

[6] ———, *Banach spaces of continuous functions*, vol. 1, Warsaw 1971.

[7] L. Waelbroeck, *Le complété et le dual d'un espace localement convexe*, Bull. Soc. Math. Belg. 16 (1964), 393-406.



Address of the author:

Instytut Matematyczny
Uniwersytet Warszawski
PKiN, IX p.
00-901 Warszawa
Poland


The paper is in final form and no similar paper has been or is being submitted elsewhere.

# FILLERS FOR NERVES

**Ross Street**
**Macquarie University**
**New South Wales 2109**
**Australia**

An adjunction between the category of $\omega$-categories and the category of simplicial sets was described in [S]. To each $\omega$-category $A$, the right adjoint assigns a simplicial set $\Delta A$ called *the nerve of* $A$. An element $f$ of $\Delta A$ of dimension $n$ is an "$n$-simplex drawn in $A$": each $m$-dimensional face simplex amounts to an $m$-cell in $A$ with source and target obtained by appropriate pasting of the lower dimensional face cells.

Nerves are generally not Kan. It is a strong condition on an $\omega$-category to ask all horns in the nerve to have fillers. Yet certain horns do have fillers. For example, the horn on the left below is filled by the triangle on its right. Notice furthermore that the 2-cell in the triangle is an identity and as such is unique.

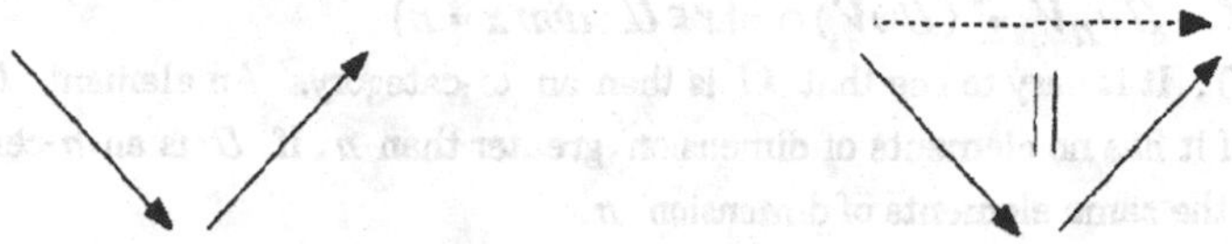

The nerve supports an extra structure which cannot be captured purely from the simplicial set. There are certain distinguished elements called *hollow*. An element $f$ is hollow when the cell of the top dimensional face is an identity (that is, is a lower dimensional cell). For example, the triangle above is a hollow element of dimension 2. In [S], certain horns, called *admissible*, were defined in terms of this hollowness structure.

The purpose of this paper is to prove that every admissible horn in the nerve of an $\omega$-category has a unique hollow filler. The question of characterizing nerves in terms of filler conditions ( there are conjectures of **John E. Roberts** and of the author [S]) will be addressed thoroughly in the forthcoming thesis of **Michael Zaks** (Macquarie University). Of course, when every horn (admissible or not) in the nerve has a unique hollow filler, every cell in the $\omega$-category must be invertible; this situation was analysed by **M.K. Dakin** [D] who used the word "thin" instead of "hollow".

Let $\omega$ denote the ordered set $\{0,1,2,...\}$ of finite ordinals.

Recall from [S] that an *$\omega$-category* $A$ is a set equipped with, for each $n \in \omega$, unary operations $s_n$, $t_n$, and a binary operation $*_n$ for which $a *_n b$ is defined precisely when $t_n(a) = s_n(b)$, satisfying a list of equational axioms (which we omit ). An *$n$-cell* of $A$ is an element $a$ with $s_n(a) = a$. An *$\omega$-functor* $f: A \to B$ is a function which preserves the operations.

Let $\Omega$ denote the set of finite subsets $x$ of $\omega$. Each $x \in \Omega$ orders itself as $x = (x_0, x_1, ..., x_n)$ with $x_0 < x_1 < ... < x_n$. The *dimension* $n$ of $x$ is one less than its cardinality

$\# x$. Write $x\partial_j$ for the set obtained from $x$ by deleting $x_j$.

Suppose $S$ is a subset of $\Omega$. Put $\neg S = \{x \in \Omega : x$ is not in $S\}$,

$S^+ = \{x\partial_i : x \in S$ and $i$ is odd$\}$, and $S^- = \{x\partial_i : x \in S$ and $i$ is even$\}$.

Define $S$ to be *well formed* when it contains at most one element of dimension 0 and, if $x, y \in S$ with $x\partial_i = y\partial_j$ and $i, j$ of the same parity, then $x = y$.

Define $\mathcal{O}$ to be the set of pairs $(\mathcal{U}_+, \mathcal{U}_-)$ of non-empty finite well-formed subsets $\mathcal{U}_+$, $\mathcal{U}_-$ of $\Omega$ satisfying the condition

$$\mathcal{U}_\varepsilon = (\mathcal{U}_\eta \cup \mathcal{U}_\zeta^{\varepsilon}) \cap \neg\,\mathcal{U}_\zeta^{\eta}$$

for all choices $\varepsilon, \eta, \zeta$ of signs $+, -$ with $\varepsilon, \eta$ opposite. We identify each such pair with the union

$$U = \mathcal{U}_+ \cup \mathcal{U}_-$$

since we can regain the pair of subsets using the equations

$$\mathcal{U}_+ = U \cap \neg U^- \quad \text{and} \quad \mathcal{U}_- = U \cap \neg U^+.$$

Define source and target operations by the equations

$$s_n(U) = \{x \in U : \dim x < n\} \cup \{x \in \mathcal{U}_+ : \dim x = n\},$$

$$t_n(U) = \{x \in U : \dim x < n\} \cup \{x \in \mathcal{U}_- : \dim x = n\}.$$

It is proved in [S] that $\mathcal{O}$ is closed under the operation

$$U *_n V = (U \cup V) \cap \neg\{x \in \mathcal{U}_- : \dim x = n\}$$

for $t_n(U) = s_n(V)$. It is easy to see that $\mathcal{O}$ is then an $\omega$-category. An element $U$ is an $n$-cell if and only if it has no elements of dimension greater than $n$. If $U$ is an $n$-cell, notice that $\mathcal{U}_+$, $\mathcal{U}_-$ have the same elements of dimension $n$.

For $x \in \Omega$ and $i \in \omega$, put $x/i = \{j \in x : j \leq i\}$. For $z \in \Omega$, it is fairly easy to see that the sets

$$\langle z \rangle_+ = \{x \in \Omega : i \in z \text{ and } i \notin x \text{ imply } \# x/i \text{ odd}\},$$

$$\langle z \rangle_- = \{x \in \Omega : i \in z \text{ and } i \notin x \text{ imply } \# x/i \text{ even}\}$$

constitute an element $\langle z \rangle$ of $\mathcal{O}$.

The main theorem of [S] is that $\mathcal{O}$ *is the free $\omega$-category generated by the elements $\langle z \rangle$ for $z \in \Omega$*: that is, an $\omega$-functor $f : \mathcal{O} \to A$ is uniquely determined by prescribing each $f(\langle z \rangle) \in A$ recursively subject to compatibility with $s_{n-1}$, $t_{n-1}$ for $z$ of dimension $n$.

**Henceforth, let $n$ and $k$ be fixed positive integers with $0 < k < n$.**

For brevity, put $N = \langle (0,1,\ldots,n) \rangle$ and $K = \langle (0,1,\ldots,k-1,k+1,\ldots,n) \rangle$.

Put $M\eta = \{x \in \Omega : k-1, k, k+1 \in x$ and $x \cap \neg\{k-1, k, k+1\} \in N_\eta\}$ for $\eta$ both $+$ and $-$. Let $M = M+ \cup M-$.

**Combinatorial Lemma.** $(N_\eta \cup (M\eta)^{\varepsilon}) \cap \neg(M\eta)^{\eta} = M \cup K_\eta$ *where* $\eta$ *and* $\varepsilon$ *are opposite signs.*

The proof of this result is postponed until the end of the paper.

**Proposition.** *If $k$ is odd, there is a decomposition*

$$s_{n-1}(N) = L(n-1) *_{n-2} \ldots *_1 L(1) *_0 K *_0 R(1) *_1 \ldots *_{n-2} R(n-1)$$

*in the $\omega$-category $\mathcal{O}$ in which $L(r)$, $R(r)$ are $r$-cells whose elements of dimension $r$ are the elements of dimension $r$ in $M+$, $M-$, respectively. If $k$ is even, there is a similar decomposition of $t_{n-1}(N)$.*

**Proof.** Consider the case of $k$ odd. For $1 \le r \le n$, define $r$-cells $L(r)$, $R(r)$ and an $(n-1)$-cell $K(r)$ as follows:

for each of these the elements of dimension less than $r-1$ agree with those of $N$,

the elements of dimension $r-1$ in $L(r)_+$ are those of $N_+$, in $L(r)_-$ and in $K(r)_+$ are those of $(N_+ \cup (M+)^-) \cap \neg(M-)^+$, in $R(r)_+$ and in $K(r)_-$ are those of $(N_- \cup (M-)^+) \cap \neg(M+)^-$, and in $R(r)_-$ are those of $N_-$,

the elements of dimension $r$ in $L(r)$ are those of $M+$,

the elements of dimension $r$ in $R(r)$ are those of $M-$, and

the elements of dimension greater than $r-1$ in $K(r)$ are those of $K$.

Then $K(n) = s_{n-1}(N)$ and $K(1) = K$. By the Combinatorial Lemma,

$$K(r+1) = L(r) *_{r-1} K(r) *_{r-1} R(r).$$

Substituting these equations into each other, we obtain the desired decomposition.□

Let $\Omega_n$ denote the set of elements of $\Omega$ of dimension no greater than $n$. Let $O_n$ denote the sub-$\omega$-category of $O$ consisting of those $U$ which are subsets of $\Omega_n$. Recall from [S] that *the nerve of an $\omega$-category $A$* is the simplicial set whose elements $f$ of dimension $n$ are the $\omega$-functors from $O_n$ to $A$ and whose face and degeneracy functions are given in the obvious way. An element $f$ is called *hollow* when it takes $N$ to an $(n-1)$-cell.

Write $D$ for the simplicial set $\Delta[n]$ whose elements of dimension $r$ are the order-preserving functions from $(0,1,\ldots,r)$ to $(0,1,\ldots,n)$. Write $H$ for the *k-horn* $\Lambda^k[n]$ which is the sub-simplicial set of $D$ consisting of the functions whose image does not contain all of $0,1,\ldots,k-1,k+1,\ldots,n$. Define the *hollow* elements of $H$ to be those functions whose images are in $M$.

For a simplicial set $C$, simplicial maps $f$ from $D$ to $C$ are in natural bijection with elements of $C$ of dimension $n$, while simplicial maps $g$ from $H$ to $C$ are in natural bijection with $k$-horns in $C$ of dimension $n-1$. Call an element of $C$ *a filler for a horn* when the corresponding $f$ restricts to the corresponding $g$. For a simplicial set $C$ with hollowness, a $k$-horn is called *admissible* when the corresponding map $g$ preserves hollowness.

**Theorem.** *In the nerve of an $\omega$-category, each admissible $k$-horn has a unique hollow filler.*

**Proof.** Let $\mathcal{K}$ denote the sub-$(n-1)$-category of $O_n$ generated by all the $(n-1)$-cells of the form $\langle(0,1,\ldots,i-1,i+1,\ldots,n)\rangle$ for $i \ne k$, and all the $m$-cells for $m < n-1$. A $k$-horn $H \to \Delta A$ in the nerve of an $\omega$-category $A$ amounts to an $(n-1)$-functor $f: \mathcal{K} \to A$. The $k$-horn is admissible precisely when, for all $m < n$, the $(n-1)$-functor $f$ takes each $m$-cell $\langle x\rangle$ with $x \in M$ to an $(m-1)$-cell in $A$. A filler for the $k$-horn amounts to an $n$-functor $g$: $O_n \to A$ extending $f$. There will be a unique hollow filler provided we can find such a $g$ satisfying

$$(*) \qquad g s_{n-1}(N) = g t_{n-1}(N).$$

Notice that $t_{n-1}(N)$, $L(r)$, $R(r)$ are all in $\mathcal{K}$. Since all $r$-dimensional elements of $L(r)$ are in $M$, the decomposition of $L(r)$ into generators involves only $r$-cells $\langle x\rangle$ with $x$ in $M$

and other lower-dimensional cells; so $fL(r)$ is an $(r-1)$-cell. In order to have equation $(*)$, the Proposition forces us to take

$$gK = ft_{n-1}(N).$$

As this is consistent with $s_{n-2}$, $t_{n-2}$, it uniquely defines $g$ as a hollow element of the nerve of $A$ of dimension $n$ and extending $f$. □

**Proof of Combinatorial Lemma.** We shall use the fact that the elements of $M+$ are those $x \in \Omega$ satisfying the conditions

(a) $k-1, k, k+1 \in x$,

(b) $\# x/k$ is odd, and,

(c) $m \in (0,1,...,n) \cap \neg x$ implies $\# x/m$ odd for $m < k-1$ and $\# x/m$ even for $m > k+1$.

A similar description holds for $M-$: just interchange odd and even in the above.

The proof breaks up into the following eight parts.

(1) $(M+)^+ \cap M+ = \emptyset$.

Suppose $y \in M+$ and $x = y\partial_i$ with $i$ odd. If $y_i < k-1$ then $\# x/k = (\# y/k)-1$ which is even; so (b) fails for $x$. If $y_i = k-1$, $k$, or $k+1$ then (a) fails for $x$. If $y_i > k+1$ then $\# x/y_i = i$ which is odd; so (c) fails for $x$. Hence $x$ is not in $M+$.

(2) $(M+)^+ \cap M- = \emptyset$.

The proof is similar to that of (1).

(3) $(M+)^+ \cap K_+ = \emptyset$.

If $y \in M+$ and $x = y\partial_i \in K_+$ then $y_i = k$ and, since $i+1 = \# y/k$ is odd, $i$ is even. So $x$ is not in $(M+)^+$.

(4) $M+ \subseteq N_+ \cup (M+)^-$.

Take $x \in M+ \cap \neg N_+$. Then there exists a largest $m > k+1$ with $m \in (0,1,...,n) \cap \neg x$. Put $y = x \cup \{m\}$. Since $\# x/m$ is even, $x = y\partial_i$ with $i$ even. The choice of $m$ ensures $x/j = y/j$ for all $j < m$; this together with $x \in M+$ implies $y \in M+$. So $x \in (M+)^-$.

(5) $K_+ \subseteq N_+ \cup (M+)^-$.

Take $x \in K_+$. Then $k$ is not in $x$. If $\# x/k$ is odd then $x \in N_+$. So suppose $\# x/k$ is even. Then $x = y\partial_i$ with $i$ even and $y_i = k$. If $k-1$ is not in $x$ then $\# x/(k-1) = \# x/k$ is odd, contrary to $\# x/k$ even. So $k-1$ is in $x$. Similarly, $k+1$ is in $x$. So $k-1$, $k$, $k+1$ are in $y$. So $y \in M+$. So $x \in (M+)^-$.

(6) $M- \subseteq N_+ \cup (M+)^-$.

The proof of this is similar to that of (4).

(7) $N_+ \subseteq M+ \cup K_+ \cup (M+)^+$.

Take $x \in N_+$ so that $\# x/m$ is odd if $m \in (0,1,...,n) \cap \neg x$.

Consider the case where $k-1$, $k$, $k+1$ are all in $x$. If $x$ is not in $M+$ then the first $m > k+1$ with $m \in (0,1,..., n) \cap \neg x$ has $\# x/m$ odd. Put $y = x \cup \{m\} \in M+$ so that $y\partial_i = x$, $i$ is odd, and $y_i = m$. Thus $x \in (M+)_+$.

If $k$ is not in $x$ then $x$ must be in $K_+$.

If $k$ is in $x$ but $k-1$ is not then $\# x/k+1 = \# x/k$ is even so $k+1$ is in $x$ (using $x \in N_+$). So $y = x \cup \{k-1\} \in M+$ and $x = y\partial_i \in (M+)^+$.

The final case where $k$ is in $x$ but $k+1$ is not is similar to the last yielding $x \in (M+)^-$.

(8) $(M+)^- \subseteq K_+ \cup M \cup (M+)^+$.

Take $x$ in the left-hand side so that $x = y\partial_i$ with $i$ even and $y \in M+$.

Consider the case where $k$ is not in $x$. Then $k$ is in $y$. So $y_i = k$. So $k-1$, $k+1 \in x$. If $m < k-1$ then $\#x/m = \#y/m$ which is odd. If $m > k+1$ then $\#(x/m)+1 = \#y/m$ which is even so again $\#x/m$ is odd. So $x \in K_+$.

So we may suppose $k \in x$. Put $m = y_i$. Then $m$ is not in $x$ and $\#x/m$ is even. Certainly $x$ is not in $K_+$. The cardinality of $x/p$ is $\#y/p$ for $p < m$ and is $\#(y/p)-1$ for $p \geq m$.

Suppose $m < k-1$. Then $\#x/m$ is even and $m \in (0,1,...,n) \cap \neg x$; so $x$ is not in $M+$. Suppose $x$ is not in $M-$. Then there is a $p \in (0,1,...,n) \cap \neg x$ with either $p < k-1$ and $\#x/p$ odd or $p > k+1$ and $\#x/p$ even. If $m \leq p < k-1$ then $\#x/p$ is even; a contradiction. If $p > k+1$ then $\#x/p = \#(y/p)-1$ is odd; a contradiction. So the only possibility is $p < m$. Consider the largest $p \in (0,1,..., n)$ not in $x$, $p < m$, and $\#x/p$ odd. Put $z = x \cup \{p\}$ which is in $M+$ since $y$ is. So $x \in z^+ \subseteq (M+)^+$.

The case where $m > k+1$ is similar to the last paragraph.

Suppose $m = k-1$. Then $\#y/k = \#y/(k-1)+1 = \#x/(k-1)$ which is even, contrary to $y \in M+$.

Similarly the case $m = k+1$ is impossible. □

## References


[S] Ross Street, The algebra of oriented simplexes, *Journal of Pure and Applied Algebra 49* (1987)283-335.

[D] M.K. Dakin, *Kan Complexes and Multiple Groupoid Structures*, PhD Thesis, University of Wales, 1974.


THIS PAPER IS IN FINAL FORM AND WILL NOT BE PUBLISHED ELSEWHERE.

# SIMULTANEOUS REPRESENTATIONS IN CATEGORIES

++++++++++++++++++++++++++++++++++++++++++++++

Věra TRNKOVÁ

## I. Introduction

Consider the following three statements apparently comming from different parts of mathematics.

Combinatorics. For every pair of monoids $M_0 \subseteq M_1$ there is an undirected graph G such that $M_0$ is isomorphic to the endomorphism monoid of G and $M_1$ is isomorphic to the endomorphism monoid of $G_\Delta$, where $G_\Delta$ is obtained from G by the omitting of all its edges, which are not in a triangle (proved here).

Topology. For every cardinal number m there is a rigid Tichonov space X with a *stiff* collection $\{Y_i \mid i \in m\}$ such that

$$X \subseteq Y_i \subseteq \beta X \quad \text{for all } i \in m$$

(stiff in the sense that if $f : Y_i \to Y_j$ is a continuous map, then either f is constant or i = j and f is the identity), where $\beta X$ denotes the $\beta$-compactification of X (proved in [ 24 ]).

Algebra. A submonoid $M_0$ of a monoid $M_1$ is a *dominion* in $M_1$ (for the definition of the dominion, see [12 ]) iff there exists an algebra with three unary operations, say $A = (X,\alpha,\beta,\gamma)$, such that $M_0$ is isomorphic to End A and the isomorphism can be extended to $M_1 \simeq \mathrm{End}(X,\alpha,\beta)$ (proved in [ 16 ]).

These statements have a common source in, and indeed are just corollaries of, general theorems about the simultaneous representation.

The notion of simultaneous representation is based on full and almost full embeddings of categories. Let us recall that a functor $\Phi : K \to H$ is called a *full embedding* if it is one-to-one and full. It is called an *almost full embedding* if it is one-to-one, $H$ is a concrete category (over the category of sets) and $\Phi$ is "full up to constant morphisms", i.e. if $m : a \to b$ is an $K$-morphism, then $\Phi(m)$ is a non-constant morphism from $\Phi(a)$ into $\Phi(b)$ and, vice versa, if $g : \Phi(a) \to \Phi(b)$ is a non-constant $H$-morphism, then $g = \Phi(m)$ for some (unique !) $K$-morphism $m : a \to b$.

The investigation of full embeddings as a generalization of the classical representation problems was started by J.R. Isbell in [11 ] and then developed by a number of mathematicians. From the Czech ones, let us name at least the oldest generation Z. Hedrlín, A. Pultr, V. Trnková, P. Vopěnka ; from the younger ones at least V. Koubek, L. Kučera, J. Rosický, J. Sichler. Also M. Adams, E. Fried, G. Grätzer, J. Kollár, J. Lambek, R.H. Mc Dowell, E. Mendelshon and others participated in this field of problems. More complete references are given in the monograph [ 18 ], where the results about full and almost full embeddings are presented in a systematic way (however, we quote here the papers, where the result just mentioned appears the first time). Let us recall that every small category can be fully embedded into the category Graph of all directed graphs and all compatible maps and into the category Alg(1,1) of all universal algebras with two unary operations and all homomorphisms, by [ 8 ].

These two categories further can be fully embedded in a number of categories, e.g. in the category Und Graph ofall undirected graphs ([ 7 ]), Smg of all semigroups ([ 9]), Rings of all rings ([ 5]), IntDom of all integral domains ([ 6]), so that we obtain a representation of any small category as a full subcategory of Und Graph, Smg, Rings IntDom and so on.

Also, the category Graph can be *almost* fully embedded in the category of metrizable spaces and all continuous maps ([ 20]) and the oposite category $\mathrm{Graph}^{op}$ of can be almost fully embedded in the category of compact Hausdorff spaces ([ 21 ]). This gives a topological representation of all small categories.

The investigation of full (and almost full) embeddings is, roughly speaking, the studying "how structures select morphisms". The simultaneous representations investigated here describe "how constructions, performed on structures, influence the selecting of morphisms". Let us explain this on the combinatorial assertion : let $G = (V,E)$ be an undirected graph (i.e. $V$ is the set of its vertices, $E$ is the set of its edges ; we suppose that never $(x,x) \in E$), let $G_\Delta$ be the graph obtained from it by the omitting of all edges which are not in a triangle, i.e. $G_\Delta = (V,E_\Delta)$, where $E_\Delta = \{(x,y) \in E \mid$ there exists $z \in V$ such that $(x,z) \in E$ and $(y,z) \in E\}$. If $f$ is an endomorphism of $G$, i.e. $f$ maps $V$ into itself and $(x,y) \in E \Rightarrow (f(x),f(y)) \in E$, then $f$ is also an endomorphism of $G_\Delta$ evidently. Hence the endomorphism monoid End $G$ is always a submonoid of the endomorphism monoid End $G_\Delta$. Is there some other connection (possible non-trivial) between End $G$ and End $G_\Delta$ ? By our result, the answer is negative, every pair of monoids $M_0 \subsetneq M_1$ can be obtained in this way.

Our approach based on full and almost full embeddings makes it possible to obtain also other applications than the representability of monoids, as it is shown e.g. in the topological statement.

Let us present the *Main Definition* : Let D be a diagram scheme (i.e. a small category ; it is always finite here), let $\mathcal{C}$ and $\mathcal{D}$ be diagrams over D such that

for every $\sigma \in$ obj D, $\mathcal{C}(\sigma)$ and $\mathcal{D}(\sigma)$ are categories (not necessarily small),

for every $m \in$ morph D, $\mathcal{C}(m)$ and $\mathcal{D}(m)$ are functors.

A natural transformation

$$\Phi = \{\Phi_\sigma \mid \sigma \in \operatorname{obj} D\} : \mathcal{C} \to \mathcal{D}$$

is called a *simultaneous representation of* $\mathcal{C}$ *in* $\mathcal{D}$ if $\Phi_\sigma : \mathcal{C}(\sigma) \to \mathcal{D}(\sigma)$ is

*an almost full embedding* whenever $\mathcal{D}(\sigma)$ is a concrete category with constants

and

*a full embedding* else.

In our field of problems, we suppose that the diagram $\mathcal{D}$ is given and it describes some natural construction. We investigate which diagrams $\mathcal{C}$ in

the category $\mathbb{C}$at of all *small* categories

have simultaneous representations in the given $\mathcal{D}$. In the discussed example, D consists of two objects and one nonidentical arrow, say

$$D : \sigma_1 \xrightarrow{m} \sigma_2 , \tag{1}$$

$\mathcal{D}(\sigma_1)$ is the category Und Graph of all undirected graphs, $\mathcal{D}(\sigma_2)$ is its full subcategory Und $\text{Graph}_\Delta$ consisting of those graphs in which each edge is contained in a triangle and $\mathcal{D}(m) = F_\Delta$ is the coreflector which forgets the edges which are in no triangle. Which diagrams $\mathcal{C}$ in $\mathbb{C}$at

$$\mathcal{C} : k_0 \xrightarrow{\psi} k_1 \tag{2}$$

have simultaneous representations in

$$\mathcal{D} : \text{Und Graph} \xrightarrow{F_\Delta} \text{Und Graph}_\Delta \ ? \tag{3}$$

This is fully solved in the part III.

Let us describe briefly the content of this paper. In II., we formulate and prove four necessary conditions for the existence of simultaneous representations. We present here also a survey of the known results whether the necessary condition in question is or is not also sufficient. Let us notice that the proofs of the sufficiency are always more involved than the simple proofs of the necessary conditions. We show such proofs in the part III. As already mentioned, we give here a full characterization of the diagrams(2) in $\mathbb{C}$at, which have a simultaneous representation in (3). We also characterize here all the diagrams in $\mathbb{C}$at

$$\mathcal{C} : k_0 \begin{array}{l} \xrightarrow{\psi_1} k_1 \\ \xrightarrow{\psi_2} k_2 \end{array}$$

which have simultaneous representations in

$$\mathcal{D} : \text{Bitop} \begin{array}{l} \xrightarrow{F_1} \text{Top} \\ \xrightarrow{F_2} \text{Top} \end{array}$$

where $F_1$ forgets the first topology of a bitopological space and $F_2$ forgets the second one. The simultaneous representations in the diagrams

$$\text{Bi Graph} \begin{array}{l} \xrightarrow{F_1} \text{Graph} \\ \xrightarrow{F_2} \text{Graph} \end{array} \qquad \text{and} \qquad \text{Bi Und Graph} \begin{array}{l} \xrightarrow{F_1} \text{Und Graph} \\ \xrightarrow{F_2} \text{Und Graph} \end{array}$$

are also fully clarified (where $F_1$ again forgets the first graph structure and $F_2$ the second one - the directed graph structures in the former diagram and the undirected ones in the later diagram).

In the part II, several open problems are stated.

## II. The necessary conditions

Let us present four necessary conditions for the existence of a simultaneous representation. In all of them, $\mathcal{C}$ and $\mathcal{D}$ are diagrams of categories and functors, both over a scheme D. (The conditions (FC) and (SC) are also in [ 23 ].)

1. The faithfulness condition (FC). Let there be a simultaneous representation $\Phi = \{\Phi_\sigma \mid \sigma \in \text{obj } D\} : \mathcal{C} \to \mathcal{D}$. If $m : \sigma \to \sigma'$ is a D-morphism such that $\mathcal{D}(m)$ is faithful, then $\mathcal{C}(m)$ is also faithful.

**Proof**. Since $\Phi_\sigma$ is one-to-one, $\mathcal{D}(m) \circ \Phi_\sigma$ is faithful. Hence $\Phi_{\sigma'} \circ \mathcal{C}(m) = \mathcal{D}(m) \circ \Phi_\sigma$ is faithful so that $\mathcal{C}(m)$ must be faithful.

This easy necessary condition is often also sufficient, we mention several examples. Let us denote by Metr (or Unif or Tich or Comp) the category of all metric spaces and all nonexpanding maps (or of all separated uniform spaces and all uniformly continuous maps or of all Tichonov spaces and all continuous maps or of all compact Hausdorff spaces and all continuous maps). Then a diagram in $\mathbb{C}$at

$$k_1 \xrightarrow{\psi_1} k_2 \xrightarrow{\psi_2} k_3 \tag{4}$$

has a simultaneous representation in the diagram

$$\text{Metr} \xrightarrow{F_1} \text{Unif} \xrightarrow{F_2} \text{Tich}$$

(where $F_1, F_2$ are the forgetful functors) iff the functors $\psi_1, \psi_2$ are faithful. They must be faithful, by (FC), because the functors $F_1$, $F_2$ are faithful. The proof that (FC) is also sufficient is given in [23].

Analogously, a diagram (4) in $\mathbb{C}$at has a simultaneous representation $\Phi = \{\Phi_1, \Phi_2, \Phi_3\}$ in the diagram

$$\text{Unif} \xrightarrow{F_2} \text{Tich} \xrightarrow{\beta} \text{Comp}$$

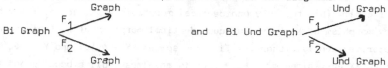

iff the functors $\psi_1, \psi_2$ are faithful. The necessity is by (FC) again, the sufficiency is proved in [ 24 ]. Let us notice that this implies the topological statement mentioned in the introduction : if $k_3$ consists of a unique object and $k_2$ has m + 1 objects $\{b_i \mid i \in m\} \cup \{b\}$, the only nonidentical morphisms of $k_2$ are $m_i : b \to b_i$ and $\psi_2$ sends every morphism of $k_2$ to the unique identical morphism of $k_3$ (this is a faithful functor, evidently), then the Tichonov spaces $X = \Phi_2(b)$ and $Y_i = \Phi_2(b_i)$ fulfil $X \subseteq Y_i \subseteq \beta X$ and, since $\Phi_2 : k_2 \to \mathrm{Tich}$ is an almost full embedding and there are no nonidetical morphisms in $\{b_i \mid i \in m\}$, $\{Y_i \mid i \in m\}$ is a stiff collection of spaces.

It is not known which diagrams in $\mathbb{C}$at

$$k_1 \xrightarrow{\psi_1} k_2 \xrightarrow{\psi_2} k_3 \xrightarrow{\psi_3} k_4$$

have simultaneous representations in the diagram

$$\mathrm{Metr} \xrightarrow{F_1} \mathrm{Unif} \xrightarrow{F_2} \mathrm{Tich} \xrightarrow{\beta} \mathrm{Comp}.$$

By (FC), the functors $\psi_1$, $\psi_2$, $\psi_3$ must be faithful, but this is *not* sufficient ; however, a counterexample is known *only* in a set theory without measurable cardinals. Also, (FC) is not sufficient in the question which diagrams (2) in $\mathbb{C}$at have a simultaneous representation in the diagram

$$\mathrm{Alg}(1,1,1) \xrightarrow{F} \mathrm{Alg}(1,1),$$

where Alg(1,1,1) (or Alg(1,1)) denotes the category of all universal algebras with three (or two) unary operations and F forgets the last operation. By [ 16 ], a simultaneous representation exists iff $\Psi$ can be decomposed as

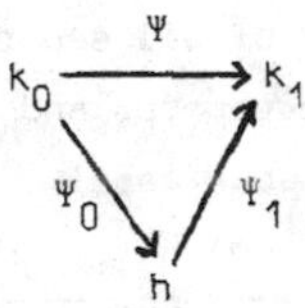


where $\Psi_0$ is an equalizer in $\mathbb{C}$at and $\Psi_1$ is a full faithful functor (let us call the functors $\Psi$, which can be decomposed in this way, *regularly faithful*). The algebraic statement mentioned in the introduction is a corollary of this theorem : we obtain it if $k_0$ and $k_1$ both have precisely one object.

Finally, let us mention that (FC) *is* sufficient for the existence of a simultaneous representation of a diagram (2) in the diagram

$$\mathrm{Und\ Graph} \xrightarrow{F_\Delta} \mathrm{Und\ Graph}_\Delta ,$$

this is proved in the part III of the present paper.

Open problem. As already mentioned in I., every small category can be fully embedded into Smg and into Rings (see [5] and [6]). Which diagrams (2) in $\mathbb{C}$at have a simultaneous representation in the diagram

$$\text{Rings} \xrightarrow{F} \text{Smg} \ ?$$

Here, F is the functor which forgets the addition, i.e. it sends each ring on its multiplicative semigroup.

## 2. The subpullback condition.

A commutative square (of categories and functors)

$$\begin{array}{ccccc} & & k_1 & & \\ & \nearrow^{\Psi_1} & & \searrow^{\Gamma_1} & \\ k_0 & & & & h \\ & \searrow_{\Psi_2} & & \nearrow_{\Gamma_2} & \\ & & k_2 & & \end{array}$$

is called a *subpullback* if, for every $a,b \in \text{obj}\,k_0$ and every $k_i$-morphisms $p_i : \Psi_i(a) \to \Psi_i(b)$, $i = 1,2$, such that

$$\Gamma_1(p_1) = \Gamma_2(p_2),$$

there exists a unique $p : a \to b$ in $k_0$ such that $\Psi_i(p) = p_i$, $i = 1,2$. Hence, if the square is in $\mathbb{C}$at and we form a pullback of $\Gamma_1$ and $\Gamma_2$ in $\mathbb{C}$at, say

$$\begin{array}{ccccc} & & k_1 & & \\ & \nearrow^{\overline{\Psi}_1} & & \searrow^{\Gamma_1} & \\ \overline{k_0} & & & & h \quad , \\ & \searrow_{\overline{\Psi}_2} & & \nearrow_{\Gamma_2} & \\ & & k_2 & & \end{array}$$

then the unique functor $\Psi : k_0 \to \overline{k_0}$ with $\overline{\Psi}_i \circ \Psi = \Psi_i$, $i = 1,2$, is full and faithful.

The subpullback condition (SC). Let

$$\begin{array}{ccccc} & & \sigma_1 & & \\ & \nearrow^{m_1} & & \searrow^{q_1} & \\ \sigma_0 & & & & \sigma_3 \\ & \searrow_{m_2} & & \nearrow_{q_2} & \\ & & \sigma_2 & & \end{array}$$

be a commutative square in D. If there exists a simultaneous representation $\Phi = \{\Phi_\sigma \mid \sigma \in \text{obj}\,D\} : \mathcal{C} \to \mathcal{D}$ and

$$\begin{array}{ccccc} & & \mathcal{D}(\sigma_1) & & \\ & \nearrow^{\mathcal{D}(m_1)} & & \searrow^{\mathcal{D}(q_1)} & \\ \mathcal{D}(\sigma_0) & & & & \mathcal{D}(\sigma_3) \\ & \searrow_{\mathcal{D}(m_2)} & & \nearrow_{\mathcal{D}(q_2)} & \\ & & \mathcal{D}(\sigma_2) & & \end{array}$$

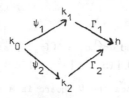

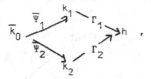

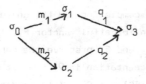

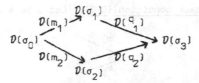

is a subpullback and if either

(i) $\Phi_{\sigma_0} : \mathcal{C}(\sigma_0) \to \mathcal{D}(\sigma_0)$ is full or

(ii) $\Phi_{\sigma_0}, \Phi_{\sigma_1}, \Phi_{\sigma_2}$ are almost full (hence $\mathcal{D}(\sigma_0), \mathcal{D}(\sigma_1), \mathcal{D}(\sigma_2)$ are concrete categories with all constant maps being morphisms !) and $\mathcal{D}(m_1)$, $\mathcal{D}(m_2)$ preserve the constant morphisms, then

$$\begin{array}{ccccc} & & \mathcal{C}(\sigma_1) & & \\ & \overset{\mathcal{C}(m_1)}{\nearrow} & & \overset{\mathcal{C}(q_1)}{\searrow} & \\ \mathcal{C}(\sigma_0) & & & & \mathcal{C}(\sigma_3) \\ & \underset{\mathcal{C}(m_2)}{\searrow} & & \underset{\mathcal{C}(q_2)}{\nearrow} & \\ & & \mathcal{C}(\sigma_2) & & \end{array}$$

is also a subpullback.

Proof. Let a,b be in obj $\mathcal{C}(\sigma_0)$ and let $p_i : [\mathcal{C}(m_i)](a) \to [\mathcal{C}(m_i)](b)$ be $\mathcal{C}(\sigma_i)$-morphisms, i = 1,2, such that $[\mathcal{C}(q_1)](p_1) = [\mathcal{C}(q_2)](p_2)$. Then $\overline{\Phi}_{\sigma_i}(p_i) = \overline{p}_i$ fulfil $[\mathcal{D}(q_1)](\overline{p}_1) = [\mathcal{D}(q_2)](\overline{p}_2)$. Since the $\mathcal{D}$-square is a subpullback, there exists a unique $\mathcal{D}(\sigma_0)$-morphism $\overline{r} : \Phi_{\sigma_0}(a) \to \Phi_{\sigma_0}(b)$ such that $[\mathcal{D}(m_i)](\overline{r}) = \overline{p}_i$, i = 1,2. If (i) or (ii) is fulfilled, then $\overline{r} = \Phi_0(r)$ for a (unique !) $\mathcal{C}(\sigma_0)$-morphism $r : a \to b$. Then $[\mathcal{C}(m_i)](r) = p_i$ for i = 1,2, evidently.

Let us mention at least one example. Denote by Compl Metr (or Compl Unif) the full subcategory of Metr (or Unif) generated by all the complete spaces. One can see easily that the square

$$\begin{array}{ccccc} & & \text{Compl Metr} & & \\ & \overset{c_m}{\nearrow} & & \overset{\overline{F_1}}{\searrow} & \\ \mathcal{D} : \text{Metr} & & & & \text{Compl Unif} \;, \\ & \underset{F_1}{\searrow} & & \underset{c_u}{\nearrow} & \\ & & \text{Unif} & & \end{array}$$

where $c_m$ and $c_u$ are the metric completion and the uniform completion and $\overline{F_1}$ is the domain-range-restriction of the forgetful functor $F_1$, is a subpullback consisting of faithful functors, moreover, $c_m$ and $F_1$ preserve the constant morphisms. If a square in $\mathbb{C}$at has a simultaneous representation in $\mathcal{D}$, then it must be a subpullback consisting of faithful functors, by (FC) + (SC). This condition is also sufficient, the proof is contained in [ 23 ].

3. The completed-subpullback condition (CSC). Let D be a span

$$\begin{array}{ccc} & & \sigma_1 \\ & \overset{m_1}{\nearrow} & \\ \sigma_0 & & \\ & \underset{m_2}{\searrow} & \\ & & \sigma_2 \; . \end{array}$$

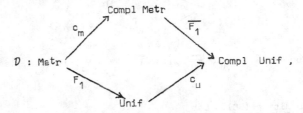

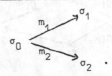

Let there exist a simultaneous representation

$$\Phi = \{\Phi_{\sigma_0}, \Phi_{\sigma_1}, \Phi_{\sigma_2}\} : \mathcal{C} \to \mathcal{D}.$$

If $\mathcal{D}$ can be completed to a subpullback consisting of faithful functors and either (i) or (ii) of (SC) is fufilled, then $\varphi$ also can be completed to a subpullback consisting of faithful functors.

<u>Proof</u>. Let

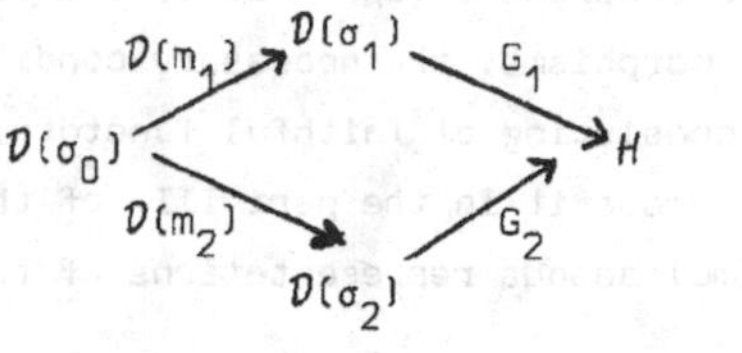


be a subpullback, let $\mathcal{D}(m_i)$, $G_i$, $i = 1,2$, be faithful. Then [(i) or (ii)] easily implies that the square

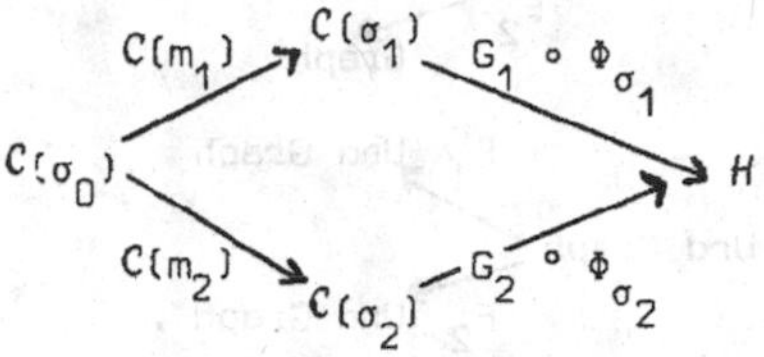


is also a subpullback. The functors $\mathcal{C}(m_1)$, $\mathcal{C}(m_2)$ are faithful, by (FC) ; the functors $G_1 \circ \Phi_{\sigma_1}$, $G_2 \circ \Phi_{\sigma_2}$ are also faithful because $G_1$, $G_2$ are faithful and $\Phi_{\sigma_1}, \Phi_{\sigma_2}$ are one-to-one.

<u>Remark</u>. If $\mathcal{C}$ is a diagram in $\mathbb{C}$at and we wish to have the completed subpullback also in $\mathbb{C}$at, we replace $H$ by its full subcategory generated by

$$[G_1 \circ \Phi_{\sigma_1}](\mathcal{C}(\sigma_1)) \cup [G_2 \circ \Phi_{\sigma_2}](\mathcal{C}(\sigma_2)).$$

The (CSC) is suitable for the situations, where $\mathcal{D}(\sigma_0)$ is the category of sets with two structures, $\mathcal{D}(m_1)$ forgets the first one and $\mathcal{D}(m_2)$ forgets the second one. For example, which spans of small categories and functors

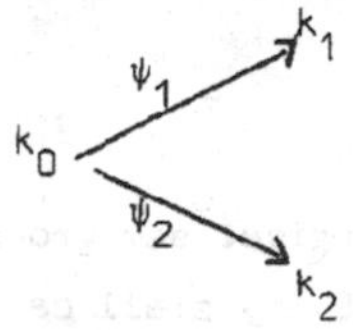

(5)

have a simultaneous representation in the diagram

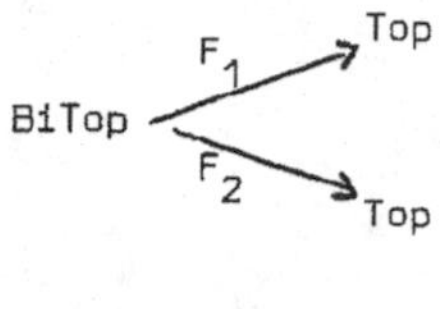

(6)

(where BiTop is the category of all bitopological spaces $(X,t_1,t_2)$ and all bicontinuous maps, $F_1$ forgets the topology $t_1$ and $F_2$ forgets the topology $t_2$) ? Since the diagram (6) can be completed to the subpulback

$$\begin{array}{ccccc} & & \mathrm{Top} & & \\ & \overset{F_1}{\nearrow} & & \overset{G}{\searrow} & \\ \mathrm{BiTop} & & & & \mathrm{Set} \\ & \underset{F_2}{\searrow} & & \underset{G}{\nearrow} & \\ & & \mathrm{Top} & & \end{array}$$

consisting of faithful functors (where $G : \mathrm{Top} \to \mathrm{Set}$ is the forgetful functor) and since $F_1$, $F_2$ preserve constant morphisms, the necessary condition if that (5) has to be completed to a subpullback consisting of faithful functors, by (CSC). This condition is also sufficient - we prove it in the part III. of this paper. An analogous situation is concerning the simultaneous representations of the diagrams (5) in $\mathbb{C}$at in the diagrams

$$\begin{array}{ccc} & & \mathrm{Graph} \\ & \overset{F_1}{\nearrow} & \\ \mathrm{BiGraph} & & \\ & \underset{F_2}{\searrow} & \\ & & \mathrm{Graph} \end{array} \qquad (7)$$

and

$$\begin{array}{ccc} & & \mathrm{Und\ Graph} \\ & \overset{F_1}{\nearrow} & \\ \mathrm{Bi\ Und\ Graph} & & \\ & \underset{F_2}{\searrow} & \\ & & \mathrm{Und\ Graph}\ , \end{array} \qquad (8)$$

where BiGraph is the category of all directed bigraphs $(X,R_1,R_2)$, $F_1$ forgets the first graph structure $R_1$ and $F_2$ forgets the second $R_2$, Bi Und Graph is the category of all undirected bigraphs and $F_1$, $F_2$ forget the first and the second structure. The both diagrams can be completed to a subpullback consisting of faithful functors, by means of the forgetful functors Graph → Set and UndGraph → Set. Hence by (CSC), the necessary condition for the existence of a simultaneous representation of (5) into (7) (or into (8)) is that (5) can be completed to a subpullback consisting of faithful functors. This condition is also sufficient, we prove it in III, of the present paper.

<u>Open problem</u>. Which diagrams (5) in $\mathbb{C}$at have a simultaneous representation in

$$\begin{array}{ccc} & & \mathrm{Smg} \\ & \overset{F_1}{\nearrow} & \\ \mathrm{Top\ Smg} & & \\ & \underset{F_2}{\searrow} & \\ & & \mathrm{Top} \end{array} \qquad ?$$

Here, Top Smg is the category of all topological semigroups, $F_1$ forgets the topology and $F_2$ forgets the semigroup structure. (Every small category can be fully embedded into Smg, by [ 9 ] ; and every small category can be *almost* fully embedded in Top, by [ 20 ].)

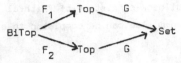

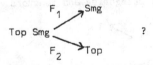

## 4. The admissible enlarging condition.

Let $\mathcal{D}$ be a diagram of categories and functors over a poset $D = (P,\leq)$. We say that a diagram $\mathcal{D}'$ of categories and functors is its *admissible enlarging* if its scheme is a poset $D' = (P',\leq)$ containing $D = (P,\leq)$ as a order ideal (i.e. if $p \in P$, $q \in P'$ and $q \leq p$, then $q \in P$) and $\mathcal{D}'$ is an extension of $\mathcal{D}$.

We say that a diagram $\mathcal{D}$ is *nice* if it consists of faithful functors, its scheme is a poset $D = (P,\leq)$ with a last element t and, for every $p_1,p_2 \in P$ such that the meet $p_1 \wedge p_2$ exists in D, the square

$$\begin{array}{ccccc} & & \mathcal{D}(p_1) & & \\ & \nearrow & & \searrow & \\ \mathcal{D}(p_1 \wedge p_2) & & & & \mathcal{D}(t) \\ & \searrow & & \nearrow & \\ & & \mathcal{D}(p_2) & & \end{array}$$

is a subpullback.

The admissible enlarging condition (AEC) : If $\Phi = \{\Phi_\sigma \mid \sigma \in \mathrm{obj}\, D\} : \mathcal{C} \to \mathcal{D}$ is a simultaneous representation such that all the functors $\Phi_\sigma$ are full embeddings and if $\mathcal{D}$ has an admissible enlarging to a nice diagram, then $\mathcal{C}$ has also an admissible enlarging to a nice diagram over the same scheme.

**Proof.** Let $\mathcal{D}'$ over $D' = (P',\leq)$ be a nice diagram which is an admissible enlarging of $\mathcal{D}$ (the scheme of $\mathcal{D}$ is a poset $D = (P,\leq)$). Let $\Phi = \{\Phi_p \mid p \in P\} : \mathcal{C} \to \mathcal{D}$ be a simultaneous representation such that all the functors $\Phi_p$ are full. We extend $\mathcal{C}$ to a diagram $\mathcal{C}'$ over $D'$ such that for $p' \in P' \setminus P$, we put $\mathcal{C}(p') = \mathcal{D}(p')$ and if $p \leq p'$, then

$$\mathcal{C}'(m(p,p')) = \mathcal{D}'(m(p,p')) \quad \text{if} \quad p,p' \in P' \setminus P,$$

$$\mathcal{C}'(m(p,p')) = \mathcal{D}'(m(p,p')) \circ \Phi_p \quad \text{if } p \in P,\ p' \in P' \setminus P,$$

where by $m(p,p')$ is denoted the unique $D'$-morphism from p to p'. Since $\mathcal{D}'$ is nice, $\mathcal{C}'$ is also nice (the verification is straightforward).

The (AEC) can be applied e.g. in the following situation : which diagrams

$$\begin{array}{ccc} & \xrightarrow{\Psi_1} & k_1 \\ k_0 & \xrightarrow{\Psi_2} & k_2 \\ & \xrightarrow{\Psi_3} & k_3 \end{array} \tag{9}$$

in $\mathbb{C}\mathrm{at}$ have a simultaneous representation in the diagrams

$$\text{(10)} \quad \begin{array}{ccc} & \xrightarrow{F_1} & \mathrm{Alg}(1,1) \\ \mathrm{Alg}(1,1,1) & \xrightarrow{F_2} & \mathrm{Alg}(1,1) \\ & \xrightarrow{F_3} & \mathrm{Alg}(1,1) \end{array} \qquad \begin{array}{ccc} & \xrightarrow{F_1} & \mathrm{Palg}(1,1) \\ \mathrm{Palg}(1,1,1) & \xrightarrow{F_2} & \mathrm{Palg}(1,1) \\ & \xrightarrow{F_3} & \mathrm{Palg}(1,1) \end{array} \quad \text{(11)}$$

Here, Alg(1,1,1) and Alg(1,1) are as in II.1., Palg(1,1,1), Palg(1,1) are the categories of partial algebras with three or two unary operations and $F_i$ always forgets the i-th operation, i = 1,2,3. The diagram (10) has the admissible nice enlarging

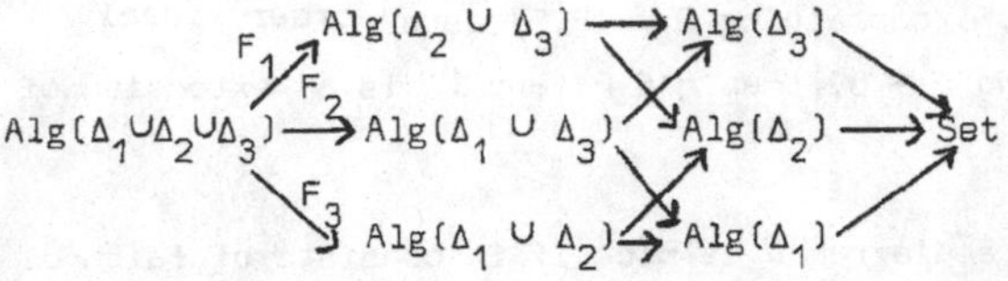

(11)

(where $\Delta_i$ denotes the type consisting of the i-th operation) consisting of the forgetful functors ; analogously, the diagram (11) has an admissible nice enlarging over the same scheme. Hence, by (AEC), if a diagram (9) in $\mathbb{C}$at has a simultaneous representation in the diagram (10) or (11), it has an enlarging to a nice diagram

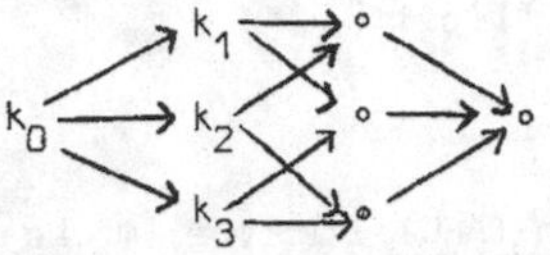

(12)

For the diagram (11) of partial algebras, this condition is also sufficient, by [16 ]. However a necessary and sufficient condition for the diagram (9) to have a simultaneous representation in (10) requires, moreover, that all the arrows in (12) are regularly faithful functors. This is proved in [ 16 ], where also more complex schemes and other types of algebras are investigated.

## III. The sufficiency

1. The aim of this part is to prove the following two theorems.

**THEOREM 1.** *A diagram (2) in* $\mathbb{C}$at *has a simultaneous representation in the diagram (3) iff the functor* $\Psi$ *is faithful.*

**THEOREM 2.** *The following properties of a diagram (5) in* $\mathbb{C}$at *are equivalent :*

*(a) it has a simultaneous representation in the diagram (6) ;*
*(b) it has a simultaneous representation in the diagram (7) ;*
*(c) it has a simultaneous representation in the diagram (8) ;*
*(d) it can be completed to a subpullback consisting of faithful functors.*

2. The necessity in Theorem 1. was shown in II, it is by the (FC). Also the implications (a) ⇒ (d), (b) ⇒ (d), (c) ⇒ (d) in the Theorem 2. were shown in II, they follow by the (CSC). Hence we have to prove the sufficiency. In the proofs of the both theorems (and in the proof of all the implications (d) ⇒ (a), (d) ⇒ (b), (d) ⇒ (c) in Theorem 2.), we use the following auxiliary lemma proved in [22 ].

**LEMMA.** *Let* D = (P,⩽) *be a poset with a last element* t, *let* C *be a diagram in* $\mathbb{C}$at *over* D *consisting of faithful functors. Then* C *has a simultaneous representation*

$\Phi = \{\Phi_p \mid p \in P\}$ *in the diagram* $G$ *defined as follows* :

a) $\mathcal{G}(t)$ is the category of all directed graphs $(X,R)$ (i.e. $X$ is a set and $R \subseteq X \times X$) which are

(i) connected (i.e. for every $x,y \in X$ [ not necessarily distinct ] there are $x_0 = x, x_1, \dots, x_n = y$ in $X$ such that $(x_i, x_{i+1}) \in R \cup R^{-1}$) and

(ii) without loops (i.e. never $(x,x) \in R$)

and all compatible maps ;

b) if $p \in P$, $p < t$, then $\mathcal{G}(p)$ is the category of all $(X,\{R_d \mid d \in P,\ d \geqslant p\})$ (i.e. $X$ is a set and $R_d \subseteq X \times X$ for each $d \geqslant p$) such that $(X,R_t)$ is an object of $\mathcal{G}(t)$ and

$$\text{if } p \leqslant d_1 \leqslant d_2, \text{ then } R_{d_1} \subseteq R_{d_2}$$

and all their compatible maps (i.e. $f : (X,\{R_\alpha \mid d \geqslant p\}) \to (X', R'_d \mid d \geqslant p\})$ is a morphism of $\mathcal{G}(p)$ iff, for all $d \geqslant p$,

$$(x,y) \in R_d \Rightarrow (f(x),f(y)) \in R'_d \ ;$$

c) if $p \leqslant p'$ (and $m(p,p')$ denotes the unique D-morphism from $p$ into $p'$), then $\mathcal{G}(m(p,p')) : \mathcal{G}(p) \to \mathcal{G}(p')$ is the forgetful functor, i.e.

$$[\ \mathcal{G}(m(p,p'))]\ ((X,\{R_d \mid d \geqslant p\}) = (X,\ R_d \mid d \geqslant p'\}).$$

Moreover, if the meet $p_1 \wedge p_2 = p$ exists in D and the square

$$\begin{array}{ccccc} & & \mathcal{C}(p_1) & & \\ & \nearrow & & \searrow & \\ \mathcal{C}(p) & & & & \mathcal{C}(t) \\ & \searrow & & \nearrow & \\ & & \mathcal{C}(p_2) & & \end{array}$$

is a subpullback, then $\Phi_p$ sends any object of $\mathcal{C}(p)$ to an **object** $(X,\{R_d \mid d \geqslant p\})$ such that

$$R_p = R_{p_1} \cap R_{p_2}.$$

**Remark**. No category $\mathcal{G}(p)$ has all constants as morphism so that (by the definition of the simultaneous representation) all the functors $\Phi_p : \mathcal{G}(p) \to \mathcal{G}(p)$ are full embeddings.

3. To prove Theorem 1, we investigate $D = (\{0,1\},\ 0 < 1 = t)$. We consider the diagram $\mathcal{G}$ over D, i.e. $\mathcal{G}(1)$ is the category of all connected directed graphs without loops and $\mathcal{G}(0)$ is the category of all bigraphs $(X,R_0,R_1)$ with $R_0 \subseteq R_1$ (and

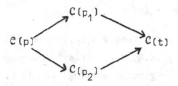

$(X,R_1)$ being connected and without loops). It is sufficient to find a simultaneous representation of $G$ in our diagram (3).

Let us recall how a full embedding of the category $G(1)$ into the category Und Graph is made in [18] : each arrow $r = (x,y)$ in a directed graph $(X,R) \in \text{obj } G(1)$ is replaced by a copy $A_r$ of the undirected graph A bellow such that the vertex a of this copy is identified with x and the vertex b of this copy is identified **with** y.

A :

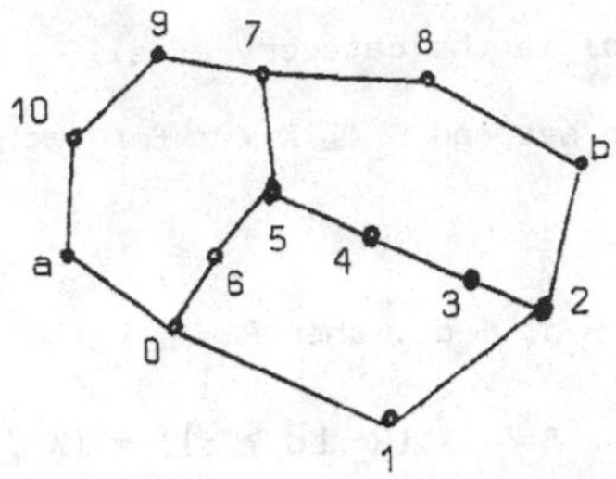


(13)

As one can see, the graph A consists of three 7-cycles (0,1,...,6), (a,0,6,5,7,9,10) and (b,8,7,5,4,3,2), having one or two or three edges in common ; its precise description is given in the monograph [18], p. 68. If $A_r$ is a copy of A, we denote by $z_r$ the vertex of it corresponding to a vertex z in A. The process of the replacing of the arrows in (X,R) by copies of A is as follows : for each $r \in R$, we take a copy $A_r$ of A, supposing them pairwise disjoint, and identify

$a_r$ with $a_{r'}$, whenever $\pi_1(r) = \pi_1(r')$,

$b_r$ with $a_{r'}$, whenever $\pi_2(r) = \pi_1(r')$,

$b_r$ with $b_{r'}$, whenever $\pi_2(r) = \pi_2(r')$,

in the sum of all these $A_r$, $r \in R$ (where $\pi_1(x,y) = x$, $\pi_2(x,y) = y$). The obtained undirected graph is denoted by $(X,R) * A$. This "arrow-construction" is described with all details (and in a more general setting) in [18], pages 105-106. If $f : (X,R) \to (X',R')$ is a $G(1)$-morphism, we define a map

$$f * 1_A : (X,R) * A \to (X',R') * A$$

such that it sends each vertex $z_{(x,y)}$ of each copy $A_{(x,y)}$ in $(X,R) * A$ on the vertex $z_{(f(x),f(y))}$ in the copy $A_{(f(x),f(y))}$ in $(X',R') * A$.

Clearly, the map

$$(X,R) \rightsquigarrow (X,R) * A, \quad f \rightsquigarrow f * 1_A$$

is a one-to-one functor $G(1) \to$ Und Graph, say M. The proof that M is full uses the fact that A consists of the three 7-cycles and $(X,R) * A$ contains no shorter cycles. Hence any compatible map $g : (X,R) * A \to (X',R') * A$ sends each 7-cycle in $(X,R) * A$

on a 7-cycle in $(X',R') * A$, consequenly a copy $A_r$, $r \in R$, on a copy $A_{r'}$, with $r' \in R'$. Since $(X,R)$ is connected, in each vertex $x \in X$ an arrow $r \in R$ either starts or terminates so that one can find a compatible map $f : (X,R) \to (X',R')$ such that $g = f * 1_A$. A more detail version of the proof that M is really a full embedding is contained in [18], p. 107.

To prove our Theorem 1, we also use the arrow construction. We define a functor $\Lambda_1 : \mathcal{G}(1) \to$ Und Graph$_\Delta$ by

$$(X,R_1) \rightsquigarrow (X,R_1) * B, \quad f \rightsquigarrow f * 1_B$$

where B is the graph obtained from A (see (13)) such that we add, for each of the three 7-cycles, one vertex and we join it with each vertex of the 7-cycle in question. Hence B looks as follows.

B:

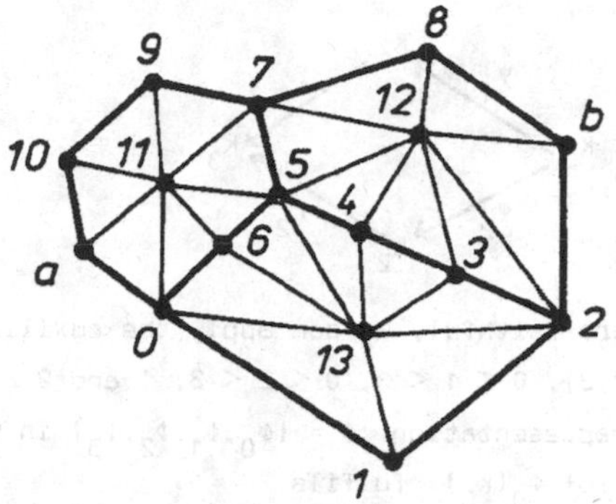


(14)

We make the identifications as in the definition of the functor M so that M is a subfunctor of $\Lambda_1$. Clearly, $\Lambda_1$ is one-to-one. We show that $\Lambda_1$ is full. If $G = (V,E)$ is an undirected graph and $x \in V$, let us denote by $G_x$ the full subgraph of G generated by the set $\{y \in V \mid (x,y) \in E\}$. (If $f : G \to G'$ is a compatible map, then necessarily $f(G_x)$ is a subgraph of $G'_{f(x)}$.) In our graph B, $B_{11}$, $B_{12}$ and $B_{13}$ are 7-cycles and $B_x$ contains no shorter odd cycle for any vertex x of B. This implies easily that any compatible map $g : (X,R_1) * B \to (X',R_1') * B$ is an extension of a compatible map $\bar{g} : (X,R_1) * A \to (X',R_1') * A$. Since M is full, $\bar{g} = f * 1_A$ for a compatible map $f : (X,R_1) \to (X',R_1')$. Then clearly $g = f * 1_B$.

To define $\Lambda_0 : \mathcal{G}(0) \to$ Und Graph, we add one edge to the graph B chosen such that it does not create a triangle with the edges of B and its ends are distinct from a and b, e.g. the edge (2,9), let us denote the obtained graph by $B^1$. If $(X,R_0,R_1) \in$ obj $\mathcal{G}(0)$, we define $\Lambda_0(X,R_0,R_1)$ such that

each arrow r in $R_0$ is replaced by a copy $B^1_r$ of $B^1$ and
each arrow r in $R_1 \setminus R_0$ is replaced by a copy $B_r$ of B

and the identifications of $a_r$, $b_r$ with $a_{r'}$, $b_{r'}$ are as above (as in the definition of M). Clearly, the omitting of the edges which are not contained in a triangle turns $\Lambda_0(X,R_0,R_1)$ into $\Lambda_1(X,R_1) = (X,R_1) * B$. If $f : (X,R_0,R_1) \to (X',R_0',R_1')$ is a $G(0)$-morphism, $f * 1_B$ has an evident extension to the newly added edges. Clearly, $\Lambda_0$ is a one-to-one functor. If $g : \Lambda_0(X,R_0,R_1) \to \Lambda_0(X',R_0',R_1')$ is a compatible map, it maps each triangle on a triangle again ; since $\Lambda_1$ is full necessarily $F_\Delta(g) = \Lambda_1(f)$ for some $G(1)$-morphism $f : (X,R_1) \to (X',R_1')$. Since g sends each edge $(2_r, q_r)$ on some $(2_{r'}, q_{r'})$, necessarily f is also $R_0R_0'$ - compatible so that $\Lambda_0$ is a full functor. We conclude that $\Lambda = \{\Lambda_0,\Lambda_1\}$ is a simultaneous representation of $G$ in the diagram (3) so that the composition $\Lambda \circ \Phi$ is a simultaneous representation of a diagram (2) in the diagram (3).

4. Now, *we prove theorem 2.* First, we present a proof of the implication (d) ⇒ (a) (which is rather long) : If a diagram $C$ can be completed to a subpullback $S$ in $\mathbb{C}$at, say

$$S : k_0 \begin{array}{c} \xrightarrow{\Psi_1} k_1 \xrightarrow{\Gamma_1} \\ \xrightarrow[\Psi_2]{} k_2 \xrightarrow[\Gamma_2]{} \end{array} k_3 ,$$

such that $\Psi_n$, $\Gamma_n$, $n = 1,2$, are faithful, we can apply the auxiliary Lemma to the diagram $S$, i.e. $D = (\{0,1,2,3\}$, $0 < 1 < 3$, $0 < 2 < 3$, 1 and 2 are incomparable). Hence $S$ has a simultaneous representation $\Phi = \{\Phi_0,\Phi_1,\Phi_2,\Phi_3\}$ in $G$ over D such that every object $(X,R_0,R_1,R_2,R_3)$ of $\Phi_0(k_0)$ fulfils

$$R_0 = R_1 \cap R_2 , \quad R_1 \cup R_2 \subsetneq R_3$$

(and $(X,R_3)$ is a connected directed graph without loops).

a) In what follows we construct a simultaneous representation $\Lambda = \{\Lambda_0,\Lambda_1,\Lambda_2\}$ of the diagram

$$G(0) \begin{array}{l} \nearrow G(1) \\ \searrow G(2) \end{array} \tag{15}$$

in our diagram (6). We use a kind of "arrow construction" again. We construct three topological spaces $Q_1$, $Q_2$, $Q_3$ on the same underlying set, say Y, such that the identity map $Y \to Y$ is continuous iff it is considered as one of the following maps :

$$1_{Q_n} : Q_n \to Q_n,\ n = 1,2,3 \text{ and } Q_3 \to Q_2,\ \ Q_3 \to Q_1 .$$

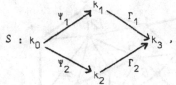

Moreover, three distinct points $a_+$, $a_-$, $e$ are given in Y. For n = 1,2, denote by $Q^r_n$ a copy of the space

$$Q_n \text{ whenever } r \in R_n,\ Q_3 \text{ whenever } r \in R_3 \setminus R_n.$$

We form a space $Z_n$ as follows : we replace each $r = (x,y) \in R_3$ by $Q^r_n$ (assuming the copies pairwise disjoint) and identify $a_-$ of this copy with x, $a_+$ of this copy with y and glueing the points e of all these copies together (for all $x \in R_3$). Since all the spaces $Q_1$, $Q_2$, $Q_3$ have the same underlying set, the spaces $Z_1$ and $Z_2$ have the same underlying set so that $(Z_1,Z_2)$ is really a bitopological space. If, for n = 1,2,

$$f : (X,R_n,R_3) \to (X',R'_n,R'_3)$$

is a *Gy*(n)-morphism, denote by $\overline{f} : Z_n \to Z'_n$ the continuous map which send each $Q^{(x,y)}_n$ in $Z_n$ onto $Q^{(f(x),f(y))}_n$ carried by "the identity" on the underlying set Y. Clearly,

$$\Lambda_n(X,R_n,R_3) = Z_n,\quad \Lambda_n(f) = \overline{f}$$

is a one-to-one functor of *Gy*(n) into Top. And $\Lambda_0$ sending $(X,R_0,R_1,R_2,R_3)$ to $(Z_1,Z_2)$ and f to $\overline{f}$ (which is bicontinuous if f is a *Gy*(0)-morphism) is a one-to-one functor of *Gy*(0) into BiTop. Clearly, $\Lambda = \{\Lambda_0,\Lambda_1,\Lambda_2\}$ is a natural transformation of the diagram (15) into (6). If we construct the spaces $Q_1$, $Q_2$, $Q_3$ in a special way, presented below, we are able to prove that $\Lambda_0,\Lambda_1,\Lambda_2$ are almost full embeddings.

b) Let K be a *Cook continuum*, i.e. a compact connected metric space, non degenerate (i.e. with more than one point) such that

> if K' is a subcontinuum of K and $f : K' \to K$ is a continuous map, then either f is constant or $f(x) = x$ for all $x \in K'$.

A continuum with this property was constructed by H. Cook in [4 ] (a more detail version see also Appendix A in [18 ]). Let J be the set of all integers, let

$$\mathbb{A} = \{A_i \mid i \in J\setminus\{0\}\},\ \mathbb{B}_n = \{B_{n,i} \mid i \in J\setminus\{0\}\},\ \mathbb{C}_n = \{C_{n,i} \mid i \in J\setminus\{0\}\},$$

$$\mathbb{D}_n = \{D_{n,j} \mid j \in J,\ j \geq 0\},\ \mathbb{E}_n = \{E_{n,j} \mid j \in J,\ j \geq 0\},\ \mathbb{F}_n = \{F_{n,j} \mid j \in J, j \geq 0\},$$

with n = 1,2,3, be disjoint systems of distinct nondegenerate subcontinua of K such that the system

$$\mathbb{X} = \mathbb{A} \cup \bigcup_{n=1}^{3} (\mathbb{B}_n \cup \mathbb{C}_n \cup \mathbb{D}_n \cup \mathbb{E}_n \cup \mathbb{F}_n)$$

is pairwise disjoint. Hence

(16) $\left\{\begin{array}{l}\text{if } X,X' \in \mathbb{X},\ K' \text{ is a subcontinuum of } X \text{ and } f : K' \to X' \text{ is a continuous} \\ \text{map, then either } f \text{ is constant or } X = X' \text{ and } f(x) = x \text{ for all } x \in K'.\end{array}\right.$

We create a space P as follows : we choose two distinct points in each $X \in \mathbb{X}$ and glue the sum of X as visualized in the picture below.

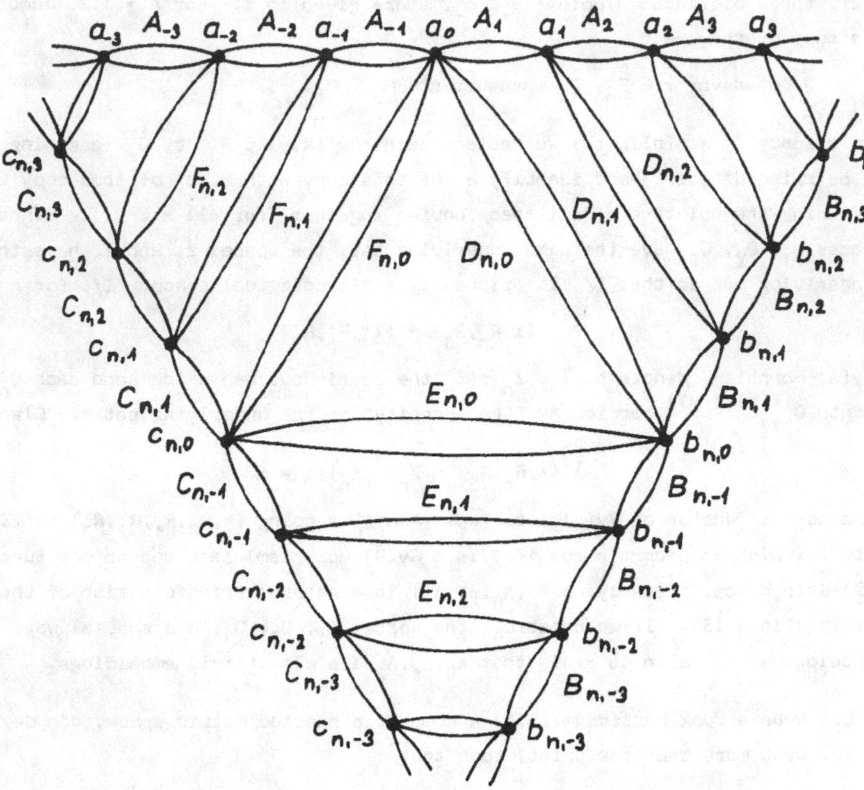


On the picture, only "one triangle" is drawn, our space P consists of three such triangles having the side consisting of $\bigcup_{i \in J \setminus \{0\}} A_i$ in common. The space P is also used in [23], where the "glueing construction" is described with all details. The glueing of one triangle is also described with all details in the monograph [18], pages 223-224.

To simplify the notation, let us suppose that the continua $A_i, B_{n,i}, C_{n,i}, \ldots, F_{n,j}$ from $\mathcal{X}$ are those contained in P (so that they are no more disjoint subcontinua of the Cook continuum ; however (16) remains valid for them and this is what we need). We also need the following statement (which really can be seen from the picture) :

(17) $\begin{cases} \text{If } X \in \mathcal{X},\ x_0 \in P \setminus X, \text{ then, for every } X' \in \mathcal{X} \text{ not containing } x_0, \\ \text{there exists a chain } X_0 = X, X_1, \ldots, X_s = X' \text{ of members of } \mathcal{X} \text{ such that} \\ \text{none of them contains } x_0 \text{ and } X_j \text{ intersect } X_{j+1} \text{ for each } j = 0,1,\ldots,s-1. \end{cases}$

Since the Cook continuum is a metric space, we can reach (by a multiplying of the metric of its subcontinua by a suitable constant) that the diameters of all $X \in \mathcal{X}$ are as on the picture so that P is a metric space, the completion cP of which is obtained by the adding of five points, namely

$$a_+ = \lim_{i\to\infty} a_i\ , \qquad a_- = \lim_{i\to-\infty} a_i$$

$$e_n = \lim_{i=-\infty} b_{n,i} = \lim_{i=-\infty} c_{n,i} \quad \text{for } n = 1,2,3.$$

c) Now, we modify the topology of cP to obtain the spaces $Q_1,Q_2,Q_3$ : all the sets open in cP are open in $Q_1,Q_2,Q_3$ and, for n = 1,2, the local basis of $e_n$ is changed, the space $Q_n$ looks as follows :

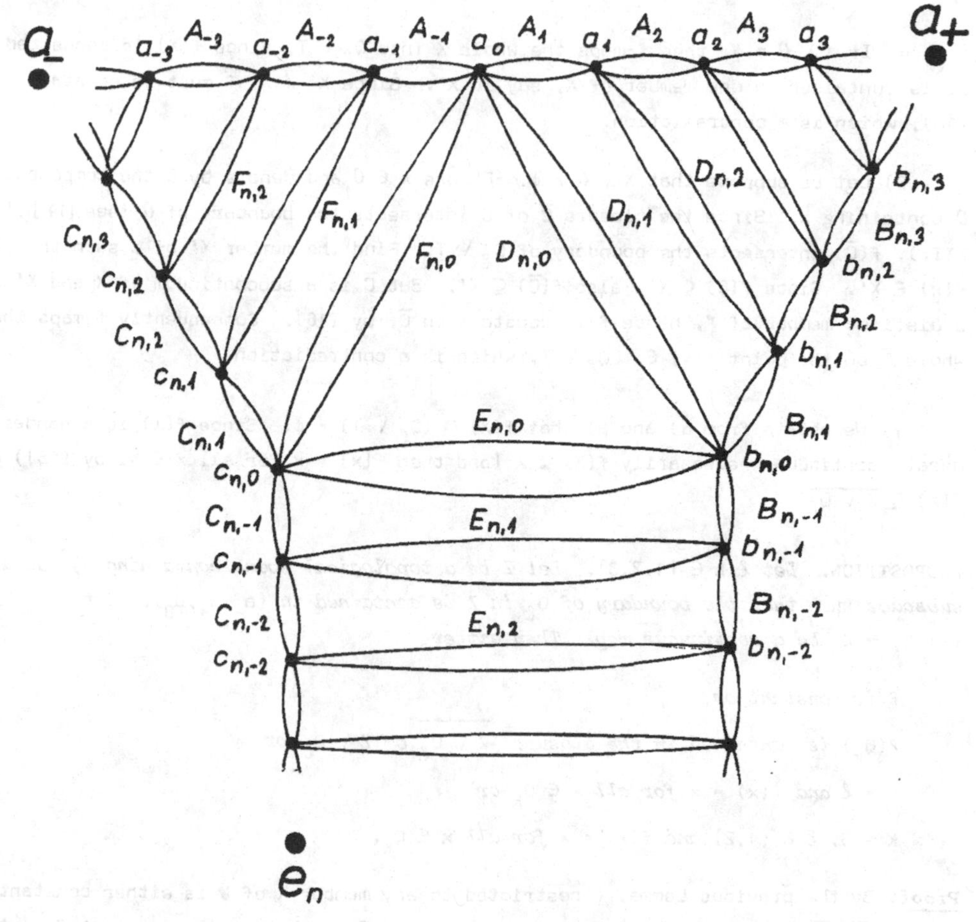


The topology of $Q_3$ is the infimum of the topologies of $Q_1$ and $Q_2$. Now, we use the spaces $Q_1,Q_2,Q_3$ in the "arrow construction" described in a), by means of the points $a_-,a_+$ and $e = e_3$. (Let us mention that making the identifications in the "arrow construction" within the category Metr, we can reach that the spaces $Z_1,Z_2$ are metric spaces.)

d) For the proof that the functors $\Lambda_0,\Lambda_1,\Lambda_2$ constructed in a) are almost full embeddings, we need the Lemma and the Proposition below. The spaces $Q_1,Q_2,Q_3$ are as in c).

**LEMMA.** *Let* $\ell \in \{1,2,3\}$, *let* Z *be a topological space containing* $Q_\ell$ *such that the boundary of* $Q_\ell$ *in* Z *is contained in* $\{a_-, a_+, e_3\}$. *Let* X *be a continuum in* $\mathcal{X}$, *let* $f : X \to Z$ *be a nonconstant continuous map. Then either* $f(X) \subseteq Q_\ell$ *and* f *is the inclusion (i.e.* $f(x) = x$ *for all* $x \in X$) *or* f(X) *is contained in the closure* $\overline{Z \setminus Q_\ell}$ *of* $Z \setminus Q_\ell$.

__Proof.__ Put $T = X \cup \{a_i \mid i \in J\} \cup \{b_i \mid i \in J\} \cup \{c_i \mid i \in J\} \cup \{a_+, a_-, e_1, e_2, e_3\} \subseteq Q_\ell$. Let us suppose that f(X) intersect $Q_\ell \setminus T$. Then $\mathcal{O} = f^{-1}(Q_\ell \setminus T)$ is nonempty and open.

α) If $X \setminus \mathcal{O} = \phi$, then f maps the whole X into $Q_\ell \setminus T$, since f(X) is connected, it is contained in one member of $\mathcal{X}$, say in X'. Since $X' \neq X$, f must be constant, by (16), which is a contradiction.

β) Let us suppose that $X \setminus \mathcal{O} \neq \phi$. Choose $x \in \mathcal{O}$ and denote by C the component of $\mathcal{O}$ containing x. Since the closure $\overline{C}$ of C intersects the boundary of $\mathcal{O}$ (see [14], §42, III.), $f(\overline{C})$ intersects the boundary of $Q_\ell \setminus T$. Find the member X' of $\mathcal{X}$ such that $f(x) \in X'$. Since $f(C) \subseteq X'$, also $f(\overline{C}) \subseteq X'$. But $\overline{C}$ is a subcontinuum of X and X' is a distinct member of $\mathcal{X}$, hence f is constant on $\overline{C}$, by (16). Consequently f maps the whole $\overline{C}$ on the point $f(x) \in Q_\ell \setminus T$, which is a contradiction.

γ) We obtain from α) and β) that $f(X) \cap (Q_\ell \setminus T) = \phi$. Since f(X) is a nondegenerate continuum, necessarily $f(X) \subseteq X$ (and then $f(x) = x$ for all $x \in X$, by (16)) or $f(X) \subseteq \overline{Z \setminus Q_\ell}$.

**PROPOSITION.** *Let* $\ell, k \in \{1,2,3\}$. *Let* Z *be a topological space containing* $Q_\ell$ *as a subspace such that the boundary of* $Q_\ell$ *in* Z *is contained in* $\{a_-, a_+, e_3\}$. *Let* $f : Q_k \to Z$ *be a continuous map. Then either*

f *is constant or*

$f(Q_k)$ *is contained in the closure* $\overline{Z \setminus Q_\ell}$ *of* $Z \setminus Q_\ell$ *or*

$k = \ell$ *and* $f(x) = x$ *for all* $x \in Q_k$ *or*

$k = 3$, $\ell \in \{1,2\}$ *and* $f(x) = x$ *for all* $x \in Q_k$.

__Proof.__ By the previous Lemma, f restricted to any member X of $\mathcal{X}$ is either constant or $f(X) \subseteq \overline{Z \setminus Q_\ell}$ or $f(X) = X$ and $f(x) = x$ for all $x \in X$. Let us suppose that f restricted to some X in $\mathcal{X}$ is constant, say $f(X) = \{x_0\}$. Let T be as in the proof of the Lemma.

α) If $x_0 \in Q_\ell \setminus T$, then every member of $\mathcal{X}$, which intersects X, has to be mapped by f on $x_0$. We can continue to the next members of $\mathcal{X}$ (i.e. those members of $\mathcal{X}$ which intersect some $X' \in \mathcal{X}$ with $X' \cap X \neq \phi$). Finally, we obtain (by (16)) that f maps the whole $Q_k$ on $x_0$.

β) Let us suppose that $x_0 = a_i$ or $x_0 = b_{n,i}$ or $x_0 = c_{n,i}$ for some $i \in J$, $n = 1,2,3$ or $x_0 \in X$.

β1) $x_0 \notin X$ : For every $X' \in \mathcal{X}$ not containing $x_0$ there exists a chain $X_0 = X, X_1, \dots, X_s = X'$ of members of $\mathcal{X}$ such that none of them contains $x_0$ and $X_j$ intersect $X_{j+1}$ for $j = 0,1,\dots,s-1$, by (17). By the previous Lemma, f maps $X_0, X_1, \dots, X_s = X'$ to $x_0$. Hence f maps any member $X' \in \mathcal{X}$ not containing $x_0$ on $x_0$. If $X' \in \mathcal{X}$ contains $x_0$, we can find $X'' \in \mathcal{X}$ not containing $x_0$ with $X' \cap X'' \neq \phi$, say $x_1 \in X' \cap X''$. Then f maps X" on $x_0$ so that $f(x_1) = x_0$. By the previous Lemma again, f maps X' on $x_0$. We conclude that f maps the whole $Q_k$ on $x_0$.

β2) $x_0 \in X$ : Let X' be a member of $\mathcal{X}$, which intersect X in a point distinct from $x_0$ ; then $x_0 \notin X'$. By the Lemma, f maps X' on $x_0$. Now, we use the case β1) for X'.

γ) We conclude from α) and β) that if $f(Q_k)$ intersects $L = Q_\ell \setminus \{a_-, a_+, e_1, e_2, e_3\}$, then either f is constant or $f(x) = x$ for all $X \in \mathcal{X}$ and all $x \in X$. In the later case, necessarily $f(a_-) = a_-$, $f(a_+) = a_+$ and $f(e_n) = e_n$ for $n = 1,2,3$. Since f is continuous, $f(e_n) = e_n'$ is possible only when $k = \ell$ or $k = 3$ and $\ell \in \{1,2\}$. If $f(Q_k)$ does not intersect L, then either f is constant or $f(Q_k) \subset \overline{Z \setminus Q_\ell}$.

e) Now, we prove that the functors $\Lambda_0, \Lambda_1, \Lambda_2$ constructed in a) are almost full embeddings. Let $(X, R_n, R_3)$ and $(X', R_n', R_3')$ be objects of $Gy(n)$ and let $\Lambda_n(X, R_n, R_3) = Z_n$, $\Lambda_n(X', R_n', R_3') = Z_n'$, $n = 1,2$. Let $g : Z_n \to Z_n'$ be a continuous map. We have to show that either g is constant or $g = \Lambda_n(f)$ for some $Gy(n)$-morphism $f : (X, R_n, R_3) \to (X', R_n', R_3')$. By the Proposition, g restricted to any $Q_n^r$, $r \in R_3$, is either constant or it sends it (carried by "the identity" on the underlying set) onto some $Q_n^{r'}$ with $r' \in R_3'$ [ hence if $r \in R_n$, then r' must be in $R_n'$ ]. We have to show that if g restricted to some $Q_n^r$ is constant, then it is constant on the whole $Z_n$. This is precisely the rôle of the glueing of the points $e = e_3$ of all the copies $Q_n^r$, $r \in R_3$ : if g restricted to some $Q_n^{(x,y)}$ is constant, then $g(a_-) = g(e) = g(a_+)$ ; then g must be constant also on each $Q_n^r$ with r starting or terminating in x or in y ; since the graph $(X, R_3)$ is connected, g must be constant on the whole $Z_n$. We conclude that either g is constant or, for every $r \in R_3$ (or $r \in R_n$) there exists $r' \in R_3'$ (or $r' \in R_n'$) such that g maps $Q_n^r$ onto $Q_n^{r'}$. Since $(X, R_3)$ is connected, we can find, for every $x \in X$ an arrow $r \in R_3$ which starts or terminates in x, say $r = (x,y)$. Then we put $f(x) = x'$ where $x' \in X'$ is the vertex in which the corresponding r' starts. Clearly, $f : (X, R_n, R_3) \to (X', R_n', R_3')$ is a $Gy(n)$-morphism and $g = \Lambda_n(f)$ ; hence $\Lambda_1$ and $\Lambda_2$ are almost full embeddings. If $g : (Z_1, Z_2) \to (Z_1', Z_2')$ is bicontinuous, then either g is constant or $g = \Lambda_1(f)$ and $g = \Lambda_2(f)$ for the same map $f : (X, R_n, R_3) \to (X', R_n', R_3')$. Since $R_0 = R_1 \cap R_2$ and $R_0' = R_1' \cap R_2'$, $f : (X, R_0, R_1, R_2, R_3) \to (X', R_0', R_1', R_2', R_3')$ is a $Gy(0)$-morphism, hence $\Lambda_0$ is also an almost full embedding.

5. Now, *we prove the rest of Theorem 2.* To prove the implication (d) ⇒ (c), we show that the diagram (15) has a simultaneous representation in (8). Let B be the graph (14), let $B^1$ be as in 3, i.e. B with the edge (2,9) added. Moreover, denote by $B^2$ the graph B with the edge (1,8) added. Starting with $(X,R_n,R_3)$ we make the arrow construction as in 3, replacing each arrow in $R_n$ by a copy of $B^n$ and each arrow in $R_3 \setminus R_n$ by a copy of B. One can see that this arrow construction really gives a simultaneous representation of (15) in (8). To prove the implication (d) ⇒ (b), it is sufficient to choose one (fix) orientation of the graph B and of the edges (2,9) and (1,8) and proceed as in the proof of the implication (d) ⇒ (c).

Concluding remark.

Finally, let us mention that there are several papers devoted to similar topics : [ 3 ], where the connection between the automorphism group and the vertex coloring group of a graph is studied ; [ 19 ], where the connection between the automorphism group and the weak automorphism group of a unary algebra is described ; [ 2 ], where the automorphism groups of two graphs which differ in one edge are studied ; [ 13 ], where the connection between the group of all isometries and the group of all autohomeomorphisms of a metric space is studied ; [ 1 ], where the endomorphism monoids of the unions of chains of lattices are investigated ; [ 10], [ 15 ], where the homomorphisms of graphs and of their orientations are investigated.

## REFERENCES
++++++++++

[ 1 ] M.E. ADAMS, D. PIGOZZI and J. SICHLER, Endomorphisms of direct unions of bounded lattices, Arch. Math. (Basel) 36 (1981), n°3, 221-229.

[ 2 ] L. BABAI, Vector representable matroids of given rank with given automorphism group, Discrete Math. 24 (1978), 119-125.

[ 3 ] V. CHVÁTAL and J. SICHLER, Automorphisms of graphs, J. of Comb. Theory 14(1973), 209-215.

[ 4 ] H. COOK, Continua which admit only identity mapping onto non-degenerate subcontinua, Fund. Math. 60(1967), 241-249.

[ 5 ] E. FRIED and J. SICHLER, Homomorphisms of commutative rings with unit element, Pacific J. Math. 45(1973), 485-491.

[ 6 ] E. FRIED and J. SICHLER, Homomorphisms of integral domains of characteristic zero, Trans. Amer. Math. Soc. 225(1977), 163-182.

[ 7 ] Z. HEDRLÍN and A. PULTR, O predstavlenii malych kategorij, Dokl. AN SSSR 160(1965), 284-286.

[8] Z. HEDRLÍN and A. PULTR, *On full embeddings of categories of algebras*, Illinois J. Math. 10 (1966), 392-406.

[9] Z. HEDRLÍN and J. LAMBEK, How-comprehensive is the category of semigroups, J. Algebra 11 (1969), 195-212.

[10] P. HELL and J. NEŠETŘIL, Homomorphisms of graphs and of their orientations, Monatsh. für Math. 85 (1978), 39-48.

[11] J.R. ISBELL, Subobjects, adequacy, completeness and categories of algebras, Rozprawy matematyczne XXXVI(1964).

[12] J.R. ISBELL, Epimorphisms and dominions, Proc. Conference on Categorical Algebra, La Jolla, 1965, Springer-Verlag 1966, 232-246.

[13] V. KANNAN and M. RAJAGOPALAN, Constructions and applications of rigid spaces II, Amer.J.of Math. 100 (1978), 1139-1172.

[14] C. KURATOWSKI, Topologie I, II, Monographie Matematyczne, Warsaw 1950.

[15] J. NEŠETŘIL, On symmetric and antisymmetric relations, Monatsh. Math. 76 (1972), 323-327.

[16] M. PETRICH, J. SICHLER and V. TRNKOVÁ, Simultaneous representations by algebras, in preparation.

[17] A. PULTR, On full embeddings of concrete categories with respect to forgetful functors, Comment. Math. Univ. Carolinae 9 (1968), 281-305.

[18] A. PULTR and V. TRNKOVÁ, Combinatorial, algebraic and topological representations of groups, semigroups and categories, North-Holland 1980.

[19] J. SICHLER, Weak automorphisms of universal algebras, Alg. Universalis 3 (1973), 1-7.

[20] V. TRNKOVÁ, Non constant continuous mappings of metric or compact Hausdorff spaces, Comment. Math. Univ. Carolinae 13 (1972), 283-295.

[21] V. TRNKOVÁ, All small categories are representable by continuous nonconstant mappings of bicompacta, Soviet. Math. Dokl. 17(1976), 1403-1406.

[22] V. TRNKOVÁ, Simultaneous representations in discrete structures, Comment. Math. Univ. Carolinae 27 (1986), 633-649.

[23] V. TRNKOVÁ, Simultaneous representations by metric spaces, to appear.

[24] V. TRNKOVÁ, Non constant continuous maps of a space and of its β-compactification, to appear.

Vera TRNKOVÁ
Math. Institute of the Charles University
18600 Praha 8., Sokolovská 83
CZECHOSLOVAKIA.

This paper is in final form
it will not be published elsewhere

# Uniform spaces can be represented by completely distributive lattices

Wolfgang Weiss
FB Mathematik, TH Darmstadt
Schlossgartenstr. 7
D-6100 Darmstadt

**Abstract:** The dual of the set of all uniform covers of a uniform space is characterized by lattice-theoretical properties within the completely distributive lattice of all stacks of the underlying set. This representation can be extended to an adjunction between the category of uniform spaces and uniformly continuous maps and a category of completely distributive lattices with specified bases and grills and complete homomorphisms preserving these structure-sets.

**Key words:** Uniform space, merotopic space, nearness space, completely distributive lattice, base, grill.

**AMS(MOS) Subj. Class.:** Primary 54E15; secondary 06D10, 18B30.

## 1 Introduction

Stone-duality and its generalizations have created representations of many classes of topological spaces by means of lattice-theoretical characterizations of the respective open-set-lattices (cf. [Comp 80],[Jo 82]). Similarly in this article uniform spaces are characterized by lattice-theoretical properties of (the dual of) the set of uniform covers. This is achieved by the use of terminology and results of the theory of merotopic and nearness spaces, in particular of the characterizatzion of uniform structures by near families. A representation theory for some subcategories of merotopic and nearness spaces has already been developed in [We 84;88]. Some of these results are included in the present note in order to make it self-contained.

Starting from the observation that the collection of all near families of a merotopic or uniform space is determined by its set of near stacks we investigate the lattice of all stacks of a set, which is called the scale of this set. It is a completely distributive lattice, in which every element is the limes superior of principal ultrafilters (Proposition 2.1). This property motivates the study of base-lattices. For each merotopic space the distinguished set of near stacks (called tribe) is a lattice-grill, i.e. the complement of an ideal (Proposition 2.3). The principal ultrafilters within the scale enable us to recover the original space. This procedure can be generalized to the case of arbitrary base-lattices with distinguished lattice-grills (called grill-lattices) and induces an adjunction between the category of merotopic spaces and uniformly continuous maps and the category of grill-lattices and grill-continuous homomorphisms (Theorem 2.6). In the last section the previous results are applied to the category of uniform spaces and uniformly continuous maps. The corresponding grill-lattices (called uniform lattices) are characterized internally by means of a lattice-theoretical analogue of the star-refinement property (Theorem 3.4).

With respect to the theory of uniform, nearness and merotopic spaces we shall use the following terminology and notation: For any set X let $\mathfrak{P}X$ denote the powerset of X and $\mathfrak{P}^2X = \mathfrak{P}\mathfrak{P}X$. If $\mathfrak{A}$ and $\mathfrak{B}$ are families of subsets of X one says that $\mathfrak{A}$ (co-)**refines** $\mathfrak{B}$ (notation: $\mathfrak{A} < \mathfrak{B}$ ($\mathfrak{A} > \mathfrak{B}$)) if for every $A \in \mathfrak{A}$ there exists some $B \in \mathfrak{B}$ with $A \subset B$ ($A \supset B$). The family $\mathfrak{A}$ is a **stack** if

$$\mathfrak{A} = \text{stack}(\mathfrak{A}) = \{C \subset X \mid \exists\, A \in \mathfrak{A} \text{ with } A \subset C\}.$$

$\sigma(X)$ denotes the set of all stacks of X. One immediately observes that $\mathfrak{A} > \mathfrak{B}$ if and only if $\mathfrak{A} \subset \text{stack}(\mathfrak{B})$. If $C \subset X$, let

$$\text{star}(C,\mathfrak{A}) = \bigcup\{A \in \mathfrak{A} \mid A \cap C \neq \emptyset\} \quad \text{and} \quad \text{star}(\mathfrak{A}) = \{\text{star}(A,\mathfrak{A}) \mid A \in \mathfrak{A}\}.$$

Dually let

$$\text{costar}(C,\mathfrak{A}) = \bigcap\{A \in \mathfrak{A} \mid A \cup C \neq X\} \quad \text{and} \quad \text{costar}(\mathfrak{A}) = \{\text{costar}(A,\mathfrak{A}) \mid A \in \mathfrak{A}\}.$$

For the sake of emphasis we shall write $\text{costar}_X(C,\mathfrak{A})$ and $\text{costar}_X(\mathfrak{A})$ occasionally. One says that $\mathfrak{A}$ **star-refines** $\mathfrak{B}$ ($\mathfrak{A}$ **costar-corefines** $\mathfrak{B}$) (notation: $\mathfrak{A} *\!< \mathfrak{B}$ ($\mathfrak{A} *\!> \mathfrak{B}$)) provided $\text{star}(\mathfrak{A})$ refines $\mathfrak{B}$ ($\text{costar}(\mathfrak{A})$ corefines $\mathfrak{B}$). Let

$$\mathfrak{A}^c = \{X - A \mid A \in \mathfrak{A}\}$$

denote the family of complements of elements of $\mathfrak{A}$. The principal filter generated by $A \subset X$ is denoted by $\overline{A}$, i.e. $\overline{A} = \text{stack}(\{A\})$ and for each $x \in X$ let $\dot{x} = \overline{\{\{x\}\}}$. Consider the following conditions for a collection $\mu \subset \mathfrak{P}^2 X$:

(C1) $\mathfrak{U} < \mathfrak{B},\ \mathfrak{U} \in \mu \Rightarrow \mathfrak{B} \in \mu$

(C2) $\emptyset \notin \mu,\ \mathfrak{P}X \in \mu$

(C3) $\mathfrak{U} \in \mu \Rightarrow \bigcup\mathfrak{U} = X$

(C4) $\mathfrak{U},\mathfrak{B} \in \mu \Rightarrow \mathfrak{U} \wedge \mathfrak{B} = \{U \cap V \mid U \in \mathfrak{U},\ V \in \mathfrak{B}\} \in \mu$

(C5) $\mathfrak{U} \in \mu \Rightarrow \exists\, \mathfrak{B} \in \mu : \mathfrak{B} *\!< \mathfrak{U}$

The pair $(X,\mu)$ is called **merotopic space** (**uniform space**) if it satisfies (C1)-(C4) ((C1)-(C5)). The members of $\mu$ are called **uniform covers**. A map $f: X \to Y$ between merotopic spaces $(X,\mu)$ and $(Y,\nu)$ is **uniformly continuous** if for each $\mathfrak{B} \in \mu$: $f^{-1}(\mathfrak{B}) = \{f^{-1}[V] \mid V \in \mathfrak{B}\} \in \mu$. The resulting categories are **Mer** and **Unif** with obvious forgetful functor V: **Mer** ⟶ **Set**. Turning to complements consider some merotopic (uniform) space $(X,\mu)$ and let the collection of **near families** be given by

$$\xi_\mu = \{\mathfrak{U}^c \mid \mathfrak{U} \subset \mathfrak{P}X,\ \mathfrak{U} \notin \mu\} = \{\mathfrak{A} \mid \mathfrak{A} \subset \mathfrak{P}X,\ \mathfrak{A}^c \notin \mu\}.$$

Then $\xi = \xi_\mu$ satisfies conditions (N1)-(N4) ((N1)-(N5)):

(N1) $\mathfrak{A} > \mathfrak{B} \in \xi \Rightarrow \mathfrak{A} \in \xi$

(N2) $\emptyset \in \xi,\ \mathfrak{P}X \notin \xi$

(N3) $\bigcap\mathfrak{A} \neq \emptyset \Rightarrow \mathfrak{A} \in \xi$

(N4) $\mathfrak{A} \vee \mathfrak{B} = \{A \cup B \mid A \in \mathfrak{A},\ B \in \mathfrak{B}\} \in \xi \Rightarrow \mathfrak{A} \in \xi$ or $\mathfrak{B} \in \xi$

(N5) $\mathfrak{A} \notin \xi \Rightarrow \exists\, \mathfrak{D} \notin \xi : \mathfrak{D} *\!> \mathfrak{A}$

Conversely, every collection $\xi \subset \mathfrak{P}^2X$ satisfying (N1)-(N4) ((N1)-(N5)) determines a merotopic (uniform) structure

$$\mu_\xi = \{\mathfrak{A}^c \mid \mathfrak{A} \subset \mathfrak{P}X,\ \mathfrak{A} \notin \xi\} = \{\mathfrak{U} \mid \mathfrak{U} \subset \mathfrak{P}X,\ \mathfrak{U}^c \notin \xi\}$$

and obviously

$$\mu_{\xi_\mu} = \mu \quad \text{and} \quad \xi_{\mu_\xi} = \xi\ .$$

Moreover, a function $f: X \longrightarrow Y$ between merotopic spaces $(X,\mu)$ and $(Y,\nu)$ is uniformly continuous if and only if for each $\mathfrak{A} \in \xi_\mu$ : $f\mathfrak{A} = \{f[A] \mid A \in \mathfrak{A}\} \in \xi_\nu$. Hence one may redefine **Unif** and **Mer** in terms of near families. We adopt the convention that uniform covers and near families define the same object and we shall always use the more convenient concept.

A merotopic space $(X,\mu)$ is a **nearness space** if

(C) $\quad \mathfrak{U} \in \mu \Rightarrow \operatorname{int}_\mu \mathfrak{U} = \{\operatorname{int}_\mu U \mid U \in \mathfrak{U}\} \in \mu$,

where $\operatorname{int}_\mu U = \{x \in X \mid \{U, X-\{x\}\} \in \mu\}$. In terms of near families one easily observes that a merotopic space $(X,\xi)$ is a nearness space provided

(N) $\quad \operatorname{cl}_\xi \mathfrak{A} = \{\operatorname{cl}_\xi A \mid A \in \mathfrak{A}\} \in \xi \Rightarrow \mathfrak{A} \in \xi$,

where $\operatorname{cl}_\xi A = \{x \in X \mid \{A, \{x\}\} \in \xi\}$. The full subcategory of nearness spaces is denoted by **Near**. An easy calculation shows **Unif** $\subset$ **Near**. One of the most important features of the category **Near** is the fact that it also contains the category of symmetric topological spaces and continuous maps as a full (and nicely embedded) subcategory. The associated topological interior- and closure-operators are given by $\operatorname{int}_\mu$ and $\operatorname{cl}_\xi$. Excellent sources for the theory of merotopic and nearness spaces are provided by the original papers of Katětov [Ka 65,67] and Herrlich [He 74 a,b] and the survey article [He 82].

Let L be a completely distributive lattice, $B \subset L$ and $\mathfrak{A} \subset \mathfrak{P}B$. The **limes superior** of $\mathfrak{A}$ is $\operatorname{limsup}\mathfrak{A} = \inf\{\sup A \mid A \in \mathfrak{A}\}$. The subset B is a **base** of L if every element of L is the limes superior of a family of subsets of B. A subset G of L is a **lattice-grill** if (1) $G \neq L$ and (2) for all $x,y \in L$: $x \vee y \in G$ if and only if $x \in G$ or $y \in G$. An element x of L is **$\bigwedge$-prime** (**$\wedge$-prime**) if for each (finite) subset $M \subset L$:

$$x \geq \inf M \Rightarrow \exists\, m \in M \text{ with } x \geq m\ .$$

The dual notions are **$\bigvee$-prime** and **$\vee$-prime**. The set P(L) of **principal elements** consists of all elements, which are both $\bigwedge$-prime and $\bigvee$-prime. A map $f: L \to M$ between complete lattices is a **complete homomorphism** if it preserves arbitrary inf's and sup's. In particular $f(0) = 0$ and $f(1) = 1$. Terminology and notation of Lattice Theory are in accordance with [Er 82] and [Comp 80]. For concepts of Category Theory we refer to [HS 79].

## 2 Scales, Bases, Grills and the Representation of Merotopic Spaces

The corefinement-axiom (N1) implies that merotopic structures are already determined by the collection of near stacks. This observation motivates the investigation of the collection $\sigma(X)$ of all stacks of a set X and its subset $\tau(\xi)$ of near stacks with respect to a merotopic structure $\xi$ on X. More specifically, the **scale** of X is defined as

$$S(X) = (\sigma(X), \leq),$$

where $\mathfrak{A} \leq \mathfrak{B}$ if and only if $\mathfrak{A} \supset \mathfrak{B}$ for arbitrary stacks $\mathfrak{A}, \mathfrak{B}$ in $\sigma(X)$. The partially ordered scale has many convenient properties, which can be verified by elementary calculations:

**2.1 Proposition.** [We 88, 2.2] Let X be a set, $S = S(X)$ its scale, $\mathfrak{A},\mathfrak{B} \in S$ and $\omega \subset S$.

(1) S is a completely distributive lattice with $\mathfrak{P}X = 0_S \neq 1_S = \emptyset$ and $\inf_S \omega = \bigcup\omega$, $\sup_S \omega = \bigcap\omega$.

(2) $\mathfrak{A} \vee \mathfrak{B} = \mathfrak{A} \cap \mathfrak{B} = \sup_S\{\mathfrak{A},\mathfrak{B}\}$

(3) The principal filters $\overline{A}$ ($A \subset S$) are precisely the $\bigwedge$-prime elements of S.

(4) The principal ultrafilters are the principal elements of S, i.e. $P(S) = \{\dot{x} \mid x \in X\}$.

(5) Each $\mathfrak{A} \in S$ is the limes superior of principal elements,

$$\mathfrak{A} = \inf\{\sup\{\dot{x} \mid x \in A\} \mid A \in \mathfrak{A}\},$$

i.e. P(S) is a base of S. □

A pair (L,B) consisting of a completely distributive lattice L and a base $B \subset L$ is called **base-lattice**. A complete homomorphism $f: (L,B) \to (K,C)$ between base-lattices is called **base-homomorphism** if it preserves the distinguished bases, i.e. $f[B] \subset C$. The resulting category is denoted by **BL**. "Par abus de langage" base-lattices (L,B) are sometimes denoted by L.

According to Proposition 2.1 every scale together with its set of principal elements constitutes a base-lattice. The assignments

$$X \quad \to \quad SX = (S(X), P(S(X)))$$
$$(f: X \to Y) \quad \to \quad (Sf: SX \to SY, \mathfrak{A} \to \{E \subset Y \mid f^{-1}[E] \in \mathfrak{A}\})$$

define the **scale-functor** $S: \mathbf{Set} \to \mathbf{BL}$. This functor is well-defined, since for every map $f: X \to Y$, $x \in X$ and $\mathfrak{A} \in S(X)$ the family of sets $Sf(\mathfrak{A})$ is a stack and

$$Sf(\dot{x}) = \overline{\{f(x)\}},$$

i.e. Sf preserves bases. Further canonical examples of base-lattices are provided by arbitrary completely distributive lattices together with their set of $\wedge$-primes, because each element of these lattices is the inf of $\wedge$-primes [Comp 80, I.3.15]. The category **BL** is one of the categories proved to be universal, i.e. such that each concrete category can be fully embedded into it (see [PT 80, V.4], where one can find also further references). Selection of bases obviously defines a well-defined **base-functor** $B: \mathbf{BL} \to \mathbf{Set}$ with

$$L = (L,B) \quad \to \quad BL = B$$
$$(f: (L,B) \to (K,C)) \quad \to \quad Bf = (f|_B : B \to C).$$

**2.2 Proposition.** [We 88, 2.5] The base-functor $B: \mathbf{BL} \to \mathbf{Set}$ is right adjoint to the scale functor $S: \mathbf{Set} \to \mathbf{BL}$. The unit $\eta$ of this adjunction is a natural isomorphism and is given by

$$\eta_X : X \to BSX,\ x \to \dot{x}.$$

The counit $\varepsilon$ is determined by $\varepsilon_L : SBL \to L,\ \mathfrak{A} \to \limsup \mathfrak{A}$.

**Proof.** Suppose X is a set, K a base-lattice and $f: X \to BK$ a map. It must be shown that there exists a base homomorphism $\overline{f}: SX \to K$ such that the following diagram commutes.

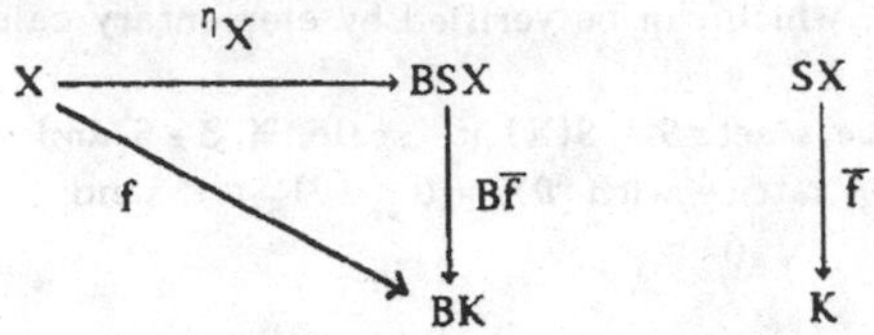


This goal is achieved by means of the following construction:

$$\overline{f}: SX \to K,\ \mathfrak{A} \to \limsup(f\mathfrak{A}) = \inf\{\sup(f[A]) \mid A \in \mathfrak{A}\}$$

Obviously $\eta_X$ is a bijective map. Finally $\varepsilon_L$ is uniquely determined by the equation

$$B\varepsilon_L \cdot \eta_{BL} = 1_{BL}.$$

Hence $\varepsilon_L(\mathfrak{A}) = \overline{1_{BL}}(\mathfrak{A}) = \limsup \mathfrak{A}$ for each stack $\mathfrak{A} \in SBL$. □

The set of near stacks of a merotopic space $(X,\xi)$ is denoted by $\tau(\xi) = \xi \cap S(X)$.

**2.3 Proposition.** [We 88, 3.2]

(1) Let $(X,\xi)$ be a merotopic space.
  - (i) $SX = (S(X), P(S(X)))$ is a base-lattice.
  - (ii) $\emptyset \neq \tau(\xi) \neq S(X)$ is a lattice-grill.
  - (iii) $P(S(X)) \subset \tau(\xi)$

(2) A map $f: (X,\xi) \to (Y,\eta)$ between merotopic spaces is uniformly continuous if and only if $Sf[\tau(\xi)] \subset \tau(\eta)$.

**Proof.** (1) is an immediate consequence of Proposition 2.1. Concerning (2) observe that f is uniformly continuous if and only if for each near family $\mathfrak{A} \in \xi$: $f\mathfrak{A} > f(\mathrm{stack}(\mathfrak{A})) > f\mathfrak{A} \in \eta$. This condition is equivalent to the property that for each near stack $\mathfrak{A} \in \tau(\xi)$: $Sf(\mathfrak{A}) > f\mathfrak{A} > Sf(\mathfrak{A}) \in \tau(\eta)$. This completes the proof. □

Motivated by these results a pair $(L,G)$ is called **grill-lattice** provided

- (i) L is a base-lattice,
- (ii) $\emptyset \neq G \neq L$ is a lattice-grill and
- (iii) $BL \subset G$.

A map $f:(L,G) \to (K,H)$ between grill-lattices is **grill-continuous** if and only if $f:L \to K$ is a base-homomorphism and $f[G] \subset H$. Let us denote the category of grill-lattices and grill-continuous maps by **GrL**. Then we define the **grill-functor** $G:\mathbf{Mer} \to \mathbf{GrL}$ by means of the assignments

$$\begin{array}{lcl} (X,\xi) & \to & GX = (SX,\tau(\xi)) \\ (f:(X,\xi) \to (Y,\eta)) & \to & (Gf:GX \to GY,\ \mathfrak{A} \to Sf(\mathfrak{A})). \end{array}$$

Other examples of grill-lattices arise from completely distributive lattices L with the set of $\wedge$-primes as bases and suitably chosen grills, for instance $G = L - \downarrow x$ for some element $x \notin \uparrow BL \cup \{1\}$. Obviously there is a forgetful functor $E:\mathbf{GrL} \to \mathbf{L}$ with $E(L,G) = L$ for each grill-lattice $(L,G)$.

Since base-homomorphisms preserve arbitrary sup's, they always have an upper adjoint, which can be used for a convenient description of grill-continuity:

**2.4 Proposition.** Let $(L,G)$ and $(K,H)$ grill-lattices, $f:L \to K$ base-homomorphism and $f^+: K \to L$ the upper adjoint of f. Then the following conditions are equivalent:

(1) f is grill-continuous.

(2) $f^+[K-H] \subset L-G$

**Proof.** Recall that for every $x \in L$ and $y \in K$: $f(x) \le y$ if and only if $x \le f^+(y)$.
(1)$\Rightarrow$(2): Let $y \in K-H$ and assume $f^+(y) \in G$. Then $f(f^+(y)) \in H$, but $f(f^+(y)) \le y$ and therefore $y \in H$, a contradiction.
(2)$\Rightarrow$(1): Let $x \in G$ and assume $f(x) \in K-H$. Then $x \le f^+(f(x)) \in L-G$, which is a contradiction.□

The foregoing duality between lower and upper adjoints generalizes the equivalence of the two approaches to merotopic spaces by means of near families and uniform covers. More precisely, for each map $F:R \to S$ between arbitrary sets the following functions constitute a Galois-connection:

$$\begin{array}{l} \mathfrak{P}^-F: \mathfrak{P}S \to \mathfrak{P}R,\ B \to F^{-1}[B] \\ \mathfrak{P}^+F: \mathfrak{P}R \to \mathfrak{P}S,\ A \to F[A] \end{array}$$

Obviously for every $A \in \mathfrak{P}R$ and $B \in \mathfrak{P}S$: $\mathfrak{P}^+F(A) \subset B$ if and only if $A \subset \mathfrak{P}^-F(B)$. Applying this connection to $F = \mathfrak{P}^-f:\mathfrak{P}Y \to \mathfrak{P}X$ for some map $f:X \to Y$ one obtains an adjunction between

$$\begin{array}{l} \mathfrak{P}^-\mathfrak{P}^-f:\mathfrak{P}^2X \to \mathfrak{P}^2Y,\ \mathfrak{A} \to \{E \subset Y \mid f^{-1}[E] \in \mathfrak{A}\} \quad \text{and} \\ \mathfrak{P}^+\mathfrak{P}^-f:\mathfrak{P}^2Y \to \mathfrak{P}^2X\ \mathfrak{B} \to \{f^{-1}[V] \mid V \in \mathfrak{B}\}. \end{array}$$

A map $f:(X,\mu) \to (Y,\nu)$ between merotopic spaces (defined in terms of uniform covers) is uniformly continuous if and only if one of the following equivalent conditions holds:

(i) $\mathfrak{P}^+\mathfrak{P}^-f[\nu] \subset \mu$

(ii) $\mathfrak{P}^-\mathfrak{P}^-f[\xi_\mu] \subset \xi_\nu$

Since $Sf = \mathfrak{P}^-\mathfrak{P}^-f|_{SX}: SX \to SY$, $\mathfrak{A} \in \xi_\mu$ if and only if $\mathfrak{A}^c \notin \mu$, and the scales are endowed with the opposite of the natural order of set inclusion, the equivalence of (i) and (ii) is contained in the above Proposition.

According to the general philosophy of this article it is desirable to endow the base of a grill-lattice (L,G) with a merotopic structure $\xi_{(L,G)}$, such that the adjunction between the categories **BL** and **Set** can be lifted to an adjunction between **GrL** and **Mer**. It would be very convenient if for each merotopic space $(X,\xi)$ the unit

$$\eta_X : X \to BSX,\ x \to \dot{x}$$

becomes an isomorphism of the merotopic spaces $(X,\xi)$ and $(BSX,\xi_{(SX,\tau(\xi))})$. This property is equivalent to the condition that for every family $\mathfrak{A} \subset \mathfrak{P}X$:

$$\mathfrak{A} \in \xi \iff \eta_X\mathfrak{A} \in \xi_{(SX,\tau(\xi))}$$

On the other hand

$$\mathfrak{A} \in \xi \iff \limsup_{SX}\eta_X\mathfrak{A} = \bigcup\{\bigcap\eta_X[A] \mid A \in \mathfrak{A}\} = \bigcup\{\overline{A} \mid A \in \mathfrak{A}\} = \operatorname{stack}(\mathfrak{A}) \in \tau(\xi)$$

The foregoing equivalences motivate the following general construction: For each grill-lattice (L,G) define $\xi_{(L,G)} = \{\mathfrak{A} \subset \mathfrak{P}BL \mid \limsup\mathfrak{A} \in G\}$.

**2.5 Proposition.** [We 88, 3.7] Suppose (L,G) is a grill-lattice. Then

$$ML = M(L,G) = (BL,\xi_{(L,G)})$$

is a merotopic space.

**Proof.** Let $\xi = \xi_{(L,G)}$.

(N1): $\mathfrak{A} > \mathfrak{B} \in \xi \Rightarrow \limsup\mathfrak{A} \geq \limsup\mathfrak{B} \in G \Rightarrow \mathfrak{A} \in \xi$

(N2): Let $A = \bigcap\mathfrak{A} \neq \emptyset$. Then $A \subset BL$ and $\mathfrak{A} \subset \overline{A}$ imply $\limsup\mathfrak{A} \geq \limsup\overline{A} = \sup A \in G$, i.e $\mathfrak{A} \in \xi$.

(N3): $\limsup\emptyset = \inf\emptyset = 1 \in G \Rightarrow \emptyset \in \xi$, $\limsup\mathfrak{P}BL = \sup\emptyset = 0 \notin G \Rightarrow \mathfrak{P}BL \notin \xi$

(N4): Suppose $\mathfrak{A},\mathfrak{B} \notin \xi$. Then $\limsup(\mathfrak{A}\vee\mathfrak{B}) = \inf\{\sup(A\cup B) \mid A \in \mathfrak{A}, B \in \mathfrak{B}\} = \limsup\mathfrak{A} \vee \limsup\mathfrak{B} \notin G$. Therefore $\mathfrak{A}\vee\mathfrak{B} \notin \xi$. □

The merotopic space ML mentioned in Proposition 2.5 is called the **base-space** of the grill-lattice L = (L,G). If $f:(L,G) \to (K,H)$ is a morphism of grill-lattices and $\mathfrak{A} \in \xi_{(L,G)}$, the relations $\limsup(f\mathfrak{A}) = f(\limsup\mathfrak{A}) \in f[G] \subset H$ imply $f\mathfrak{A} \in \xi_{(K,H)}$, i.e.

$$(Mf: ML \to MK) \to (Bf:(BL,\xi_{(L,G)}) \to (BK,\xi_{(K,H)}))$$

is a uniformly continuous map of merotopic spaces. Denote the induced functor by $M:\mathbf{GrL} \to \mathbf{Mer}$. The remarks preceding Proposition 2.5 immediately imply that each merotopic space is (isomorphic to) the base-space of a grill-lattice.

**2.6 Theorem** (The Representation Theorem for Merotopic Spaces). [We 88, 3.9]
The functor $M:\mathbf{GrL} \to \mathbf{Mer}$ is right adjoint to the grill-functor $G:\mathbf{Mer} \to \mathbf{GrL}$. The unit

$$\eta_{(X,\xi)}:(X,\xi) \to MG(X,\xi) = (BSX,\xi_{(SX,\tau(\xi))}),\ x \to \dot{x}$$

is a natural isomorphism.

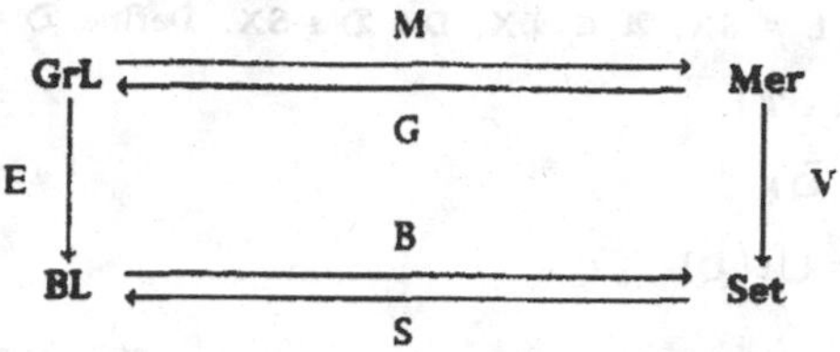


**Proof.** Let $(X,\xi)$ be a merotopic space, $(K,H)$ a grill-lattice and $f:(X,\xi) \to M(K,H)$ a uniformly continuous map. It must be shown that the base-homomorphism $\overline{f}: SX \to K$ constructed in the proof of Proposition 2.2 is grill-continuous with respect to $G(X,\xi) = (SX,\tau(\xi))$ and $(K,H)$.

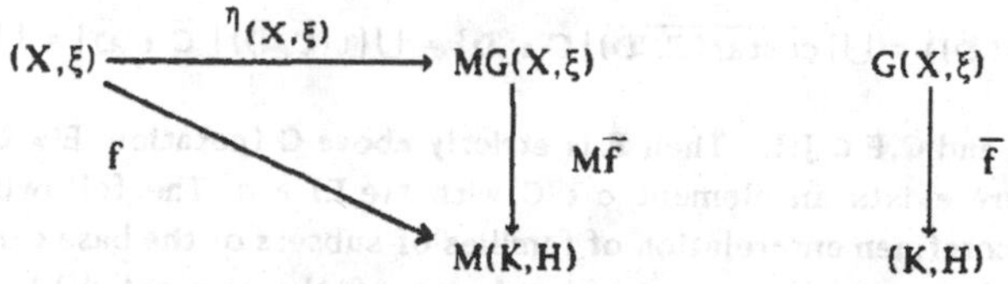


If $\mathfrak{A} \in \tau(\xi)$, then $f\mathfrak{A} \in \xi_{(K,H)}$ and therefore $\overline{f}(\mathfrak{A}) = \limsup(f\mathfrak{A}) \in H$, which completes the proof.□

The grill-lattices corresponding to nearness-spaces are characterized in [We 88]. Moreover, some categorical features of the connections between grill-lattices, base-lattices and merotopic spaces are discussed, e.g. the existence and preservation of initial sources.

## 3 Uniform Lattices represent Uniform Spaces

In order to develop a characterization of uniform spaces by suitable grill-lattices, one has to consider the algebraic counterpart of costars of stacks, which arise from intersections of subsets, i.e. elements of the powerset $\mathfrak{P}X$ of a set X. If $(L,B)$ is a base-lattice, call

$$J(L) = \{\sup A \mid A \subset B\}$$

the **join-set** of L. The powerset $\mathfrak{P}X$ of a set X admits a bijective correspondence to the join-set $J(SX)$ of the scale by means of the assignment

$$\mathfrak{P}X \to J(SX),\ A \to \overline{A}\ .$$

This bijection induces the structure of a complete lattice on $J(SX)$, where the order-relation on $\mathfrak{P}X$ is taken to be set-inclusion. More generally, if L is a base-lattice and E is a subset of $J(L)$, the supremum of E (in L) also belongs to the join-set, i.e. $J(L)$ is a complete lattice. The induced infimum may be characterized internally by

$$\inf_{J(L)} E = \sup\big((\downarrow \inf_L E) \cap BL\big) = \sup\{b \in BL \mid b \le \inf_L E\}\ .$$

If $e \in E$, define $T(e,E) = \{x \in E \mid \downarrow\{e,x\} \cap BL \neq BL\}$ and $t(e,E) = \inf_{J(L)} T(e,E)$. Let

$$t(E) = \{t(e,E) \mid e \in E\}.$$

**3.1 Lemma.** Let X be a set, $L = SX$, $\mathfrak{A} \subset \mathfrak{P}X$, $D \in \mathfrak{D} \in SX$. Define $\overline{\mathfrak{D}} = \{\overline{C} \mid C \in \mathfrak{D}\} \subset J(L)$.

(1) $\overline{\bigcap \mathfrak{A}} = \inf_{J(L)}\{\overline{A} \mid A \in \mathfrak{A}\}$

(2) $\overline{\mathrm{costar}(D,\mathfrak{D})} = t(\overline{D},\overline{\mathfrak{D}})$

(3) $\mathrm{stack}(\mathrm{costar}(\mathfrak{D})) = \bigcup t(\overline{\mathfrak{D}})$

**Proof.** (1): Identify J(L) with $\mathfrak{P}X$ as in the preceding remarks. Then $\inf_{\mathfrak{P}X}\mathfrak{A} = \bigcap \mathfrak{A}$ yields the result. (2): Consider the following equations:

$$\overline{\mathrm{costar}(D,\mathfrak{D})} = \overline{\bigcap\{C \in \mathfrak{D} \mid C \cup D \neq X\}} = \inf_{J(L)}\{\overline{C} \mid C \in \mathfrak{D}, C \cup D \neq X\} =$$

$$= \inf_{J(L)}\{\overline{C} \in \overline{\mathfrak{D}} \mid \downarrow\{\overline{D},\overline{C}\} \cap BSX \neq BSX\} = t(\overline{D},\overline{\mathfrak{D}})$$

(3): $$\mathrm{stack}(\mathrm{costar}(\mathfrak{D})) = \bigcup\{\overline{\mathrm{costar}(C,\mathfrak{D})} \mid C \in \mathfrak{D}\} = \bigcup\{t(\overline{C},\overline{\mathfrak{D}}) \mid \overline{C} \in \overline{\mathfrak{D}}\} = \bigcup t(\overline{\mathfrak{D}})$$ □

Let L be a base-lattice and $C, E \subset J(L)$. Then E is **strictly above** C (notation: $E \gg C$) if $t(E) \subset \uparrow C$, i.e. for each $e \in E$ there exists an element $c \in C$ with $t(e,E) \geq c$. The following calculations show that the costar-corefinement-relation of families of subsets of the base can be transferred into the $\gg$-relation of canonically associated subsets of the join-set J(L). The transition between the two concepts is established by means of the sup-operation

$$\mathfrak{P}^2 BL \longrightarrow \mathfrak{P}J(L), \ \mathfrak{A} \longrightarrow \sup\mathfrak{A} = \{\sup A \mid A \in \mathfrak{A}\}$$

and for the reverse direction by means of

$$\mathfrak{P}J(L) \longrightarrow \mathfrak{P}^2 BL, \ E \longrightarrow \Delta E = \{\downarrow e \cap BL \mid e \in E\}.$$

For each $j \in J(L)$ let us write $\partial j = \downarrow j \cap BL$. Then

$$\Delta E = \{\partial e \mid e \in E\}$$

for every subset $E \subset J(L)$.

**3.2 Lemma.** Let L be a base-lattice, $\mathfrak{A} \subset \mathfrak{P}BL$ and $E \subset J(L)$.

(1) For each $e \in E$ and $A \in \mathfrak{A}$:

$$t(e,E) \geq \sup A \ \Rightarrow \ \mathrm{costar}_{BL}(\partial e, \Delta E) \supset A$$

(2) $E \gg \sup\mathfrak{A} \ \Rightarrow \ \Delta E \ *\!> \mathfrak{A}$

**Proof.** (2) is an immediate consequence of (1). For the proof of (1) consider an arbitrary element $a \in A$. Then the following relations are obvious:

$$(*) \quad a \leq \sup A \leq t(e,E) = \inf_{J(L)} T(e,E).$$

Let $x \in E$ and $\partial e \cup \partial x \neq BL$. It remains to show that $a \in \partial x$. Now $\downarrow\{e,x\} \cap BL = (\downarrow e \cap BL) \cup (\downarrow x \cap BL) = \partial e \cup \partial x \neq BL$ and therefore $x \in T(e,E)$. From (*) follows $a \leq x$, i.e. $a \in \downarrow x \cap BL = \partial x$, which completes the proof. □

**3.3 Lemma.** Let L base-lattice, $C \subset J(L)$ and $\mathfrak{D} \subset \mathfrak{P}BL$.

(1) For each $c \in C$ and $D \in \mathfrak{D}$:

$$\mathrm{costar}_{BL}(D,\mathfrak{D}) \supset \partial c \Rightarrow t(\sup D, \sup\mathfrak{D}) \geq c$$

(2) $\mathfrak{D} *> \Delta C \Rightarrow \sup\mathfrak{D} \gg C$

**Proof.** For the sake of brevity write $W = \{\sup A \mid A \in \mathfrak{D}, A \cup D \neq BL\}$. We claim that

(i) $T(\sup D, \sup\mathfrak{D}) \subset W \subset J(L)$.

The second inclusion follows from the definitions. For the proof of the first inclusion consider an arbitrary $e \in T(\sup D, \sup\mathfrak{D})$. Then there exists $A \in \mathfrak{D}$ with $e = \sup A$ and $\downarrow\{\sup D, e\} \cap BL \neq BL$. Choose an element $x \in BL - \downarrow\{\sup D, e\}$. Then $x \nleq \sup D$ and $x \nleq \sup A$ and therefore $x \notin A \cup D$, i.e. $A \cup D \neq BL$. This completes the proof of (i). Thus

(ii) $t(\sup D, \sup\mathfrak{D}) \geq \inf_{J(L)} W = \sup\big((\downarrow\inf_L W) \cap BL\big)$.

Moreover we assert

(iii) $\mathrm{costar}_{BL}(D,\mathfrak{D}) \subset (\downarrow\inf_L W) \cap BL$.

Let $x \in \mathrm{costar}_{BL}(D,\mathfrak{D})$, $A \in \mathfrak{D}$ and $A \cup D \neq BL$, i.e. $w = \sup A \in W$. Then $x \in A$ and therefore $x \leq w$. Since w was chosen arbitrarily from the set W, the assertion of (iii) is clear. Finally,

$$t(\sup D, \sup\mathfrak{D}) \geq \sup\big((\downarrow\inf_L W) \cap BL\big) \geq \sup\big(\mathrm{costar}_{BL}(D,\mathfrak{D})\big) \geq \sup\partial c = c\ .$$

(2) is an immediate consequence of (1). □

The previous results motivate the following definition: A grill-lattice (L,G) is called **uniform lattice** if and only if for each subset $C \subset J(L)$ with $\inf C \notin G$ there exists subset $E \subset J(L)$ with $\inf E \notin G$ and $E \gg C$. The full subcategory of **GrL** with the uniform lattices as objects is denoted by **UniL**. The main result of the present note is contained in the following Theorem:

**3.4 Theorem.** (The Representation Theorem for Uniform Spaces)

(1) A grill-lattice (L,G) is a uniform lattice if and only if its base-space M(L,G) is a uniform space.

(2) A merotopic space $(X,\xi)$ is a uniform space if and only if $G(X,\xi)$ is a uniform lattice.

(3) $M: \mathbf{UniL} \to \mathbf{Unif}$ is right adjoint to $G: \mathbf{Unif} \to \mathbf{UniL}$. In particular, every uniform space is (isomorphic to) the base-space of a uniform lattice.

**Proof.** (1): Assume (L,G) is a uniform lattice. Let $\mathfrak{A} \subset \mathfrak{P}BL$ and $\mathfrak{A} \notin \xi_{(L,G)}$, i.e. $\limsup\mathfrak{A} \notin G$. It must be shown that there exists a collection $\mathfrak{D} \subset \mathfrak{P}BL$ with $\limsup\mathfrak{D} \notin G$ and $\mathfrak{D} *> \mathfrak{A}$. Since $\sup\mathfrak{A} \subset J(L)$, $\inf(\sup\mathfrak{A}) = \limsup\mathfrak{A} \notin G$ and (L,G) is uniform, there exists a subset $E \subset J(L)$ with $\inf E \notin G$ and $E \gg \sup\mathfrak{A}$. For each $e \in E$ the equation $\sup\partial e = e$ is valid and therefore $\limsup\Delta E = \inf E \notin G$. By Lemma 3.2(2) $\mathfrak{D} = \Delta E *> \mathfrak{A}$, i.e. M(L,G) is a uniform space. Conversely, assume that M(L,G) is a uniform space. Let $C \subset J(L)$ and $\inf C \notin G$. It must be shown that there exists a subset $E \subset J(L)$ with $\inf E \notin G$ and $E \gg C$. The relations

$$\operatorname{limsup}\Delta C = \inf\{\sup({\downarrow}c \cap BL) \mid c \in C\} = \inf C \notin G$$

show that $\Delta C \notin \xi_{(L,G)}$. Since $\xi_{(L,G)}$ is a uniform structure on $BL$, there exists a family $\mathfrak{D} \subset \mathfrak{P}BL$ with

$$\inf(\sup\mathfrak{D}) = \operatorname{limsup}\mathfrak{D} \notin G \quad\text{and}\quad \mathfrak{D} *\!> \Delta C.$$

Finally, Lemma 3.3 implies $E = \sup\mathfrak{D} \gg C$ and the proof of (1) is complete.
(2): Since the map $\eta_X : (X,\xi) \longrightarrow MG(X,\xi)$ is an isomorphism of merotopic spaces, the space $(X,\xi)$ is uniform if and only if $MG(X,\xi)$ is uniform, which in turn happens if and only if $G(X,\xi)$ is a uniform lattice by virtue of part (1).
(3): The assertions follow immediately from the above and Theorem 2.6. □

An example of a grill-lattice, which is not uniform, is provided by the following data: Let $L = [0,1]^2$ with the usual order, $a = (1/2,1/2) \in L$, $BL = \{(x,y) \in L - \{a\} \mid x + y = 1\}$ and $G = L - {\downarrow}a$. Then $(L,G)$ is a grill-lattice. In order to show that $(L,G)$ is not uniform, observe that $J(L) = {\uparrow}BL \cup \{0\}$. For any $E \subset J(L)$

$$\inf_{J(L)} E = \inf_L E, \quad \text{if } \inf_L E \in {\uparrow}BL, \text{ and } \quad \inf_{J(L)} E = 0 \quad \text{else.}$$

Consider $C = \{(x,y) \in BL \mid x < 1/2\} \subset J(L)$. Then $c = \inf_L C = (0,1/2) \notin G$. Suppose $E \subset J(L)$ satisfies $E \gg C$, i.e. $t(E) \subset {\uparrow}C \subset {\uparrow}BL$. Then for each $e \in E$

$$\inf_{J(L)} T(e,E) = t(e,E) \in {\uparrow}C \subset L - \{0\}$$

and therefore $\inf_{J(L)} T(e,E) = \inf_L T(e,E)$. We claim that then necessarily $\inf_L E \in G$. Assume $\inf_L E \notin G$. It will be shown that in this case

(*) $\quad E \subset {\uparrow}c\,.$

Assume that there exists an element $e \in E - {\uparrow}c$. Then $e \in T(e,E)$ and therefore $\inf_L T(e,E) \notin {\uparrow}c \supset {\uparrow}C$, a contradiction. The relations (*) and $\inf_L E \notin G$ imply

$$d = \inf_L E \in {\uparrow}c \cap {\downarrow}a\,.$$

Thus for arbitrary elements $e,f \in E$: ${\downarrow}\{e,f\} \cap BL \subset {\downarrow}E \cap BL \subset {\uparrow}c \cap BL \neq BL$. Hence $T(e,E) = E$ for each $e \in E$ and therefore $t(e,E) = \inf_L E \notin G$, but also $t(e,E) \in {\uparrow}C \subset G$. This contradiction shows that $E \gg C$ implies $\inf_L E \in G$, i.e. $(L,G)$ is not uniform.

It is still an open problem, whether the category of uniform lattices is a reflective subcategory of the category of grill-lattices. Moreover, separation axioms for grill-lattices and uniform lattices have not been investigated until now. One may also ask for lattice-theoretical characterizations of completions and compactifications of uniform spaces. A positive answer would link this approach to the results of Kent, who investigated representations of the Weil-completion and the Samuel-compactification of uniform spaces [Ke 67; 83].

**Acknowledgments**: Many valuable suggestions of M. Erné are gratefully acknowledged. His representation theory of closure spaces [Er 84] has very much influenced the representation theory of merotopic spaces.

**References**

[Er 82] M. Erné : Einführung in die Ordnungstheorie, (Bibliographisches Institut) Mannheim et al. 1982.

[Er 84] - Lattice Representations for Categories of Closure Spaces, in: Categorical Topology (Proc. Conf. Toledo 1983), ed. by H. L. Bentley et al., (Heldermann) Berlin 1984, 197-222.

[Comp 80] G. Gierz, K. H. Hofmann, K. Keimel, J. D. Lawson, M. Mislove and D. S. Scott: A Compendium of Continuous Lattices, (Springer) Berlin et al. 1980.

[He 74a] H. Herrlich: A Concept of Nearness, Gen. Top. Appl. 5 (1974), 191-212.

[He 74b] - Topological Structures, in: Topological Structures (Proc. J. de Groot, Amsterdam 1973), ed. by P. C. Baayen, Math. Centre Tracts 52 (1974), 59-122.

[He 82] - Categorical Topology 1971-1981, in: General Topology and its Relations to Modern Analysis and Algebra V (Proc. Fifth Prague Top. Symp. 1981), ed. by J. Novak, (Heldermann) Berlin 1982, 279-383.

[HS 79] H. Herrlich and G. E. Strecker: Category Theory, (Heldermann) Berlin 1979.

[Jo 82] P. T. Johnstone: Stone Spaces, (Cambridge Univ. Press) Cambridge et al. 1982.

[Ka 65] M. Katětov: On Continuity Structures and Spaces of Mappings, Comment. Math. Univ. Carolinae 6 (1965), 257-278.

[Ka 67] - Convergence structures, in: General Topology and its Relations to Modern Analysis and Algebra II (Proc. Second Prague Top. Symp. 1966), Prague et al. 1967, 207-216.

[Ke 67] D. C. Kent: On the Scale of a Uniform Space, Invent. Math. 4 (1967), 159-164.

[Ke 83] - On the Order Scale of a Uniform Space, J. Austral. Math. Soc. (Series A) 34 (1983), 248-257.

[PT 80] A. Pultr and V. Trnková: Combinatorial, Algebraic and Topological Representations of Groups, Semigroups and Categories, (North Holland) Amsterdam et al. 1980.

[We 84] W. Weiss: Darstellung von Merotopischen und Nearness-Räumen, DiplomThesis, Univ. Hannover 1984.

[We 88] - Representation of Merotopic and Nearness Spaces, Top. Appl. 28 (1988), 89-99.

This paper is in final form and will nt be published elsewhere.

**LECTURE NOTES IN MATHEMATICS**

Edited by A. Dold and B. Eckmann

**Some general remarks on the publication of proceedings of congresses and symposia**

Lecture Notes aim to report new developments - quickly, informally and at a high level. The following describes criteria and procedures which apply to proceedings volumes. The editors of a volume are strongly advised to inform contributors about these points at an early stage.

§1. One (or more) expert participant(s) of the meeting should act as the responsible editor(s) of the proceedings. They select the papers which are suitable (cf. §§ 2, 3) for inclusion in the proceedings, and have them individually refereed (as for a journal). It should not be assumed that the published proceedings must reflect conference events faithfully and in their entirety. Contributions to the meeting which are not included in the proceedings can be listed by title. The series editors will normally not interfere with the editing of a particular proceedings volume - except in fairly obvious cases, or on technical matters, such as described in §§ 2, 3. The names of the responsible editors appear on the title page of the volume.

§2. The proceedings should be reasonably homogeneous (concerned with a limited area). For instance, the proceedings of a congress on "Analysis" or "Mathematics in Wonderland" would normally not be sufficiently homogeneous.

One or two longer survey articles on recent developments in the field are often very useful additions to such proceedings - even if they do not correspond to actual lectures at the congress. An extensive introduction on the subject of the congress would be desirable.

§3. The contributions should be of a high mathematical standard and of current interest. Research articles should present new material and not duplicate other papers already published or due to be published. They should contain sufficient information and motivation and they should present proofs, or at least outlines of such, in sufficient detail to enable an expert to complete them. Thus resumes and mere announcements of papers appearing elsewhere cannot be included, although more detailed versions of a contribution may well be published in other places later.

Surveys, if included, should cover a sufficiently broad topic, and should in general not simply review the author's own recent research. In the case of surveys, exceptionally, proofs of results may not be necessary.

"Mathematical Reviews" and "Zentralblatt für Mathematik" require that papers in proceedings volumes carry an explicit statement that they are in final form and that no similar paper has been or is being submitted elsewhere, if these papers are to be considered for a review. Normally, papers that satisfy the criteria of the Lecture Notes in Mathematics series also satisfy this

.../...

requirement, but we would strongly recommend that the contributing authors be asked to give this guarantee explicitly at the beginning or end of their paper. There will occasionally be cases where this does not apply but where, for special reasons, the paper is still acceptable for LNM.

§4. Proceedings should appear soon after the meeeting. The publisher should, therefore, receive the complete manuscript within nine months of the date of the meeting at the latest.

§5. Plans or proposals for proceedings volumes should be sent to one of the editors of the series or to Springer-Verlag Heidelberg. They should give sufficient information on the conference or symposium, and on the proposed proceedings. In particular, they should contain a list of the expected contributions with their prospective length. Abstracts or early versions (drafts) of some of the contributions are very helpful.

§6. Lecture Notes are printed by photo-offset from camera-ready typed copy provided by the editors. For this purpose Springer-Verlag provides editors with technical instructions for the preparation of manuscripts and these should be distributed to all contributing authors. Springer-Verlag can also, on request, supply stationery on which the prescribed typing area is outlined. Some homogeneity in the presentation of the contributions is desirable.

Careful preparation of manuscripts will help keep production time short and ensure a satisfactory appearance of the finished book. The actual production of a Lecture Notes volume normally takes 6 -8 weeks.

Manuscripts should be at least 100 pages long. The final version should include a table of contents and as far as applicable a subject index.

§7. Editors receive a total of 50 free copies of their volume for distribution to the contributing authors, but no royalties. (Unfortunately, no reprints of individual contributions can be supplied.) They are entitled to purchase further copies of their book for their personal use at a discount of 33.3 %, other Springer mathematics books at a discount of 20 % directly from Springer-Verlag. Contributing authors may purchase the volume in which their article appears at a discount of 33.3 %.

Commitment to publish is made by letter of intent rather than by signing a formal contract. Springer-Verlag secures the copyright for each volume.

Vol. 1173: H. Delfs, M. Knebusch, Locally Semialgebraic Spaces. XVI, 329 pages. 1985.

Vol. 1174: Categories in Continuum Physics, Buffalo 1982. Seminar. Edited by F.W. Lawvere and S.H. Schanuel. V, 126 pages. 1986.

Vol. 1175: K. Mathiak, Valuations of Skew Fields and Projective Hjelmslev Spaces. VII, 116 pages. 1986.

Vol. 1176: R.R. Bruner, J.P. May, J.E. McClure, M. Steinberger, $H_\infty$ Ring Spectra and their Applications. VII, 388 pages. 1986.

Vol. 1177: Representation Theory I. Finite Dimensional Algebras. Proceedings, 1984. Edited by V. Dlab, P. Gabriel and G. Michler. XV, 340 pages. 1986.

Vol. 1178: Representation Theory II. Groups and Orders. Proceedings, 1984. Edited by V. Dlab, P. Gabriel and G. Michler. XV, 370 pages. 1986.

Vol. 1179: Shi J.-Y. The Kazhdan-Lusztig Cells in Certain Affine Weyl Groups. X, 307 pages. 1986.

Vol. 1180: R. Carmona, H. Kesten, J.B. Walsh, École d'Été de Probabilités de Saint-Flour XIV – 1984. Édité par P.L. Hennequin. X, 438 pages. 1986.

Vol. 1181: Buildings and the Geometry of Diagrams, Como 1984. Seminar. Edited by L. Rosati. VII, 277 pages. 1986.

Vol. 1182: S. Shelah, Around Classification Theory of Models. VII, 279 pages. 1986.

Vol. 1183: Algebra, Algebraic Topology and their Interactions. Proceedings, 1983. Edited by J.-E. Roos. XI, 396 pages. 1986.

Vol. 1184: W. Arendt, A. Grabosch, G. Greiner, U. Groh, H.P. Lotz, U. Moustakas, R. Nagel, F. Neubrander, U. Schlotterbeck, One-parameter Semigroups of Positive Operators. Edited by R. Nagel. X, 460 pages. 1986.

Vol. 1185: Group Theory, Beijing 1984. Proceedings. Edited by Tuan H.F. V, 403 pages. 1986.

Vol. 1186: Lyapunov Exponents. Proceedings, 1984. Edited by L. Arnold and V. Wihstutz. VI, 374 pages. 1986.

Vol. 1187: Y. Diers, Categories of Boolean Sheaves of Simple Algebras. VI, 168 pages. 1986.

Vol. 1188: Fonctions de Plusieurs Variables Complexes V. Séminaire, 1979–85. Edité par François Norguet. VI, 306 pages. 1986.

Vol. 1189: J. Lukeš, J. Malý, L. Zajíček, Fine Topology Methods in Real Analysis and Potential Theory. X, 472 pages. 1986.

Vol. 1190: Optimization and Related Fields. Proceedings, 1984. Edited by R. Conti, E. De Giorgi and F. Giannessi. VIII, 419 pages. 1986.

Vol. 1191: A.R. Its, V.Yu. Novokshenov, The Isomonodromic Deformation Method in the Theory of Painlevé Equations. IV, 313 pages. 1986.

Vol. 1192: Equadiff 6. Proceedings, 1985. Edited by J. Vosmansky and M. Zlámal. XXIII, 404 pages. 1986.

Vol. 1193: Geometrical and Statistical Aspects of Probability in Banach Spaces. Proceedings, 1985. Edited by X. Fernique, B. Heinkel, M.B. Marcus and P.A. Meyer. IV, 128 pages. 1986.

Vol. 1194: Complex Analysis and Algebraic Geometry. Proceedings, 1985. Edited by H. Grauert. VI, 235 pages. 1986.

Vol.1195: J.M. Barbosa, A.G. Colares, Minimal Surfaces in $\mathbb{R}^3$. X, 124 pages. 1986.

Vol. 1196: E. Casas-Alvero, S. Xambó-Descamps, The Enumerative Theory of Conics after Halphen. IX, 130 pages. 1986.

Vol. 1197: Ring Theory. Proceedings, 1985. Edited by F.M.J. van Oystaeyen. V, 231 pages. 1986.

Vol. 1198: Séminaire d'Analyse, P. Lelong – P. Dolbeault – H. Skoda. Seminar 1983/84. X, 260 pages. 1986.

Vol. 1199: Analytic Theory of Continued Fractions II. Proceedings, 1985. Edited by W.J. Thron. VI, 299 pages. 1986.

Vol. 1200: V.D. Milman, G. Schechtman, Asymptotic Theory of Finite Dimensional Normed Spaces. With an Appendix by M. Gromov. VIII, 156 pages. 1986.

Vol. 1201: Curvature and Topology of Riemannian Manifolds. Proceedings, 1985. Edited by K. Shiohama, T. Sakai and T. Sunada. VII, 336 pages. 1986.

Vol. 1202: A. Dür, Möbius Functions, Incidence Algebras and Power Series Representations. XI, 134 pages. 1986.

Vol. 1203: Stochastic Processes and Their Applications. Proceedings, 1985. Edited by K. Itô and T. Hida. VI, 222 pages. 1986.

Vol. 1204: Séminaire de Probabilités XX, 1984/85. Proceedings. Edité par J. Azéma et M. Yor. V, 639 pages. 1986.

Vol. 1205: B.Z. Moroz, Analytic Arithmetic in Algebraic Number Fields. VII, 177 pages. 1986.

Vol. 1206: Probability and Analysis, Varenna (Como) 1985. Seminar. Edited by G. Letta and M. Pratelli. VIII, 280 pages. 1986.

Vol. 1207: P.H. Bérard, Spectral Geometry: Direct and Inverse Problems. With an Appendix by G. Besson. XIII, 272 pages. 1986.

Vol. 1208: S. Kaijser, J.W. Pelletier, Interpolation Functors and Duality. IV, 167 pages. 1986.

Vol. 1209: Differential Geometry, Peñíscola 1985. Proceedings. Edited by A M. Naveira, A. Ferrández and F. Mascaró. VIII, 306 pages. 1986.

Vol. 1210: Probability Measures on Groups VIII. Proceedings, 1985. Edited by H. Heyer. X, 386 pages. 1986.

Vol. 1211: M.B. Sevryuk, Reversible Systems. V, 319 pages. 1986.

Vol. 1212: Stochastic Spatial Processes. Proceedings, 1984. Edited by P. Tautu. VIII, 311 pages. 1986.

Vol. 1213: L.G. Lewis, Jr., J.P. May, M. Steinberger, Equivariant Stable Homotopy Theory. IX, 538 pages. 1986.

Vol. 1214: Global Analysis – Studies and Applications II. Edited by Yu.G. Borisovich and Yu.E. Gliklikh. V, 275 pages. 1986.

Vol. 1215: Lectures in Probability and Statistics. Edited by G. del Pino and R. Rebolledo. V, 491 pages. 1986.

Vol. 1216: J. Kogan, Bifurcation of Extremals in Optimal Control. VIII, 106 pages. 1986.

Vol. 1217: Transformation Groups. Proceedings, 1985. Edited by S. Jackowski and K. Pawalowski. X, 396 pages. 1986.

Vol. 1218: Schrödinger Operators, Aarhus 1985. Seminar. Edited by E. Balslev. V, 222 pages. 1986.

Vol. 1219: R. Weissauer, Stabile Modulformen und Eisensteinreihen. III, 147 Seiten. 1986.

Vol. 1220: Séminaire d'Algèbre Paul Dubreil et Marie-Paule Malliavin. Proceedings, 1985. Edité par M.-P. Malliavin. IV, 200 pages. 1986.

Vol. 1221: Probability and Banach Spaces. Proceedings, 1985. Edited by J. Bastero and M. San Miguel. XI, 222 pages. 1986.

Vol. 1222: A. Katok, J.-M. Strelcyn, with the collaboration of F. Ledrappier and F. Przytycki, Invariant Manifolds, Entropy and Billiards; Smooth Maps with Singularities. VIII, 283 pages. 1986.

Vol. 1223: Differential Equations in Banach Spaces. Proceedings, 1985. Edited by A. Favini and E. Obrecht. VIII, 299 pages. 1986.

Vol. 1224: Nonlinear Diffusion Problems, Montecatini Terme 1985. Seminar. Edited by A. Fasano and M. Primicerio. VIII, 188 pages. 1986.

Vol. 1225: Inverse Problems, Montecatini Terme 1986. Seminar. Edited by G. Talenti. VIII, 204 pages. 1986.

Vol. 1226: A. Buium, Differential Function Fields and Moduli of Algebraic Varieties. IX, 146 pages. 1986.

Vol. 1227: H. Helson, The Spectral Theorem. VI, 104 pages. 1986.

Vol. 1228: Multigrid Methods II. Proceedings, 1985. Edited by W. Hackbusch and U. Trottenberg. VI, 336 pages. 1986.

Vol. 1229: O. Bratteli, Derivations, Dissipations and Group Actions on C*-algebras. IV, 277 pages. 1986.

Vol. 1230: Numerical Analysis. Proceedings, 1984. Edited by J.-P. Hennart. X, 234 pages. 1986.

Vol. 1231: E.-U. Gekeler, Drinfeld Modular Curves. XIV, 107 pages. 1986.

Vol. 1232: P.C. Schuur, Asymptotic Analysis of Soliton Problems. VIII, 180 pages. 1986.

Vol. 1233: Stability Problems for Stochastic Models. Proceedings, 1985. Edited by V.V. Kalashnikov, B. Penkov and V.M. Zolotarev. VI, 223 pages. 1986.

Vol. 1234: Combinatoire énumérative. Proceedings, 1985. Edité par G. Labelle et P. Leroux. XIV, 387 pages. 1986.

Vol. 1235: Séminaire de Théorie du Potentiel, Paris, No. 8. Directeurs: M. Brelot, G. Choquet et J. Deny. Rédacteurs: F. Hirsch et G. Mokobodzki. III, 209 pages. 1987.

Vol. 1236: Stochastic Partial Differential Equations and Applications. Proceedings, 1985. Edited by G. Da Prato and L. Tubaro. V, 257 pages. 1987.

Vol. 1237: Rational Approximation and its Applications in Mathematics and Physics. Proceedings, 1985. Edited by J. Gilewicz, M. Pindor and W. Siemaszko. XII, 350 pages. 1987.

Vol. 1238: M. Holz, K.-P. Podewski and K. Steffens, Injective Choice Functions. VI, 183 pages. 1987.

Vol. 1239: P. Vojta, Diophantine Approximations and Value Distribution Theory. X, 132 pages. 1987.

Vol. 1240: Number Theory, New York 1984–85. Seminar. Edited by D.V. Chudnovsky, G.V. Chudnovsky, H. Cohn and M.B. Nathanson. V, 324 pages. 1987.

Vol. 1241: L. Gårding, Singularities in Linear Wave Propagation. III, 125 pages. 1987.

Vol. 1242: Functional Analysis II, with Contributions by J. Hoffmann-Jørgensen et al. Edited by S. Kurepa, H. Kraljević and D. Butković. VII, 432 pages. 1987.

Vol. 1243: Non Commutative Harmonic Analysis and Lie Groups. Proceedings, 1985. Edited by J. Carmona, P. Delorme and M. Vergne. V, 309 pages. 1987.

Vol. 1244: W. Müller, Manifolds with Cusps of Rank One. XI, 158 pages. 1987.

Vol. 1245: S. Rallis, L-Functions and the Oscillator Representation. XVI, 239 pages. 1987.

Vol. 1246: Hodge Theory. Proceedings, 1985. Edited by E. Cattani, F. Guillén, A. Kaplan and F. Puerta. VII, 175 pages. 1987.

Vol. 1247: Séminaire de Probabilités XXI. Proceedings. Edité par J. Azéma, P.A. Meyer et M. Yor. IV, 579 pages. 1987.

Vol. 1248: Nonlinear Semigroups, Partial Differential Equations and Attractors. Proceedings, 1985. Edited by T.L. Gill and W.W. Zachary. IX, 185 pages. 1987.

Vol. 1249: I. van den Berg, Nonstandard Asymptotic Analysis. IX, 187 pages. 1987.

Vol. 1250: Stochastic Processes – Mathematics and Physics II. Proceedings 1985. Edited by S. Albeverio, Ph. Blanchard and L. Streit. VI, 359 pages. 1987.

Vol. 1251: Differential Geometric Methods in Mathematical Physics. Proceedings, 1985. Edited by P.L. García and A. Pérez-Rendón. VII, 300 pages. 1987.

Vol. 1252: T. Kaise, Représentations de Weil et $GL_2$ Algèbres de division et $GL_n$. VII, 203 pages. 1987.

Vol. 1253: J. Fischer, An Approach to the Selberg Trace Formula via the Selberg Zeta-Function. III, 184 pages. 1987.

Vol. 1254: S. Gelbart, I. Piatetski-Shapiro, S. Rallis. Explicit Constructions of Automorphic L-Functions. VI, 152 pages. 1987.

Vol. 1255: Differential Geometry and Differential Equations. Proceedings, 1985. Edited by C. Gu, M. Berger and R.L. Bryant. XII, 243 pages. 1987.

Vol. 1256: Pseudo-Differential Operators. Proceedings, 1986. Edited by H.O. Cordes, B. Gramsch and H. Widom. X, 479 pages. 1987.

Vol. 1257: X. Wang, On the C*-Algebras of Foliations in the Plane. V, 165 pages. 1987.

Vol. 1258: J. Weidmann, Spectral Theory of Ordinary Differential Operators. VI, 303 pages. 1987.

Vol. 1259: F. Cano Torres, Desingularization Strategies for Three-Dimensional Vector Fields. IX, 189 pages. 1987.

Vol. 1260: N.H. Pavel, Nonlinear Evolution Operators and Semigroups. VI, 285 pages. 1987.

Vol. 1261: H. Abels, Finite Presentability of S-Arithmetic Groups. Compact Presentability of Solvable Groups. VI, 178 pages. 1987.

Vol. 1262: E. Hlawka (Hrsg.), Zahlentheoretische Analysis II. Seminar, 1984–86. V, 158 Seiten. 1987.

Vol. 1263: V.L. Hansen (Ed.), Differential Geometry. Proceedings, 1985. XI, 288 pages. 1987.

Vol. 1264: Wu Wen-tsün, Rational Homotopy Type. VIII, 219 pages. 1987.

Vol. 1265: W. Van Assche, Asymptotics for Orthogonal Polynomials. VI, 201 pages. 1987.

Vol. 1266: F. Ghione, C. Peskine, E. Sernesi (Eds.), Space Curves. Proceedings, 1985. VI, 272 pages. 1987.

Vol. 1267: J. Lindenstrauss, V.D. Milman (Eds.), Geometrical Aspects of Functional Analysis. Seminar. VII, 212 pages. 1987.

Vol. 1268: S.G. Krantz (Ed.), Complex Analysis. Seminar, 1986. VII, 195 pages. 1987.

Vol. 1269: M. Shiota, Nash Manifolds. VI, 223 pages. 1987.

Vol. 1270: C. Carasso, P.-A. Raviart, D. Serre (Eds.), Nonlinear Hyperbolic Problems. Proceedings, 1986. XV, 341 pages. 1987.

Vol. 1271: A.M. Cohen, W.H. Hesselink, W.L.J. van der Kallen, J.R. Strooker (Eds.), Algebraic Groups Utrecht 1986. Proceedings. XII, 284 pages. 1987.

Vol. 1272: M.S. Livšic, L.L. Waksman, Commuting Nonselfadjoint Operators in Hilbert Space. III, 115 pages. 1987.

Vol. 1273: G.-M. Greuel, G. Trautmann (Eds.), Singularities, Representation of Algebras, and Vector Bundles. Proceedings, 1985. XIV, 383 pages. 1987.

Vol. 1274: N. C. Phillips, Equivariant K-Theory and Freeness of Group Actions on C*-Algebras. VIII, 371 pages. 1987.

Vol. 1275: C.A. Berenstein (Ed.), Complex Analysis I. Proceedings, 1985–86. XV, 331 pages. 1987.

Vol. 1276: C.A. Berenstein (Ed.), Complex Analysis II. Proceedings, 1985–86. IX, 320 pages. 1987.

Vol. 1277: C.A. Berenstein (Ed.), Complex Analysis III. Proceedings, 1985–86. X, 350 pages. 1987.

Vol. 1278: S.S. Koh (Ed.), Invariant Theory. Proceedings, 1985. V, 102 pages. 1987.

Vol. 1279: D. Ieşan, Saint-Venant's Problem. VIII, 162 Seiten. 1987.

Vol. 1280: E. Neher, Jordan Triple Systems by the Grid Approach. XII, 193 pages. 1987.

Vol. 1281: O.H. Kegel, F. Menegazzo, G. Zacher (Eds.), Group Theory. Proceedings, 1986. VII, 179 pages. 1987.

Vol. 1282: D.E. Handelman, Positive Polynomials, Convex Integra Polytopes, and a Random Walk Problem. XI, 136 pages. 1987.

Vol. 1283: S. Mardešić, J. Segal (Eds.), Geometric Topology and Shape Theory. Proceedings, 1986. V, 261 pages. 1987.

Vol. 1284: B.H. Matzat, Konstruktive Galoistheorie. X, 286 pages 1987.

Vol. 1285: I.W. Knowles, Y. Saitō (Eds.), Differential Equations and Mathematical Physics. Proceedings, 1986. XVI, 499 pages. 1987.

Vol. 1286: H.R. Miller, D.C. Ravenel (Eds.), Algebraic Topology Proceedings, 1986. VII, 341 pages. 1987.

Vol. 1287: E.B. Saff (Ed.), Approximation Theory, Tampa. Proceedings, 1985–1986. V, 228 pages. 1987.

Vol. 1288: Yu. L. Rodin, Generalized Analytic Functions on Rieman Surfaces. V, 128 pages, 1987.

Vol. 1289: Yu. I. Manin (Ed.), K-Theory, Arithmetic and Geometry Seminar, 1984–1986. V, 399 pages. 1987.